# IBM &

# COMPUTING

The Story of the
Iconic Company Amid
Technological Progress
from the Abacus to AI

William C. Shaffer

Cover design by Adri Rossow

Editing by Michelle Asakawa

Indexing by Douglas Easton

First printing

Library of Congress cataloging in publication data

Name: Shaffer, William Caldwell, author

Title: IBM and Computing: The Story of the Iconic Company Amid Technological Progress from the Abacus to AI

Description: First Edition | Boulder, CO : The Widsten Press : [2024] | includes index

Identifiers: LCCN 2024910140 | ISBN 978-1-7358078-1-2 (paperback) | ISBN 978-1-7358078-2-9 (Kindle/EPUB ebook)

Subjects: LCSH: International Business Machines - History | Computer industry - United States - History

Classification: LCC HD9696.2 U6 S53 2024 | DDC 338.7/61004 – dc23

LC record available at https://lccn.loc.gov/2024910140

Printed in the USA

# CONTENTS

# PREFACE

My first day at IBM was quite a wake-up call. I rolled into the branch office wearing a sport coat, open shirt, and two-tone shoes. I was quickly hustled into my new manager's office. He tactfully explained that I needed to make a few adjustments. The revised uniform included a conservative suit, white dress shirt, diagonal-pattern tie, and ten-pound wingtip shoes. Once I returned to the office properly dressed, I went out on a customer call with a IBM Systems Engineer. The company was using IBM's punched-card equipment. "Unit record" was the IBM jargon. The IBM 407 Accounting Machine was the formal name. They called the hulking machine the "gray elephant". It was a mechanical monster that clicked and clacked as it read, sorted, and punched 80-column cards. At this point, I was thinking—did I make a mistake joining IBM?

I had just graduated from college and had a gap year before going on to graduate school, I needed a temporary job, and the IBM office in Miami made me an offer. What I desperately needed at that point was the book you now have in your hands, something to educate me on computing and IBM's place in it. Still, I was able to make it past the first few days. It was actually exciting. This IBM office focused on small and medium businesses. A few of these companies used punched-card machines while most were completely manual. They were open, even hungry, for someone to guide them into computers. I had taken one computer course in college, so I could definitely be that person. To my great surprise, at the end of the gap year I elected to stay. Actually, I stayed for another 42 years.

I continued to work in that very exciting part of IBM, its small and midrange system operations. This area, formally called the General Systems Division, would have been the second-largest computer company in the world if it had emerged from under the shadow of IBM as a stand-alone entity. It would be well ahead of DEC, Data General, and all of the "Seven Dwarfs" classical computer companies. After retiring, I turned to writing, a lifelong passion. I finished a book on automotive history. As that project was winding down, a new history of IBM hit the bookstores. I read it cover to cover and discovered to my chagrin that I was not in it. There was perhaps a sentence or two within its 800-plus pages on IBM's smaller systems. The rest of the book was about big, bad mainframe IBM.

It turns out that this particular book was not an outlier. There is a veritable library of books about IBM. For years, it was the most admired company in the world and attracted mountains of press. As I read more of these books, a pattern emerged. They all told the story of IBM as if all the company ever produced was large computers. They were missing the smaller IBM story, my story. Puzzled, I tracked down the author of that new book and asked him why this story was missing. A fellow IBMer, it came as no surprise that his experience had been large mainframe computers. He suggested I take up the pen and write my own story. And so, here we are.

I wanted to produce the complete chronicle of the company, including its place in the broad span of computing history. It would touch all the bases about the epic adventure that was computing, but I would have free license to talk about my unheralded IBM. More than that, I was interested in going beyond what happened and when, and delve into how and why each step in the journey fit into the overall picture. Though I had just that one computer course in college, I spent a good deal of my time in IBM in technical, engineering, and programming pursuits. I rose to technical "top gun" for IBM's leading small system. I naturally wanted to understand the underpinnings of each technology that emerged along the way.

As an added bonus to the reader, I can guarantee that this will be the first book on IBM that will have a small part for a fifty-six-foot tall chicken. Oh, I had one other reason to write this story. My older son joined IBM. It was quite a shock. I am sure that he would want to read this book.

# INTRODUCTION

One of the very early lessons one learns at IBM is how to make an effective presentation. It is a simple three-step process. You start by describing what you are going to cover, then actually cover it, and end by going over what you covered. An IBM presentation would typically involve a sizeable stack of foils or slides. We will limit ourselves to the first two steps of the process here and dispense with any foils.

IBM plays a central role in the story of computing. Yet, computing began well before the company rolled out its first punched card machine. We have to go back a little further, perhaps even several thousand years. Early calculating devices appeared as early as 2700 BC in Mesopotamia. They were simple counting aids a step or two removed from using fingers and toes. Eventually they took the form of a framed tool with beads or stones to represent counting digits, the abacus. By the 17th century, advances in producing automated devices such as clocks led to the first mechanical calculators. Blaise Pascal applied a maze of gears, cogs, and levers to the process of decimal addition. His Pascaline was widely recognized as the first mechanical calculator.

The coming of industrialization and the steam age in the 19th century quickened the pace of computing invention. Inspired by the economic theories of Adam Smith, Charles Babbage sought to make calculating a manufacturing operation. He witnessed a manual version of calculating, where hundreds of human "computers" with pen and paper converted the French property tables to metric following the French Revolution. He aimed to accomplish this by applying steam and massive machinery to the task. He enlisted a colleague, the Lady Ada Lovelace, to program his "Analytical Engine." Alas, both came along a century too soon. They had 20th-century ideas constrained by 19th-century technology.

The early devices focused on computation, mainly addition and subtraction, and sometimes the far more demanding multiplication and division. Greater and greater speeds were welcome. In many cases, the demands of scientific analysis or the production of vast mathematical tables drove the development of such devices. Soon, another computing challenge loomed. With the Industrial Revolution, companies got much bigger and the processing of data, not calculations, became a central issue. It was the once a decade task that provided a central push in this new direction—the United States Census. By 1880, manual processing of this population data took up to ten years. Herman Hollerith, an engineer and former Census summer hire, invented a new system that recorded the information on 63 million Americans into holes punched in paper cards. His machine read the holes and tabulated the results.

Meanwhile, an ambitious lad named Thomas Watson from the small burg of Painted Post, New York, found his way through luck and perseverance to the highest heights of one of the largest and most successful corporations of the time. Run imperiously by John Patterson, National Cash Register (NCR) was a forerunner of the modern corporation. Patterson understood the value of research to design new products to go along with his innovations in sales. National Cash Register was also a prime example of a dominate company in the age of trusts and monopolies. Patterson worked hard at cornering the market for cash registers, by means fair and foul. The capricious John Patterson also had a penchant for disposing of his executives at his slightest whim. Watson would join the growing ranks of ex-NCR managers.

While Patterson's NCR was crushing the competition and extending its hold on the cash register industry, a serial monopolist named Charles Flint was busy doing similar kind of work. Deemed the "father of trusts", Flint aimed to gain market power by buying up all the companies in a particular industry. He constructed groupings, or "trusts," in rubber and chewing gum before organizing an odd combination of companies that included Hollerith's punched-card company. He called it the "Computing, Tabulating, and Recording Company" as it also contained meat scale and time recorder companies in addition to Hollerith's tabulators.

Watson was on the street looking for a job at the same time that Charles Flint was looking for someone to run his unique trust. A high-ranking executive from NCR was a good catch indeed for Flint, and Watson did not shy away from demanding the autonomy and compensation he needed to make the most of his career turn. A quick study, he knew almost from the outset that Hollerith's punched-card technology was the crown jewel of his new company. Watson applied the lessons learned at NCR and pushed the company to preeminence in the rapidly growing industry of data processing.

His company fought off several competitors, and by 1924 Watson was ready to rechristen the enterprise with a name to match his ambitions—"International Business Machines." Driven by outstanding engineering and sales, his company flourished. He did not let up as the Great Depression immersed the country in darkness. He hired more salespeople and kept the manufacturing plants humming. His company would grow and prosper through this bleak period thanks in part to an assist from Franklin Roosevelt. Roosevelt's alphabet soup of new government agencies would need considerable amounts of data processing machinery—machinery that Watson's plants just happened to have sitting in inventory.

World War II effectively ended the Great Depression. The war would be a watershed period in the history of computing. The conflict brought dramatically escalated requirements for computing, including insatiable demand for IBM's punched-card equipment to manage the massive and far-flung production of America's "Arsenal of Democracy." There was also ever increasing need for raw computing speed to calculate the flight of artillery shells, decode the enemy's secret communications, and simulate the explosion of an atomic bomb. This resulted is rapid technological progress, driven by electronics.

However, IBM would miss a critical technological shift. High speed computing machines using electronic vacuum tubes would soon render IBM's electromechanical devices obsolete. Wartime's fastest machine broke the encrypted communications of the German armies. However, it would remain shrouded in secrecy until 1975. The second-fastest computer created during the conflict, the ENIAC, would get all the accolades. The designers of the ENIAC would soon form a company to commercialize their invention. Their UNIVAC computer was well-positioned to dominate the emerging computing industry.

As for IBM, it decided to build Charles Babbage's 1820 Difference Engine in 1940 and went on to produce a more elaborate version of Babbage's design in 1951. Meanwhile, UNIVAC became interchangeable with the word "computer." A UNIVAC would successfully call the winner of the 1952 presidential election. To add insult to injury, UNIVAC replaced IBM machines at the U.S. Census. An IBM company tied to punched-card machines was in mortal danger.

By the war's end, Thomas Watson was, at age 72, the "Old Man" at IBM. He had poured vast sums into the Babbage projects at a time when computing technology had moved on. However, there was another Thomas Watson potentially available to take over and lead the company across the electronic divide. Tom Watson Jr. had been a poor student, a screw-off, and a playboy. However, he had grown up and matured as a bomber pilot in the war. He came back from Europe wanting to be an airline captain, not a captain of industry. He would change his mind. Returning to IBM, he accepted the UNIVAC challenge and accelerated the imperative to transform IBM from electromechanical to electronics. With a series of larger and larger computers, IBM soon took over the new industry and regained its former dominance, leaving UNIVAC and the "Seven Dwarfs" competitors far behind.

By 1959, a potentially fatal flaw had emerged in the company's strategy. IBM had produced computer after computer, but they were all incompatible with each other. Maintaining so many separate lines was slowly killing the company. Watson Jr. bet the company on a complete transformation, throwing out all of IBM's existing computer lines and developing one broad new composite line, the System/360.

If IBM was dominant before the new line debuted, it was doubly dominant after. General Motors was the standard for market dominance when it controlled 50 percent of the car market for a brief time in the 1950s. Riding the wave of System/360, IBM would capture over 70 percent of the worldwide market for computers. The company was just as dominant with its culture. It was the most-admired company in the world for years, a great place with unbounded opportunity in an exciting new industry.

However, there was a "small" problem. The System/360 was not small enough. There was plenty of room below IBM's entry computers. This space could easily be filled with inexpensive but powerful machines based on a new technology—transistors. Moreover, there was not any shortage of companies angling for this market, with DEC and Wang just the first of many. They would establish themselves with small machines and soon expand upward to encroach on IBM's big computers. IBM risked gradual erosion from below.

Watson Jr. recognized this shift. He created a new division to address the challenge. The "small" operation soon had manufacturing sites amid the cornfields of Rochester, Minnesota, and along Interstate 95 in Boca Raton, Florida. They produced a small system that rented for less than $1,000 a month. Watson located the division's headquarters in Atlanta, away from the bureaucracy in New York. The new division was just a few miles south of that fifty-six-foot chicken. The IBM branch office that I joined in Miami was one of the early sales outposts for the new division.

IBM was on top of the computing world. With its mainframes, small systems, and soon the IBM PC, it was growing even faster, firing on all eight cylinders. Watson Jr. retired from the company in 1971, and executive hubris would soon follow. His successors assumed that the company would continue to grow at the same explosive rate. That would require building a slew of new plants and hiring many thousands of new employees. The growth did not materialize, and the overhang of the massive expansion came crashing down. IBM enlisted an outsider to save the company. It was painful, but IBM survived and even thrived, for a time.

The emergency financial measures used to pull the company back from the abyss became ongoing strategies adopted by subsequent IBM

CEOs. The focus on financial targets instead of innovation and growth led to a key technology shift being missed—the advent of cloud computing. It was not long before a second existential moment was at hand. The result was a far smaller, less relevant IBM struggling to regain its former prominence. In the midst of this effort, a second challenge erupted -- the meteoric rise of Artificial Intelligence (AI). As I write this, I do not know how it will end. I leave that story for my son.

Now, I have told you what I am going to tell you. It is time now to tell you.

# PART ONE
# EARLY COMPUTING

# COUNTING BEADS AND SPINNING WHEELS

The story goes that Richard Feynman, a future Nobel laureate in Physics and master of IBM punched-card machines, faced a computing contest in Rio de Janeiro. After working on the Manhattan Project during the war, Feynman ended up as a professor at Cornell University. He quickly soured on the cold winters of Ithaca, New York, and readily accepted a part-time position in 1951 as a lecturer in physics at Brazil's Center for Physical Research in Rio. He lived just a few miles from famed Ipanema Beach. One day, he was having lunch at a local eatery, the only customer in the place. A Japanese man entered the restaurant wanting to sell his abacus calculators. The salesman challenged the waiters to see who could add up a customer's bill faster. Instead, they suggested he challenge one of their customers. Feynman agreed to the contest.

He did not have the usual arsenal of computing machines that he had at Los Alamos—the IBM machines, the high-speed desk calculators, and the pioneering computers. Rather, he had only a pencil and paper, plus the ten symbols of decimal arithmetic and a celebrated physicist's deep knowledge of mathematics.

The two settled down for the match. The first round was addition and subtraction. With the abacus, you can start counting immediately. Feynman had to first write the numbers down and then start adding.

In the meantime, the abacist was a Whirling Dervish with the counting beads. He won handily. Feynman suggested a second round, this time with multiplication. The abacist won again, but his margin of victory was narrower. Feynman proposed a third round, this time with division, knowing the harder the problem, the better chance he had. The division contest ended up in a tie. It was at this point that the abacist made a fatal error by dramatically increasing the mathematical difficulty with a problem in cube roots. The challenge was to determine the cube root of 1,729.03. The abacist immediately started a flurry of frantic bead moves. Feynman did not even write the number down. Instead, he had already divined the first two digits (12) and scribbled that down. He continued working in his head on the remainder digits and had the cube root to five digits within seconds. The abacist was soundly defeated. Feynman recounted his approach:

> "The number was 1,729.03. I happened to know that a cubic foot contains 1728 cubic inches, so the answer is a tiny bit more than 12. The excess—1.03—is only one part in nearly 2,000, and I had learned in calculus that for small fractions, the cube root's excess is one-third of the number's excess. So all I had to do is find the fraction 1/1728 and multiply by 4 (divide by 3 and multiply by 12).[1]"

The abacus salesman had picked the wrong customer in Feynman. A similar contest had pitted the Japanese abacus expert Kiyoshi Matsuzaki against Thomas Wood, a private in the U.S. Army. Wood was using an electric desk calculator. The abacist won the event 4-1, losing only in the multiplication round[2]. What is most remarkable about these two computing episodes is not that the abacus won both times but that the device was still around in the 20th century. It would take transistors to apply the coup de grace to the ancient device.

The abacus dates back to the Sumerians, in the third century BC. The Romans used it extensively. A Chinese version of the counting frame was called the suanpan and dates to the second century. The suanpan made its way to Japan in the 14th century and was renamed the soroban. Until the 1970s, the teaching of the soroban was standard in Japanese education. Until the advent of transistor calculators, the device was a common tool in business. There are still 6,500 soroban schools in the country with competitions that culminate in national championships[3].

A typical abacus is composed of a wooden frame with beads mounted on vertical metal rods. The rods represent the columns in base ten, starting at the rightmost rod with ones, then tens, hundreds, thousands, and so forth. The abacus further divides each rod into upper and lower sections. The lower section on each rod typically has four beads that facilitate counting one through four. The upper section has one or two beads, with each bead having the value of five. Moving a bead to the center of the abacus frame activates it value. For example, the number eight is expressed by moving three beads from the lower section to the center (counting three) and moving one bead from the upper section (counting five). Though the abacus has survived, it does have several limitations that put it outside of the mainstream of computing history. The unusual combination of two number bases, with base five in the upper frame, results in an awkward carryover sequence. If, for example, we wanted to add three to the number eight already entered, the resulting sum would have one lower-frame bead activated in the tens column (ten) and one lower-frame bead in the ones column (one). We need to reset all the other beads. In the hands of a skilled abacus user, these steps could be blindingly fast but at the same time, very difficult to automate.

The abacus used base ten in a roundabout manner, likely a result of the counting process with fingers and toes. This changed with the invention of the zero. Indian mathematicians came up with the concept before the 4th century. Arab mathematicians pick up the zero through trade in the 9th century. Pope Sylvester II introduced the Roman abacus with the zero in the year 999, though it took some time before the ten numeric symbols assumed their present form.

## PASCAL, SCHICKARD, AND LEIBNIZ

Calculating with the new base ten and its newly minted zero would open up new vistas in computing. A far simpler and potentially automatic approach could replace the abacus counting process. Accepting this challenge was Blaise Pascal, a Renaissance man who was a towering figure across many disciplines, including mathematics (probability theory, fluid dynamics), philosophy, and religion. In 1639, Cardinal de Richelieu named his father Etienne finance head for Normandy province, centered in Rouen, 75 miles northwest of Paris. Etienne's most important task was to calculate property taxes. His staff did this work by hand calculation or the abacus. Blaise started work on a calculating machine at age 19 in 1642. The core challenge would be that carryover, resetting the current base-ten column and adding one to the next higher column. With each column now containing the equivalent of ten abacus beads, it sounds relatively simple. This is until you contemplate the addition of one to the number "99,999." This would require a cascade of four carryovers in order to arrive at answer "1,000,000." Blaise would go through 50 prototypes before he perfected a working machine, dubbed the "Pascaline."

**Figure 1.** Blaise Pascal's Pascaline calculator (1642). This model is in the IBM collection and one of an estimated eight Pascalines that have survived. Reprint Courtesy of IBM Corporation ©.

His original machine had five sets of gearing, enabling addition up to 99,999. The mechanism was ingenuous. Pascal had adapted the gear mechanism of large clocks. For each digit, three different wheels, each with ten ratchet positions, moved using a set of spring-loaded levers, pins, and pawls. He used a local clockmaker as his artisan. Each counter wheel assembly was independent and moved by gravity. Pascal would note that a machine of his design with 10,000 wheels would work just as well[4].

He gave his first public demonstration in 1645. His device was good for addition but awkward for subtraction as the gears only moved in one direction. It was not useful for multiplication, as you had to run repetitive additions. In addition, the operator needed to reset the machine after every operation by setting the dials to all nines and adding one. That moved all the carryover gears to the start position. The completed machines were incredibly ornate, with accomplished craftsmanship. While his first models had the five wheels, he had "accounting models" built with ten wheels. Pascal received the equivalent of a patent for his device from King Louis XIV. Pascal produced only twenty Pascalines. They were expensive to build, and it was hard to find enough craftsmen with the requisite skills. Pascal moved on from his calculator work to study religion and philosophy. He entered a Jansenist convent in 1655 at age 32. Frail all his life, he died seven years later. Nine Pascalines survive to this day[5]. IBM holds one of them in its computing collection.

Pascal's calculator was considered the first automated calculator, at least for four centuries. In 1935, a biographer of the famed astronomer Johannes Kepler came upon a series of letters and sketches to Kepler detailing a calculation box. The correspondence was from a German professor of astronomy named William Schickard. Another Renaissance man who also taught Oriental languages, mathematics, and geography, Schickard worked at Tubingen University[6]. Kepler, famed for his three laws of planetary motion, had spent six years at Tubingen. He had met Schickard there and encouraged him to build a device to ease astronomical calculations. Schickard had to take some time to free his mother, who was on trial as a witch, before he could start on the computing device. He would build two machines around 1623, one that he sent to Kepler and the other he kept for himself. The year 1623 was the year that Blaise Pascal was born, so Schickard's work preceded the Pascaline by at least two decades. Both of his machines were lost: the Kepler machine in the crossfire of the Thirty Years War, and Schickard's own machine in the fires set by the combatants.

Schickard designed his device to go beyond the simple addition and subtraction of the Pascaline. It combined a geared system for addition and subtraction in conjunction with Napier rods for multiplication and division. Also called "Napier's bones," the Scottish mathematician John Napier translated the multiplication process into a compressed series of additions. In broad terms, Napier created a device (the "bones") that simulates how one does multiplication by hand. His "bones" were simply embedded single-digit multiplication tables (i.e. 3x3). Consider the simple case of multiplying 43 (the multiplicand) by 12 (the multiplier). The "bones" provide the values of the first multiplier digit: "2x3=6" and "2x4=8" or "80" with the required tens shift. Adding these partial values, we have "86". Moving to the next digit of the multiplier, we have "1x3 = 3" and "1x4 = 4 or "40" with the tens shift. Adding these partial values, we have "43" but the multiplier digit is in the tens place so the result is actually "430". Adding "86" and "430" just as one would in manual calculation yields the correct answer (516).

Napier published his "bones" concept in 1617. He died the same year. Though an actual "bones" device failed to catch on, his algorithm to simplify and automate multiplication became widely adopted. Schickard's calculator performed addition, subtraction, and partial multiplication. He added a bell whenever there was an overflow in the computation. In 1960, a lecturer at the University of Tubingen constructed a reproduction on Schickard's machine, based on the diagrams and descriptions found in his notes to Kepler. Schickard and his entire family died of the bubonic plague that the war brought to town[7].

The discovery of the Schickard machine ignited a debate as to who invented the automatic mechanical calculator. Schickard was undoubtedly the first, but he did not commercialize the device. It was lost to history for 400 years. In addition, the design of his gearing was too fragile to have resulted in a reliable calculator. His key innovation was the use of the Napier bones to produce a machine that could plausibly do multiplication.

With the loss of Schickard's machine, the next advance in "computers" would fall to another polymath from Germany, Gottfried Wilhelm von Leibniz. Though there is no record that the "von" in his name was the result of an appointment to nobility, he achieved that exalted status in science and philosophy. He invented calculus at the same time as Isaac Newton. It is his notation that survives. He became interested in mechanized calculation and was able to work with and assess a Pascaline device. He invented a special gear, the "Leibniz wheel", to simplify the carry function. He called this notched wheel the "Stepped Reckoner." Incorporating the wheel, he developed a calculator that would perform addition, subtraction, multiplication, and division.

He demoed the machine to the Royal Society in London in 1673 and they immediately made him a member. Though an impressive engineering feat, his device did not work that well. His design outpaced the technology and craftsmanship that was available at the time. Yet, his design was a major influence on the calculating devices that followed. He also thought about far more advanced concepts for computing machines. He predated the work of George Boole by declaring, "All truths of the reason would be reduced to a kind of calculation," meaning binary operations and logic[8]. His ideas on automation would influence the concepts of Charles Babbage and Ada Lovelace 150 years later.

John Napier produced another invention three years before his Napier "bones." He discovered logarithms, the inverse of the exponent function. Logarithms serve to transform multiplication and division operations into addition and subtraction. Both his Napier "bones" and logarithms aimed at reducing the labor involved in arithmetic computation. Mathematician Henry Briggs published the first table of logarithms, which computed the log values of the first 1,000 integers to 14 decimal places[9]. A few years later, the Reverend William Oughtred invented the slide rule, an analog device that uses logarithms to make calculations. Oughtred laid out the logarithmic scale on two parallel rulers. The answer reflects the relative distance across the two scales. The issue with the slide rule is its accuracy. As with all analog devices, it produces an approximation of the result. Bigger slide rules can improve accuracy to a point. In contrast, digital devices such as the Pascaline or Leibniz's calculator produce an exact answer.

The works of Pascal, Schickard, and Leibniz developed the concepts of a mechanical calculator but did not result in a commercial product. That changed when Charles Xavier Thomas de Colmar focused on the task. He was a French mathematician who had joined the French army and would rise to manager of the army's entire supply function. This role required a significant amount of computation. He decided to develop a machine. De Colmar used the Leibniz wheel design and received a French patent on the desk-size unit he created in 1820. He shopped his design to the French and English government but received no interest. The English government was already financing Charles Babbage's Difference Engine and had enough problems with that venture.

## THE ARITHMOMETER

De Colmar left the army and went into the insurance business, a vocation that required even more computation. He continued to experiment with his design as he built up his business. He was very successful, building the largest insurance company in France. All the while, he continued to tinker with his mechanical calculator. After 30 years, he was ready to produce a commercial model. Coming 200 years after Pascal, he had the advantage of major advances in technology and mass production. He also had the funds to produce the device on his own. De Colmar's new calculator was handheld and called the Arithmometer. He demoed it at the Great Exposition of 1851, held at the Crystal Palace in London[10]. He was to be disappointed with the trip though, as a calculator invented by Izrael Staffel took the gold prize. He tried again at the 1855 Paris Exhibition, this time with a far more elaborate, piano-size model. Again, he came away discouraged as the gold medal was awarded to Georg Scheutz, who had succeeded in producing Charles Babbage's computing machine. De Colmar persisted and soon produced the world's first commercially successful calculator.

It was a revolutionary machine with dials, cylinders, cogs, gears, teeth, levers, springs pawls, and a carriage. De Colmar simplified the calculator's workings over the previous designs. For addition, the operator enters the first number via six slide bars and turns the crank once. The operator then enters the second number, turns the crank again, and views the answer on the 12-position display at the top of the unit. Multiplication involves a few more cranks. To multiply 1,599 by 235, the operator would first enter 1,599 and turn the crank five times. After shifting the platen to the right for the tens column, the operator cranks three times. Another platen shift for the hundreds column and two cranks produces the answer. The base ten values of the multiplier determine the total number of cranks. Where raw addition would require 235 cranks, De Colmar's platen scheme requires just ten cranks (2+3+5). He also included a switch to reverse the gearing for subtraction and division and a knob to reset the machine in one motion[11].

The De Colmar Arithmometer was virtually the only automatic calculator on the market for the next 40 years. De Colmar built 1,500 machines, including some expensive models for European royalty. He also licensed his design, which resulted in a total production of roughly 5,500 machines. De Colmar died in 1870, leaving behind a huge fortune. He also left a significant legacy, moving the pace of calculating technology forward ever faster. His Arithmometer would greatly influence the future designs from Monroe, Friden, and Marchant.

**Figure 2**. Platine Arithmometer, circa 1822. Faceplate covers a maze of cogs, gears, levers, springs, and pawls. Numbers are dialed in with top row. A crank produces the carry mechanism for multiplication (from the book celebrating the 100th anniversary of the insurance company "Le Soleil" 1829-1929, and its founder Thomas de Colmar).

Inventors in the United States adopted and enhanced the Arithmometer. In 1872, Frank Baldwin modified the Leibniz wheel into a pinwheel and started the Monroe Calculator Machine Company. Dorr Felt invented a similar calculator in 1885, marketed by the Felt and Tarrant Company. It was the first machine to use a keyboard to enter numbers. William Burroughs introduced the Burroughs Adding Machine in 1892. His key innovation was the printed tape, showing the numbers entered and the totals. Burroughs started in business as the American Arithmometer before adopting the Burroughs name. His company would grow rapidly to 2,500 employees. In 1893, Otto Steiger invented the "Millionaire" calculator. It performed multiplication with just one turn of the crank for each digit in the multiplier. The Dalton machine arrived in 1902. It was the first machine with the familiar ten-key entry format.

Scientists and engineers, with serious computing challenges, considered many of the early machines as pedestrian. For this group, it had to be a Marchant or a Friden. Rodney and Alfred Marchant founded the Marchant Calculating Machine Company in 1911 in Oakland, California. They used a pinwheel design invented by Willdogt Odhner, a Swede who worked in Saint Petersburg. Engineers valued the Marchant calculators for their speed and reliability. They added gearing for each digit that enabled calculations to run at different speeds. The resulting Marchant calculator had hundreds of gears under the covers and three separate drive shafts. It was the Cadillac of desk calculators, with Cadillac pricing. In 1915, they added an electric version.

Carl Friden was a young engineer for Swedish Match Trust who had immigrated to the United States after World War I[12]. He joined Marchant and modified their existing pinwheel gearing to avoid impinging on existing German patents. He went on to design the top-end Marchant calculator. He received $100,000 plus one dollar for every machine sold. In a patent dispute with Monroe, he netted an additional $225,000. The Marchant Company let him go in the midst of the Depression. By this time, he had only $27,000 left. He used these funds and added investment monies to start his own his own company. His Friden calculator was an absolute monster. Addition was sedate enough, a single grind of the gearing. Multiplication was a different story. The machine came to life as it clicked, clanked, and rumbled through the digits, with the heavy typewriter-like platen moving digit by digit to an answer. Anything short of a stout desk would shake in time with the machine. Division was more complicated but no less noisy. The platen first had to clunk to the starting point of the divisor and then it was off to the races. If that was not enough, the machine had a bell (shades of Schickard) that would ring when there was an error such as a negative result or numeric overflow.

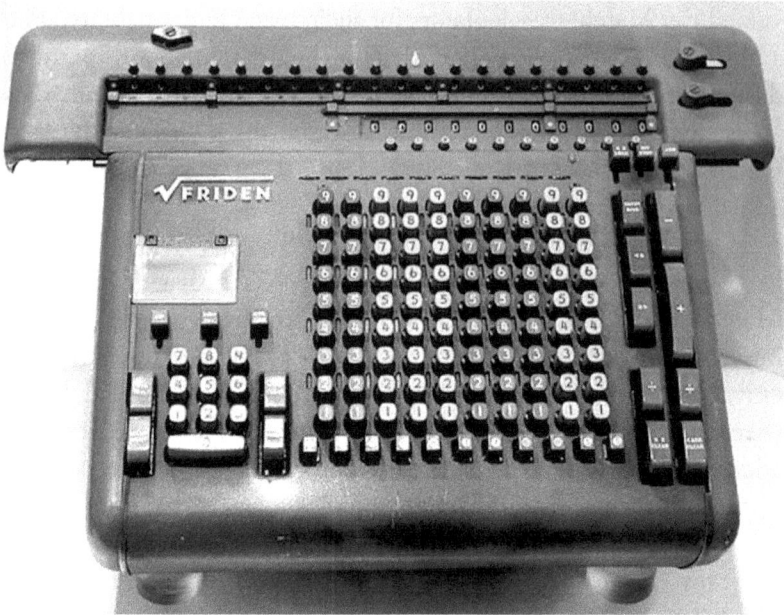

**Figure 3.** Friden mechanical calculator (Tokyo University of Science).

# BABBAGE AND "STEAM" COMPUTING

B laise Pascal developed the Pascaline to ease the computation work of his tax assessor father. The French Revolution from 1789 to 1799 presented a computing challenge on a completely new level. France under Napoleon had adopted a new scheme for property taxation based on the metric system. This system defined the meter as one ten-millionth of the distance of the meridian from the North Pole passing through Paris and on to the equator. This distance was ten million meters, or 5,195 miles.

Metric measurements required new tables of sine and logarithmic values that were decimal-based. The existing property tables used the sexagesimal system, or number base 60. Time segments and angles also used base 60. The change to metric divided a circle into 100 sections as opposed to 360 degrees of arc. The inventor of the slide rule, mathematician William Oughtred, had pressed for such a change. It was met with substantial resistance, as all current land tables would become obsolete. Napoleon's metric conversion promised a fairer system for property taxation. Francois Callet produced tables to seven places for each 10,000th segment of the Paris meridian. However, his tables were not granular enough for taxation purposes[1].

A new organization, the Bureau de Cadastre, was set up to administer the property system. Cadastre mapping defined the bounds and area of a given piece of real property. Baron Gaspard de Prony, a French mathematician and engineer who at the time was the chief engineer for roads and bridges, was the director of the property bureau. Prony came into his new position with grand ambitions, far beyond the taxation mandate. He viewed his challenge an "imposing monument of computation." De Prony wanted far more gradations of the existing trigonometric and logarithmic tables and twice the accuracy, with some values calculated to 29 decimal places[2].

De Prony organized his calculation project as one would organize a factory. He got inspiration from the writings of Adam Smith on the mass production methods of the steam age. De Prony would "manufacture logarithms as one manufactures pins.[3]" He would set up his computation "factory" with three hierarchical groups of employees. The first group consisted of five to six expert mathematicians who would define the necessary formulas. They would use the method of finite differences, a mathematical algorithm that reduced complex calculations involving multiplication and division into simple additions and subtractions. This method was rooted in the solving of differential equations in the calculus invented by Newton and Leibniz. A second group consisted of individuals that were familiar with calculus. They would develop the first 10,000 values and define the calculation process for the third group. The third group consisted of 100 "computers," individuals who would each calculate about 1,000 additions or subtractions per day. This translated into 200 logarithms[4]. Many of these "computers" were former hairdressers, as fancy hair styles went out of style with the French Revolution.

The output of the De Prony "factory" was prodigious. It included the natural sines of the 10,000 segments of the Paris meridian calculated to 25 places and the logarithmic sines of each quadrant calculated to 14 decimals. It went on to compute the ratios of the sines to their arcs, the logarithmic tangents, and several other ratio tables. The process calculated the simple logarithms of the numbers from one to 10,000 to 19 decimals and the numbers from 10,000 to 200,000 to 14 places. Charles Babbage later estimated that just a couple of these undertakings involved over eight million individual calculations.

De Prony's teams posted the completed values into massive books of tables. They produced seventeen hefty volumes before the project ended far short of completion in 1801 and French authorities dissolved the entire bureau[5]. The handwritten volumes were never typeset and printed. De Prony tried to get the British government to produce the tables, but Britain was still using the base 60 system and not interested in conversion. The level of detail that De Prony had pursued was wildly extravagant. Later computation efforts produced tables to just seven or eight decimal places. The French Revolutionary government adopted the metric system in 1795, but the law did not explicitly require the angle values that De Prony's teams were producing. Still, De Prony pressed on for six more years.

The De Prony project seems fantastical, yet it was not the largest table project of the era. Edward Sang, a Scottish mathematician and civil engineer, spent over 40 years, from 1848 to 1890, creating a mammoth set of tables that ran to 47 volumes with logarithms computed to 28 places. Sang and his two daughters, Jane and Flora[6], solely produced this unbelievable and mind-numbing result. Later table makers could argue about the relative accuracy of the De Prony and Sang tables, but errors in the process would diminish the value of both efforts. This included errors in the computations and mistakes in the transcription to the printed volumes. Despite the greater scope of Sang's work, de Prony is the one remembered, and that is because of the unique division of labor he employed.

## ENTER CHARLES BABBAGE

The work of Baron de Prony would be a footnote in history if it not for a British scientist who would take note of de Prony's methods. That scientist was Charles Babbage. He entered Trinity College at Cambridge in 1810. Disappointed by the mathematics he encountered, he formed the Analytical Society with two fellow undergraduates and lobbied the college to upgrade the mathematics curriculum. They wanted Trinity to switch from the calculus notation of Isaac

Newton to that Leibniz. After graduation, Babbage struggled to secure a teaching position, losing several times to other candidates. He married early and had to rely on the financial support of his father for several years. He eventually became a professor at Cambridge and would hold the Lucasian chair of mathematics previously occupied by Isaac Newton. Babbage spent most of his time in London and never delivered a single lecture.

Babbage was instrumental in founding the Astronomical Society in London. In 1821, Babbage and John Herschel embarked on the production of new star tables. Once the summer of work was completed, they noticed many errors in both the calculations and the transcription. During this time, there was acceleration in the demand and production of such tables, with errors being the dirty little secret of such efforts. At one point Babbage is reported to have remarked, "I wish to God these calculations had been executed by steam!"

With the end of the Napoleonic Wars, scientific interchange between France and England was renewed. Charles Blagden, a British scientist and member of the Royal Society, lobbied for the printing of the de Prony tables. Though his effort did not come to fruition, it pushed Babbage in a direction he was already considering—how to automate such work.

**Figure 4**. Charles Babbage, 1871, from a wood engraving of the
Wellcome Collection.

Babbage had learned of the de Prony table project work when he
traveled to France in 1819. He was later able to view the de Prony
volumes. As a mathematician, he was very familiar with the method of
finite differences that de Prony had employed.

Babbage later commented on the moment his idle remark about
producing computations by steam became a real plan of action: "One
evening I was sitting in the rooms of the Analytical Society at
Cambridge, my head leaning forward on the table in a kind of dreamy

mood, with a table of logarithms laying open before me." Another member, coming into the room, and seeing me half asleep, called out, 'Well Babbage, what are you dreaming about?' to which I replied, 'I am thinking that all these tables (pointing to the logarithms) might be calculated by machinery.[7]' He would call his machine the Difference Engine. It married a mechanism of cogs and gears with the method of finite differences to calculate the desired tables. In de Prony "factory" terms, he would create a device that would mechanize the labors of his third group of adders and subtractors. Babbage was specifically interested in scientific computations. The method of finite differences reduced a polynomial equation to additions and subtractions that his Difference Engine could calculate. Babbage set down his ideas in a paper he presented to the Astronomical Society in 1821 titled, "Observations on the Application of Machinery to the Computation of Mathematical Tables." The Society awarded his paper its gold medal.

Babbage produced a small tabletop working model in 1822 that could solve any polynomial equation to six figures of accuracy. He actively promoted the concept, as he needed significant funding to move forward. He received assistance from the Royal Astronomical Society, which supported his petition to the British government. Interested in the potential of the machine for use in such projects as updating the British nautical tables, the government agreed to fund of 1,700 British pounds[8]. When Thomas de Colmar sought British investment in his Arithmometer, he discovered that the government had already made the commitment to Babbage.

Babbage started the project in earnest in 1823, working out of his house. His plans detailed a machine that was ten feet high, five feet deep, and twenty feet long. Composed of 25,000 parts, it would weigh up to 15 tons. The precise machining of the thousands of intermeshed gears challenged the techniques of metal fabrication available at the time. Babbage engaged a top engineer and craftsman, Joseph Clement, to design the manufacturing tools and construct the rods and gears for the mechanism. In 1827, Babbage inherited 100,000 British pounds from his father's estate. Despite this fortune, he still needed additional funds to build a fully functional machine. Two years later, he was again asking the government for additional monies, and he received 7,500 pounds[9]. The government's outlay would eventually reach 17,000 pounds.

By 1832, now roughly ten years into the project, Babbage completed one section of the machine but continually fought with his engineer. Clement grossly abused the relationship, charging the project for tools he developed for other projects in his shop. He extorted more funds by holding back his detailed design drawings. Babbage's project stalled for several years as he attempted to get out from underneath Clement's various schemes. Babbage certainly did not help his cause, as he was a perfectionist. Even when Clement was cooperating, progress was slow. By 1834, he finally succeeded in securing the in-process Difference Engine and design drawings from Clement. The myriad frustrations had led to Babbage's near breakdown, and he took some time off. It was during this period that Babbage developed the concept of a far more advanced machine, a design he called the Analytical Engine.

Twelve years into the Difference Engine and seeking even more funds from the British government, Babbage made a fatal mistake. He mentioned he had conceived of a better machine. He recommended scrapping the Difference Engine and starting a new project. The British government, having wasted thousands of pounds and many years with nothing to show for it, ended its support.

Babbage designed the Difference Engine for a single function, to solve polynomial-based mathematical problems. He equipped the Analytical Engine with a programming capability to solve an entire range of mathematical problems. He arrived at the new concept when he realized that his Difference Engine would only approximate the result over a specific range of values. The new machine could loop back whenever it veered outside the range, adjust the settings, and proceed. To control this loopback function, he adopted the punched-card approach used in the Jacquard loom. Joseph Jacquard, a French weaver, improved on previous loom designs by using cards punched with holes to program the loom's operation. He completed his programmable loom around 1790. By 1812, there were an estimated 11,000 Jacquard looms in France[10].

Babbage completed the basic design of his Analytical Engine in 1836. The punched cards would instruct the machine's mechanics through changes in the standard processing sequence such branching and looping. He defined a section to the machine where numbers could be saved, which he termed the "store." The "store" had metal counter wheels to represent up a thousand 50-digit numbers. He had a separate section to perform the actual arithmetic functions. He called this the "mill." The set of controlling instructions in punched-hole cards was the controlling program.

Babbage struggled with the Difference Engine, and now his Analytic Engine was at least an order of magnitude more complex. There were certainly valid reasons why the Difference Engine fell short. He was constantly short of funds. He struggled with his craftsman, Joseph Clement. He also struggled with himself. The Difference Engine was important to him, but it was not his only passion. His other interests were wide-ranging. He authored a book on economics. He enjoyed the verbal parrying at the Royal Society. He consulted with the Post Office. He studied the British railways, advocating a shift to broad-gauge tracks. He invented the "cow catcher", the frame at the front of the locomotive to clear the tracks. He pursued a long campaign against noise in London streets[11]. He was in charge of a mapping survey of Ireland. He translated a French textbook on calculus. He wrote about theology, including a defense of divine miracles. He was interested in cryptography and solved the seemingly unbreakable Vigenere cipher during the Crimean War[12]. He ran for Parliament twice, both times unsuccessfully. In this light, it is easier to understand how he failed to finish either machine or give a single lecture at Cambridge.

Babbage had an additional avocation. He enjoyed hosting parties at his home, where the rich and famous in science, politics, and business joined him. Babbage delighted in showing his guests the operation of the small working model of the Difference Engine. It was at one of these affairs in 1833 that he met Ada Lovelace. She was the daughter of famous poet Lord Byron. Byron had numerous affairs, and eventually both his wife and Ada left him. Ada married William King, the Earl of Lovelace, in 1835, making her the Countess of Lovelace. Fascinated by mathematics, she engaged Augustus De Morgan as her math tutor. She quickly mastered trigonometry and the calculus. Babbage's machine intrigued her, and the two started corresponding. Although many others saw Babbage's working model and were entertained by the flurry of gears, Ada appreciated the potential of the Difference Engine and, even more so, the Analytical Engine.

The English government had lost patience with Babbage, but scientists in Italy were very interested in his machines. Babbage traveled to Turin in 1842 and presented the Analytical Engine design to a group of mathematicians and engineers. One of the mathematicians, Luigi Menabrea, would later become the president of Italy. Babbage encouraged him to write a report about the Analytical Engine, which Menabrea published in French. Ada volunteered to translate the description of the Babbage machine into English. She also decided to add her own section at the end. This "Notes by the Translator" ballooned to 20,000 words, or seven times the length of Menabrea's text. Her "Notes" became famous for the four concepts she defined in the Analytical Engine. First, it was a general-purpose machine, unlike the Difference Engine. Second, the machine worked in numbers but its use was not limited to numbers. Third, the machine ran by a series of instructions, or a program, that could include algorithms and subroutines. Finally, she ended with the caveat that despite all of the things the machine could do, it could not think. It only followed the holes in the inserted punched cards. Ada also developed a program to run on the planned machine, the computation of Bernoulli numbers. She listed the steps in the calculation algorithm, including defining the needed "variables." In this 20,000-word addendum, she essentially described the art of computer programming.

Babbage eventually realized his Analytical Engine was beyond the technology of the day. He returned to the Difference Engine in 1847. Borrowing some of the ideas from the Analytical Engine, he worked on Difference Engine No. 2. It would handle 31-digit numbers while calculating up to seven orders of finite differences. He displayed a partial version at the 1862 International Exposition in London. Babbage would continue to tinker on both the Difference and Analytical engines until his death in 1871. His youngest son, Henry, continued the work, starting with Difference Engine No. 1. He produced six additional sections. He also built a portion of the Analytical Engine, the "mill" or central processing unit. Henry completed that unit in 1910 when he was 86 years old. It handled all four arithmetic functions.

In 1991, the Science Museum in London completed construction of Babbage's Difference Engine No. 2. The completed machine stands seven feet high, 11 feet long, and weighs five tons. It actually works. Cranking by hand, not by steam, sends thousands of gears whizzing and clicking as it calculates and prints the answer. With the success of this project, Doron Swade of the Science Museum organized a second project in 2011 to construct the Analytical Engine. It is perhaps a fitting tribute to Babbage that the plan fell short. However, the patient reader will soon learn that IBM built two versions of Babbage's concept, the Automatic Sequence Controlled Calculator (ASCC) in 1944 and the Selective Sequence Electronic Computer (SSEC) in 1948.

As Charles Babbage struggled to complete his Difference Engine, another parallel effort emerged. His book on mass production, *Economy of Machinery and Manufactures*, included descriptions of his ideas to automate mathematical calculations. Dionnysus Lardner, a popularizer of the latest in science, wrote a review of the book in 1834 and the Difference Engine in the *Edinburgh Review*[13]. Georg Scheutz, a Swedish engineer who invented a printing press and steam turbine, read both Babbage's book and the review. He was intrigued[14]. Assisted by his son Edvard, they constructed several experimental models. In 1837, Edvard proposed building a full-size working Difference Engine. It was smaller in scope than the Babbage design, featuring just three orders of differences and printed results to five decimal places. However, the machine worked, despite the fact that Edvard built the unit in a wooden frame with hand tools.

Georg and Edvard tried to sell the completed machine to the British government. The government was still recovering from the Babbage experience and wanted nothing to do with another Difference Engine. The Scheutz father and son secured funding from the Swedish government to construct a full-size Difference Engine, one constructed of metal. Completed in 1853, it handled four orders of differences and up to 15-digit numbers. Crank operated, its maze of gears that turned banks of decimal wheels.

The Scheutz father and son took the machine to London in 1854 and then on to Paris in 1855 for the World Exposition. It won the show's gold medal, beating out Thomas de Colmar's Arithmometer. Babbage took his Difference Engine prototype to the Great Exhibition at the Crystal Palace in London in 1851, with far less success. When in London, the Scheutz father and son met with Babbage, and to their surprise, he encouraged them and promoted their machine. Babbage also visited the Paris show and saw the Scheutz Difference Engine in operation. The British government would finally order a Scheutz machine in 1859. The device added forty-four 32-digit numbers within 60 seconds. The government put it to work calculating life insurance tables[15]. The success of the Scheutz duo in creating the Babbage Difference Engine certainly begs a question about Babbage. Was he undone by the desire to create a more complex machine or did perfectionism lead to his failure?

Benjamin Gould, an American astronomer, purchased the Paris Scheutz machine in 1857 and shipped it to the Dudley Observatory in Albany, New York. Gould knew Babbage and he coordinated with fellow astronomers to make the purchase and put it to work computing astral orbital positions for the *Nautical Almanac*. Over time, disputes over the cost and effectiveness of the machine led to its disuse. The advent of more accomplished desk calculators spelled the end of mammoth devices like the Babbage and Scheutz Difference Engines. The Dudley Observatory machine collected dust for decades until Dorr E. Felt, the inventor of the Comptometer, purchased in 1924 for his collection. The Smithsonian Institute acquired the machine in 1963.

It was the Colmar Arithmometer, commercialized in 1850, that was the successful computing technology during Babbage's time. Despite all of the varied accomplishments of his life, his reputation rests on an invention he was never close to completing. His Analytical Engine was simply a century ahead of its time. Historians would call him the "father of the computer" despite his never building a working one. Babbage did make one other contribution to computing. He was asked, "Pray, Mr. Babbage, if you put into the machine wrong figures, will the right answers come out?[16]" He answer was, of course, that the right answers would not magically appear. He was thus describing the familiar "GIGO" concept: garbage in, garbage out—again a century early.

As for Ada, Lady Lovelace, she died young of ovarian cancer. She has been called the first computer programmer. It turns out that Ada would be a good name for a computer programming language, but that would have to wait.

# HOLLERITH AND DATA PROCESSING

From Pascal to Babbage, the focus was on mathematical calculations. What about those situations where the computation was minimal but the volume of data was overwhelming? In the late 19th century, there was a surge in the consolidation of companies. Far larger enterprises, with railroads and insurance companies in the vanguard, dealt with the vast amounts of transactions required to keep their operations running. The typical response was to multiply what had worked when the organizations were small. A small insurance company with a few accountants would grow by adding banks and banks of accountants until the work was complete. It was the manufacturing approach of De Prony but applied to the data-driven tasks. Though there were large companies with such challenges, one accounting task dwarfed all the others. That was the United States Census. It did not get much bigger than accounting for 50 million people.

The Constitution required running this immense data gathering and counting task every ten years. The first census in 1790 dealt with a total population of 3.9 million. It limited the data gathered to several questions, including the name of the head of the household, the number of individuals living in the household, the number of free white males, free white females, and the number of slaves[1]. The handwritten forms ended up in Washington for compilation. Census processing took 18 months. In 1840, with the U.S. population at 17 million, the compilation required 28 clerks. By 1870, with the population at 40 million, there were 438 clerks and the Census report ballooned to 3,473 pages[2]. Census clerks used large sections of paper called tallying sheets that spread the data elements across the top and individuals down the side. Each tally sheet contained 100 individuals. The clerks "counted" with groups of five tally marks, combining four right-slanting slashes with a left-slanting slash[3]. As the number of questions increased, these sheets became unwieldy, despite an escalating number of clerks.

Charles Seaton was the chief clerk for the 1870 Census. He created a rudimentary wooden roller device that enabled tally counts up to 160 entries per row. The innovation brought widely separated data columns closer together for easier tabulation. It was an improvement to be sure, but one that did not dramatically reduce the overall scope of the work. Seaton received an award of $15,000 for his invention. A typical Census clerk had to work 29 years to earn that amount of money. It is roughly equivalent to $360,000 in 2024 dollars. The Census used the Seaton device for the latter part of the 1870 Census and the entire 1880 Census. In 1881, Seaton became superintendent of the Census.

The scope of the 1880 Census increased substantially. In addition to the basic accounting of 50 million individuals, census takers asked far more questions. These included questions on race, occupation, real estate, personal estates, birth and marriage dates, school attendance, and literacy. Though Congress passed a law in 1850 limiting the Census to 100 questions, the pressure for more data kept working against that target. The processing of the 1880 Census required 1,495 clerks armed with massive tally sheets and the wooden Seaton boxes[4]. The tabulation took nine years to complete, with the last volume published in 1889. The 1890 Census was just around the corner, and prospects for its processing were daunting. The Census expected a population increase by at least ten million, and a doubling of the amount of the surveyed data.

The laborious task of the 1880 Census was not a surprise for one individual who had worked on the project as a summer job. His name was Herman Hollerith. Born in 1860 in Buffalo, New York, he attended the City College of New York and then the Columbia School of Mines, where he studied engineering. He graduated in 1879 at the age of 19. One of his professors at Columbia, William Trowbridge, was an agent for the Census. He offered Herman a job working for him in Washington, compiling a report from Census data on the use of steam and power. Hollerith had the opportunity to see and use Seaton's tally sheet mechanism.

While Hollerith was in Washington, the new Census director Francis Walker enlisted Dr. John Shaw Billings to work on the Vital Statistics section of the Census. Billings was a major in the U.S. Army and the Army Surgeon General. Trowbridge connected Hollerith to Billings and Hollerith began work on the Vital Statistics project. By happenstance, Hollerith ran into Billings' daughter Kate at a Potomac River rowing party. Kate took notice when Hollerith enjoyed the chicken salad served at the party. She invited him to her house for dinner and naturally had chicken salad on the menu. During the course of that dinner, Hollerith talked with Dr. Billings about the inefficiency of the Census tabulating process. Billings suggested using notches on cards to represent the data. That got Hollerith to thinking. Years later, when asked where he got the idea for punched-card data processing, Hollerith would respond, "Chicken salad![5]"

Francis Walker left the Census in 1881 to become president of MIT. A year later, he persuaded Hollerith to join the MIT faculty as a mechanical engineering instructor. While at MIT, he started to experiment with card-based tabulation. He needed to register the data on a card and read the information back mechanically. After a year at MIT, Hollerith concluded he preferred inventing to teaching. He went to work at the Patent Office in Washington DC, a good place to start for a budding inventor.

**Figure 5.** Herman Hollerith, circa 1888. Library of Congress
collection (original photograph by Charles Milton Bell).

Hollerith continued to work on the concept of a card-based
capture system. Seaton's huge award for such a simple invention
spurred him on. While traveling on the New York Railway, Hollerith
observed the conductor using punched holes on passengers' tickets to
profile that person. The punches represented such characteristics as

male or female, light or dark hair. As a mechanical engineer, he was familiar with the use of holes punched in cards to control the Jacquard loom. Wooden pegs in the loom sensed the holes in the cards. Hollerith also had the example of the telegraph. Invented twenty years earlier, the device used punched holes in paper tape to record the incoming message. Hollerith's first design also used paper tape as an input to the tabulating device. Heeding the advice of Dr. Billings, he switched from paper tape to hard stock cards. His evolving design included a card punch, a card tabulator, and a sorter.

Hollerith was at the Patent Office for a year. He filed a patent application for his tabulation system titled "Certain new and useful improvements in the art of compiling statistics." While at the Patent Office, he worked with Tolbert Lanston, who was working on an adding machine based on Thomas de Colmar's design. Hollerith made a mental note of Lanston's design. He already knew that his tabulator would eventually need to do more than just count. It would also need to add. Hollerith spent some time as a patent consultant, but by 1885, he was short of funds and took a position as an engineer for Mallinckrodt Brake Company, a Missouri railroad equipment manufacturer.

Hollerith continued to fine-tune his design. He borrowed $2,500 from his brother to help finance the development of his machinery. He decided that his cards would be 6-5/8 inches by 3-1/4 inches—the size of the current U.S. bank note[6]. He selected this size because the Treasury Department could provide organizing trays for that size at no cost. He clipped one corner of the card in order to ensure the proper orientation, a feature that remained in all future IBM cards. The card design that he patented in 1884 had 20 columns and 10 rows. He expanded the card to 24 columns for the 1890 Census.

Operators punched data from the field recoding sheets using a device called a pantograph. His first prototype was a levered unit that punched one hole at a time before repositioning. In early tests, this proved to be far too labor-intensive. With at least 13 million census sheets to be processed, he needed a better design. Hollerith revamped his pantograph to punch all the positions in the card at once. In tests with trained users, typical production times were 700 cards per day per user, with experts achieving 1,100 cards[7]. That represented 18,700 total data punches. He contracted with Pratt and Whitney to build these keypunches with electrics supplied by Western Electric. The initial cards were blank. Those working day in and day out with the cards could tell you the identity of any data element just by the position of the hole.

Hollerith designed a sorting box to sequence the punched cards for the tabulation run. Once a deck of cards was ready, the operator placed each card on the tabulator-sensing platform. Metal pins protruded through the punched holes, connected with a mercury cap, and completed an electrical circuit. It sensed all the holes in the card in one operation. The electrical connections registered on forty counting dials that incremented to 9,999 before being manually reset. Hollerith also deployed telephone relays to link several holes in a set. An example was counting white individuals who were female. A bell would ring if the card did not register properly. A skilled operator could process about 40 cards per minute. Hollerith graduated from Columbia ten years before electrical engineering was taught there. It took Edison's invention of the light bulb and the power system to exploit it before this technology made the curriculum. Hollerith's use of electricity in the tabulator was all self-taught.

The initial machinery was in place by 1886, and Hollerith looked for a test case. He found it with the City of Baltimore. He would be working again with Dr. Billings, who was managing the tabulation of mortality statistics for the city. The key to success in this first project was Billings' decision to consolidate all of the information for an individual on a single card. After a successful test, Hollerith implemented his system at the Bureau of Vital Statistics in New Jersey and the Health Department in New York City. In 1889, Dr. Billings installed Hollerith machines in the offices of the Army Surgeon General. This was a permanent installation and led to Hollerith's decision to rent his machines instead of selling them. The equipment required a considerable amount of service to stay operational and this was easier to administer with a rental contract.

Certainly in the back of Hollerith's mind at this time was the colossus of data tabulation, the upcoming 1890 Census. Robert Porter was the new superintendent for this census. Hollerith had worked for him on the 1880 Census, and they remained on friendly terms. Porter wanted to implement a far better system than the one in use. He decided to evaluate several promising alternatives by conducting a contest. He received three proposals, from Charles Pidgen, William Hunt, and Hollerith. All three submissions assumed a new process with one card per individual, replacing the tally sheets. Pidgen and Hunt developed a manual system for the 1885 Massachusetts census. Now, each was proposing variations of that process. They used the so-called "chip" system, where Census clerk transcribed the input data onto different-colored cards, or "chips." This made sorting easier. Tabulation remained a manual process. Porter selected the census data from one county for the test, comprising 10,491 individuals from St. Louis. The Hollerith system took 78 hours start to finish while the Pidgen approach took 155 hours and the Hunt system 200 hours. Though the timings for data capture and sorting were closer, the Hollerith machines shined in tabulation, completing that phase in six hours while the other systems were at least eight times slower[8]. Since the Hollerith process captured the data on machine-readable punched cards, additional tabulation runs would maintain and extend that dramatic edge.

As the clear winner, Hollerith received a Census contract in December 1889. Superintendent Porter ordered 56 machines at an annual rent of $1,000 per machine. He added six machines for the Department of Vital Statistics. Perhaps anticipating the use of the Hollerith invention, Census administrators decided to increase the information recorded on each household, virtually doubling the data to 235 separate criteria.

**Figure 6**. The Hollerith-designed punched card electrical counting machine. Drawing from Scientific American, August 30, 1890.

The 1890 Census used 46,804 enumerators to collect the data of 18 million households. The average clerk punched 7,000 to 8,000 cards per day. On a typical day, eighty-one clerks produced 556,346 cards. [9] The Census went through over 100 million cards. Just six weeks into the count, the Hollerith machines calculated the total U.S. population count at 62,622,250. This created a minor controversy as the *New York Herald* felt that number was low and concluded that Hollerith's machines must be at fault. Superintendent Porter defended Hollerith and the accuracy of the newly automated Census process.

Census results were available in 18 months, comprising 26,000 pages of information. The 1890 Census cost $5.8 million, roughly double the cost of the 1880 Census[10]. This reflected the near doubling of the information processed and the 26 percent increase in the U.S. population. The permanent capture of the census data on punched cards also enabled additional sorting and tabulation at minimal incremental cost. Hollerith received $750,000 in equipment rental fees.

The 1890 Census saved an estimated $5 million in expense. This was ten times the estimate going into the work.

The machine rental fees were not the only recompense Hollerith received. He demonstrated his machines at the 1889 Paris Exposition and received the gold medal. The following year he received the Franklin Institute invention award. Columbia awarded him a doctorate. Hollerith was able to complete doctorate requirements away from campus. His dissertation was, of course, a detailed description of his electric tabulation system.

Not all was goodness though. When the 1890 tabulation was complete, Census returned the machines and the rental payments abruptly ended. He could not survive solely by winning the U.S. Census contract every ten years. Hollerith first operated his business as the Herman Hollerith Tabulating System. He formally incorporated in 1896 as the Tabulating Machine Company. Hollerith retained 50.2 percent of the company while trusted friends bought the remaining shares.

He needed additional customers, and other censuses were the logical low-hanging fruit. He traveled to Europe, stopping in in Austria, Italy, Russia, England, and Switzerland. He won agreements for census processing in Norway, Canada, Austria, England, Germany, and Russia. The most important of these was Russia. Czar Nicholas II approached Hollerith in 1895 ahead of Russia's planned first census in 1897. At the time, Austria had invalidated the Hollerith patents and was working on its own machines. In an ill-advised remark, Hollerith told the Russians that they could go ahead and use the Austrian machines if they did not want Hollerith's quality. Russia decided to explore this option. Meanwhile, Hollerith was running short on cash to operate his business and needed significant capital if the Russian deal came through. He cut back all operations and expenses. In the end, Russia decided to go with the Hollerith machines. The rental fees to conduct the Russian 1897 Census, surveying 125 million people[11], would help Hollerith stay afloat until the 1900 U.S. Census. As his business expanded, his data processing machines became widely known as the Hollerith System and the punched card as the Hollerith Card.

Hollerith realized it would be feast or famine with census work, even with multiple contracts with a number of countries. He turned to commercial accounts. He developed a sales brochure for life insurance companies, describing tabulation applications for "wage administration, sales analysis, and cost accounting.[12]" His first customer was Prudential Life Insurance, which installed the Hollerith System in 1891. The New York Central Railroad approached him in 1894, wanting to process freight waybills. Railroads had hundreds of accountants calculating these freight invoices, totaling four million a year[13]. Each invoice contained the route-miles and the customer charge.

However, there was a problem. Both companies wanted to enter and add numeric fields. They were not interested in just simple counting, as was the case with census data. Hollerith remembered the Lanston adding machine from his days at the Patent Office. He retrofitted his basic machine with three separate adding machines that mounted vertically within the tabulator frame to support addition up to eight decimal digits. He enhanced the implementation with his version of the Leibniz "step reckoner." He installed a pilot system at the New York Central Railroad in 1895. The railroad was initially not happy. Hollerith redesigned the add function and the railroad put the machines into full production in 1897.

The Census staged another contest to select the vendor for the 1900 Census. Hollerith won again. He contracted for 311 tabulators, 1,021 keypunches, and 20 sorters, and was paid $550,000 in rent[14]. There were several major changes ahead of the 1910 Census. First, the Census organization was no longer a quadrennial function. It was now a permanent bureau. Second, the new Census Bureau had funding for an in-house development operation called the Machine Shop. Hollerith had raised his rental prices. He now charged a fee for every card tabulated and every card sorted. The card tabulation fee was 65 cents per thousand cards, which alone came to about $65,000 in added cost to the bureau[15]. Furthermore, the Bureau was required to purchase all from his company. Card sales were highly profitable.

In 1901, Teddy Roosevelt became President, after McKinley's assassination. The new administration replaced Census superintendent Robert Porter, a Brit, with Simon North. Porter returned to England, and after securing patent and royalty rights from Hollerith, set up The Tabulator, Limited, in 1904. This would later become the British Tabulating Machines Company. Porter's company did the British census in 1911. New superintendent North, a former newspaper editor and lobbyist for the wool industry, had been the chief statistician of the 1900 U.S. Census. North felt the cost of the Hollerith equipment was exorbitant. In addition, there was a sense that the departure of Porter along with patent rights was not entirely above board. North secured $40,000 in additional funding in 1904 and decided that the bureau would develop its own machines. Hollerith voided the existing Census contract and pulled his machines. He was flooded with business elsewhere, so it was an easy decision.

Meanwhile, North staffed his expanded Machine Shop with three engineers that had previously worked for Hollerith, including his key foreman. Hollerith was not happy. The Census shop staff would grow to sixteen. One of those engineers was James Powers. Born in Odessa, Russia, he immigrated to the United States in 1889. He worked briefly for Western Electric before the Census hired him to work on 1890 Census. North asked Powers to design new sorters and keypunches. Powers was determined to improve on the Hollerith machines. He designed a whole card keypunch with a 240-key keyboard. He automated the card feed and ejection with a foot-actuated mechanism. He retained the existing Hollerith card, now with 45 columns and 12 rows. The Powers keypunches proved to be unreliable, forcing the Census Bureau to overhaul their existing Hollerith machines.

By this time, a number of Hollerith's original patents had lapsed. As he enhanced his machines, he added new patents. Patent rights still protected the electric sense mechanism of the tabulator. Powers designed a replacement method, pushing spring-loaded rods through the holes in the card to count mechanically. His new tabulator, even with its purely mechanical operation, was significantly faster than the Hollerith machine. Hollerith sued the government for patent infringement and even wrote letters to the President. He would spend seven years on litigation, all to no avail. The new Census machines also added a significant new feature, a printing mechanism that replaced the manual process of recording and resetting the counters. North loaned the new machines to Cuba for its census as a test, then used the machines for the 1910 and 1920 Censuses.

The Census Bureau and Hollerith's Tabulating Machine Company (TMC) parted ways in 1905. Commercial accounts became a full-time focus. Hollerith's company signed Aetna Life in 1910. By 1931, nearly every department at Aetna had a "Hollerith" department[16]. New York Edison adopted the Hollerith system in 1903 for sales analysis. Taft-Peirce, the company in Rhode Island manufactured Hollerith's early tabulators, started using the machines in 1905 for cost accounting. The other key partner, Western Electric, installed Hollerith equipment in 1906. Eastman Kodak in Rochester, New York, added machines that same year. Metropolitan Life approached Hollerith early on, but he said he was too busy to design custom machines for them. Still, they became a customer in 1911. Marshall Fields was the first retail account, installing equipment in 1906 for sales analysis. Other early accounts included National City Bank, Merrill Lynch, ATT, and Swift and Company. By 1908, TMC had 30 commercial customers.

The railroads continued to be a strong customer segment for TMC. Following the New York Central, the Long Island Railroad— part of the Pennsylvania Railroad system—was the second line to sign up, in 1903. The parent railroad followed the next year. The Rock Island Railroad, Union Pacific, and Southern Pacific installed TMC equipment. By 1928 most, U.S. railroads were using Hollerith machines. The Union Pacific had waited two years to receive its machines. Part of the delay was manufacturing problems at the Taft-Peirce Company. Hollerith had purchased the company in 1901, when business was strong and the combination made sense. However, Taft-Peirce soon ran into financial problems and became a drag on TMC income. Hollerith sold the company in 1905.

Though Hollerith had a near-monopoly with his tabulation system in the early years, the experience with the Census Bureau showed that this might not continue. James Powers left the Machine Shop to start his own tabulating company. This meant that TMC needed to continue to innovate to stay on top. This actually played to Herman Hollerith's core strength -- invention. As a manager of a fast-growing business, Hollerith was difficult to work with. He had poor interpersonal skills and a frequently irritable temperament. As his business grew, he lacked the essential skills to adapt and, most importantly, to delegate. He continued to be a micro-manager. He was not a fan of salesmen and preferred to secure new business by word of mouth and his own "selling" efforts. He was also not big on advertising. He would turn down new customers if he was too busy.

One positive aspect of the Taft-Peirce relationship was Eugene Ford. He was the company's principal engineer, an inventor based in Uxbridge, Massachusetts. In 1897, Ford developed an innovative typewriter and moved to New York City to drum up interest in the machine. When that effort fell short, he signed on with Taft-Peirce. By 1904, he was working in tandem with Hollerith on tabulation engineering. Once the company moved past the simple U.S. Census requirements, the need for enhancements exploded. Hollerith replaced the mercury caps of the early tabulators with wire brushes, resulting in faster and more reliable reading. The keypunch was electrified. The year 1906 saw a major change in setting up the tabulator. The early tabulators physically connected wires from the punched hole sensors to either the counter wheels or relays. Since the Census had one application and it rarely changed, this was more than sufficient. As companies embraced the Hollerith system, they soon had additional applications. Each new application required renting an additional tabulator or rewiring an existing machine. The answer to this problem was the plugboard. It was a removable panel that used jumper wires to connect card positions to registers. It enabled one machine to handle multiple applications simply by swapping the plugboard. Hollerith released the enhanced tabulator as the Hollerith Type 1.

# CHARLES FLINT AND HIS TRUSTS

General Antonio López de Santa Anna, famed for the siege of the Alamo in Texas in 1836, served 11 terms as the president of Mexico. The terms were not consecutive, as he was forced into exile from 1855 to 1874[1]. In 1866, he improbably ended up in Staten Island, New York, attempting to sell chicle for the manufacture of tires. Chicle was the sap of a Mexican tree prized as a gum for its flavor and high sugar content. The gum was also strong and flexible enough to be a potential substitute for rubber. Santa Ana imported a ton of chicle to the United States in the hopes of testing that proposition. He retained Thomas Adams, a scientist and inventor, to explore the market for his chicle[2]. Adams was unsuccessful in developing chicle as a rubber substitute and decided to try to commercialize the material as a chewing gum. In 1871, he developed a machine to produce chicle in sticks. His chicle chewing gum in convenient sticks caught on and before long, a new market emerged with a raft of chewing gum competitors.

This was the state of affairs at the end of the 19th century when Charles Flint got involved. Flint was a stock manipulator who was very much at the top of his game. Born in 1850 in a small town in Maine, he would carve a broad path through many of the country's industries before he was done[3]. He had many interests before he got involved in chicle. He embraced the automobile in its infancy, starting with a De Dion-Bouton in 1896, and followed by a Stanley Steamer and Locomobile[4]. He was one of the first individuals to get an automobile license and was a founder of the Automobile Club of America. He was also part of a group that built the yacht that defended the America's Cup. He brokered naval ships to Japan and then to Russia, which used them against each other in the 1905 war. He was a business partner of the Wright brothers and sold their airplanes. In his time, Flint was a force in the world. He knew all the important players, in business and in politics.

**Figure 7.** Charles R. Flint, 1907. Photograph from the Library of Congress collections.

However, despite these myriad endeavors, he was known as the "Father of Trusts," a title bestowed by the Chicago newspapers. A trust was a consolidation of companies in a single industry. The companies were brought together to gain monopoly market power.

The practice was not unique to Flint. It was rampant starting in the 1880s. Flint shied away from the use of the term "trusts." He called his creations industrial consolidations, and he would set forth the many advantages of greater size, scope, and efficiencies of large companies over smaller ones. He avoided any mention of the key benefit of monopolistic pricing, except to his financial backers.

Flint had a slow start in trusts. His first foray into the field was United States Electric Lighting. It failed to prosper as it competed with Thomas Edison and General Electric[5]. In the early 1890s, Flint acted as a consultant for the American Bicycle Company, a conglomeration headed by Colonel Albert Pope that tried to consolidate the industry just as the bicycle craze was ending and the automobile era was beginning. In 1892, Flint took aim at the rubber industry. He used his considerable stock skills to combine a number of smaller rubber manufacturers to form U.S. Rubber. As was his typical approach, the stock of the newly formed trust wildly overstated the value of the constituent companies.

By 1899, Flint had finally gotten his trust formula down. With his burgeoning reputation, trust-to-be companies came to him rather than Flint seeking them out. He stitched a number of New England woolen mills together to form American Woolen. Moving to a related industry, he created United States Bobbin and Shuttle. Then, he improbably turned to chewing gum. He bought Thomas Adams's pioneer company, Adams Chewing Gum, and then added Chiclets, Dentyne, and Beemans to form American Chicle. He capitalized the new gum trust at $14 million though the true value was likely only $500,000[6]. He maintained that most of the difference was the "goodwill" valuation of each chewing gum's brand. American Chicle started with about 50 percent of the market. There was a small competitor in Chicago that escaped his grasp. Wrigley's Chewing Gum had barely a 1 percent share of the market, but it embraced advertising while American Chicle did not. In short order, American Chicle's market share sank to 15 percent while Wrigley's soared to over 50 percent[7].

Despite the setback with chewing gum, Charles Flint was on to additional industries and more trust combinations. The Sherman Antitrust Act of 1890 did not appear to slow him down. He set his sights on the time-clock industry. The industry leader was Bundy

Manufacturing. Willard Bundy was a clock maker in Auburn, New York, who developed a clock-based recorder for his employees to punch in and out during the workday. A novel invention, customers called the company's time clocks, "Bundys." Willard's brother Harlow set up Bundy Manufacturing in Binghamton, New York. George Fairchild, a six-term U.S. congressman and friend of Harlow, joined the company in 1896. After securing Bundy, Flint looked for other time-clock companies for his "time trust." He added Standard Time, Willard and Frick, Chicago Time Register, Dey Time Register, and Syracuse Time Record. He moved the headquarters of the new enterprise, now christened International Time Recorder Company, from Binghamton to nearby Endicott.

Starting in 1901, Flint sought to duplicate the formula with weighing scales and meat and cheese slicers. He acquired the Computing Scale Company in Dayton, Ohio, and added Money-weight Scale, Detroit Scale, and American Automatic Scale to form the much larger trust that retained the name Computing Scale Company.

Flint departed from his normal formula in 1908 when he made a run at Hollerith's Tabulating Machine Company. It was an odd divergence from his normal practice. The only common denominator was monopoly power, as TMC created a new industry and was its own market trust. Hollerith rebuffed Flint's offer. Three years later, Flint upped his offer by a factor of twenty and approached Hollerith again. The timing was better. Hollerith was increasingly anxious to retire. He had suffered a heart attack and his doctor had recommended rest. He was still trying to run his far larger operation by himself. Flint and Hollerith reached an agreement. Flint would pay $2.3 million for the company[8]. Hollerith would pocket slightly more than half of that, with his investors receiving the balance. In addition, Flint retained Hollerith as a consultant at an annual salary of $20,000 for a period of ten years. Hollerith still wanted a hand in the ongoing improvement of his tabulating system, including veto power over any new designs.

Flint decided to combine the tabulating company with his time clock and computing scale trusts. It was clearly not a pure industry trust that had worked so well before. The three companies all produced electromechanical devices with similar working components such as dials and gears. Flint felt that they could also share administrative and sales resources. More than likely, the real impetus was in stock manipulation with a payoff in capital gains.

Flint called the combined enterprise C-T-R, for "Computing-Tabulating-Recording." He capitalized the company at $19 million, with $12 million in stock and $7 million in bonds. He financed it with $4 million in loans from Guaranty Trust of New York[9,10]. He installed Fairchild as the President. The new company had roughly 1,200 employees, with a head office in New York City and major plants in Endicott, New York, and Dayton, Ohio. Net income for the first year was $800,000. Charles Flint was especially proud of C-T-R. He was an active participant, attending every biweekly board meeting for the next eight years.

**Figure 8.** 1911 C-T-R (Computing-Tabulating – Recording) logo in the then-popular Rococo style. Reprint Courtesy of IBM Corporation ©.

Flint did not see George Fairchild in an ongoing hands-on role. In 1912, Flint tapped Frank Kondolf, the lead executive of International Time Recorder, to run the three companies. Yet, neither Fairchild nor Kondolf performed to Flint's satisfaction. He wanted a rapid integration of the companies and he was not going to wait. He put out feelers that he was looking for a replacement executive.

PART TWO
# THE WATSON SENIOR YEARS

# THOMAS WATSON, SR. AND NCR

Thomas J. Watson, a man who would become one of the highest paid individuals in the entire country, started out life quite modestly. He was born in 1874 in Painted Post, a small town in upstate New York, 75 miles west of Binghamton. His father owned a farm and ran a small lumber operation. Watson attended District School Number Five, better known as the "Little Red Schoolhouse," in a rural, one-room building. In his teens, he attended Addison Academy. His father wanted him to go to college at Cornell, located in Ithaca 45 miles to the northeast. Instead, Watson spent one year studying accounting and business at the Miller School of Commerce. He left in 1891, taking a bookkeeper position at Risley's Meat Market in town for $6 a week[1]. He developed a ledger-card accounting system for the shop. However, he found himself out of work when business declined.

The following year, he worked for an itinerant salesman named George Cornwell, selling pianos and organs out of the back of a wagon. When Cornwell moved on, Watson took over the wagon and the territory. His salary was $12 a week, which seemed like a lot until another salesman pointed out that he could make far more if he was on commission. However, a lapse in judgment derailed his new venture. He left his traveling rig outside a saloon while he had more than a few drinks inside. When he emerged, the horse, wagon, and all the demonstration goods were gone. The company fired him and billed him for the lost property[2].

Watson moved to Buffalo, where he hoped to get back on his feet. He sold sewing machines for Wheeler and Wilcox. He lasted less than a year before he was fired. Still short on good judgment, he fell in with a slick and seemingly good salesman named C. R. Barron. Barron had arranged to sell shares in the Buffalo Building and Loan Company in exchange for commissions. Barron and Watson discussed plans to use their earnings to open up a chain of butcher shops. The plan came crashing down when Barron ran off with all of their monies. The bank fired Watson. His father stepped in, perhaps unwisely, and loaned his son the money to open a butcher shop. The new business proved a struggle, and Watson had to sell the shop's assets in order to pay off outstanding debts. It was yet another crushing defeat for the young Watson. However, he was driven. He needed to be successful and have others recognize him as successful.

The closeout of the shop left him with just one asset, the store's NCR cash register. That was no surprise, as National Cash Register under John Henry Patterson dominated the cash register industry. Patterson was born in 1844 to a pioneer family near Dayton, Ohio. He attended Miami University and Dartmouth College. After graduating, he worked on the family farm and as a toll collector on the Miami and Erie Canal. He moved on to the Southern Coal and Iron Company and rose to general manager. The company store continually lost money. Patterson discovered that his clerks were stealing from the cash drawer. He heard about a Dayton saloonkeeper named James S. Ritty who, with his brother John, had invented a device to record sales. The cash drawer remained locked until a sale was made and a very audible bell rang as the drawer was opened. The brothers called the device "Kitty's Incorruptible Cashier"[3] and sold it for $50. Patterson bought two for his company store. In 1882, he purchased shares in the company, got on the board of directors, and eventually bought the company. He changed the name to National Cash Register. He bought a rundown factory in Dayton in order to expand operations. He had second thoughts about the whole venture and tried to back out, but it was too late.

Patterson set about improving on the Ritty design. The resulting machine was indeed better but also quite a bit more expensive. It had many advantages over the standard cash register that were not readily apparent. This required selling. He decided to reinvent the practice of sales. He knew that there was a huge untapped market of businesses that needed an NCR cash register but did not realize it. The person to address that opportunity, his NCR sales representative, would not be a George Cornwell or C. R. Barron. He would dress formally, have a defined territory and quota, and be skilled in the art of selling. Patterson had his top salesmen record their standard sales pitch along with the typical flow of a sales call. He codified it in the NCR *Primer*, a booklet that included opening sales call statements such as, "I am visiting you to find out why you don't want a cash register." It included the advantages of the NCR machine, techniques to handle objections, and closing phrases such as "how soon do you want delivery?" Patterson had a fixation on the number "5"so naturally there were five steps to every sales call. Salesmen who used the guidelines in the *Primer* simply sold more cash registers.

*The NCR Primer* became the basic text for the sales training school that Patterson created in Dayton. He called it the Hall of Industrial Education, or simply the "Schoolhouse." Every prospect was termed a "probable purchaser." Patterson augmented the formal sales training with a summer event called Sugar Camp. He also started the Century Point Club, for those salesmen that had achieved their annual quotas. It included additional training in Dayton followed by three days in New York City. NCR sales representatives made very good money for the time.

Patterson did stop with his sales programs. He constructed a modern company facility outside of town on the family acreage. Besides the manufacturing and office areas, it included lockers, showers, restrooms, dining rooms, and an infirmary. The company, known locally as "the Cash," implemented a range of progressive programs not commonplace at the time, including employee training and shorter working hours. Patterson also focused on continual improvement of the company's cash register. He hired a future giant of American invention to lead his engineering team. Charles Kettering went straight from graduation at Ohio State University to head the NCR research department. The company's current cash registers were slow, hand-cranked devices. Kettering took up the challenge and his R&D team designed and produced an electric cash register, a momentous advance in the industry.

For all his good works, Patterson had a darker side. Like Charles Flint, he wanted to dominate his industry. He would acquire some eighty competitors in order to amass control of 95 percent of the cash register market. To protect his now monopolistic empire, he created a web of patents, combining the patents that he registered in-house with those he gained through acquisition. He was aggressive in suing for patent infringement.

Back in Painted Post, Thomas Watson took the butcher shop cash register to the NCR office in Buffalo. He purchased it on installment and needed to transfer ownership. While there, he decided to apply for a job. The NCR sales manager, John Range, took some convincing but finally agreed to hire him. It was 1895 and Watson was just 21. He had no success in his first few days of selling. He had learned all the wrong lessons in selling from Cornwell and Barron. Range accompanied him on several sales calls and used that disarming introduction from The NCR Primer, "I've visited to find out why you don't want a cash register.[4]" It was a way to understand the prospect ("probable purchaser") really needed one. Watson absorbed the sales lessons from Range and soon became a very successful salesman, selling 60-80 cash registers a week. His commissions were substantial, including a week when he earned more than $1,200, a princely sum in that era. He had finally found his calling.

After four years selling in Buffalo, NCR promoted Watson to sales manager in Rochester, a struggling branch office 75 miles to the east. He now earned a cut on each salesman's commissions. He had also developed a few sales techniques of his own, and not all were completely above board. Watson would have his salesmen tail the competition on their prospecting calls, then walk in directly afterward and make a competing presentation. Such sales tactics and the branch's success drew the attention of Hugh Chalmers, the lead sales executive in Dayton. Chalmers became his advocate and urged his promotion. Watson arrived in Dayton in 1903, with a very important project waiting for him.

John Patterson chaffed at any threats to his monopoly on cash registers. His 95 percent share of the cash register market was not enough. He chaffed at a cottage industry of small storefronts that bought and sold used NCR machines. They could offer them at a lower price than new NCR registers. Patterson had used a variety of tricks to beat competition, including creating a low-cost "knock-out" register to undercut a competitor's price. However, these shops continued to be successful.

Patterson and Chalmers outlined a scheme to Watson to force these used sellers out of business. Watson was to work undercover in each of the cities where a used cash register operation had surfaced. He was to set up a fake storefront nearby and begin selling used NCR registers at an even lower price. Chalmers gave Watson a budget of $1 million and a team of 56 men to carry out the plan. The first target was New York City, where Watson opened his first "store," called Watson's Cash Register and Second Hand Exchange. It was just down the street from the real used register shop. After weeks of under-cutting the target shop's prices, Watson would describe a bleak future for the competitor's shop and offer to buy him out for $21,000[5]. If the offer was accepted, they simply closed the shop. Watson's team repeated the deceit in city after city until most of the major used shops were gone.

Watson did an exceptional job with the scheme. Though it was patently illegal, Patterson rewarded him with a continued fast track at the Cash. In 1908, he became the assistant sales manager. Two years later, he was the national sales manager, overseeing 900 salesmen in more than 200 cities. He replaced his mentor, Hugh Chalmers, who Patterson had summarily fired, likely for taking too much credit for the company's success. Watson attacked his new sales mission with gusto. In his first year (1911), his sales team topped 100,000 cash registers for the first time[6]. The company reached the milestone of one million total cash registers sold by the following year. Patterson was apprecia-tive and rewarded Watson with a new house and a luxury Pierce-Arrow automobile.

The good times did not last, however. The "scorched earth" campaign Watson led was an open secret in the industry. Hugh Chalmers was bitter about his treatment by Patterson. He jumped at the chance to provide testimony on the scheme to the Taft administration's attorney general, George Wickersham. Chalmers walked the government through the scheme city by city, recounting the details. In 1912, the Justice Department indicted Watson and 26 other NCR executives, including Patterson, for violating the Sherman Antitrust Act by using illegal means to monopolize market share. It was a criminal indictment, meaning that jail time was possible if convicted. The following year, the trial concluded with all executives, including Watson and Patterson, found guilty. Watson had pled not guilty and vowed to appeal the verdict.

That March, the "Great Easter Flood" struck Dayton. The entire city was under water, with 90,000 people homeless[7]. Patterson sprang to action. He put his entire company at the city's disposal. He had his employees build rowboats in the factory and served thousands of meals at the company cafeteria. For his part, Watson chartered a train in New York City, had it loaded with relief supplies, and sent it to Dayton. The dramatic and welcome response brought calls for pardons for the NCR executives.

In the midst of all this turmoil and despite his now rehabilitated reputation in Dayton, Patterson fired Watson. As benevolent as Patterson was with most of his employees, he was brutal with his senior executives, especially those he viewed as amassing too much power. He was not a nice man but was particularly cruel with his executives. He would fire them on a whim and apparently delighted in the manner of each dismissal. An unforgiving transgression was a failure to memorize word for word Patterson's beloved 16-page *NCR Primer*.

Watson was in esteemed company as a Patterson target. Between 1910 and 1930, roughly a sixth of the top executives in the United States came from NCR. It was the golden age of automobiles, and ex-NCR men gravitated to the car companies in droves. Chalmers formed the Chalmers automobile company. He would merge with Maxwell and would later be acquired by Walter Chrysler, forming the #3 automobile maker after Ford and General Motors. Alvan Macauley, an engineer and patent attorney at NCR, would take over American Arithmometer, which would become the Burroughs Adding Machine Company. He would go on to run the Packard car company.

Edward Jordan, who managed advertising at NCR, worked for the Jeffrey Automobile Company and then founded the Jordan automobile company in 1916. This was no consolation for Watson. He was out on the street and still facing jail time. His only consolation was that former NCR executives were highly regarded and he could afford to be choosy.

# THE ODD TRUST OF COMPUTING-TABULATING-RECORDING

Watson had attended a speech by Charles Flint when he was the sales manager in Rochester. He learned that Flint was looking for someone to manage the Computing-Tabulating-Recording Company and gave him a call[1]. Watson viewed the opportunity as interesting, but he made it clear to Flint that he wanted a share of the profits and free rein to invest in growing the companies. George Fairchild was opposed to his hiring, in part because Watson was a convicted felon and could end up in prison. However, Fairchild also had another reason. His focus was on the company's stock price and timely dividends, which Watson's growth ideas would likely disrupt. Flint sided with Watson and hired him in May 1914 at a starting salary of $25,000 and options on C-T-R stock. But most importantly, he agreed to a profit sharing of 5 percent of earnings after dividends were paid. Flint had one key condition: Watson's conviction had to be resolved. Watson was hired on probation and when the case was closed by consent decree, Flint removed the interim label.

Watson's first order of business was to assess what he had in C-T-R. The time recorder company was the big profit driver. It was a near monopoly with solid earnings but limited growth potential. The story at Recording Scale was similar, but again, he was skeptical that any substantial investment would lead to greater results. He realized quite early on that Hollerith's operation was the jewel of the trust.

Watson saw the Hollerith equipment running at Eastman Kodak in 1904 when he was the NCR branch manager in Rochester. They used the tabulating machines to track salesmen production. He was impressed. When he dug into Hollerith's business model with rental equipment, maintenance, and very profitable card sales, he was doubly impressed. The card revenue from TMC and International Time Recording represented 75 percent of overall C-T-R revenue the year after he took over[2]. Watson was not shy about his assessment where further investment should go. The management teams at International Time Recording and Computing Scale were less than pleased.

Charles Flint had loaded the new company with debt. Watson needed to address this or it would drag down his plan for growth. He decided to reinvest earnings in 1914 and 1915 and issued no dividends. Total revenue was $4.1 million with margins at roughly 30 percent. Revenue would grow from 1915 to 1920 at a 25 percent compounded growth rate, reaching $20 million in 1920. With his profit sharing, Watson was also doing well. He could afford two Packards, a Model T, and electric cars for his wife and mother. The company moved to a new 20-story skyscraper in 1915 at 20 Broad Street. This was just fifty yards from the New York Stock Exchange, something that would please Charles Flint. By 1918, TMC had an inventory of 1,400 tabulators on rental installed at 650 different locations. The U.S. government had used hundreds of punched-card machines during World War I, most of which were from TMC.

Watson did not hesitate to change the structure and culture of his new company. He borrowed liberally from his experience at NCR. He took over a company with 400 salesmen. He provided extensive sales training and implemented defined territories. He expected his salesmen to be professionals, starting with professional dress—white shirt, stiff collar, and dark suit. With the far more complex tabulating products, he needed an extensive solution-selling program. He established a recognition event in 1915 for salesmen that made their quota. He formalized this as the 100% Club the following year. Watson would often remark about his fondness for sales. He could not help himself when it came to hiring more[3]. Watson also followed the Patterson NCR playbook by providing good benefits, competitive salaries, cafeterias, paid vacations, and ample promotions, especially for his top salesmen.

The Tabulating Machine Company had been a virtual monopoly, for a time. Hollerith had invented the industry. In 1911, James Powers left the Census Bureau and incorporated the Powers Accounting Machine Company. Having "Accounting Machine" in the company name was a not-so-subtle indication that he would focus on commercial applications. Powers hired engineers to assist him in developing a tabulation system that would outperform the Hollerith machines. One of those engineers was William W. Lasker. Lasker would develop the adding and printing functions for the Powers tabulator. The adding component was relatively easy, as the numerous Leibniz-style calculator mechanisms had blazed the path. The printer was Lasker's seminal development. With the Hollerith system, you still had to stop the machine and manually record the totals. The Powers tabulator would read a specially designed punched card, the stop card, and print the interim totals while automatically resetting the accumulators to zero. In addition to the printer, the Powers system had an easier keypunch, a simpler card-sensing mechanism, and was faster than the Hollerith machines.

On the brink of a breakthrough in the nascent industry, Powers abruptly resigned from his own company in 1918. He much preferred a role as inventor to managing a rapidly growing company. Several years later, he would later seek to return but his former company had moved on. William Lasker took over as the lead inventor role and continued making enhancements. In 1919, he added the flexible connection box, a linkage box that eased the switching from one application to the next. The box was a precursor of the programming plugboard. The Lasker design, however, required that any changes had to be made at the Powers factory. In 1919, the Prudential Assurance Company in London bought the Powers subsidiary in the United Kingdom. It expanded the numeric-only capability of the tabulating system by adding full alphabetic function. The Powers Company in the United States imported this innovation.

The tabulating technology leadership went back and forth during these years. However, market success was remarkably one-sided. The Powers Company did not hire a sales manager until 1920. Meanwhile, Watson applied everything he learned in his extensive sales career to honing the best sales organization. It was really no contest.

Besides the frontal assault by Powers, there was competition of the company's own making, dating to Hollerith's early efforts to expand. Former U.S. Census director Robert Porter secured the rights to the Hollerith technology for the British market. Hollerith had also licensed a German company called DeHomag in 1910. This would complicate Watson's eventual desire to expand internationally.

It was clear the TMC division could not stand still and remain competitive. Though Watson would always have a soft spot for sales-men, he had learned that the product side of the house, with robust research and development, was equally important. This lesson came from Charles Kettering. The two young men had arrived at NCR in Dayton about the same time, bonded, and would influence each other over time. Kettering attended the famous NCR sales school. He embraced sales, as he wanted to be a practical engineer. He would note, "I lived with the sales gang. They had some real notion of what people wanted."[4]

Patterson hired and fired Kettering five times, a common shared experience at the Cash. Kettering left NCR for good in 1909 to join in the automotive adventure sweeping the country. He started Dayton Engineering Laboratories Company (Delco). He was already working on an ignition system for automobiles when Henry Leland of Cadillac called. A close friend of Leland's in the automobile business died while starting his car when the engine backfired and the crank kicked back, striking him[5]. Leland was determined to eliminate the starter crank. When his engineers at Cadillac were unsuccessful at coming up with a solution, he contacted Kettering, a known expert in electric motors. Kettering developed the modern starter/generator for Leland. Leland had the prototype installed in the 1911 Cadillac Roadster.

Watson was still at NCR at the time. Kettering learned that Watson was returning to Dayton from a business trip. He drove the new Cadillac to the rail station and parked out front. When Watson emerged, he offered him a ride to the NCR offices. Once Watson got in the car, Kettering just sat there. Watson asked him if he should get out of the car and hand crank the engine. Instead, Kettering pushed a button on the dash and the car roared to life. Watson was apparently one of the first individuals to witness the new invention. It reinforced the importance of R&D; a lesson not lost on Watson. The move to invest in R&D was a source of contention in board meetings. Charles Flint continued to back Watson. Though C-T-R was already in debt to the New York Guarantee Trust, Watson called the bank and asked for $40,000 more. It would be a starter fund to get his new department going. With his typical sales flourish, he characterized the loan as a shrewd investment in the future of C-T-R. He got the money.

Herman Hollerith was ostensibly in charge of product development for TMC. When Watson arrived, he and Hollerith immediately clashed. Watson held salesmen in the highest regard while Hollerith disdained the profession and felt any machine that did not sell itself was not worth producing. Hollerith had decided to remain in Washington, far from the center of company activity in New York City and Endicott. He resigned from the board of directors and was no longer actively involved in management, which was fine with him.

Hollerith was still a major factor in design and engineering in the three years before Watson joined the company. Watson located his R&D group in New York City. Though Hollerith would communicate from Washington by letter, management in New York and Endicott progressively marginalized him. Watson moved forward without him. He moved Eugene Ford from Uxbridge, Massachusetts, to New York City and asked him to build an engineering staff. Watson helped the effort by hiring two colleagues from NCR. The team set about analyzing the current Hollerith machines and compiling a list of enhancements. Ford helped Watson recruit a young automobile engineer named Clair Lake. A designer for the Locomobile steam and gasoline automobile company, Lake had no prior experience in anything relating to tabulation. Lake would eventually become head of tabulator development, the plant manager at Endicott[6], and the holder of hundreds of patents.

Watson also hired James Bryce in 1917 as head engineer at International Time Recording. Bryce would become chief engineer of the entire company in 1922. He was a brilliant engineer and inventor who would direct every tabulator project over the next two decades and retire from the company with over 400 patents. Ford moved his engineering team to Endicott to be close to manufacturing. Watson and the executive team remained at the new C-T-R headquarters on Broad Street.

Hollerith had all new machines in place by 1907, with the suite of tabulator, sorter, and keypunch. Eugene Ford had worked with him on the renovation of the line. New patents replaced the patents that had expired. The focus turned to the Powers challenge with its new edge in printing and alphanumeric tabulators. Eugene Ford's team acquired a Powers tabulator and broke down its function in order to avoid any patent infringement. Clair Lake joined Ford in designing the new TMC printing tabulator, which debuted in 1920. The following year the engineers quickened their pace. They designed a new electric keypunch. They enhanced the printing tabulator with another function that Powers had pioneered—group subtotals. Fred Carroll, another engineer hired from NCR, designed the Carroll press[7], a device to automate the production of punched cards. Starting with a roll of card stock, the rotary machine churned out 460 cards per minute, four times the speed on the platen-based press it replaced. Carroll upgraded the press to 850 cards per minute. Card sales represented a quarter of TMC's profits.

Watson had no sooner come on board when he faced a difficult request from James Powers. Powers' company needed a licensing agreement from TMC for several of the Holleriths's newer patents. If Powers continued without the licenses he sought, TMC could sue for patent infringement. Powers and his team had already explored designs to circumvent the patents, without success. Since the desired licenses were central to their machines, lacking a license would essentially put them out of business. This was 1914 and Watson was still a felon, with the conviction for monopolistic practices. However, old habits die hard. Though TMC was very close to a monopoly in its nascent industry, Watson took full advantage of the situation. He dictated a particularly onerous royalty arrangement, requiring Powers to pay 25 percent of its machine revenues and 18 percent of card sales revenues. In addition, the agreement prevented the Powers Company from modifying its equipment for electromechanical operation for twenty years. It was an arrangement that could have easily brought in the Justice Department. Watson was unrepentant, but in time, the axe would fall.

While the Powers Company was the most formidable competitor, it was not the only one. John Royden Peirce, an inventor in New York City, also entered the field. He focused on bookkeeping applications. His major insight was using the punched card as the originating document. He designed different cards for different applications, such as restaurant invoicing, electricity consumption, bank accounting, and sales analysis. He started with a 43-column card but converted it to a double-decker 86-column card in order to support applications with alphanumeric fields. He added a printer, one that could also print the three lines of an address. His machines were all mechanical, like Powers equipment, to circumvent the Hollerith patents. Peirce was also careful to patent his own designs. TMC would infringe on his patents for sorting and tabulating, forcing Watson to pay $212,000 to the licenses. Several years later, Watson decided to go ahead and purchase Peirce's company for $261,000. The acquisition brought Peirce into the company along with his engineering team. They were set up as Invention Department No. 1 on Varick Street in New York City. Fred Carroll headed up Invention Department No. 2 and Eugene Ford headed Invention Department No. 3, both in Endicott.

With the beefed-up engineering group, the pace of new machines quickened. Eugene Ford worked with Hollerith on the 070 sorter, which would read 250 cards per minute. He upgraded the machine, announced as the famed IBM Type 80 sorter, running at 450 cards per minute. The Type 80 would sell over 10,000 units. The tabulator was the main focus of the engineering teams. C-T-R introduced the Hollerith Type III in 1920. The Hollerith Type IV tabulator, also known as the IBM 301, debuted several years later. It was the first tabulator to offer subtraction. This was not a trivial feature. Clair Lake designed the mechanism to react to a specific punch in the card that designated a negative number or subtraction. That would set in motion a reversal of the accumulating gears, levers, and ratchets while preserving the carrying and borrowing operations.

Providing group totals was more involved. Hollerith had originally implemented the function by using blank "stop cards" inserted into the card stream at the point where subtotals were needed. When the tabulator read a stop card, it stopped the machine. The operator manually recorded the current totals, reset the accumulators, and restarted the machine. In 1921, Clair Lake developed "automatic group totals."[8] He modified processing to read two successive cards and compare the fields controlling the group number. If the field changed, the tabulator printed the totals, and reset the group accumulators.

By the late 1920s, the existing punched card was proving inadequate. The Hollerith card had 45 columns and 12 rows. This was the standard, one also adopted by Powers. Watson agreed to the revamp of the card but employed a new stratagem in its development. He pitted his engineering teams, the Lake group in Endicott and the Peirce group in New York City, against one another in a design contest. Both teams retained the Hollerith card's physical dimensions. The Peirce group doubled the number of card positions to 90 by recording multiple digits in a given column. Lake's team designed a card with smaller, rectangular "holes" that yielded enough space to fit 80 columns of data across the card while preserving the 12 rows for complete alphanumeric flexibility. Watson asked James Bryce, as chief engineer, to choose the best option. Bryce selected the Lake solution, as it would be easier to implement on existing machines.

The new 80-column card had three top rows, called the zone area, and ten bottom rows, the numerics area. The bottom row was the "9" row which led to the now-famous "9 edge first" directive to insert the card correctly into the reader. Watson patented the new card, right down to the use of square holes. It would become the "IBM card" and would soon have the ubiquitous warning at the bottom, "Do not bend, fold, spindle, or mutilate." The Type IV tabulator, also known as the IBM 301, was the first tabulator to use the new card. Remington Rand, which had acquired the Powers Company, responded with a 90-column card in 1930 that included a full alphanumeric encoding. It used the double-decker Peirce approach. Bryce's team responded in 1931 with alphanumerics added to the 80-column card by using a combination of a zone punch and a numeric punch in the same column for the letters and special characters.

Charles Flint listed C-T-R on the New York Stock Exchange in 1915. By 1918, Watson had doubled revenues to $9 million. The stream of enhancements at TMC and the high demand for punched-card equipment with World War I contributed to the increase. The card business was booming, growing with the increase of tabulating equipment on rent. International Time Recording continued as a solid profit center. Earnings doubled to $2 million. The size of the company's workforce also doubled, rising to 2,731 in 1920[9]. Watson continued to invest heavily in R&D while minimizing dividends. The end of the war resulted in a surplus of tabulating machines on the market. The U.S. government returned the equipment it had rented during the conflict. Watson would rely on the new designs that his engineering team had completed plus an aggressive sales force to keep the momentum going. Buoyed by the continuing growth and a strong performance in 1920, he bumped his sales quotas by over 30 percent for 1921. He expanded the workforce to ensure capacity for the higher volume of machines.

Then the recession of 1921 struck. It was in fact closer to a depression than a recession. It followed recession-type conditions that had set in when the war ended. The downturn was a result of a number of converging factors, including the transition from a wartime economy to peacetime production and consumption and the need for the economy to absorb the returning soldiers and all those that had directly supported the war effort. On top of all that, the Spanish flu pandemic arrived in 1918 and took two years to run its course. Finally, the Wilson administration hewed to conservative policies instead of using monetary and fiscal actions to spur the economy forward.

C-T-R, and especially the tabulating division, was in a somewhat recession-proof industry. The railroads were not going back to calculating freight bills manually. The company had a built-in cushion with its inventory of machines on rent, but revenue still dropped by a third. The planned new business built into the raised quotas did not materialize. Though the company managed a small profit in 1921, it was dangerously overextended and in the midst of a cash-flow crisis. Fairchild and others on the board exacerbated the crisis by insisting on paying dividends. Watson laid off a number of employees and shut down his prized "future plans" department. He took away some valuable lessons from the recession experience: keep dividends low, cash high, and continue to invest in growth.

The recovery from 1921 was slow. The year 1922 was essentially flat, with revenues of just $10.7 million and earnings of $1.4 million. The company would not rebound to 1920 levels until 1929. The year 1926 brought another dip in the economy, though not as severe as 1921. As income gradually improved, the company declared dividends of 5 percent in 1928 and continued even into the Great Depression. George Fairchild, the board chairman and largest stockholder, died in late 1924. Though Watson had already reduced his deference to Fairchild, his death further cemented Watson's ability to run the company as he saw fit, with limited board oversight. Despite the setbacks, he remained very bullish on the company's prospects. He noted that the Canadian and South American subsidiaries went by "International Business Machines" or IBM for short. Watson liked the name as it reflected his high ambitions. In 1924, he rechristened the Computing-Tabulating-Recording company as IBM. As 1929 progressed, prospects looked bright. It was a year of 30 percent growth for IBM.

**Figure 9.** 1924 IBM logo, with the then "modern" san-serif font. The globe arrangement hints at Watson's ambitions for the company. Reprint Courtesy of IBM Corporation ©.

In the 1920s, the top four business machine suppliers were Remington Rand, NCR, Burroughs, and trailing those top three by a substantial margin, IBM. Burroughs blossomed after William Burroughs licensed patents on the Comptometer and designed the Burroughs Adding Machine. The Comptometer from Felt and Tarrant failed to make the transition to electromechanical models and faded. Burroughs died in 1898, but his company continued to grow, expanding into accounting machines. John Patterson died in 1922, the year National Cash Register sold its two-millionth cash register. Patterson had carved out a monopoly and supercharged it with his revolution in salesmanship.

The Powers Company slid into receivership in 1920. The onerous royalty arrangement with Watson did not help, and, in fact, Powers was in arrears on payment. The company struggled until purchased by Remington Rand in 1927. The new owners had the size and financial muscle to challenge IBM in tabulators. Remington Rand started business as Rand Ledger. James Rand Sr. had developed a unique system of tabbed indexed cards for business. His son, James Rand Jr., joined the firm but soon clashed with his father. He proceeded to set up his own competing tabbed-card company. Father and son eventually reconciled, and combined the two companies in 1925. James Rand Jr. then went on a buying spree, adding seven companies, including Powers and Remington Typewriter. Rand renamed the new combination Remington Rand. It was the world's largest business machine company.

The Powers Company would not survive the acquisition. It suffered from control, financing, and focus issues and never fully recovered. The royalty agreement between Powers and TMC expired in 1929 with $350,000 still owed to IBM. Remington Rand now assumed that obligation. Another round of negotiations ensued, resulting in a five-year extension. The two companies made a separate agreement on pricing, which set in motion a U.S. Justice Department antitrust action. A viable and independent Powers Company would have absolved IBM of any antitrust implications, but Watson charged ahead. By 1931, Powers was no longer driving the technology but reacting to IBM innovations.

Overseas, an Austrian engineer and inventor named Gustav Tauschek attempted to re-create the Hollerith magic. In 1921, he worked for the Austrian National Bank, where he encountered Hollerith machines supplied by the German licensee Dehomag[10]. Watson had purchased 90 percent of Dehomag in 1923. Tauschek designed a number of improvements for the machines at the bank. He learned English so that he could research the Hollerith and Powers patents and avoid infringement. In 1926, he went to work for the Rheinmetal Company, where he designed a complete punched-card system with keypunch, sorter, and tabulator. The company formed a subsidiary to market his system. Dehomag determined that IBM was infringing on Tauschek's patents. In 1928, before any manufacturing started with Tauschek's designs, IBM purchased the company, paying

Tauschek $250,000. IBM also hired him as a consultant on a five-year contract based in New York City[11]. Tauschek would go on to file 169 patents during his time with IBM.

Watson continued to mold IBM to his liking. Many of the additions to the company culture came from his days at NCR. This included the 100% Club, family dinners, sales school, and of course, the "Think" sign. He had created that during his sales training sessions at NCR, making the point that sales success involved using your head. Originally, he just wrote "Think" on flipcharts, but he graduated to producing desktop "Think" signs after John Patterson approved.

Watson got other ideas from George Johnson. Watson had met him in 1914 on an early visit to Endicott. The Endicott-Johnson shoe company was the biggest employer in town when Watson moved C-T-R there. Johnson told him his company had no unions because he treated workers well. He was actually on a first-name basis with most of his employees. He provided a range of benefits including profit sharing. He called his management approach the "square deal" though some derided it as "welfare capitalism." When union organizing came to Endicott-Johnson, 15,000 employees voted it down.

Watson took note. He added the Quarter Century Club in 1924 to honor long service. His Open Door policy encouraged any employee with an issue to take it up with any manager up to and including Watson. The Suggestion Plan debuted in 1928. Other additions included employee sports teams, a company band, and an employee newspaper.

As Watson assumed more control over the company, he added a very unique element to the culture—the company song. Songs would be an ever-present feature of Watson's stewardship. Employees sung them at sales schools, recognition events, and family dinners. Many of the songs were about Watson, extolling his great leadership. The lyrics would describe him as "head and soul of our splendid IBM," "our leader fine," or "inspiring all the time." The nearest "competitor" to having multiple songs was the manufacturing vice president Otto Braitmayer and he had only two. He had joined Herman Hollerith and TMC at age 15, and spent the next 40 years with the company. He was serenaded with "we adore you, Otto Braitmayer, our great pioneer." Fred Nichol, who had started in 1909 at age 17 as Watson's aide at NCR and was now his executive assistant, was praised as "a real Go-Getter" in his tune. Division executives had songs. The men running the Canadian and European subsidiaries had songs. The key engineers in the company -- Claire Lake, James Bryce, J. Royden Peirce, Fred Carroll, and Eugene Ford -- all had songs. The Watson song "You're Our Leader Fine" was sung to the tune of "Auld Lang Syne." The lyrics[12]:

*T. J. Watson, you're our leader fine, the greatest in the land,*
*We sing your praises from our hearts—we're here to shake your hand.*
*You're IBM's bright guiding star—you're big and square, and true.*
*No matter what the future brings, we all will follow you.*
*You've made our IBM so great in every land and clime*
*Our service meets all needs of men on Weights, Accounts, and Time.*
*You've brought us through to victory, with leadership so fine.*
*We'll always love and honor you for the sake of Auld Lang Syne.*

# IBM AND THE DEPRESSION

The year 1929 was a good one for IBM. The company recorded $18 million in revenues and $7 million in income with a workforce that had grown to 6,000. Starting after the recession in 1921, the stock market went on a nine-year bull run. The Dow Jones average rose by a factor of ten during this time. Record company profits fueled a continual surge in the market. At some point, valuations became unhinged from underlying conditions and stock purchases became pure speculation. There were ample warning signs and even discussion on actions that might tamp down the stock mania. The market reached its peak on September 3, 1929, when the Dow Jones Industrial Average hit 381. It would not rebound to that level until 1954. The market drifted down until October 24, Black Thursday, when 12 million shares traded, resulting in an 11 percent drop in the market. A 13 percent drop followed on October 28 and a 12 percent drop the next day -- Black Tuesday, October 29.

The U.S. economy spiraled down from the market collapse. The nation's GDP declined by 50%. Prices fell 32 percent. The unemployment rate went from 4 percent to 25 percent, meaning 15 million Americans were out of work. Nearly half of the country's banks failed. Many had used depositors' money to invest in the stock market. At a time when a quarter of the U.S. population was still on the farm, falling demand cratered farm prices. The early sense that the downturn would unfold like the recession of 1921 soon faded. The Great Depression took hold, and conditions continued to deteriorate until reaching the nadir in 1932. The collapse of demand created a downward spiral where each ratchet downward led to job losses and bankruptcies, further pushing demand down, creating a vicious cycle. Auto sales crashed 75 percent from 1929 to 1932. As with the 1921 recession, Republican administrations viewed the decline as a normal part of the business cycle. They would not entertain any discussion of governmental intervention.

IBM stock fell from $241 a share before the collapse to $129. Yet, IBM stock had been at $93 a share in August 1927[1]. Anyone buying at that time would still have appreciation and dividends to show for the purchase. IBM paid 5 percent dividends each year from 1928 to 1931. The IBM board watched as the stock fell by 50 percent even though earnings were up and Watson continued to earn his 5% profit-sharing commitment. There was some discussion on the board about who should be at the helm at this time, though they took no action. Watson's protector, Charles Flint, retired in 1930, but Watson was safely in charge by that time. IBM revenues suffered less of a drop than other companies did as the business had the cushion of equipment on rent. Revenues would only fall from $18 million in 1929, which was a good year, to $17 million in 1932.

Watson pushed the company forward despite the deteriorating conditions. He was determined to maintain employment and production, to be in a position to benefit as conditions improved. He continued to run the plants, producing machines that would simply end up in inventory. He actually increased production by 30 percent in the first few years of the crisis. He tried to avoid layoffs, though there were some in 1930 and 1931. His penchant for salesmen was unabated, and he continued to hire them into the depths of the Depression. He also hired IBM's first women sales representatives. He also continued to invest in R&D, spending $1 million to build a new lab in Endicott in 1933. That amount was equal to the total company earnings for the year.

Watson sold the Computing Scale division in 1933 to Hobart. He kept the Time Recording division as it used punched cards. By 1930,

IBM was producing four billion cards each year. He purchased the Electromatic Typewriter Company in 1933, in part to reengineer their machines for keypunch and tabulator keyboards.

In 1930, Watson moved IBM headquarters from the New York Financial District to a new 28-story building at 270 Broadway. The building became famous not for being IBM's home office but for the occupants on the 18th floor during World War II. General Leslie Groves set up administrative offices there to manage his far-flung military enterprises. Groves named his team the "Manhattan Engineer District," leading to the name "Manhattan Project" for the atomic bomb program.

Watson continued his focus on employee training. His education department was up and running in 1916. Education did not end with just the sales team. There were programs for engineering, and tool and die classes for factory workers. In 1933, Watson erected the IBM Schoolhouse in Endicott. It was a magnificent Art Deco structure with the words "Think Observe Discuss Read Listen" etched into the front steps. Newly hired sales trainees stayed at the Hotel Frederick. That is, except for women, who stayed at the luxurious IBM Homestead[2], the former Spaulding estate Watson acquired in 1935. All new hires attended classes at the Schoolhouse in groups of fifty. Those classes began each day with fifteen minutes of song[3].

Besides the impressive new IBM School building, Watson built the new Research and Development building in Endicott. He consolidated his three engineering teams, working under the direction of Chief Engineer James Bryce. A project of central importance was a new tabulator. It would have significant new capabilities, prompting a change in the name to "Accounting Machine." Introduced in 1934, the IBM 405 Alphabetical Accounting Machine would be the flagship of the company's product line for well into the 1950s. It was an electromechanical marvel, with 55,000 parts[4]. It included group subtotals, direct subtraction, and full printing. Engineering added a third card read station to support three lines of address. With so much going on within the machine, timing was essential. A central rotating shaft provided the machine's clock. Every operation synchronized to the shaft's rotation. Engineering divided each 360-degree rotation of the shaft into 20 intervals of 18 degrees of arc each[5]. This arc segment was the basic timing unit and would correspond to one discrete machine function—read one card row, add one digit, or print one character. The full rotation of the shaft corresponded to a complete card read cycle. For example, one segment might read the sixth row from the card. The next segment would read the fifth row after the card moved forward a fourth of an inch. The machine contained hundreds of relays controlling the logic of these physical operations.

Addition and subtraction counters used an ingenious counter mechanism, a contemporary implementation of the Leibniz step reckoner. Each single digit of a number corresponded to a combination of a rotating wheel with associated cams, magnets, pawls, clutches, and relays. The counter wheel had 20 teeth or cogs that synchronized with the main shaft timing segments. The counter wheels operated in groups of eight to represent eight-digit decimal values. The 405 had 16 of these banks of counter wheels, resulting in 128 intricate assemblies. This function became registers when implemented in computers.

The built-in printer ran 100 lines per minute and used vertical type bars, one for each of the 88 print positions across the platen. The 43 type bars on the left were alphabetical while the 45 bars on the right were numeric. Each bar contained all of the needed alphabetic or numeric characters. The bars moved up or down to select the designated character before shaft timing initiated the strike against the ribbon and paper. There were so many metal moving parts inside the machine that it required a lubrication system with an oil pump to keep everything running smoothly[6]. The machine weighed in at roughly 2,500 pounds. It was far more technically advanced than competitors' machines.

This complex machinery of rotating shafts, whirling counters, and reciprocating print bars needed programming to generate a sales analysis report or a utility bill. The answer was the plugboard. Hollerith's original tabulators simply connected a particular card punch to its wired counter. The 1906 Type 1 Tabulator added a rudimentary wiring panel. None of these small steps proved satisfactory. The engineers focused on creating a programming board that was relatively easy to set up and immediately swappable, from application to application.

The IBM plugboard arrived in 1934 with the 405 Accounting Machine. It was a rectangular panel 24" by 18" in dimension and an inch thick, weighing 25 pounds. The board contained a maze of holes termed "hubs." There were 1,600 hubs, and they matched hub receptors located inside the machine. Patch cords with a plug at each end completed the electrical connection, hence the name plugboard. A plug inserted into the programming side of the plugboard protruded out the other side of the board, making contact with the machine's hub receptor.

The plugboard hubs had different sections that were clearly marked on the board face. The sections included hubs for the 80 columns of card read data, print functions, counters, and logic or compare operations. As a simple example, plugboard cables that ran from the hubs of card columns 60-65 to the hubs of printer columns 2-6 resulted in that card field being printed on the current print line at those positions. A simple application might read a transaction card, add some fields, print a detail line, and then accumulate group and final totals. One might immediately sense that a completed plugboard business application would result in an utter maze of plugboard wires. That certainly was the case. As antiquated as this whole process sounds, it was more than high-tech in its time. IBM set up a methods research department that would roam the field looking for promising customer plugboard applications and make those examples available to other customers.

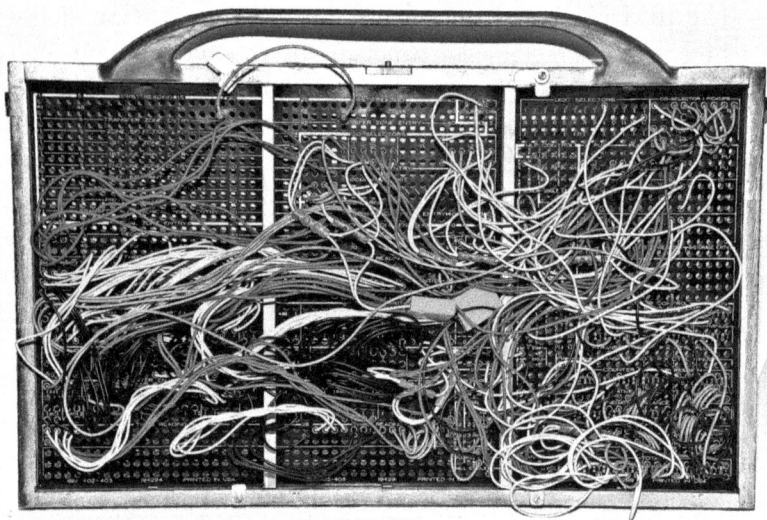

**Figure 10**: IBM 402 plugboard (Glenn Henry's Computer Museum).

An obvious question soon arose – what to do about multiplication? Early tabulator applications had to make do with addition. However, more applications surfaced where multiplication was an essential requirement, such as telephone billing. As we have seen through the progression from the Pascaline through Napier's bones and on to the early desk calculators, the key to effective multiplication was a combination of addition with built-in algorithms to reduce the number of calculator cycles to a reasonable number. The standard method multiplied the digits of the multiplicand and multiplier one at a time coupled with shifts in the base ten columns, replicating the process when one multiplied by hand.

Paul Breedveld[7] provides an example with the multiplicand 7354 times the multiplier 87. The process is similar to that of the De Colmar Arithmometer. The value "7454" is added 7 times followed by a shift to the tens place. The "7354" is added eight times for the tens place. The two partial totals are then added together to produce the answer. There are 16 total additions instead of the 87 with repetitive addition.

In the 1920s, a Marchant or Friden calculator would whir and clunk its way at the problem, perhaps for up to a minute, in order to multiply two numbers. With the tabulator reading punched cards at a rate of 150 cards per minute or higher, that kind of delay would simply not work. In 1926, Watson asks his engineers to come up with a solution for multiplication. With IBM's electromechanical technology, adding multiplication to existing tabulators would dramatically increase the number of components and result in an even more mammoth machine. The engineering team, led by Bryce, decided to develop multiplication in a separate machine.

The first attempt, the 600 Multiplying Punch, was short-lived as Bryce was soon unhappy with it. He set about designing an improved multiplication process. He employed a novel technique to reduce the number of additions. Novel is probably not quite the right word as his technique was straight out of John Napier's famous "bones." Using relays, he built tables of single-digit multiplications up to 9x9 into the machine. For example, his table would show that 7x8 = 56[8]. In a precise sequence, Bryce had the product of each digit of the multiplier and multiplicand recorded from the relay tables and shifted into four accumulators. Using the earlier example of 7354 multiplied by 87, his approach ran the "7" multiplier through the multiplicand digits using his table and arranged the products diagonally as 49-21-35-28. He then performed the same routine with the "8" multiplier, yielding 56-24-40-32 with the necessary offsets. He summed these four partial products. This reduced the total number of additions to the number of digits in the multiplier plus one. In this case, the number of additions needed is just three, versus eight on the 600 Multiplier. The following diagram shows the four partial products that are summed for the total. You can see how the "bones" multiplications are arranged diagonally (note the "28", for example) and placed in the proper power of tens column[9].

IBM released Bryce's invention as the IBM 601 Multiplying Punch in 1933. The 601 read the numbers from an input card and punched the multiplication results back onto the same card.

**Figure 11.** IBM 601 Multiplying Punch (1931). The machine read
two numbers from a punched card, multiplied them, and punched
the result in the same card. Reprint Courtesy of IBM Corporation ©.

During this period, Bryce was chief engineer as well as a one-man
research and patent department. As the engineering work accelerated,
he knew he needed help. Fortunately, he knew a recent MIT graduate, and actually knew him pretty well. Arthur Halsey Dickinson lived
nearby in Binghamton, New York[10]. Bryce and Dickinson had known
each other since 1920. In fact, Bryce kept his car in the Dickinson
garage. They exchanged Christmas cards.

Dickinson graduated from Union College in Schenectady in 1928
with BS and BE degrees. He earned a graduate degree in fuel and gas
engineering at MIT in 1930 and stayed on to work on a public utility
project. The Depression hit the utilities industry hard. Bryce
extended an invitation for Dickinson to come to New York City for an

interview. Liking what he heard about working for IBM, Dickinson spent some of his accrued vacation time in Endicott learning punched-card machines. The newly released 601 Multiplier was a highlight of his stay there. Bryce hired him in 1932 as his assistant, increasing the size of the applied science and patent department to two.

Dickinson's role was to prepare patent applications, a job that required detailed understanding of the machines submitted for review. He also was responsible for turning Bryce's ideas, many of which came directly from Watson, into engineering designs. He attended the annual 100% Club meetings in New York, an excellent opportunity to meet with salesmen and learn what new enhancements their customers were asking for, a technique used to great success by Charles Kettering. Despite Bryce's commanding position within the company, Dickinson would describe him as a very shy individual.

Though most ideas came from Watson based on his conversations with customers and salesmen, one major new idea came from a statistics professor at Columbia. Ben Wood, an Army psychologist in World War I, headed the Collegiate Educational Research group at the school. He was in charge of conducting and analyzing achievement tests given to college seniors. The tests achieved a high profile as they indicated there were wide variances between colleges in results. As the number of tests conducted ramped up from 35,000 a year to several million[11], Wood began to think about automated approaches to scoring. At the time, the statistical work-up of these tests relied on human computers. An earlier professor at Columbia (Harold Jacoby) mentioned "Miss Harpham, our chief computer" meaning she was in charge of the Burroughs, Brusnviga, and Millionaire machines used for these calculations[12].

Wood sent letters to ten business machine companies. He received only one response, and it was a telephone call from Watson. He agreed to meet and allocated exactly one hour in his busy schedule for Wood. It helped that Watson was a friend of Nicholas Butler, the President of the University[13]. Watson was intrigued by Wood's proposal, and the one-hour meeting ran to almost four hours[14]. Watson was already "big" in New York, in business, in the art world, and at the opera. He was interested in working with universities on research, in part to enhance his reputation beyond being a successful businessman. Columbia was an obvious choice as it was less than five miles uptown from IBM headquarters in Manhattan.

Watson agreed to look into a test-scoring machine, but the conversation also led to more immediate moves. Wood sketched out how he might use existing IBM tabulating equipment in his statistical work. Watson agreed, and within two days, three truckloads of IBM tabulating gear arrived. Watson also provided a technician to assist in getting the machines up and running. That technician was John McPherson, who would be a future giant in IBM. Born in Short Hills, New Jersey, in 1908, McPherson went to Hotchkiss Preparatory School and then on to Princeton University. Before McPherson graduated in 1929 with an electrical engineering degree, Watson gave a lecture at the school. Knowing that McPherson was from his hometown of Short Hills, he encouraged him to join IBM.

McPherson started as a sales trainee, going through the training program in Endicott and learning punched-card equipment. He was from a family with a railroad background and, as such, gravitated to a railroad territory. He spent three years as the sales representative for the Reading Railroad and then seven years in New York City with the Railroad Group marketing team. He authored the treatise "Machine Methods for Railroad Accounting.[15]" Before all that transpired, he helped Ben Wood install and run the new IBM machines at Columbia.

The equipment would form the basis of the Columbia Statistical Bureau and started a long and fruitful collaboration between the University and IBM. Once the equipment was operational, Wood had a conversation with Watson about the speed of the machines. He was careful not to imply that IBM's machines were slow. Rather, he impressed upon Watson that future machines of this sort would be bound only by the speed of light[16]. That was a prescient forecast in 1928 and one that Watson would remember when the time came.

With more experience under his belt by 1930, Wood asked Watson if he could build a custom tabulator designed specifically for statistics, one that would calculate exponentials and the sum of squares. Watson asked James Bryce to lead the effort, and a huge one-off tabulator was the result. It was called the "Columbia Machine," or "Ben Wood's Machine," or even the "Packard," the latter a reference to the very large luxury automobile. The *New York World* newspaper simply called it the "Super Computing Machine.[17]" It was also known as the "Difference Tabulator," as the mathematicians at Columbia used the machine to implement a polynomial-based algorithm that was based on the method of finite differences, the same concept that Charles Babbage employed for his Difference Engine. Using this approach, the tabulator was able to calculate the equivalent of 1.2 million multiplications in a single 42-hour period. On the 601 Multiplier, which ran at the speed of one multiplication every six seconds, the processing would have taken more than 800 hours. Though Hollerith had originally designed his tabulators for the massive data processing at the U.S. Census, some of IBM's current machines were now re-missioned for scientific tasks.

Ben Wood's statistical department running Watson's machines sparked interest from Wallace Eckert, a professor in the astronomy department. Eckert had degrees from Amherst College and worked at the University of Chicago and Columbia before getting his PhD in astronomy at Yale in 1931. He worked the computations for his doctoral thesis on celestial mechanics on Monroe desk calculators[18]. Once at Columbia, he stopped by Pupin Hall to see Wood's installation. He asked if Wood would discuss a similar setup with Watson for astronomy. Watson agreed and asked James Bryce to have Dickinson meet with Eckert to assess his needs. Another truckload of IBM punched-card equipment arrived at Columbia. It included the new Model 285 tabulator, a numeric-only version of the 405, and one of Bryce's new 601 Multipliers.

To adapt punched card processing for scientific computation, Eckert did not have to start from scratch. He could lean on a famed astronomer in London. Leslie J. Comrie was a New Zealander who became head of His Majesty's Nautical Almanac Office at the Royal Greenwich Observatory. Comrie's expertise was in table making using Brusnviga desk calculators. He graduated to punched-card devices, renting them from British Tabulating Machines, the Hollerith licensee set up by former Census superintendent Robert Porter. Comrie wanted to calculate the positions of the moon, a set of tables called an ephemeris. A fellow Brit, Professor Ernest Brown of Haverford College in the United States, spent twelve years producing such tables using desk calculators. He published his "Tables of the Moon" in 1919. Comrie wanted to tackle the task with more granularity and accuracy but without the tedium of desk machines. This required punching 20 million holes in 500,000 cards to set up the computation[19]. He calculated the positions of the Moon from 1935 to 2000. He published a complete account of the tabulator process in 1932 in a Royal Astronomical Society paper.

Eckert took Comrie's methods and dramatically expanded them. He received custom modifications from IBM to calculate tables of logarithms and LaGrangian interpolation, which he applied to solving the differential equations required to compute the orbits of the planets. Such computation was unimaginable when he was plugging away on the desk calculator at Yale. As he honed his use of the IBM equipment, Eckert concluded that he needed a switch between the 601 Multiplier and the tabulator in order to eliminate the intermediate card punch step and fully automate his calculations. He discussed this need with his IBM customer engineer.

Upon hearing of Eckert's concept, Bryce assigned a new trainee to work with Eckert on the switch. Steve Dunwell had started with IBM in 1933 in Endicott as a co-op student while working toward his electrical engineering degree at Antioch College in Ohio. He never graduated, as he decided to hire on with IBM full-time. Dunwell set up a patch cable between the 601 Multiplier and the Type 285 tabulator with a switch enabling 12 separate operations for each card read[20]. Ben Wood's "Columbia" machine had since been retired, allowing Dunwell to steal a number of relays for his switch. The completed modification enabled the tabulator to use the 601 Multiplier as an on-demand calculator. Eckert's astronomy lab would eventually have fourteen 601 machines humming away. This rather Rube Goldberg style arrangement predated many future IBM systems where near automatic, high-speed calculations were required.

Eckert documented his use of punched-card equipment in the book *Punched Card Methods in Scientific Computing*, published in 1940[21]. Also called the "Orange Book" due to its bright orange cover, it provided the introduction to scientific computing for a number of future computing luminaries, including Vannevar Bush, J. Presper Eckert, Howard Aiken, John Backus, and Ted Codd. In 1937, the

Eckert computing facility was renamed the Thomas J. Watson Astronomical Computing Bureau. John von Neumann, Hans Bethe, and Richard Feynman (all future Nobel laureates) were on hand. Feynman wowed Eckert with his knowledge of tabulator machines, including the programming of plugboards, a skill he would soon put to use at Los Alamos. Watson reveled in the attention.

# IBM ACCELERATES TO THE LEAD

B y the middle of the Depression, Watson was in complete control of the company, with little board oversight to worry about. However, he would soon discover that he was also subject to public oversight. His compensation surfaced as a major issue. He started with C-T-R in 1914 at a $25,000 salary plus his negotiated 5 percent share of profits after dividends. In the early 1920s, the company increased his salary to $60,000, and then increased it again to $100,000 heading into the Depression. His total compensation averaged $184,000 per year from 1914 to 1935. It was in 1934 that his earnings became a problem. That year, he made $264,432 in profit sharing in addition to his $100,000 salary (this was a kingly sum, equivalent to roughly $8.5 million in 2024 dollars). However, the Revenue Act of 1934 required the Treasury Department to publish the names of all corporate officers earning more than $15,000.

The inaugural list had 50,000 names, and Watson was at the very top of the list. Watson's take was reportedly greater than a number of high-profile personalities. Will Rogers was #2 at $324,000; Alfred P. Sloan, the chairman of General Motors, made $202,000; and Walter Chrysler was close behind, at $197,000[1]. Watson's NCR colleague Charles Kettering was also high on the list. Watson's total of $365,358 unfortunately included "365" in the amount. The press labeled

Watson "the $1,000-a-Day Man.[2]" Though his earnings success would normally be a rags-to-riches tale for someone who started out selling pianos from the back of a wagon, the "$1,000-a-Day" label was not a distinction one would want in the middle of the Great Depression. That year, the average wage earner, the lucky ones who still had jobs, earned about $1,000 for the entire year. It was little consolation that a year-end Treasury report had Watson far down the list, after William Randolph Hearst at $500,000 and Mae West at $450,000. However, the bad press had already gone out. It would turn out that 1934 was not even Watson's best year. That would be 1940, when he earned $546,000.

Watson was not big on company stock, the normal path to corporate riches. In case the economy really disintegrated and he was out of job, he purchased a farm in Indiana. Yet he was sensitive to the bad press. In 1942, he reduced his profit sharing from 5 percent to 2.5 percent. When this proved insufficient because of wartime business, he capped his total earnings below the pre-war level, at an average of just over $400,000 each year. Watson was handsomely paid, but he earned it given IBM's continual outstanding performance.

Watson's issue with compensation was part of a somewhat darker side. He had an insatiable need for recognition, for praise, for even exaltation, and it frequently got him in trouble. The numerous company songs extolling his leadership were the most innocent manifestation. He liked to dominate company events such as the 100% Club and family dinners, where he might drone on for hours[3]. He had the longest entry in *Who's Who in America*, the compendium of the famous and near famous. His entry comprised 16.5 inches of achievements, including positions, boards, and awards. His entry ran to 155 total lines, an all-time record[4]. The list included over 20 honorary degrees from colleges, something Watson's father would have found quite ironic.

Watson also liked the trappings of running the IBM Company. He purchased the Homestead estate in 1935 to use for important meetings. He had two very nice suites added upstairs to accommodate his frequent visits. His flat organizational structure, with most managers reporting directly to him, was a reflection of his need to be the center of the IBM universe. Despite a huge sales force, he micromanaged the group as if he was still the branch manager in Rochester. He had a culture of yes men around him, with Charley Kirk at the lead. Watson, upon returning from a sales office visit, might idly mentioned to Kirk that he talked with a salesman who did not seem in the IBM mold. It was Kirk's tacit understanding that he travel to the sales office and either demote or fire the man. This was without any kind of review or any probing as to the man's real record.

Though he was generous and supportive of the company's work-force, he had a terrible temper and typically reserved his outbursts for managers who had disappointed him. His ego would get him into even greater trouble than the compensation scandal. In one of his frequent trips to Europe, he lauded the "great leader" Benito Mussolini and then received the German Eagle award from Adolf Hitler. Watson attended the 1937 meeting of the International Chamber of Commerce where he succeeded in having the group adopt "World Peace Through World Trade" as its motto. Then, he met with Hitler for tea[5].

IBM's German subsidiary, Dehomag, had close ties to the Nazi regime. The general manager, Willy Heidinger, was a Nazi Party member and enthusiastic supporter of Hitler. The company provided tabulation machines for the 1925 Prussian census and the 1939 German census. Historians raised the issue whether IBM machines had a role in the administration of the Holocaust. Race and religion were not fields collected and punched into cards in the 1939 census. There was a separate effort to collect that information later but there is no record of a merger of the two data sets[6]. Nevertheless, Watson was incredibly naïve about fascism in Europe. He returned the Nazi medal but the situation cried for a much stronger stand. Both Henry Ford and Charles Lindbergh received Hitler's Eagle award and did not do much more than Watson did when they realized their mistake.

Though Watson had this darker side, his bright side was very bright indeed. The company shined during the Depression. While Watson did especially well, his employees also benefitted. Until 1938, Watson clocked in every day just like everyone else in the company. This was no surprise as the company was still in the time clock business. He gave freely to universities and charities. He put the company at the country's service whenever war intervened. He was quick to act in time of disaster, as he had done with the Dayton floods. In 1939, he planned to have a huge coming-out party for the company at the New York World's Fair. He funded the company's presence at the fair and arranged to have one day declared "IBM Day". He paid for 6,000 IBM employees, along with their spouses, to be there. Furthermore, it was not just for the day but included a three-day stay at New York City hotels. He reserved ten special trains to take IBMers from Endicott to New York City. All told, this was a $1 million expense for the company, or about 10 percent of that year's annual income[7].

Traveling at night, two of the reserved trains collided with one another at Port Jervis, New York. Upwards of 250 IBM employees were injured, some critically. Word reached Watson in New York City at 2:00 am. He mobilized the company immediately and with his assistant Fred Nichol and daughter Jane, headed by car to Port Jervis. He spoke to each person in the hospital. He had flowers delivered before breakfast the following morning. He had a makeshift hospital set up for those less seriously injured. His quick and forceful actions in such dire circumstances were not soon forgotten.

Watson not only maintained full employment during the Depression but also grew the IBM workforce, doubling in size from 1929 to 1940, from roughly 6,000 to 12,000. The success of the company helped fund employee policies and programs. In 1933, he standardized a 40-hour workweek. The following year he eliminated those employees still on piecework pay. All IBMers would be on salary. In 1935, he added group life and survivor benefits. In 1937, he added paid holidays and paid vacation days. He would turn 66 at the end of the decade. More and more IBMers affectionately referred to him as the "Old Man."

The downward spiral of the Depression continued unabated until 1932. The country needed an injection of demand, which would stabilize prices and increase jobs. President Herbert Hoover did not grasp the situation. He believed in the Republican business orthodoxy that downturns would eventually right themselves. Republicans had controlled the White House since before the Civil War with the exception of the terms of Grover Cleveland and Woodrow Wilson. Hoover had crushed his Democratic opponent, Alfred E. Smith, in

1928 by a five-to-one margin. However, business conditions in 1932 were far different than 1928.

In the run-up to the 1932 presidential election, Franklin Roosevelt promised a flurry of actions to jump-start the economy. The result was an eight-to-one electoral landslide for FDR. He pressed forward immediately after his inauguration in March 1933. He oversaw an alphabet soup of new organizations and programs under the aegis of the New Deal to address the nation's economic problems. Programs like the Agricultural Adjustment Act sought to address the farm situation by a combination of price supports and reduced production. He pushed a series of programs to create jobs, including the Civilian Conservation Corps (CCC), Works Progress Administration (WPA), Civil Works Administration (CWA), Federal Writers Project (FWP), and the National Youth Administration (NYA). In concert, they made the U.S. government the largest employer in the country. Other programs addressed banking (the Federal Deposit Insurance Corporation, or FDIC), home mortgages (the Federal Housing Administration, or FHA), labor relations (the National Labor Relations Board, or NLRB), and oversight of the stock market (the Securities and Exchange Commission, or SEC).

One other New Deal program would have a dramatic impact on IBM. That was the Social Security Administration, instituted in 1935. The program set aside a percentage of an employee's wages to fund a basic retirement pension. Great Britain, France, and Germany had already implemented old-age pensions. The new program would fall under the purview of Labor Secretary Frances Perkins. While discussions with foreign governments stressed the difficulty of such a program in the United States, Roosevelt and Perkins were committed to moving forward aggressively. To administer such a program, the federal government would need tabulating machines, and a very great number of them.

The newly established Social Security Administration (SSA) asked ninety business equipment manufacturers to bid on the contract. Only eight submitted bids. The three finalists were Burroughs, Monroe, and IBM. Remington Rand failed to submit a bid. Its new line of alphanumeric machines would not be available until 1938. The contest among the finalists boiled down to IBM with its market-leading punched-card automation versus the modified book-keeping machines from Burroughs and Monroe. The IBM sales representative, H. J. MacDonald, led a team that worked for two years on the opportunity. IBM won the contract in September 1936. It was worth 400 accounting machines and over 1,200 keypunches[8].

There were several reasons for IBM's selection. First and foremost, IBM had the machines. Watson's decision to maintain full employment and continue producing equipment, even if they went unsold, meant the company could respond quickly to the SSA timetable. A second key to the win was IBM's commitment to tailor the machines to SSA requirements. This meant creating a new machine, the company's first collator, the IBM 077. It featured two input card readers and five output hoppers with programming provided by a plugboard. It could merge two card decks while searching for any duplicates. For the SSA, merging the weekly or monthly earning cards with the employee's master Social Security record was critical.

Watson was no doubt helped by aligning with Roosevelt and his administration. He had once been a staunch Republican, but his prosecution by the Taft administration Justice Department while at NCR changed that. Watson built a worldwide operation as the company's name suggested, and he was naturally interested in international relations. Roosevelt would at times ask Watson to entertain foreign dignitaries when they visited New York. For IBM, it did not hurt to have a friend in the White House.

The Social Security Act required that earnings reporting begin by 1937, including compulsory compliance by employers. The SSA took delivery of over 400 machines in 1936. Due to the combined weight of the machines and mountains of punched cards, a suitable and very sturdy site needed to be located. It turned out to be a former Coca-Cola plant located along the harbor in Baltimore. At the onset, the SSA enrolled 26 million Americans, creating an 80-column master card for each individual sequenced by their newly assigned Social Security number. The number of individuals enrolled grew to 46 million by 1945. The Social Security law required any company with eight or more employees to report earnings of its employees. The logical choice for larger companies was to adopt IBM machinery to produce the earnings cards for the SSA in Baltimore. In addition, IBM's success did not end with the Social Security Administration. Many of the other agencies created in the New Deal also had huge data processing requirements, and most tended to follow the lead set by the SSA. The success with the SSA, related agencies, and adopting companies produced an estimated 81 percent increase in IBM revenue from 1935 to 1939[9].

It was not all good news for the company during the Depression years. Watson's monopolistic practices were bound to catch up with him at some point, and that turned out to be 1936. At issue was IBM's virtual monopoly on 80-column cards. Endicott was still using the Fred Carroll presses to produce up to ten million cards per day, selling three billion cards annually. That amount was over 80 percent of the total market for punched cards. Remington Rand was a distant second at 19 percent[10]. Card sales represented 10 percent of total revenue but 30-40 percent of profits. As Herman Hollerith had demanded from the start, IBM required users of IBM equipment to use IBM cards.

The situation finally prompted the Justice Department to act, filing an antitrust suit against IBM and three other companies, including Remington Rand, for monopolistic "tie-in" practices. The Justice Department announced their decision in April, 1936. The ruling forced IBM to accept the use of competitive cards in its machines. It also required the company to publish the specifications for its cards and to assist in the creation of competitors to enter the market. IBM appealed the decision, arguing that it had patents both for the cards and for their use in its machines. The company's appeal went all the way to the Supreme Court, which upheld the Justice Department decision. Watson was unrepentant in his views on monopoly, and this result only increased his animosity toward any antitrust action.

IBM emerged from the 1930s a far bigger and stronger a company. The first couple of years were flat with revenues around $20 million and income in the $5-7 million range. The low point was 1933 with $17.6 million in revenue and $5.7 million income. From 1934 on, it was mostly double-digit growth to 1940, when revenue exploded to $46 million and income grew to $9 million. In 1928, IBM had risen to #3 in the office equipment industry, behind Remington Rand and

NCR[11]. NCR suffered under mismanagement by John Patterson's underachieving son and sank in the ratings. By 1939, IBM was slightly behind Remington Rand in revenue but earned six times the profit. Six years later, Remington was in the rear view mirror and IBM was solidly in front. It had grown to dominate the punched-card industry, with a roughly 80 percent share. Though the SSA contract was a very big boost, the company was doing well elsewhere. A prime example was Metropolitan Life. The company nearly went with Peirce machines in 1914. By 1940, Metropolitan had 300 machines on lease from IBM[12]. This would continue to grow, reaching over 1,000 machines in the early 1950s.

With the rapid growth, Watson commissioned a new world headquarters for the company at 590 Madison Avenue, just south of Central Park. It replaced the former seven-story headquarters at 310 Fifth Avenue[13]. The 20-story structure, completed in 1936, was much closer to Columbia University, where Watson served as a trustee and the company had built a substantial relationship.

# VANNEVAR BUSH AND AN ANALOG INCURSION

I BM was riding high, dominating the industry of punched-card data processing while making some forays into scientific applications. Brute force calculations with desk calculators like the Friden had proved far too slow. Those experimenting with punched cards like Comrie and Eckert were constantly looking for faster and faster computing. The desk calculators and IBM machines were digital. At the same time, most scientists and engineers were still using the analog device invented by Reverend William Oughtred—the slide rule.

A student at Tufts College built a rudimentary analog computing device called the "profile tracer." He mounted a tracking stylus within a box mounted between two bicycle wheels. As the device traversed a section of land, it would trace and plot the contours of the property. It used a similar approach to that taken in the 1870s by Lord Kelvin (William Thomson) to predict the tides[1]. Kelvin's device, called a harmonic analyzer, used a series of gears and pulleys to plot out a tide report for a given seaport location. The sheer amount of calculations required to approximate this kind of function discouraged digital methods. Using embedded sun and moon positions along with local tide information, it could produce tidal estimates for an entire year in about four hours[2]. The Allies used an updated version of Kelvin's harmonic analyzer in World War II to predict the tides for the Normandy landing and the Pacific island assaults. Lord Kelvin's machine was still used to predict Liverpool tides as late as the 1960s[3].

The young man with the profile tracer was Vannevar Bush. He left Tufts in 1913 with both bachelor and master's degrees. His master's dissertation described the profile tracer. Bush worked for a short time at General Electric in Schenectady before returning to Tufts as a mathematics instructor. He saved just enough money to attend MIT for one year to secure his doctorate. At the time, the engineering doctorate was a joint program of MIT and Harvard. His 1916 electrical engineering thesis concerned oscillating-current circuits. He struggled with the sheer amount of computation involved. His faculty advisor refused to certify him for a doctorate, stating that he had done insufficient work. However, the young Bush signaled at an early age that he would be a force to be reckoned with. He escalated to a faculty advisor and the professor was overruled[4]. Bush received his PhD and returned to Tufts as an assistant professor. He moved back to the MIT Electrical Engineering Department in 1919.

In 1924, he lined up financial backing for physicist Charles G. Smith, who invented a voltage regulator called the S-tube. It was a rectifier, a specialized vacuum tube that would convert alternating current to direct current. This enabled radio sets to use standard household sockets. Bush and Smith started the American Appliance Company to market the device. The company would later become Raytheon. It would make Bush very wealthy[5].

As a professor at MIT, Bush supervised engineering graduate students. Most of their dissertation projects involved endless computations. Bush suggested to one graduate student, Herbert R. Stewart, that he look into an analog device that could perform continuous integration. The integral of a mathematical function was the area between the curve of the function and the underlying x-axis. Bush, working with Stewart and F. D. Gage, devised the Product Intergraph in 1925. It approximated the area under a curve by mechanical means, an analog approach to integration. A key inspiration for the design came from an unlikely source, a spinning disk used for aiming naval guns invented by mechanical engineer Hannibal Ford[6]. The initial MIT device handled only first-order (linear) differential equations, a capability of limited value.

Looking to perfect the design, Bush enlisted several others (Harold Hazen and Samuel Caldwell) to develop a far more advanced version, one that would handle higher order differential equations, equations that described curves. Completed in 1931, Bush called it the differential analyzer. The device was completely mechanical, a maze of gears, rods, pulleys, disks, and wheels. Like Bush's profile tracer, the differential analyzer would trace out the curve on graph paper, an analog rendering of the differential equation. Each equation required a different setup for the device. Given the mammoth size of the machine, this step required a wrench and a bit of "elbow grease."

Though this second version would handle second-order equations, it was still of limited use[7]. Bush's key innovation was designing a way to interconnect six complete sets of these mechanical integrators together in order to solve up to six-order differential equations. It was an engineering marvel to watch in operation. Taking up an entire room and "resembling a giant 100-ton foosball set[8]", eighteen drive shafts propelled by a small electric motor would trace the curve of the target function and compute the integral. The essence of the device was the interplay of a fixed spinning disk with the follower wheel that rested upon it. The follower wheel would rotate at different speeds depending on where it was on the timing disk, in effect producing the desired curve.

Bush's device would provide an analog approximation of the answer, in the same manner that a slide rule provided an approximation of a multiplication result. A larger differential analyzer would improve on the analog result, in a similar way a larger number of digital multiplications would yield a more accurate answer. However, given the slow speed of Bryce's 601 Multiplier (one multiplication every six seconds), the Bush analog machine would have a short window of opportunity until digital computing speeds increased.

It was not long before news of the Bush differential analyzer got around. Irven Travis heard about it in 1928. Travis was an assistant professor at the Moore School of Engineering, an adjunct to the

University of Pennsylvania. The Moore School had just started operation two year earlier, running out of a modest three-story brick
building on campus. Travis had undergraduate degrees at Drexel and
Penn, and then completed his master's degree at Moore, where he
decided to stay on. Interested in solving nonlinear differential equations, he considered ganging together a whole series of desktop calculators but soon discarded the idea as fanciful. He talked with
Vannevar Bush and discussed the feasibility of building a Differential
Analyzer at Moore[9]. Harold Pender, the dean of the Moore School, an
accomplished electrical engineer, and very funny guy with an outsized
IQ, supported Travis Pender secured agreement from Bush and negotiated a Public Works Administration grant to build one machine at
Moore and oversee production of a second machine for the U.S.
Army's Ballistic Research Laboratory (BRL) in Aberdeen, Maryland.
He also agreed to let the BRL use the Moore School machine in the
event of war.

The Ballistic Research Lab's ongoing mission was to compute
firing tables for the Army's artillery guns. Such tables were essential
for the guns to hit their target. The firing table allowed for such variables as range, altitude, air temperature, and wind speed. A separate
firing table was required for each gun and shell combination, and there
were upwards of 4,000 individual trajectories computed in a single
table. A human computer working with a desk calculator could
produce one trajectory in about two hours. The Army had set up a
large group of human computers at Aberdeen to calculate the tables in
this manner. With war looming, the need for firing tables escalated
dramatically as new guns and new shells were being developed. The
Army decided to fund an additional staff of 100 human computers, to
work out of the Moore School.

Meanwhile, Travis began development of the two Bush differential analyzers. He tapped into Works Progress Administration (WPA)
funds to staff this project. He assembled a construction group of 150,
far more than needed but he wanted to provide jobs in the midst of the
Depression. Travis's group completed the Moore machine in late
1934. They delivered the second differential analyzer to the BRL four
years later. Tests run on the Moore Differential Analyzer resulted in

an 80-fold improvement in the time to calculate a shell trajectory over hand computation.

The news of the Differential Analyzer at the Moore School made its way to General Electric. GE had a transmission-engineering problem and thought the machine at Moore could solve it. Travis assisted GE in setting up the problem and producing the desired results. Impressed, GE lobbied for its own Differential Analyzer. The company retained Travis as a consultant to assist in building the machine. He drafted a memo suggesting that they should leapfrog the Bush mechanical technology and build a fully digital, electronic Differential Analyzer. However, GE ended up building the same mechanical design. They would go on to build four Differential Analyzers at a cost of $125,000 each[10].

Back at MIT, Bush sought a far more powerful differential analyzer. In 1931, he turned to Warren Weaver, a director for the Rockefeller Foundation. Weaver was a professor at the University of Wisconsin who joined the foundation in 1930. The Foundation approved an initial grant of $85,000, though the new machine would eventually cost $230,500[11]. Eliminating the laborious manual setup of the analyzer was a central impetus for the new machine. It would use paper tape control. MIT began construction of the Rockefeller differential analyzer in 1935 and completed it in 1941, just in time for war service computing firing tables for the U.S. Navy. It was a monster, weighing in at 100 tons with 2,000 vacuum tubes, thousands of relays, and 150 electric motors. It was a unique design marrying digital electronics with the analog and mostly mechanical differential analyzer design.

Douglas Hartree at the University of Manchester copied and enhanced the Bush machine in 1934. He had visited Bush at MIT in 1932[12]. Hartree constructed his test version entirely with Meccano parts. Meccano was a toy construction set developed in England, the British equivalent of an Erector set. The Manchester machine cost just twenty British pounds. It handled first order differential equations. A second Meccano construction supported up to third-order equations. The project was a proof of concept rather than an attempt to create a device capable of calculating firing tables. The success of these simple models enabled Hartree to secure funding for a full-scale differential analyzer. He gave a lecture at Cambridge on the differential analyzer, using the Meccano model. Maurice Wilkes attended the lecture and found the machine "irresistible". He assumed responsibility for managing it[13]. A Meccano test machine at Cambridge would also lead to full-scale differential analyzer. It would also springboard Wilkes into computing.

**Figure 12**. Differential Analyser at Cambridge, 1938. Courtesy of the Department of Computing Science, University of Cambridge.

Vannevar Bush was never short of new ideas. With the advent of the war, there were virtually unlimited funds available to develop devices for use in the conflict. His Comparator was a decryption machine funded by the U.S. Navy. He was unable to get it to work, and the Navy soon pulled out of the project. His Rapid Selector was a high-speed microfilm reader. It was never completed. And then there was the Memex. It was a visionary concept that posited a vast hypertext system of information. Though Bush never translated into a working device, the concept was a precursor of the World Wide Web.

Bush aspired to make MIT a center for computation. Despite his analog work, his new goal was a digital device that would eclipse the tabulators of IBM. He would call it the Rapid Arithmetic Machine. He fine-tuned his ideas in a 1940 memo where he indicated that the machine would be similar in ambition to Babbage's Analytical Engine. It would be completely automatic, with data and program instructions on paper tape. The registers and computation would all be electronic. It would read data from paper tape, store the decimal numbers in registers, perform arithmetic operations digitally via vacuum tubes, and print the results. For someone so steeped in analog computing, his concept was surprisingly digital and quite visionary.

Bush planned to secure funding from NCR. The project was under the direction of his former graduate student Samuel Caldwell. However, work on the Rockefeller differential analyzer, a proven technology at the time, took valuable cycles away from this project. Then the war intervened. Two engineers, William Radford and Wilcox Overbeck, took over the project[14]. Overbeck was "requisitioned" for the Manhattan Project and the computation project died. It was somewhat puzzling that Bush never established a relationship with either IBM or Remington Rand, the two leaders in computing devices. Watson chose to work with Columbia and later Harvard. The Rapid Arithmetic Machine, had Bush been able to follow through with it, would have put MIT at the forefront of the coming computer revolution. As it turned out, that role would go to the Moore School.

There was another reason why a possible digital breakthrough escaped Bush. Vannevar Bush emerged as a central figure in mobilizing the entire scientific community in the war effort. In 1938, he was named president of the Carnegie Institution for Science, an organization that endowed research projects. That same year, the National Advisory Committee for Aeronautics, a forerunner of NASA, selected Bush as its chairman. With the German invasion of France in 1939, the likelihood of the United States involvement in the conflict increased substantially. This signaled the urgency of organizing scientific research. Bush was able to use his connections to meet with Roosevelt and propose a super-agency to coordinate all research. Roosevelt agreed, and the National Defense Research Committee (NDRC) was underway by June 1940. Bush was the logical person to head this group. He would have an estimated 30,000 scientists and engineers working under his purview.

A year into the work of the NDRC, Bush chafed at the lack of coordination between the research his organization was doing and the implementation. Bush went back to Roosevelt and argued for the expansion of the NDRC to include development. Roosevelt agreed again and authorized a new group, the Office of Scientific Research and Development, again with Bush at the head. Bush was strong, forceful, decisive, and well respected. His office oversaw a huge range of critical projects during the war including radar, antiaircraft fire control, the proximity fuse, and, of course, the Manhattan Project. When FDR died in 1945 and Truman became president, it fell to Bush to explain the atomic bomb to him.

Though the Rapid Arithmetic Machine was a bust, Bush was still involved in the development of the digital computer—but it was through one of his graduate students. Claude Shannon received degrees in mathematics and electrical engineering at the University of Michigan, graduating in 1936. He went onto graduate study at MIT. He worked on the Rockefeller differential analyzer, which proved to be somewhat ironic given his subsequent total focus on digital computation. He held summer jobs at Bell Labs, working with the electromagnetic relays used in telephone networks. He connected the on-off nature of relays to the logical design of electronic circuits. Working with Bush as his faculty advisor, his doctorate thesis was "A Symbolic Analysis of Relay and Switching Circuits[15]." Completed in 1937, it was a landmark paper, linking the binary concepts of Boolean algebra to the task of electronic logic design. Shannon also described how the Boolean logic elements of and, or, and not could be designed into computing. Taking his analysis a critical step further, he suggested that this naturally led to digital computation in binary (base two).

At the time, binary was prominent in mathematical circles, but decimal reigned supreme in calculators. As the common joke about binary goes, "There are 10 types of people in the world—those who understand binary, and those who don't." Both Blaise Pascal and Gottfried Leibniz studied number systems, including binary. Babbage considered alternative number systems for his Analytical Engine but ended up with decimal. In 1837, Samuel F. B. Morse somewhat accidentally brought the idea of binary to the forefront as his eponymous telegraph code used a short electrical pulse for a dot and a long electrical pulse for a dash. The implication was it required only two elec-trical pulses or voltage levels to represent numbers electronically in binary. Shannon even coined the term for the basic unit of informa-tion in binary, the bit. The bit represented the value of the flip-flop or on-off circuit. He went on to calculate the number of bits required for various repositories of information if done in binary. A phonograph record took 300,000 bits. The *Encyclopedia Britannica* would require over a billion bits. One hour of television was ten billion bits, and the entire Library of Congress would take up 100 trillion bits.

Given that Shannon had worked at Bell Labs and linked tele-phone relays to binary, it was logical that relay-based calculation first surfaced there. The Bell engineer with this idea was George Stibitz. He had grown up in Dayton, Ohio, and attended the advanced high school established by Charles Kettering of NCR, Delco, and GM fame[16]. He went to Dennison and Union College, spent a year with General Electric, and then obtained his PhD in mathematical physics from Cornell in 1930. He joined Bell Labs the same year at their West Street lab in Manhattan. Within a year, he put two and two together, so to speak, coming to Shannon's position on binary. In 1937, Stibitz connected a series of relays at home to create a simple adder. He called the device the Model K, as he put it together in his kitchen. It contained two relays representing two binary digits. He asked Bell to fund a full-size version. Bell initially balked but soon backtracked, realizing the potential to handle advanced calculations. Working with S. B. Williams, Stibitz constructed the Complex Number Calculator in 1939. It used 450 relays to enable the multiplication of eight-digit numbers and used paper tape to input instructions. Wanting to create a big splash, he ran a demo of his new calculator for the 1940 meeting of the American Mathematical Society, held in McNutt Hall at Dart-mouth College. He ran the demo remotely from his offices at Bell Labs. John von Neumann, a famous mathematician and polymath, was among the attendees at Dartmouth.

Stibitz then asked Bell for $50,000 to take the next step but the company declined. However, the war quickly changed the decision and he got the funding. He would build four relay calculators during the war years. His original machine was renamed the Bell Model I, and he followed it with the Models II to V. All of them found their way to the Army's Ballistic Research Lab to do firing tables. The early models provided a tenfold improvement over desk calculation. His final machine, the Model V, contained more than 9,000 relays and could execute a multiplication in just under a second. Installed in 1946, it weighed in at ten tons[17]. It was a technology, however, completely outmoded by war's end.

# THE FIRST COMPUTER

Though Vannevar Bush and Claude Shannon would signal the path to the coming revolution in computing, the real breakthrough would not come in the hallowed halls of MIT or the research center of Bell Labs. It would occur 1,300 miles west, in Ames, Iowa. Actually, the brainstorm would occur in a roadhouse near Rock Island, Illinois, and the ideas perfected at Iowa State College in Ames.

John Vincent Atanasoff, a physics professor at the college, had been thinking about ways to speed up scientific calculations for a number of years. He had experimented with the school's IBM punched-card equipment. He flirted with developing an analog device such as the Bush differential analyzer. He became convinced that a digital approach was the right way forward, but he struggled with the details.

One cold December night in 1937[1], Atanasoff was once again trying to piece together the elements of his digital device. Nervous and unable to think clearly, he decided to go for a drive to clear his mind. He headed east out of Ames and onto the Lincoln Highway, then south on Highway 6. He drove his Ford V-8 "hard," as was his custom. Since driving his father's Model T as a boy, he was fascinated with automobiles. Now with a professor's salary, he bought a new car each year. The Ford V-8 was the hot car of the 1930s, so driving it "hard" would have been going very fast, at least 90 mph. Iowa roads at the time had no speed limit. Lost in his thoughts, he was jolted when he crossed the Mississippi River and realized he was in Illinois. He had been driving over two hours and was now 200 miles from campus.

A bit thirsty, he exited the highway near Rock Island and pulled into a roadhouse. Iowa was a dry state while Illinois was not. He ordered a bourbon and soda. Soon he was relaxed, his mind was clear, and he focused on his planned calculator. For the next three hours, he sketched out the key parameters of his digital "computer" on cocktail napkins. He made four key decisions. First, the mechanism would be electronic using vacuum tubes. Second, it would operate in binary. Third, it would have a storage device for memory that used capacitors. And fourth, it would compute digitally, using Boolean logic on-off circuitry[2].

Atanasoff had grown up in Polk County, Florida, in the center of the state. His father had emigrated from Bulgaria and was a self-taught electrical engineer. His mother was a schoolteacher. He was a precocious child. His father bought a highly prized Dietzgen slide rule[3]. Young John mastered its operation at the age of ten, and then sought to understand the mathematics behind it, learning logarithms and trigonometry in the process. He learned different number bases in his early math experiences, including binary base two. He worked at the phosphate surface mines in central Florida to earn money for college. He attended the University of Florida starting in the fall of 1921. He would graduate with a BS in electrical engineering, recording the highest grade point average in school history[4].

While at the university, Atanasoff also taught science at Gainesville High where he was de facto head of the department. Iowa State College offered him a teaching position and a path to a master's degree. He received a late acceptance from Harvard but chose to honor his commitment to Iowa State. He received his master's in physics and mathematics in 1926 and stayed on as an instructor. He decided to further his studies and enrolled at the University of Wisconsin to seek a doctorate in physics[5]. He started in the middle of the school term. Professor John Van Vleck, the only theoretical physicist on staff, was not happy to have a student walk into his quantum mechanics class in the middle. Atanasoff quickly caught up and was one of the few students who aced the final examination. Both Van Vleck and visiting professor Paul Dirac would go on to win Nobel Prizes in physics. Van Vleck also taught John Bardeen and Walter Brattain, two of the three inventors of the transistor. Atanasoff would later muse that with his electronics and radio crystal background, the transistor idea should have come to him. Warren Weaver, who would later join the Rockefeller Foundation, taught courses on electricity and magnetism. Weaver had graduated from Throop College of Technology, which was later renamed Caltech[6]. Atanasoff would have an ongoing relationship with Dr. Weaver. It was at Wisconsin that Atanasoff got his first serious dose of massive calculations. His doctoral thesis, "The Dielectric Constant of Helium[7]", required eight weeks of intense Monroe desk calculator work, cementing in his mind the need for a better way.

Atanasoff returned to Iowa State in 1930 as a professor of theoretical physics, the lone professor in the discipline. Iowa State College was a land grant college more focused on applied science than pure science. As the institution was located in the midst of broad cornfields, the term "applied" often meant something agricultural. Atanasoff did a study in 1937 entitled "Measurement of the Viscosity of Eggs by the Use of a Torsion Pendulum[8]." It was a legitimate physics problem but one that you would be hard pressed to make up. Clearly, Ames was a backwater when it came to physics. Atanosoff's 1940 paper on "Generalized Taylor Expansions," linked him to Babbage and the method of finite differences. On campus, Atanasoff was simply "JV." He was amiable and widely admired. He could also come across as a "know it all," even domineering at times. He would sometimes abruptly end a conversation if he lost interest. In other words, he was not unlike many other physics professors.

The struggles of Atanasoff's graduate students with intense desk computation amplified his own experiences. The slide rule was woefully insufficient. To achieve the desired right level of accuracy, one would need "a slide rule the length of a football field.[9]" Still looking at analog approaches, he teamed up with graduate student Glenn Murphy to build a small analog device called the Laplaciometer, designed to solve Laplace differential equations by mechanical means. The limitations of the device reinforced the need for digital accuracy, but he wrestled with how?

Iowa State had a shop with IBM punched-card equipment. Like Comrie, Wood, and Eckert, Atanasoff experimented with them. From 1935 to1937, he did linear programming calculations on the IBM tabulators. Working with his IBM customer engineer, he adjusted the equipment to speed up the calculations. He wrote up his work in 1936 with the paper "Application of Punched Card Equipment to the Analysis of Complex Spectra." He wrote IBM about the possibilities of an enhanced tabulating machine, incorporating his ideas. IBM was not interested and was much more concerned about his "renegade" use of its rental punched-card equipment. An IBM memo turned up later that stated, "keep Atanasoff off of the tabulator[10]." At one time, he considered ganging desk calculators together to create a vector-calculating tandem. Atanasoff realized the cost would be prohibitive and that the contraption would be a behemoth.

Though disappointed by IBM's lack of response, he had already concluded that the IBM machines would not work because of their limited memory. Given his electrical engineering, mathematics, and physics background, he naturally gravitated to the latest electronic devices, vacuum tubes and condensers (now known as capacitors). His graduate students faced computing problems that could be solved through a set of simultaneous linear equations. This led him to consider a device that would be able to do this automatically. The four decisions that he had made at the roadhouse in Rock Island would coalesce into a mechanism that could solve up to 29 simultaneous linear equations with 29 variables[11]. The machine would resolve the variables one at a time until the entire equation set completed. The mathematical name of the process was Gaussian elimination. Charles Babbage designed his Analytical Engine to work in a similar manner.

This was an age-old problem in calculation. Solving ten simultaneous equations with ten variables required about 3,000 multiplications and subtractions. The computation needed to be extended ten decimal positions to assure accuracy. The number of calculations required increased dramatically as the number of simultaneous equations and variables grew. Atanasoff estimated his computer would take upwards of 125 hours to solve 20 equations with 20 variables. The speed at which computers solved these types of problems would later become a common benchmark for supercomputers. In 2010, for example, rating of the fastest supercomputer involved solving a million simultaneous linear equations with over a million variables.

At the heart of Atanasoff's design was the vacuum tube. This electrical device was based on a phenomenon observed by Thomas Edison in his light experiments, where electrons would flow from a heated element to a cooler metal plate. Naturally, it was called the "Edison Effect." In 1904, John Ambrose Fleming invented the diode vacuum tube, a device that restricted current flow to one direction. The diode soon found a place in radio. Lee de Forest, who was already heavily involved in radio, added a third wire to the diode in 1907 and called the enhanced device the "grid audion." This triode had the new property of amplifying the incoming signal, which proved very useful to boost radio transmissions. The amplification effect was slight at the outset as his triode used only a partial vacuum. Harold D. Arnold of Bell Labs perfected the device by simply converting the tube to a perfect vacuum. This improved vacuum tube was now useful for radio, television, and telephone purposes.

For digital applications, it was the speed and not the amplification that was the attraction. To count digitally, the vacuum tube needed to act as a flip-flop, or on-off, switch, much like the relay. The circuit was either conducting or not conducting. The solution was to simply divide the voltage ranges of the vacuum tube into two, with a high voltage denoting a "1" or "on", and a low voltage denoting a "0" or "off." The key was the speed at which the vacuum tube made this transition. The electromechanical relay could switch roughly 100 times per second, and because of its composition with mechanical moving parts, it could not maintain that high rate for very long. In contrast, the vacuum tube could switch at 1,000 times per second. Since it had no moving parts, it could operate continually, only subject to overall heat buildup.

After returning from his drive to the roadhouse in Rock Island, Atanasoff constructed a test setup in to prove his basic concept. It was a very simple device with only eleven vacuum tubes. With success, he drafted a proposal to fund a more complete model. In the spring of 1939, he received $650 from Iowa State and another $110 from the science dean[12]. Later in the year, the college kicked in an additional $700. The next step was to find a bright graduate student to assist on the project. A fellow professor recommended Clifford Berry. Berry was an Eagle Scout, well liked, with an aptitude for putting electronics together. He earned his BA in electrical engineering and was working on his master's in physics when he joined the project. Atanasoff would later say that Berry was "one of the best things that could have happened to the project.[13]"

The core of Atanasoff's "computing machine" was the vacuum tube-based calculating unit. Atanasoff called it the Add-Subtract Mechanism (ASM). It was a significant expansion of his small test device. The complete unit contained 280 vacuum tubes, arranged in 30 individual modules. Each module performed binary addition or subtraction, including the carry and total functions. Multiplication was repetitive additions with the James Bryce "601" type shortcuts.

Atanasoff developed the vacuum tube logic circuitry for the computation through trial and error. Though Claude Shannon had defined the binary and Boolean logic approach for such a task in 1937, Atanasoff was unaware of his paper. He simply set up a table of addition with binary numbers, determined the result of each operation (zero or one), and translated the table into his vacuum tubes. The standard Gaussian method of solving linear equations required many multiplications. He designed a method that would only require additions and subtractions, with one division at the end. He decided to do that final division on a desk calculator[14]. Though Atanasoff called his computing unit the Add-Subtract Mechanism, later computers would adopt the name Arithmetic and Logic Unit (ALU).

In addition to his core computing mechanism, Atanasoff needed a storage method to hold the variables of two sets of equations as the computation progressed. Each equation could have up to 29 variables. He wanted accuracy of the variable values to 15 decimal digits. He converted that number to binary, resulting in the need for 50 bits for each binary number. He considered a number of options for the short-term storage, including electromechanical and ferromagnetic components. Vacuum tubes were the ideal solution was but this would require 3,000 tubes. At a minimum of $4 each, this was unaffordable given his limited funding[15]. He considered using capacitors selling for

5 cents but he eliminated them because they only retain their charge, or bit value, for a limited time. However, he soon came back to capacitors when he devised a process to "jog" or refresh the capacitor value.

His next hurdle was to find a method to read and write the charge of each capacitor at a speed that was in sync with the rest of the machine. He decided to mount the capacitors on two rotating drums. Working with Cliff Berry, they affixed 1,600 capacitors with wax on the inside of each drum, with the metal stud of the capacitor protruding on the outside surface[16]. That provided the needed 3,000 bits of storage. A small electric motor spun the drums at one rotation per second, the equivalent of a clock cycle of 60 Hertz. The drum rotation synchronized the machine's operation. The machine wrote the coefficients of the 30 variables from two equations onto the capacitors of the two drums. With each drum rotation, up to 3,000 binary bits (thirty 50-bit binary numbers) registered in the 30 Add/Subtract vacuum tube modules. The Add/Subtract modules would run in parallel, performing up to 30 binary additions or subtractions with each clock cycle. The device wrote the results onto the second drum and then posted them on Atanasoff's punched-card format.

The finished "computing machine" was a compact desk-sized unit that weighed about 700 pounds. Input was in 80-column cards with the equation coefficients punched in decimal notation. A third purely mechanical drum would convert the input decimal numbers to the 51-bit binary format. Programming was done by buttons and switches on the console. There was no stored program. The application needed an operator and that was Clifford Berry.

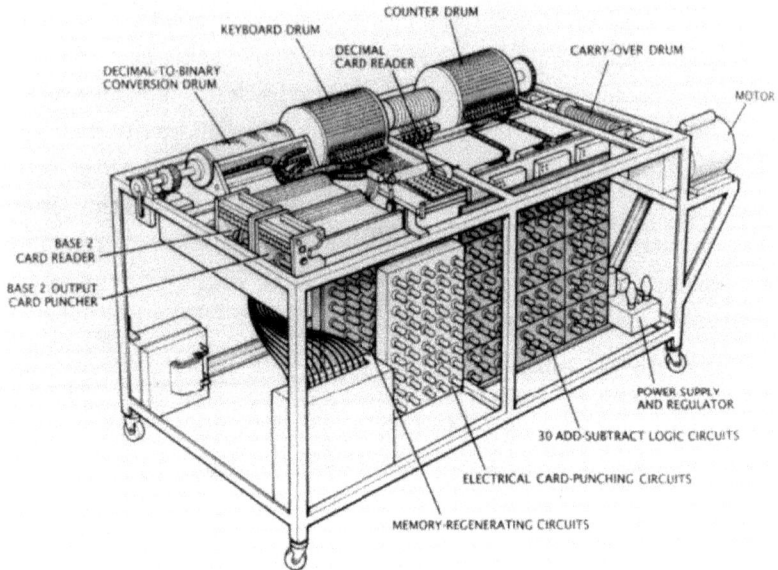

**Figure 13**. Atanasoff-Berry Computer. Diagram courtesy of Iowa State University Archives.

Once operational, the system could solve a full set of 29 linear equations in a day. Atanasoff was very familiar with the manual process[17]. Hand computing of eight equations with eight variables took eight hours and that assumed no errors, which was unlikely. He estimated that his device would calculate "at a speed about 30 times that of the largest tabulator." Atanasoff and Berry wrote a 35-page manual called the "green book[18]" covering all aspects of the design and operation of the machine, which would later be called the Atanasoff-Berry Computer, or ABC. Berry would write an extensive description of the workings of the computer as his PhD thesis.

Atanasoff designed his machine to prove the concept, within the bounds of his funding. He planned significant enhancements to improve the operation. The immediate task would be to automate the flow so Clifford Berry did not have to monitor each step. Beyond that, Atanasoff wanted to turn the machine into a more general-purpose computer. Though designed for the specific purpose of solving simultaneous linear equations, the ABC pioneered the three essential elements in a digital computer. It was binary-based, it performed the logic and computation with electronics, and it had a separate memory. Atanasoff's goal was to take the machine to the next level as soon as he secured more funding.

With a working machine completed, Atanasoff reached out to a number of companies to solicit their interest. He sent letters to IBM, General Electric, Remington Rand, and Raytheon. Despite the earlier snub of his punched-card ideas, IBM remained his first choice. He met with Clement Ehret, the IBM manager of market research and an IBMer that rated a song, which included, "he's a high-speed dynamo, with ideas he's all aglow.[19]" Ehret was interested but not without a patent in place. He urged Atanasoff to start immediately on the patent application. In the meantime, IBM provided 500 electric contact brushes at no charge. GE, Raytheon, and Remington Rand also were interested but reiterated the patent requirement. Raytheon provided 100 vacuum tubes at no charge.

Atanasoff met with Vannevar Bush in 1940 in Washington, DC. Bush showed little interest in Atanosoff's computer as he was still very analog focused. Atanasoff also met with Warren Weaver, his former physics professor, in New York City. Weaver was now a director with the Rockefeller Foundation, tasked with evaluating scientific projects. At the time, he was focused on medical research. He did not have $5,000 to spare for Atanasoff, though he had recently approved over $1 million for Ernest Lawrence to develop a cyclotron device at Berkeley. Lawrence was more experienced in the foundation grants game and had framed his physics machine as medical research when it was decidedly not.

Atanasoff also met with Theodore Brown, a Harvard professor and consultant to IBM. Brown stressed the need for a patent. He had introduced fellow professor Howard Aiken to IBM's James Bryce to look at Aiken's proposal for a Babbage-type machine. Atanasoff met with Aiken, who was in the early stages of discussions with Bryce and not very receptive to the Atanasoff design. Atanasoff considered Aiken's heavily electromechanical design as "primitive and not very valid . . . the last throb of the old era.[20]" It was an early but accurate assessment of Aiken's future.

Atanasoff also met with Samuel Caldwell, the graduate student who worked with Bush at MIT on the differential analyzer. Caldwell was now working for Warren Weaver. Caldwell would visit Ames to see the ABC and would recommend that the Research Corporation, the patent agency for MIT, provide funding. In early 1941, the Research Corporation did in fact grant $5,330[21].

Atanasoff was well aware of the need to patent his machine. Iowa State was not initially interested. With very little word of mouth, Atanasoff had secured a sizeable grant, demonstrating the potential value of his invention. Now, the President of Iowa State was interested. However, he wanted 90 percent of the patent proceeds to go to the college. Atanasoff pushed back, and they agreed to a 50/50 split with ten percent of Atanasoff's share going to Berry[22]. The college selected Richard Trexler, a Chicago attorney, to file the patent application. Atanasoff worked with Trexler, first on the required patent search and then on the application itself. Atanasoff and Berry sent detailed materials and drawings to Trexler. Atanasoff and Berry also traveled to the patent office in Washington to search for any conflicting patents. They passed on their assessment of any patent exposures to Trexler. At this point, Atanasoff felt that he had supplied the necessary information and it was now up to the lawyer to run with it.

Atanasoff was naturally interested in who else might be working on computing devices. In September 1940, he had traveled to Dartmouth College for the meeting of the American Mathematical Society. He saw the demonstration by George Stibitz of his Model I relay-based Complex Number Calculator[23]. In December 1940, Atanasoff was back East again, this time with his family. The day after Christmas, he attended a meeting of the American Association for the Advancement of Science, on the campus of Ursinus College, just outside of Philadelphia. One of the presenters at the meeting was John Mauchly, a physics professor at the school. Mauchly gave a lecture on an analog device he had created, a "harmonic analyzer" that he planned to use in weather modeling. After the lecture, Atanasoff started a conversation with Mauchly. He simply could not resist telling a fellow physics professor about his computer. He described a digital calculating machine, one that he had constructed for just $2 per storage bit. He invited Mauchly to come to Ames and see his machine. He did caution Mauchly, "We're afraid that IBM might get hold of my ideas and so we're careful not to spread them around.[24]"

Mauchly had grown up in Chevy Chase, Maryland. His father, a German émigré, was head of the Terrestrial Electricity section at the Carnegie Institution. Mauchly entered the engineering program at Johns Hopkins in 1925, but after two years, he switched to physics. He enrolled in an accelerated program that allowed one to skip the undergraduate degree requirements and go straight to graduate work. He worked summers at the Bureau of Standards and received a stipend as an instructor in physics. He used a Marchant calculator in his physics studies and relied on an Erector set to build some of his experimental devices[25]. His doctorate dissertation was in molecular spectroscopy, an area considered old physics by many of the schools he applied to for a faculty position. He ended up remaining at Johns Hopkins as a research assistant for a year. It was the utter bottom of the Depression and all jobs, including college professor, were hard to come by. He eventually secured an appointment as a professor of physics at Ursinus College, where he was the entire physics department. Ursinus was far from the center of theoretical physics, without the funding or equipment of the leading research universities.

Mauchly decided to take Atanasoff up on his offer to see the ABC machine. In June 1941, he drove out to Ames. As money continued to be a problem for him, a neighbor helped with gas money and he had paying passengers that he dropped off along the way[26]. For some reason, he brought his son along. He apparently expected Atanasoff's wife, Lura, to tend to his son despite her having three children of her own. Once more, Atanasoff let down his guard in his enthusiasm to show off his new machine. He made himself and Clifford Berry completely available over a period of five days, providing a detailed demo followed by many hours of discussion on the inner workings of the system. It remains a mystery why Atanasoff was so open with an invention that lacked patent protection. He did become suspicious when Mauchly asked for more details about certain ABC functions. He provided a copy of the 35-page "green book" for Mauchly to review but gave it with the caveat that he could not keep it. His wife was far more suspicious[27]. She witnessed Mauchly making extensive notes from the manual. Atanasoff was either incredibly naïve or concluded that Mauchly did not have the background or capabilities to steal his ideas and build his own computer.

Sometime after his visit to Ames, Mauchly wrote to Atanasoff about the possibility of building an ABC-like computer. Atanasoff asked Mauchly to hold off on any discussion or action until his patent came through. At the time, he did not know that the patent lawyer, Richard Trexler, had failed to take any action on it. He assumed that Iowa State was overseeing the process and if they needed his attention, they would contact him. Iowa State lost their patent files during the course of the war. Trexler would later claim that no one authorized him to proceed. He moved on to other cases. Once Atanasoff got to Washington for war service, he was swamped. In the end, no patent was ever filed.

# WORLD WAR

Though the Great Depression substantially eased by 1940, another threat loomed on the horizon. The rise of Adolf Hitler and the Nazi Party, German rearmament during the 1930s, and the German annexation of the Sudetenland in 1938 made it clear that another European war was likely. In 1936, Hitler had provided German dive-bombers in support of fellow fascist General Francisco Franco during the Spanish civil war. The Japanese invasion of Manchuria in 1937 further darkened the war clouds. Meanwhile, American isolationism was in full bloom, thanks in part to the efforts of Henry Ford, Charles Lindbergh, and the America First Committee. U.S. military preparedness was at low ebb. In 1940, the United States ranked 19th in armed forces capability in the world, after Romania and just ahead of Holland. Hitler struck Poland on September 1, 1939, with an overwhelming "blitzkrieg" military invasion.

Roosevelt had sought all manner of low-key ways to prepare for war. He listened to Vannevar Bush's advice and agreed to the establishment of the National Defense Research Committee. He asked his advisors for the name of the best manufacturing man in the country. The answer was "Big Bill" Knudsen, who ran manufacturing for Henry Ford and then Chevrolet before rising to president of General Motors[1]. Knudsen would lead the Office of Production Management, the key arbiter between the country's manufacturers and the needs of the armed forces. Roosevelt delivered his famous "Arsenal of Democracy" speech in December 1940. When Japanese forces attacked the U.S. bases at Pearl Harbor on December 7, 1941, any lingering doubt about America's role in the global conflict disappeared.

In 1939 during the ABC development, Atanasoff had met with Warren Weaver, who was working for Vannevar Bush at the National Defense Resources Council. Weaver was in charge of the fire control section and overseeing a project for the U.S. Navy called Project X to improve the tracking and targeting of aircraft by anti-aircraft guns. The degree of difficulty was high, as any device developed would need to weigh the aircraft's speed, future position, and the transit time of the fired shell, among many factors. The Naval Ordnance Lab (NOL), located at the U.S. Naval Gun Factory on the Anacostia River in Washington, DC, ran the program. Atanasoff agreed to work on the program part-time. In 1941, Atanasoff had a number of graduate students, a full course load, and this Navy project. Something had to give, and it was the continued work on the ABC.

Atanasoff hoped that Clifford Berry would be able to carry through on the ABC enhancements. He sought a draft deferral to keep Berry in Ames. However, in June 1942, Berry married Atanasoff's secretary and left for war service at Consolidated Engineering in California. Atanasoff petitioned the NOL to work out of his Ames office but he was ordered to report Washington in September 1942, where he became chief of the acoustics division. With anti-aircraft fire control, as with most wartime NRDC projects, the research mantra was simply "just try something." Duplication of effort was encouraged in order to speed the development. The urgency of fire control increased dramatically with the war in the Pacific. The program attracted quite a few luminaries in computing besides Atanasoff. Irven Travis would leave the Moore School to work in this area, as would Julian Bigelow. Claude Shannon left a fellowship at Princeton to join the effort. George Stibitz and Bell Labs signed on. The various teams pursued digital approaches but the technology was simply not mature enough before the war was over. The focus was on analog methods. Two related research programs, radar at MIT and the proximity fuse at Johns Hopkins, would combine to produce an effective anti-aircraft gun, the Mark 51.

Watson, as he had done in World War I, immediately pledged the resources of the IBM Company to the war effort. He wrote a letter to Roosevelt to that effect. He also announced that the company would accept only a one percent profit on war production. The company secured a $57 million defense contract to manufacture Browning automatic rifles, M-1 carbines, pistols, and ammunition. A separate company, the Munitions Manufacturing Corporation, was set up to keep war work separate from commercial work. The government contract included funds to add a building in Endicott and build a new plant on the Hudson River at Poughkeepsie. To get off to a fast start, IBM bought the former location of the Delapenha pickle factory near the new plant location. By early 1942, the new plant was well under construction. The company christened it as IBM Plant No. 2, as Endicott was IBM Plant No. 1. There would be 10,000 employees working at Poughkeepsie by war's end.

IBM expanded the Endicott site significantly, adding 700,000 square feet[2]. Staffing there would increase from 4,130 employees in 1940 to over 9,000 in 1943. Managing the war effort would require the widespread use of punched-card machines. IBM and Remington Rand would combine to ship 9,000 tabulators in 1942 and 10,200 more in 1943[3]. The Endicott factory and lab would also be responsible for the highly secret Norden and Sperry bombsights. Watson and his board later decided to pay back the government for all the funds provided for plant construction.

Beyond the core wartime commitments, Watson was also open to special projects. John McPherson had distinguished himself with his work in sales and the railroad marketing team. In 1940, Watson asked him to start a new department called Future Demands to assess the new ideas that came in. McPherson nominally reported to general manager Fred Nichols, but he had unfettered access to Watson. He became the conduit during the 1940s and early 1950s, passing ideas up and handing out Watson's decisions down. He might deal with two or more requests a day. He received all manner of suggestions, but perhaps the wildest one came from Madame Chiang Kai-Shek. She wanted to know if IBM could electrify the suanpan, the Chinese version of the abacus. McPherson actually put a couple of engineers on it[4]. The war put his Future Demands role on hold, but he continued with essentially the same role as director of engineering.

IBM employees also went to war. Watson announced that every IBM soldier would receive 25 percent of his or her pre-war salary. He authorized the erection of panels at 590 Madison headquarters that listed every one of the 2,000 IBMers in uniform. Many IBM engineers ended up in war positions. Steve Dunwell, who had helped Ben Wood at Columbia, worked in the army's intelligence service modifying punched-card machines for code breaking. Ralph Palmer, manager of the new Poughkeepsie lab, led cryptographic work for the Navy, working out of the NCR plant in Dayton, Ohio. "Dick" Watson, Watson's younger son, dropped out of Yale to enlist. The Army assigned him to the Signal Corp, where he pressed the concept of field data centers. Called Mobile Records Units (MRU), they consisted of tabulating equipment mounted in trucks. The Army received 274 MRU trucks during the war[5]. Thomas Watson Jr., the older son, became an Army Air Corps pilot.

In 1940, Wallace Eckert of the Watson Astronomical Bureau at Columbia received appointment as the director of the Nautical Almanac at the U.S. Naval Observatory in Washington, DC. It was a prestigious position and one he saw as a new career path. The immediate mission was to construct an accurate set of Air Almanac tables. These tables would tell aviators where to find the main celestial objects at any time in order to fix their location. He arrived at the Naval Observatory to find the current staff working with desk machines[6]. The previous director had hired people that were good with logarithms instead of pursuing automation. This would not do for Eckert. He immediately drew up a list of IBM machines to order. Because the almanac dealt in longitudes and latitudes, he asked for and received IBM 405 tabulators that were set up in base 60, enabling computation in 360 arc gradations.

He also faced the transcription issue that had plagued so many before him. The Air Almanac was comprised of many volumes, with thousands of pages of tables. The transfer of calculated values from the machines to print introduced many errors. He had the applications modified to print the computations directly on the 405 tabulator. With his updated and accurate Air Almanac tables, pilots could fix their positions within a minute of arc (about a mile). The prior hand methods only ensured a 30-mile level of accuracy. Reconnaissance pilots who spotted German U-boats could accurately radio their position. Meanwhile, back at the Watson Astronomical Bureau, Eckert loaned his punched-card equipment for calculations for the B-29 bomber and the Manhattan Project atomic bomb.

# ENIAC: "THE GIANT BRAIN"

Irven Travis completed the construction of two Bush differential analyzers during the Depression years. The agreement with the Army's Ballistic Research Lab (BRL) required turnover of the Moore School machine to the Army in case of war. The Army created by the BRL in the 1930s, with a core mission to create firing tables. It was a monumental task. A single trajectory could take up to 12 hours to calculate manually, and each firing table could have thousands of trajectories[1]. The Bush differential analyzers reduced the time for each trajectory to 10-20 minutes, yet it still took upwards of 750 hours to compute one complete firing table for a single gun and shell combination.

In 1938, Colonel Herman Zornig, the director of the BRL, engaged IBM to produce firing tables on tabulating machines. Funding was not available until 1941, at which time the BRL installed the IBM equipment and used it exclusively for firing tables. Zornig visited the Moore School and authorized an "army" of human calculators to work in conjunction with the Differential Analyzer installation. This staffing grew to more than 150, nearly all of whom were women[2]. The firing table work on the Differential Analyzer still required manual calculations to smooth the results, so Moore ordered additional Marchant and Monroe desk machines. When the war came, this extensive combination of machines and human calculators was nowhere near adequate to calculate the large number of firing tables needed. Travis was already thinking beyond the use of the Differential Analyzers. In 1938, he looked into the concept of ganging mechanical calculators in series to do computation[3]. It was the same idea that Atanasoff had considered and discarded. Travis came to a similar conclusion.

The Moore School had an additional mission. Wartime brought the urgent need for upgrading the nation's skills in engineering and science. This led the War Department to create a nationwide education program, the Engineering, Science, and Management War Training program. Running from 1940 to 1945, 227 participating colleges conducted 68,000 courses for nearly two million students[4]. The Moore School was one of the training sites for the program. When John Mauchly returned from his visit with Atanasoff, he immediately enrolled in the Defense Training in Electronics class for scientists at Moore. It was a ten-week crash study for mathematics or physics scientists, teaching them the basics of electronics for wartime research work. Mauchly was one of two PhDs in the class. The other was Arthur W. Burks. Burks had degrees in mathematics and physics from DePauw and a PhD from the University of Michigan. The two roomed together. Mauchly was the oldest person in the class. A 24-year-old graduate student named J. Presper Eckert gave some of the lectures. Mauchly and the young Eckert became lab partners. Besides the age difference, it was an odd pairing. Mauchly was undisciplined, a chain smoker, always worried about money, quite a stark departure from Eckert.

Eckert came from a wealthy family. His father was a prominent Philadelphia developer. He had his son chauffeured to his prep school in a limo[5]. Presper grew up with electronics as a hobby. His father moved in high circles in Philadelphia and arranged for Presper to spend time with Philo Farnsworth, the inventor of the television. Presper was an exceptional student. MIT readily accepted him but his parents pushed him to attend a school closer to home. He ended up at the Moore School at the University of Pennsylvania. He graduated in early 1941 with a degree in electrical engineering. He had been a student of Irven Travis, who had a very high opinion of his engineering skills.

After the defense class concluded, Mauchly sought and secured an appointment at the Moore School. The school was in a position to hire him as some members of their staff, including Travis, were leaving for war service. Mauchly did not accept the offer at Moore until he received the same pay that he received at Ursinus College. Mauchly's visit to Ames in June had shown him the path to creating a high-speed digital calculator. He discarded his prior ideas about using an analog approach. He now had a blueprint and some of the details, depending on how much of Atanasoff's "green book" he had been able to copy. He was further armed by the electronics education he received in the Moore class and a number of discussions with Irven Travis. Mauchly put his ideas down in a preliminary document titled "Notes on Electrical Calculating Devices." It was late September 1941 when Mauchly wrote to Atanasoff about building an ABC-type computer. He neglected to mention that he had already developed and turned in a recommendation[6].

Despite the negative reply from Atanasoff, Mauchly chose to ignore his wishes and proceed. He did not advise Atanasoff of his intentions. He worked on fleshing out his preliminary document into a complete proposal. He leaned heavily on his former lab partner, Presper Eckert, to ensure the electronics in his design were correct. By August of the following year, he submitted the updated plan, titled "The Use of High-Speed Vacuum Tube Devices for Calculation,[7]" to John Brainerd, the dean of the Moore School. The proposal aimed to replace the Differential Analyzer with a digital device that would calculate firing table trajectories in just 100 seconds. His proposal went nowhere. Many felt that it was too big an undertaking and that the war would be over before the project ended, if it was even possible. Though Travis supported the concept, he too felt it was too ambitious and unlikely to finish in time for war use. The Moore School was actively considering an automated Differential Analyzer, similar to Bush's Rockefeller machine.

With the war crisis deepening, Mauchly's proposal got a second look from an expert in both mathematics and ballistics. Herman Goldstine was a professor of mathematics at the University of Chicago and had taught a course on ballistics. The Army drafted him in July 1942 and assigned him to the Ballistic Research Lab. He would be directing the Army's differential analyzer operation at the Moore School. Not long after he arrived, he read a copy of the Mauchly document and immediately saw the promise. He recognized that the proposed machine could be a game-changer. He had the proposal revised and titled, "Report on an Electronic Differential Analyzer," knowing that a reference to the differential analyzer would help the proposal move forward.

The Moore School billed the proposed machine as a thousand times the speed of anything currently in use for firing tables. The BRL had a high-powered staff. Colonel Herman Zornig, Colonel Leslie Simon, and Paul Gillon were all MIT graduates. Dr. Oswald Veblen, a highly decorated mathematician who had served in the Army during World War I, had recently rejoined the Army and they posted him to the BRL. Goldstine convinced this group to entertain the concept, and in April 1943, headed a presentation to Simon and Veblen in Aberdeen, Maryland. Veblen agreed with Goldstine on the merits of the proposal and urged Simon to fund the project, saying, "Simon, give Goldstine the money[8]." Simon agreed to an initial funding of $61,700, not knowing at the time that the completed project would run to $486,000.

The BRL authorized the project despite opposition from the NDRC. Both Vannevar Bush and George Stibitz opposed the project. Stibitz favored his relay approach and believed that a device with 18,000 vacuum tubes could never function long enough to do real work. Bush and his associate Samuel Caldwell thought the project was too expensive. Bush also agreed that the project was unlikely to finish before the war ended. Bush would have a couple of additional reasons to oppose it. His Rapid Arithmetic Machine remained an active project. In addition, he was still a big believer in the power of analog computation.

Work on the program, dubbed Project PX, started in June 1943. The original name of the machine was the Electronic Difference Analyzer, hinting at both its heritage and its technology[9]. The Army renamed the planned device to ENIAC, for Electronic Numerical Integrator and Automatic Computer. Despite the funding predicated on artillery firing tables, the design enabled the machine to also solve differential equations for missile trajectories. John Brainerd was the overall project supervisor. Arthur Burks was the principal architect. Hans Rademacher provided the mathematics.

The central roles for the project went to Mauchly and Eckert. Mauchly was not an engineer. He did not have deep expertise in either electronics or hardware. His role beyond providing the original concept and being a close associate of Eckert was not clear. In contrast, Presper Eckert was the chief engineer. Though he was central to the project, he tended to lose sight of the ultimate goal to complete ENIAC in time for actual wartime work. He was a perfectionist, constantly tinkering with every detail. As a whiz with electronic circuits, he was usually quite far ahead of others working on the hardware. He could be intimidating and imperious, had a temper, and did not suffer fools. Arthur Burks, who was responsible the ENIAC multiplier, agreed that Eckert was tough to work with[10]. All of this added tension to the work environment. It was in this milieu that Herman Goldstine was the key. He drove the project and ensured that everyone, including Eckert, stayed focused and worked toward the common goal.

**Figure 14.** ENIAC team circa 1946 (U.S. army photograph from the archives of Mrs. Kay Gillon). From left, J. Presper Eckert (lead engineer), Professor John Brainerd (project leader, Moore School), Samuel Feldman (Chief Engineer, BRL), Captain Herman Goldstine (Liaison, Moore School), John Mauchly (Moore School), Dean Harold Pender (Moore School), General G. M. Barnes (Chief of Army Ordnance and Research), and Col. Paul Gillon (Chief of Research, BRL).

The design team laid out the machine in three sections: one for arithmetic operations, a second for the storage of numbers and instructions, and the third for the master programming interface. Vacuum tubes were the heart of the machine, to be used in all high-speed areas, predominantly the logic and the arithmetic computation circuits. The tubes would in effect replace the gears and wheels of the Differential Analyzer. The initial plan called for 5,000 vacuum tubes, but additional computation requirements added by the Army raised the number to 18,000.

A first order of business was to address the issue of vacuum tube reliability. Manufacturers rated vacuum tubes to last an average of 2,500 hours. With 18,000 of them, the calculated odds predicted a failure every eight minutes[11]. Eckert suspected that tube failure was a result of running them at high power or constantly powering them up and down. He ran the tubes at far lower voltages and never shut the machine down. Even with these adjustments, about 50 vacuum tubes failed each day. This resulted in the machine being up and running only about half of the time.

A related issue was the clock speed of ENIAC. This was the basic electronic cycle time, analogous to the rotation of Atanasoff's ABC drum. The design called for a cycle speed of 200 kilohertz, or 200,000 cycles per second. However, the machine would experience problems if you cranked the speed much past 70 kilohertz. A final issue was heat. With that many vacuum tubes, the machine ran very hot. The room with the ENIAC and the room below it with the school's IBM equipment were the only two rooms with air conditioning[12]. The heat problem was never fully resolved, even after the machine moved to the BRL. The eventual work-around was to eliminate 3,000 of the vacuum tubes.

In the spring of 1943, Mauchly unexpectedly showed up at Atanasoff's office at the Naval Ordnance Lab. Atanasoff was startled, as there was a very high level of security clearance required to access his lab. Mauchly would have to have that level of clearance. The Moore School had cut Mauchly's salary and he asked if there was any work available at the NOL. Atanasoff arranged for Mauchly to work part-time for Dr. Herman Ellingsen in the statistics department. He would later hear from Ellingsen that Mauchly talked a good game but did not make much of a contribution[13]. During this time there, Mauchly would talk with Atanasoff about computers but would not disclose the nature of the project he was working on at Moore. When the subject came up in 1944, he admitted that it was a computation project but said it was of a completely different design from the ABC.

There was a mid-project addition to the ENIAC program in the summer of 1944. Herman Goldstine ran into John von Neumann at the Aberdeen train station. This was not a total surprise as both did work for the BRL. Goldstine, with a PhD in mathematics and a professorship at the University of Michigan, knew of von Neumann and his outsized reputation. He was even apprehensive that von Neumann might be busy or dismissive when he approached him on the train platform. Yet von Neumann was very personable, and the two exchanged pleasantries while waiting for their trains. In fact, von Neumann was getting somewhat bored with the conversation when Goldstine mentioned that he was working on a computer at the Moore School that could compute well over 300 multiplications per second.

Von Neumann was always on the lookout for faster computing. He was present in 1937 with Wallace Eckert for the dedication of the Watson Astronomical Computing Bureau. He had joined the Manhattan Project in 1943 and was familiar with the IBM computing operation there, led by Richard Feynman, Nicholas Metropolis, and Stanley Frankel. He was also familiar with the relay computer work of George Stibitz at Bell Labs. However, Goldstine's comment about the speed of ENIAC hit him like a thunderbolt. He wanted to know more. The Moore School was only 45 miles from his office at Princeton. He started to spend time in Philadelphia with the ENIAC team, beginning in August 1944.

John von Neumann was a giant in 20th-century mathematics and physics. He was born in Budapest, Hungary, in 1903. His father was a banker and was ennobled by the Hapsburg emperor, hence the "von." Young John displayed his prodigious intellect early. He was a math prodigy, familiar with calculus at the age of eight. He had an incredible memory. He could supposedly recite word-for-word virtually any book that he had read. His memory banks would also include the largest collection of "off-color limericks[14]." He attended the University of Budapest, graduating in 1925 with a doctorate in mathematics with minors in physics and chemistry. By 1927, he was a professor at the University of Berlin, the youngest ever. He had already published 12 major papers on mathematics.

He worked with the famed mathematician David Hilbert. Hilbert, the leader of the German mathematics establishment, was certainly no slouch. He would say that von Neumann possessed "the fastest mind I ever met.[15"] Von Neumann had the ability to do elaborate mathematics in his head. Like the polymaths of the 18th and 19th centuries, his interests and expertise were wide ranging and constantly growing. They included set and game theory, linear programming, quantum mechanics, and geometry. As a world-renowned mathematician, the

Institute for Advanced Study at Princeton invited him in 1929 to accept a lifetime appointment.

Von Neumann was also a "Martian", a name given to a number of Hungarian scientists who came to the U.S. before the war. The list of Martians included Leo Szilar, Eugene Wigner, Edward Teller, and John Kemeny. They all spoke English with a "Bela Lugosi" accent. Bela was another Hungarian immigrant. Szilard had coined the name. His based the moniker on the fact that they all came from a small, nondescript European country, spoke a strange language, and had superhuman intellects. Szilard's "joke" became more elaborate over time. A later version had them as descendants of actual Martians who landed in Budapest in 1900.

The Moore School team completed the ENIAC design in June 1944, but it would not be ready for testing until November 1945. The machine, or more aptly the enormous grouping of machines, weighed in at more than 27 tons. The combined units measured eight feet high, three feet deep, and over 100 feet long. ENIAC took up 1,800 square feet of floor space. It consumed 160 kilowatts of electricity, enough to cause brownouts in the area when it was in operation. The machine contained 17,480 vacuum tubes installed in 30 massive racks, each holding 500 to 1,500 tubes. The electronics also included 70,000 resistors, 10,000 capacitors, 7,200 crystal diodes, 6,000 switches, and 1,500 relays. Putting it all together required five million soldered joints[16].

Figure 15. ENIAC computer after relocation to Building 328 at the Army's Ballistic Research Laboratory (circa 1947). Programmers Glenn Beck in the background, Betty Snyder in the foreground.

Central to the ENIAC design was the arithmetic and logic unit (ALU) that Burks designed and Eckert implemented. They rejected the binary structure of the ABC in favor of a more familiar decimal concept that mirrored the design of Monroe mechanical calculators. The ENIAC team was able to obtain input on design from other sources. It was wartime, and subject to secrecy concerns, scientists were encouraged to share information. Perry Crawford, a fellow graduate student at MIT with Claude Shannon, had written his thesis on digital design with vacuum tubes. There was also a 1937 paper by Stevenson and Getting on flip-flop circuit design[17].

The team was also able to talk with engineers at NCR and RCA, contacted through the NDRC. Both Joe Desch at NCR and I. E. Grosdorf at RCA experimented with tube-based digital counters. Mauchly, as was his nature, would claim many years later that these electronic counter ideas from RCA and NCR were not very good or relevant, and that he and Eckert invented the counter logic for

ENIAC. Their design required 20 triode vacuum tubes for each decimal number, providing ten-digit accuracy. This was more than twice the accuracy of the Differential Analyzer but far short of the 51-bit binary numbers of the ABC computer. The decimal decision and the counter design was the source of the vast number of tubes required.

Arthur Burks designed the multiplication function for ENIAC. A standard clock cycle was 200 microseconds, the time required for one addition. The machine could perform 5,000 additions per second, or one addition for each clock cycle. Burks patterned multiplication after the 1931 IBM 601 Multiplying Punch developed by James Bryce. IBM engineers would later write Burks, "You were doing electronically almost exactly what we had in general use for twelve years in the 600 and 601.[18]" Limited memory in the ENIAC required the transfer of the partial products of multiplication to slower input/output devices. Still, Burks's design enabled the ENIAC to perform 357 multiplications or 38 divisions per second[19], as Goldstine had related to von Neumann on the train platform. John Mauchly would later insist in his patent application that the multiplication design was his. He would contend that he learned it at the Remington Rand booth at the 1939 New York World's Fair by asking the show personnel how multiplication worked in their machines. With no engineering background, Mauchly's contention was dubious.

Goldstine estimated that the 2,000 to 4,000 trajectories of a firing table would take up to 750 hours on the Differential Analyzer[20]. ENIAC was certainly lightning fast, but it only mattered to the BPL how fast it could complete firing tables. The computation of a single ballistic trajectory took about 20 hours by hand, 15 minutes on the differential analyzer, and 30 seconds on ENIAC[21]. Though ENIAC was substantially faster than the Bush differential analyzer, it was not faster than the vacuum-tube-infused Rockefeller differential analyzer. The Rockefeller machine was operational in 1942 and used for missile trajectories. It was able to trace the German V-2 rocket trajectories that were hitting London from their launch pads in Germany. By war's end, however, the Differential Analyzer was obsolete while the ENIAC and similar computer technologies were just getting started.

Goldstine would call the programming process on ENIAC "unsatisfactory." The "Master Programmer" was the name given to the central programming panels. This suggested that there was much more going on than was actually there. The mathematicians conceptualized the program flow. Six programmers, all selected by Goldstine and all women, did the actual machine setup. They were Kathleen McNulty, Frances Bilas, Betty Jean Jennings, Elizabeth Holberton, Ruth Lichterman, and Marlyn Wescoff. The Ballistic Research Lab at Aberdeen assigned them to learn IBM punched card processing and the wiring of plugboard panels. John Holberton was in charge of the team and later married Elisabeth "Betty" Snyder, who became Betty Holberton. Mauchly would comment later on Goldstine's choices for the programmers. He thought some did not have sufficient skills in math. Mauchly's first wife had died tragically, and he married Kathleen McNulty. McNulty had been a math major, so she was, of course, acceptable.

The programming team set up ENIAC applications by connecting patch cables, some up to 80 feet long, and dialing in up to 3,600 rotary switches. The actual workflow was very similar to a punched-card installation. The card reader processed IBM cards containing the input numbers and stored the values in relays. The program routine executed and punched the results on IBM cards to be printed offline. The setup process for a single function took weeks. This was less a problem for the BRL as the firing table programming was set up once and then run continually. Mauchly would later insist that ENIAC was a stored program machine, presumably because the "Master Programmer" controlled the maze of switches and patch cables. He was almost belligerent about it. Years later, he was still upset about not getting credit for inventing the stored program computer.

IBM was aware that something was in the works at the Moore School. In early 1944, Watson had a visit from Colonel Gillon of the Ballistic Research Lab. Gillon wanted some help on a project he was funding at the Moore School. Watson agreed, and he sent two engineers. They were surprised by what they encountered on the first floor of the main Moore building. In addition, the Army had contacted James Bryce, asking IBM to supply custom card readers and punches and ship them to a secret location. It also turned out to be the Moore School.

By the time that ENIAC was ready for programming, the war was over. Thus, it did not make much sense to continue running artillery firing tables. One of the early programs run was called Problem A. Von Neumann had been working with the Los Alamos team to evaluate the feasibility of a hydrogen bomb. He had become an expert of fluid dynamics, something that was critical in determining the explosive force needed to trigger the bomb. He suggested using the ENIAC for some of the calculations. The Problem A test was obviously top secret. Input for this test consisted of about one million IBM cards[22]. Von Neumann was not a programmer and relied on Adele Goldstine, Herman's mathematician wife, to interpret his hand-written instructions. She had written the operating manual for the ENIAC so was well versed in setting up a program. Nicholas Metropolis and Stanley Frankel of the Los Alamos team traveled to Philadelphia to oversee the program, starting in November 1945.

The program involved solving differential equations, which entailed very high volumes of multiplications. Elizabeth Holberton did the physical machine programming. Once everything was set up and tested, ENIAC processed a partial differential equation in two minutes of machine time and 28 minutes of card operations[23]. This complex computation highlighted the inadequacies of the ENIAC computer. With only live storage for 20 ten-digit numbers, the programming team had to segment the problem into several machine steps, each of which required changing the plugboard and switch settings.

With the war over, the Army agreed to make the ENIAC secret public. The Army and Moore School decided that the reveal would be more than a simple press release. This would be ENIAC's big moment. They would spare no expense. The announcement took place on Valentine's Day 1946. The ENIAC was properly dressed up for the occasion, including the added touch of cutting Ping-Pong balls in half and placing them over the machine's many blinking lights. They aimed to show graphically that the machine was working. The press takeaway was that the machine was actually "thinking." The Moore School hosts treated the attendees to a lavish five-course dinner. None of the women programmers was invited to the affair. The accompanying press release covered all the impressive facts and features of the nearly 30 tons of electronics and machinery. It credited John Mauchly with the concept, J. Presper Eckert with the design and engineering, and Herman Goldstine with the mathematics and program management. There was no mention of Dr. John Atanasoff.

Arthur Burks handled the machine's demonstration. He started with the simple addition of 5,000 numbers. ENIAC completed that in one second. Then, he explained that he would generate a gun trajectory for a salvo that would take 30 seconds to reach its target. He told the audience that human computers "could compute this trajectory in three days, and the differential analyzer could do it in 30 minutes." ENIAC performed the calculation in just 20 seconds, "faster than the shell itself could fly.[24]"

There was an ulterior motive for all this manufactured excitement. Vannevar Bush, Howard Aiken, and others derided ENIAC as a giant waste of money. Demonstrating the power of the machine would help secure the support and funding for a follow-on machine, and there was already one in planning. After the press conference, the Army moved ENIAC to the BRL and it remained out of sight. The BRL team modified the machine in 1948 to enable program instructions to be stored in the 3,600 decimal switches (thus 3,600 digits or bytes of memory). This was a major upgrade but fell short of true stored program design. That would have to wait for the ENIAC successor.

At the time of its unveiling, ENIAC was the only known electronic computer. Its essential quality was its blistering speed. ENIAC was roughly 500 times faster than any comparable computing device. Though groundbreaking, it was clearly transitional. It was still using relays. It lacked a stored program. It used a wired board and switches to instruct the machine. Though Mauchly had obtained his basic ideas for digital computing from Atanasoff, he and Eckert left out some key elements of the ABC design, including the binary structure, Boolean logic, and the division between memory storage and ALU functions. John Brainerd recognized the contributions of Mauchly and Eckert but also lauded the many other engineers who contributed to the project. He would credit Colonel Simon and the BRL for sticking with the hugely expensive program for years.

Von Neumann's Problem A involved computations that simultaneous linear algebraic equations could solve. Given this was the essential nature of the ABC, von Neumann's problem might have run faster on Atanasoft's machine. Moreover, it would do so at a fraction of the cost. The ENIAC was certainly faster in calculating firing tables—that was what the team designed it to do. The decision to make ENIAC a decimal machine goes a long way in explaining the dramatic divergence in size and complexity. The ENIAC required 18,000 vacuum tubes to handle 20 ten-digit numbers, resulting in a room-sized machine. In contrast, the ABC was able to represent thirty 15-digit numbers (in their 51-digit binary form) with 600 vacuum tubes, all packed within a desk-size machine. The final cost of the ENIAC was well over $500,000 plus an estimated 200,000 man-hours of effort[25]. The ABC cost less than $5,000 and was built by two men. Despite its shortcomings, the ENIAC was a major step forward. The physical machine was booked solid for two years.

# A MUCH BETTER ENIAC

M auchly and Eckert were aware early on of the deficiencies of the ENIAC design. The Moore School engineered the machine in haste, and issues surfaced almost immediately. Though significant work remained to finish ENIAC, the two were anxious to move on to a new design. The key requirement was a system with sufficient memory to contain both the data and program instructions in memory at the same time. Von Neumann had been part of the ENIAC team since 1944 and participated in many discussions on a follow-on system. The core group was von Neumann, Mauchly, Eckert, Goldstine, and Burks. Whereas Mauchly and Eckert tended to discuss issues haphazardly, von Neumann continually brought the focus back to the core objective. He also injected a rigor to the process. The group held regular meetings with documenting notes taken. Eckert, for one, was thrilled to have him on the team.

They christened the new design the Electronic Discrete Variable Automatic Computer, or EDVAC. It was a general-purpose computer as it stored programs internally. EDVAC would feature fewer vacuum tubes, a larger memory, easier programming, and a smaller footprint. Eckert and Mauchly focused on the hardware. Von Neumann worked on the conceptual structure of the machine. The Moore School submitted a proposal to the BRL in August 1944. The ENIAC press conference in February 1946 was critical in securing funding. The Moore School signed a contract in April 1946 to produce three EDVAC systems for $467,000[1].

As the design for EDVAC came together, von Neumann was interested in precisely capturing the concepts embodied in the new machine. He traveled frequently to Los Alamos by train. On one of his return trips, he worked on a 101-page report called "First Draft of a Report on the EDVAC[2]." His goal was to combine all of the EDVAC concepts into a coherent story. It clarified the concept of a stored-program computer with program instructions and data resident in memory, describing a "very high speed automatic digital computing system" based on binary architecture that would fetch an instruction from memory, execute it via the arithmetic and logic unit (ALU) and accumulating registers, and advance to the next instruction. The resident stored program would enable control of the program flow, including loops and branches in the logic. It was an updated version of Lady Ada Lovelace's "Notes" description of computer programming.

Goldstine called von Neumann's June 1945 draft "a masterful analysis" and distributed copies to 24 academic colleagues[3]. Then, unbeknownst to Goldstine and von Neumann, the document found its way into many other hands. In other words, the contents became public domain. Goldstine should have classified the report, especially if there was any intention to patent inventions contained in the EDVAC definition. However, von Neumann would disagree. It was his view that basic computer concepts should be in the public domain, publicized in order to spur the computer revolution. The EDVAC stored-program definition became known as the von Neumann architecture. Von Neumann did not even cite Eckert and Mauchly in his paper. They were understandably resentful. Goldstine agreed that von Neumann should have properly recognized them, but he reiterated that he felt von Neumann was the only "indispensable" one[4] in the definition of a computer. Mauchly and Eckert had a greater issue with the publication. It robbed them of the ability to patent their inventions.

Irven Travis returned to the Moore School in 1946 as a full professor. He was also named director of research programs owing to his significant experience managing research while at MIT and during this wartime work as head of the fire control effort. He was shocked to discover that George McClelland, the university president, had signed a document in January 1946 granting patent rights for the ENIAC to Mauchly and Eckert. Travis shared John Brainerd's view that a project done during the war with U.S. Army funds could not be used for commercial gain. It was borderline unethical in his view. McClelland had also given patent rights to the U.S. government, though this was implicit in government contracts. Travis was further distressed to discover the lack of any standard agreement between the university and its employees when patentable work was being done. There was no formal process to deal with these situations.

He then discovered that Eckert had circulated a memo in September 1944 to all ENIAC engineers asking them to renounce patent rights[5]. In effect, it reserved all patent rights to himself and Mauchly. Many engineers did not understand the concept of claiming patent rights, particularly when the government funded the work. Mauchly and Eckert were also upset that John Brainerd was writing ENIAC progress reports, worried that this would also compromise their patent rights. Herman Goldstine had previously drafted a patent application for key elements of the ENIAC. He included his name and several other engineers who made significant contributions. Mauchly and Eckert amended the application, dropping Goldstine's name. In fact, the only two names left were theirs.

Travis set about correcting the situation, starting with an agreement on patent rights that he asked Mauchly and Eckert to sign. It was essentially a patent release form to the university. Because there were many potential patentable inventions with the EDVAC program, Mauchly and Eckert had sought the same arrangement that they had with ENIAC. Travis made it clear that going forward, the Moore School owned the patent rights and any deviation from that required an individual negotiation. Mauchly and Eckert refused to sign the patent release and resigned.

The Army hosted a meeting held in 1947 with the BRL and its lawyers to discuss patent rights. Mauchly, Eckert, and von Neumann attended. The Army reiterated its position that when it funded a development program, it retained nonexclusive rights to use the invention elements produced. The government lawyers also confirmed that the release of the draft report put the EDVAC content into the public domain, and thus it was not patentable.

The unveiling of the ENIAC and the dissemination of von Neumann's draft report resulted in intense interest. This led to the Moore School Lectures, a series of 48 lectures conducted over eight weeks in the summer of 1946. The Army and Navy funded the program. The ENIAC team did half of the lectures. Over eleven sessions, Eckert presented circuits, adders, and the EDVAC logic. He also demonstrated his invention for storage using mercury delay lines. Burks and Goldstine covered the mathematical methods deployed in the ENIAC and EDVAC design. Howard Aiken of Harvard described the machine he was developing in concert with IBM. Jan Rajchman of RCA presented the Selectron, an advanced vacuum tube his company was developing. George Stibitz covered relay-based approaches to computing. A highlight of the lecture series was an in-depth, hands-on demonstration of the ENIAC.

The lecture series came after Mauchly and Eckert had left Moore and formed their own company, taking many Moore engineers with them. Mauchly and Eckert were contractually bound to participate, and the preparation of their lectures came in the midst of starting their new business. However, the exposure certainly did not hurt their fledgling enterprise. It did make for a tense environment with three opposing camps that were uneasy with one another. There was Eckert, Mauchly, and the ENIAC engineers they had enticed to leave Moore. Then, there was Goldstine, Burks, and von Neumann, who would be working on a new computer von Neumann planned to construct at Princeton. Finally, there was the team under Travis that had remained to finish the EDVAC.

The attendees of these lectures would come to represent a Who's Who in the nascent computing universe. Many had been recipients of the draft report sent out by Goldstine[6]. The class included Claude Shannon from Bell Labs; Jay Forrester from MIT; David Rees from Manchester University; R. D. Elbourne, who worked with John Atanasoff at the Naval Ordnance Lab; Cuthbert Hurd of Allegheny College, who would soon become a major player at IBM; and James Pendergrass of the Navy's cryptographic organization. Maurice Wilkes of Cambridge University arrived two weeks late. He had met with Lesley Comrie, who happened to have a copy of von Neumann's report as well as news of the Moore School lectures. Wilkes decided he had to attend, but his passage on a converted cargo ship arrived late and he only caught the last part of the class, though it still included the essentials of the ENIAC design[7]. The Moore School invited a number of leading computing companies, including ATT, Eastman Kodak, General Electric, RCA, and NCR. For some reason, IBM was not one of them.

Many of the attendees returned from the lectures and embarked on their own computer projects. Wilkes would return to Cambridge, this time booking passage on the *Queen Mary*, and would lead the development of the EDSAC computer. Howard Aiken would build a series of Mark computers at Harvard. Jay Forrester would take a wartime project at MIT to create an analog flight simulator into a digital computer program called Whirlwind. David Rees would return to the University of Manchester to build the Manchester "Baby." And, there was von Neumann and his plan for a project at Princeton.

Meanwhile, EDVAC remained to be completed. Travis took over the supervision of the program after Mauchly and Eckert left. Travis had more than a few challenges. Two of the central designers had abruptly quit the project. In addition, a number of the engineers who had worked on ENIAC and EDVAC left to join Mauchly and Eckert in their new company. This left Travis short on manpower and skills, significantly extending the development time. He did not deliver EDVAC to the Army until 1949. It suffered a checkered history of consistent uptime. The Army did not recognize EDVAC as operational and reliable until 1952, at which time it remained in service until 1962.

EDVAC fully implemented the von Neumann architecture. It was a binary machine with 6,000 vacuum tubes and 12,000 diodes, with roughly 2,000 bits of internal memory. It ran with a fast one-megahertz clock (one million cycles per second), ten times the speed of ENIAC. Addition was substantially faster while multiplications were roughly on a par, perhaps a result of not having Arthur Burks on the team. Engineers whittled the instruction set down to just 12 instructions, in part to reduce the number of vacuum tubes. EDVAC slimmed down to just eight tons and the 490 square feet of floor space[8]. The EDVAC was actually closer to ABC than ENIAC. It was binary-based and used a drum for additional memory. To avoid the use of slow IBM punched-card devices for input-output, the machine used magnetic tape drives.

Main memory, the space for the stored program and data, would use a new technology called mercury delay lines. This was an invention of Presper Eckert. In 1941, John Brainerd contacted MIT's Radiation Lab and secured several projects for the Moore School. One of those was adapting the new radar concept to store information bits. This involved transmitting a pulse and receiving an echo to signify the bit value. Eckert decided to use mercury as the medium as sound traveled faster in mercury than water. Reading and writing of the memory bits involved crystal transducers at either end of a mercury column. This led Eckert to visit Atanasoff at the Naval Ordnance Lab in August 1944, as Atanasoff was an expert in crystalline physics. Supposedly, Alan Turing countered Eckert's choice of mercury, suggesting that gin would be a better way to go[9]. However, Eckert stayed with mercury. Each mercury delay line stored 1,024 44-bit words. Bits stored in this manner were roughly 1/100 the cost of using vacuum tubes. However, this technology would prove to be problematic. The negatives soon overwhelmed the positives. There was the toxicity of mercury, the mechanical and electrical complexity, and the sheer size of the units. They took up an entire room. Most importantly, the delay lines operated at the speed of sound, not the vastly faster speed of light embodied by electronics. Mauchly and Eckert would patent mercury delay lines for memories, but cathode ray tubes, magnetic drums, and magnetic disks soon became the standards for data storage.

The programming of the EDVAC was vastly easier than ENIAC, though still stone age by modern standards. The small instruction set helped. Programmers filled out coding forms that had the English symbolic operators for the instruction on the left and the machine language binary code on the right. It was a giant step forward after the patch cables and switches of the ENIAC. Yet, the programmers remained at or near the machine level with an endless string of binary values. Von Neumann was entirely happy with this. He did not see the need for such aids as assemblers or compilers. Of course, he never wrote program code. He merely designed the logic and handed it off to a programmer.

The draft report, the ENIAC and EDVAC projects, and the Moore School lectures represented the high-water mark for the Moore School and the University of Pennsylvania in computing. For whatever reason, it moved on to other endeavors after EDVAC. Its former prominence rapidly faded, and the computing baton was passed to others.

Mauchly and Eckert had left the Moore School. They had debated the merits of offering their services to a computer company or starting their own company. They soon had an offer from Watson at IBM to set them up in a lab. Mauchly was leery of IBM and preferred to go it alone. Eckert would have been fine working at IBM but soon came around to Mauchly's view. In 1944, well before the flare-up over patent rights, Mauchly had been testing the waters on going commercial. He had met with the Census Bureau, the U.S. Weather Bureau, the National Bureau of Standards, and the Army Signal Corps to assess their interest in a commercial version of the EDVAC. With significant interest in hand, they sought start-up funding from a number of companies, including NCR, Remington Rand, Philco, and Burroughs, but there were no takers. John Eckert knew his son's invention had huge commercial opportunity. He urged them to take the leap and pledged his support in getting the company going. He was also interested in having his son close to home in Philadelphia. Von Neumann intended to build an EDVAC-type computer at Princeton University and offered a significant role to Eckert. That would put him 45 miles away, and even that was too far.

Mauchly and Eckert formed their company, the Eckert-Mauchly Computer Company (EMCC) in December 1947. Though they had promised the Moore School that they would not "raid the school[10]" for engineers and programmers, they did in fact do just that. Many ENIAC and EDVAC engineers joined the new company. The ENIAC programmers—Kathleen McNulty, Betty Holberton, and Jean Bartik—came on board. At the outset, Eckert was heads down engineering. Mauchly was not a salesman, nor was he good at managing a company. He had also procrastinated on completing their patent application, a key central asset for the new company.

The wider computing community viewed the EDVAC architecture as the von Neumann architecture. Herman Goldstine, as the instigator, certainly had an opinion. He believed that von Neumann was central to the concepts. He credits von Neumann with cogently describing the stored program computer on paper and, most importantly, convincing the world of science that this was the path forward in computing. Von Neumann's stature was an essential part of this dynamic. Goldstine draws parallels with Leibniz and Newton and calculus. Leibniz knew most of what Newton knew but did not do anything with the knowledge. Newton set it all down in grand detail, transforming science. It was the same story with Einstein and relativity. Others had pieces of the puzzle, but Einstein synthesized it into a complete and fully logical framework. Julian Bigelow agreed. Von Neumann stimulated ideas and was "generous intellectually[11]."

# THE MARK I

In 1937, James Conant and Harlow Shapley of Harvard visited Wallace Eckert at Columbia to tour his Astronomy Department. Conant was president of the university, and Shapley was a noted astronomer and head of Harvard's observatory. Joining them on the trip was a physics graduate student named Howard Aiken.

Aiken had arrived in this rarefied company in a very unconventional manner. An only child, he endured a tough upbringing. As a young boy, he finally grew big enough to push his abusive father out of the house for good. He took a job installing telephones and, with glowing recommendations, graduated from a technical high school. He worked for Madison Gas and Electric while attending the University of Wisconsin[1]. Warren Weaver was his calculus professor. Upon graduation with an electrical engineering degree, a local utility hired him as chief engineer. Aiken worked in the electric power industry for ten years before concluding that he was not made to be a manager.

He decided to go back to college, starting at the University of Chicago and then switching to Harvard. He arrived on campus in 1933 at the ripe old age of 33. A graduate student in physics, his doctoral thesis was on the conductivity of vacuum tubes, a choice that would certainly prove ironic. He had the same experience as many budding scientists, stuck with endless, laborious calculations of partial differential equations. He decided to do something about it. He began work on a proposal for an automatic calculator. Knowing very early that any machine he would design would be very large; he scouted the Harvard buildings for a suitable location. During this search, the chairman of the Physics Department told him that there was some kind of mechanism collecting dust in the Science Center attic. It was a partial set of gears for the Babbage Analytical Engine. Babbage's son Henry had built it and presented it to Harvard on its 250th anniversary in 1886.

The discovery led Aiken to research Babbage. He would have a series of Babbage's books in his office and would remark, "there's my education in computers.[2]" Babbage's works inspired him to develop his calculator concept. He wrote a memo to the Harvard Physics Department indicating that he could produce a modern implementation of the Babbage machine. It was 1936 and Aiken was still a graduate student, albeit an older one. Aiken simply titled his completed proposal "Proposed Automatic Calculating Machine[3]." It took the reader some time to get to the actual proposal. Aiken first reviewed the history of calculating machines, working his way through the early devices of Pascal, Leibniz, and Napier, then on to Babbage, Scheutz, and Hollerith. Aiken then went on to discuss the differences between commercial punched card processing and scientific requirements, with the vastly greater need for repetitive arithmetic in the latter case. He reviewed the capabilities and limitations of present IBM equipment with its plugboard programming.

Aiken's goal was a machine that could perform integration, calculating the area under a curve by computing the individual areas of many thin rectangular slices. This would require many, many multiplications. He also wanted those multiplications to be automatic, hence the name he gave to his proposed machine, the Automatic Sequence Controlled Calculator. It was a dramatic extension of the IBM 601 Multiplying Punch and even surpassed the semi-automatic switch that Steve Dunwell had set up for Wallace Eckert at Columbia. Additionally, since his method of "slice" integration resulted in an approximation, he wanted the machine to work with decimal numbers up to 24 places, an echo of the excessive level of accuracy that de Prony and others had demanded before him. The concept that he ended up with was not that far afield of the Irven Travis idea, to cobble together a series of Monroe desk calculators.

Of course, it was one thing to have a concept and quite another to have the funding to actually produce the concept. That had not changed since the days of Charles Babbage. In 1937, Aiken had approached Monroe Calculators for funding but he was rejected[4]. During the meeting at Columbia, Wallace Eckert had suggested that he contact IBM. The timing was good for Aiken, as Watson was interested in courting Harvard and here was a promising opening. He also wanted to showcase IBM's talents, and he certainly did not mind if the project would burnish his own image. It did not hurt that Aiken was quite a salesman. It was also fortuitous that IBM's lead engineer, Clair Lake, was in the meeting[5].

Watson sent one of his sales executives to Boston to meet with Aiken. The executive encouraged Aiken to write a letter to James Bryce, with more details on the proposed machine. Bryce was familiar with Babbage, so the proposal resonated with him. The two met in November 1937 at the current IBM headquarters building at 270 Broadway. Bryce introduced Aiken to Halsey Dickinson[6], who mentioned that he was exploring the use of vacuum tubes to produce counting circuits. Aiken fully understood the idea and suggested that Dickinson read a recent paper by Stevenson and Getting that described the use of flip-flop circuits. Dickinson would start experimenting with vacuum tubes in earnest the following year.

Bryce referenced IBM's work at Columbia and invited Aiken to spend some time in Endicott to be educated on current IBM technology. Bryce also secured funding of $100,000 from Watson, though in the end most of the funds would come from the U.S. Navy[7]. He assigned Clair Lake to work with Aiken. Lake worked for Wally McDowell, the Endicott lab director. McDowell, a graduate of MIT in 1930, led the engineering of most of IBM's projects during the war, which would come to number nearly one hundred.

Lake, as chief engineer for the Aiken project, would add Frank E. Hamilton and Benjamin M. Durfee to assist in the design of the machine. Hamilton had joined C-T-R in 1923 as a draftsman and had advanced to engineering. Durfee had joined C-T-R in 1917 as a service technician. Work did not start on the project until February

1940. Aiken spent that summer in Endicott coming up to speed on IBM punched-card equipment and the underlying mechanics[8]. In turn, he schooled the IBM engineers on the mathematical functions he wanted automated. In addition to standard arithmetic operations, he wanted the hardware to calculate sines, logarithms, and exponentials. There was a significant educational gap between Aiken and many of the engineers. Aiken did not help the situation as he treated IBM's expert inventors as mere mechanics. The project very quickly became unpopular. On top of this friction, there was a misunderstanding on who would do what. It was IBM's belief that it would provide the components for Aiken's machine but Harvard would do all the assembly. That did not happen, and IBM ended up assuming the entire load.

In 1941 and in the midst of this turmoil, the Navy drafted Aiken. Robert Campbell, a graduate student in physics at Harvard, took Aiken's place in working with the IBM team. He would spend significant portions of his time in Endicott during the period from 1942 to 1944. Aiken was able to join him during the summers. Once the war began, progress slowed, as higher-priority work took precedence. The first tests of a work-in-process machine occurred in January 1943. Campbell developed several test programs. One traced the path of Neptune's orbit. Another calculated the shape of suspension cables for New York City's Williamsburg Bridge. The Endicott team demoed a nearly complete machine to the Harvard faculty in December. In May 1944, IBM moved the completed system to Harvard. IBM had spent $500,000 plus thousands of uncounted engineering hours in developing the machine. Though the IBM engineers were not math experts, they proved very adept at translating Aiken's concepts into a massive combination of wheels, counters, shafts, and relays. Vacuum tubes were conspicuously missing from the design.

The machine, known as the Harvard Mark I, was a monster. It weighed five tons and took up 4,000 square feet of floor space. It was comprised of roughly 750,000 parts[9], with most being off-the-shelf components. Conceptually, it was the equivalent of 72 adding machines linked together. Under the covers were 2,225 counters, 3,500 relays, and 60 sets of dial switches. It was held together with 500 miles of wire and three million welded connections.

**Figure 16**. IBM Automatic Sequence Controlled Calculator, also known as the Harvard Mark I (1944). Concept by Howard Aiken. Invention, engineering, and development by IBM. Reprint Courtesy of IBM Corporation ©.

The machine would have been much larger but for two inventions by Clair Lake: the Lake relay and the Lake counter. Lake had developed these two innovations with extra monies that Watson provided his engineers to work on any project they wished (a forerunner of the IBM Fellow program). Relay technology seemed to be long past its peak, but the Lake relay design was five times the speed of the Bell Labs relays and far more compact. The Lake counter was a wonder. A modern equivalent of the Leibniz wheel, it was a sprocketed wheel controlled by ratchet and clutch mechanisms with the overall function managed by relays. There were 3,000 of these counters, arranged to create 70 accumulation registers. Each register had 24 of the counter wheels, in order to support a 23-digit decimal value plus the sign.

**Figure 17.** The Claire Lake-designed electromechanical counter. One of 3,000 that went into the Mark I at Harvard. Photograph courtesy of Ken Sherriff from the Computer History Museum collection.

The multiplier on the machine was a separate unit comprised of relays. James Bryce enhanced his IBM 601 multiplication process by performing the partial products operations two digits at a time. This cut the multiply time in half, from six seconds to three. In addition, you were free to run additions using separate registers while the multiplication was in process. Per Aiken's design requirements, there were also separate hardware registers for the built-in sine, logarithm, and exponential functions. As with the case with all IBM tabulators, a rotating shaft provided overall synchronization. With the mammoth size of the Mark I, that shaft was 55 feet long and ran the entire length of the machine. A five-horsepower electric motor rotated the shaft. Data was input via IBM cards or through the machine dials. Programmers punched instructions on paper tape.

With the Mark I, the rotation time of the drive shaft dictated performance. One cycle was three-tenths of a second. Additions, subtractions, and register transfers took that one cycle. Multiplications averaged ten cycles, or three seconds. Division could take up to 52 cycles, which would be nearly 16 seconds[10]. The special functions that Aiken had specified took significantly longer. Logarithms took 90 seconds while exponentials and sines required a minute apiece.

Though the Mark I was dramatically faster than existing IBM machines, it was no match for the ENIAC. The difference in machine cycle time between vacuum tubes and electromechanical counters and relays showed up in the performance numbers. The ENIAC produced 5,000 additions or 333 multiplications each second. To solve a partial differential equation, the Mark I would run for 80 hours while the ENIAC took two minutes of processing plus some 28 minutes of card processing. The multiplication progression went from five minutes by hand, 15 seconds by deck calculator, three seconds for the Mark I, one second for the Stibitz Bell Model III, and one-thousandth of a second for ENIAC. The key advantages of the Mark I would be in programming and automation.

Watson provided a final touch for the machine. He hired Norman Bel Geddes, the most famous industrial designer of the period, to enhance the look of the Mark I before IBM moved it to Harvard. Bel Geddes forte was modern, streamlined, even futuristic designs during the age of Art Deco. His work included the classic Chrysler Airflow automobile and the super-streamlined Locomotive No. 1.

Watson traveled by train to Cambridge, Massachusetts for the August 1944 dedication. He expected a reception at the station and a limo to the hotel. Instead, Frank McCabe, the IBM Boston branch manager, picked him up in his dowdy Chevrolet. Once in the car, McCabe showed him the press release that Harvard had produced for the event. It extolled Harvard and Aiken's amazing machine, with just a minor mention of IBM at the end of the eight-page missive. The same basic story appeared in all of the Boston newspapers. The *Boston Daily Globe* heralded the "world's greatest mathematical calculator" and proceeded to profile "the inventor, Commander Howard H. Aiken[11]."

Watson was beyond irate. It was certain that the IBM engineers who had designed and built the machine shared his ire. The dedication itself included the governor of Massachusetts (Leverett Salton-stall), Harvard's president (James Bryant Conant), Harvard officers, U.S. Navy admirals, and the key IBM engineers responsible for the machine (Clair Lake, Frank Hamilton, and Benjamin M. Durfee). Despite his fury, Watson joined the dedication and gave his usual strong performance. He also presented Harvard with a check for $100,000 to provide for ongoing maintenance of the Mark I. This public announcement of the new system was a year and a half prior to the ENIAC press conference.

After the dedication, the Harvard dean wrote a letter of apology to Watson. Aiken also wrote a letter of apology, but not for a full month later, and the text had the look of a Harvard administrator. Watson continued to stew about how Harvard had treated IBM. He felt that Aiken and Harvard had not only robbed IBM inventors of their central role in the machine but also blunted Watson's goal to have IBM seen as a leader in scientific circles. The university had even appropriated the machine's name.

Watson authorized an extensive review of who did what in the development of the Mark I. The review lasted six months, with Watson personally conducting some of the interviews with his engineers. The resulting report went through over 40 revisions, according to John McPherson, who ended up in the middle of the controversy. The report's conclusion recognized Aiken's role with the general concept and explanation of the mathematics required, with the balance of the credit and invention going to the IBM engineering team. Watson had his team develop its own manual for the Mark I. At the front, it included a statement of the machine's inventions. It read:

- Multiplying machine, invented by James Bryce in 1934
- Dividing machine, invented by James Bryce in 1936
- Multiplying and dividing machine, invented by James Bryce and Halsey Dickinson in 1937
- Unit counter invented by Fred Carroll in 1925

- Double card feed, invented by Clair Lake in 1921
- Electromatic typewriter, adapted by Clair Lake and Frank Hamilton in 1936
- Counter readout and emitter, invented by James Bryce in 1928
- Commutator totaling mechanism, invented by George Daly in 1926

The bitterness never abated. Most considered Aiken's actions and position untenable. Robert Campbell agreed, "IBM had not been treated fairly.[12]" Watson vowed to create another computer to put the Mark I to shame.

Robert Campbell was the initial programmer on the Mark I. Richard Bloch and Grace Hopper soon joined him. Bloch emigrated from Germany. His parents perished in the Holocaust[13]. Bloch graduated from Harvard in 1943 with degrees in mathematics and physics. He signed up for the Navy's V7 program, which provided an accelerated path to the rank of ensign, or as some characterized a graduate as the "90-day wonder." The Navy assigned Campbell to the Naval Research Lab at Anacostia in Washington, DC. He had the opportunity to escort Howard Aiken on a tour of the facility. When Aiken found about his Harvard background, he offered him a position in the Mark I lab. Ensign Bloch would become the principal programmer. He would also lay claim to teaching Grace Hopper how to program, starting her on an illustrious career.

Grace Hopper received her doctorate from Yale and was a professor of mathematics at Vassar when the war began. She tried to join the Navy but they considered her too old. She settled for the Naval Reserves. In the summer of 1944, the Navy assigned Lt. Grace Hopper to work with Aiken. When she reported to Harvard, Aiken gave her a copy of Babbage's 1864 book, *Passages from the Life of a Philosopher*, as an introduction to the underpinnings of the Mark I. Hopper was initially a programmer until Aiken ordered her to write a manual of operation for the new machine. That is correct, he ordered her. As a lieutenant commander, he was her immediate superior. And, Aiken ran a tight ship. Hopper dove into what would become a 561-

page tome, with the obligatory first chapter for Aiken on computer history and Charles Babbage in particular. Harvard published the manual in 1946 with Hopper and Aiken listed as the authors.

In his earlier days as a manager in the electric power industry, Aiken was ill at ease with management. However, now as a naval commander, he did not have to worry much about such things as employee empowerment or management style. He could be imperious. He adopted a strict Navy protocol at Harvard. Subordinates addressed him as "Commander" or "Sir." Even after war's end, he ran the shop like a military operation. He was a large man, 6' 4" tall, with a commanding physical presence. He would form quick first impressions of someone, good or bad, and that impression usually stuck. One staff officer apparently suggested that he read Dale Carnegie's *How to Win Friends and Influence People*, with little success[14]. Others would describe Aiken as a "tough hombre," or a "hard man.[15]" Later, his students would call him "the tiger." Maurice Wilkes met with Aiken in 1946 and again in 1950. He would admit that he was "rather afraid of Aiken." He concluded that "Aiken relished conflict," noting "a look which would come into his eyes that I can now see was an invitation, almost a pleading, for his adversary to close in for a good sparring match."[16] Wilkes avoided such conflicts but characterized Aiken as "technologically backward."[17]

Though the Mark I was not a stored-program computer, it resembled modern computers in that it ran instructions in sequence via program control, unlike the ENIAC. With a reasonably complete arithmetic instruction set and plenty of accumulators, the machine could solve virtually any problem, as long as the programmer framed the problem in mathematical language and translated it into machine codes.

A typical program sequence involved laying out the required math, sketching the steps in a flowchart, translating the steps into machine code, and punching a paper tape. Programmers could also instruct the Mark I with plugboard wiring and setting switches. Hopper would describe Bloch as "the Mozart of the computer" as he could draft a program in ink and have it run successfully the first time[18]. For lesser mortals, there was at least one additional step, and that was the iterative process of fixing program errors. The eventual term for this was "debugging," a name attributed to Grace Hopper. While working on the Mark I, she discovered a moth that had landed on a relay and shorted out the circuit. The term "bug" became a generic word for any program error.

Given the tedium sometimes involved in developing a program, the team gravitated early to capturing standard program routines for reuse. Bloch and Campbell kept a notebook of repetitive code. The routines could also be punched in paper tape and run from a Mark I reader in conjunction with a mainline program. The team discovered that the special hardware IBM engineers produced for the sine, logarithm, and exponential calculation ran too slow. Bloch wrote separate machine code routines for those functions and added them to the notebook.

Program E was one of the first programs tested on the Mark I. Requested by the Navy group working on anti-aircraft firing control, Problem E produced firing tables for the Navy's five-inch anti-aircraft gun. The computation took into account a slew of variables, including bearing, elevation, range, and the ship's pitch and roll. The program required 15 subroutine segments. In late 1944, John von Neumann was working on another implosion problem with the Manhattan Project. The program needed to model spherical shock waves created by the implosion by solving partial differential equations. It was very complex, so naturally, Aiken assigned the task to Bloch. That program ran on the Mark I for most of the month of September 1944.

Unfortunately, these critical programs did not represent the major use of the Mark I. Instead, Commander Aiken wanted to create massive tables of Bessel values. Bessel functions represented differential equations for circular or cylindrical analysis, often used to describe planetary orbits. They are highly used in applied mathematics. Perhaps because he was channeling de Prony and Babbage, Aiken decided to use the calculation power of the Mark I to produce complete Bessel tables. This work would run for five years. The Bessel program ran any time there were no higher-priority jobs ready. This worked out to some 90 percent of the available Mark I processing time. In fact, the machine was nicknamed "Bessie" because of preponderance of this work[19].

Aiken's office was right next to the Mark I. The machine had a unique sound, created by the combination of the rotating main shaft and the maze of gears, relays, and wheels. Richard Bloch would describe the Mark I in operation where "the action of the mechanism's steel clutch engaging and releasing at a frequency of some three plus times per second emitted a distinct sound" like a horse clattering on a stone street. Whenever that sound stopped, Aiken came out of his office and wanted to know why the machine was not running. He wanted to be constantly "makin' numbers," by which he meant running Bessel calculations. Harvard published twenty volumes of Bessel tables. According to Herb Grosch, Aiken was managing one of the only high-performance computing engines in the world and was "wasting it on a ridiculous project to calculate Bessel functions.[20]" It would not be long before computers could compute Bessel functions on demand, making printed tables obsolete.

Aiken certainly did not want the machine to stop when giving the Navy brass a demonstration. Hopper described a demo for several admirals where the machine was failing every cycle. She just kept her finger on the start button and the demo went fine[21]. Yet, reliability was the hallmark of the Mark I. It ran 24 hours a day, seven days a week. IBM had an around-the-clock service team assigned to the installation. They were dubbed the "I" specialists after the "I" in Mark I. They even had the letter "I" embroidered on their shirts. The machine just ran and ran, and would continue to do so until 1959.

With the Mark I, IBM engineers molded their existing technology into a reliable computing engine. The machine would run until forcibly stopped by an operator. It was the absolute acme of electro-mechanical design. Did it move the needle on computer progress? Technology-wise, the answer is probably "no". Yet it was very high profile. It led to a series of Mark machines and established Harvard as a center of the computing arts. Leslie Comrie called the Mark I "a realization of Babbage's project in principle.[22]" Like Babbage's Analytical Engine, the Mark I had a "store" (the accumulating registers) and a "mill" (the counters and registers). It lacked a key element of Babbage's design, the conditional branch. IBM later added a separate unit that provided a limited implementation of this function.

Babbage himself weighed in, speaking to the future and to

whoever realized his vision. In *Passages from the Life of a Philosopher*, he wrote:

> "If, unwarned by my example, any man shall undertake and shall succeed in really constructing an engine . . . upon different principles or by simpler mechanical means, I have no fear of leaving my reputation in his charge, for he alone will be fully able to appreciate the nature of my efforts and the value of their results[23]."

Aiken in turn noted that Babbage's vision was slightly premature from a technology standpoint, arguing that "if old Babbage had lived another fifty years, there wouldn't have been much left for me to do.[24]" He did feel that the Mark I was "Babbage's dream come true[25]". Aiken must have gritted his teeth when forced to send the letter of apology to Thomas Watson. He never wavered about who invented the Mark I. Furthermore, he decided that he would prove his own ability as a computer inventor. He would demonstrate that he did not need IBM and that he could develop a far better computer on his own. He did not have any qualms about using Watson's money, though. He used the $100,000 that Watson provided for Mark I maintenance to construct the Harvard computer center. This would later be renamed the Howard Aiken Computing Center.

**Figure 18.** Designers of the IBM Automatic Sequence Controlled Calculator (Mark I). From left: Frank Hamilton, Clair Lake, Howard Aiken, and Ben Durfee. Reprint Courtesy of IBM Corporation ©.

Aiken secured funding for his own computer in 1945 from the U.S. Navy, again for ballistics tables. Aiken and Robert Campbell designed the new machine, called the Mark II. It replaced much of the electromechanical components of the Mark I, including the Lake counters, with relays. By this time, relays were a semi-mechanical and outdated technology. Nevertheless, Aiken proceeded to place 13,000 of them in the Mark II. The machine broke no new ground, which would be Aiken's ongoing reputation. He avoided the stored program and continued to use decimal representation. Harvard delivered the Mark II to the Navy in 1948. It ran about eight times faster than the Mark I. Grace Hopper was unimpressed, saying of the Mark II "she was a dead duck and everybody was going electronic"[26].

Nevertheless, the Navy funded $500,000 for an additional system, also for ballistic firing tables. This would be the Mark III or as the Navy called it, the Aiken Dahlgren Electronic Calculator (ADEC). It used vacuum tubes, 5,000 of them. Aiken continue to lag in computer technology, choosing to use magnetic drums for main memory when other more advanced and faster options were available. Armed with vacuum tubes, the Mark III was dramatically faster. Multiplication improved by a factor of about fifty over the Mark II. Time Magazine featured this "Mark" in January 1950[27]. With a surreal drawing of the Mark III on the cover, it continued with the "Brainiac" theme that began with ENIAC, calling it the "thinking machine." The article quotes Aiken on the need for such large machines: "We'll have to think up bigger problems if we want to keep them busy."

The Mark IV was the last computer developed by Aiken. By 1952, Harvard was far less enthusiastic about sponsoring such projects. Funded by the newly created U.S. Air Force, the Mark IV was fully electronic, with vacuum tube logic circuitry and magnetic core and magnetic drum memories. It was another decimal machine. Designed for ballistics, there was no allocation for programming tools, not even an assembler. Programmers wrote in binary machine code and then awkwardly entered the instructions on a magnetic tape device. Aiken's grudge against IBM had not eased, and he did not want any IBM equipment such as punched-card devices near his computers. The Mark IV was only marginally faster than the Mark III, and slow compared to the other machines of its era. IBM's 650, a small departmental system, was 25 percent faster than the Mark IV, while IBM's top-end system, the 701, was nearly 30 times faster.

Watson did like the aesthetics of the Mark IV. It was a good-looking, sleek machine.

None of Aiken's Mark computers had a major influence on computer design, unlike the many von Neumann architecture computers that proliferated. Aiken was conservative, and his machines were slow. All of the Mark machines used the Harvard architecture, where program instructions and data were separate. This was antithetical to the von Neumann architecture as it eliminated the ability to modify program instruction paths dynamically based on the data, the cornerstone of computer program logic. Aiken would rail against binary machines and the von Neumann "stored program." Though he was an important figure in computer history, his ideas were consistently out of the mainstream. His major contribution was not his machines but the computing program he established at Harvard. A number of prominent computer scientists came through Aiken's program, including An Wang, Fred Brooks, and Grace Hopper.

Aiken was well known for his 1947 comment, "Only six electronic digital computers would be required to satisfy the computing needs of the entire United States[28]," or in another quote, "there will never be enough problems, enough work for more than one or two of these computers.[29]" He went on to make fun of the plans of Mauchly and Eckert to create a commercial computer company. Future events would prove his comments widely off-base.

After the war ended and the staff at Harvard no longer had to call Aiken "Sir" or follow his orders, there was quite an exodus. Robert Campbell and Richard Bloch both left for Raytheon. Campbell subsequently left Raytheon and joined Irven Travis, who had landed at Burroughs. Travis knew Aiken well and, through him, was familiar with Campbell's work. Grace Hopper would join Mauchly and Eckert at their fledgling company. Another member of Aiken's wartime staff, Rex Seeber, would leave for IBM. He had graduated from Harvard in 1932 and started a career in insurance at John Hancock. He became an expert in using desk calculators and IBM punched-card equipment. Drafted by the Navy, he had heard about the Mark I computer at Harvard and asked for an assignment there. He joined the Mark I staff in 1944, and according to Seeber, he and Aiken immediately took a dislike to one another. This would have been no surprise to anyone familiar with Aiken's penchant for instant classification, good or bad[30]. However, it was a Navy assignment and he was stuck with it. He called the office a "sweat shop.[31]" Yet, Seeber knew the Mark I inside and out and developed a list of modifications to improve the machine. Aiken ignored all of them.

Seeber relished the idea of taking his ideas to IBM, and as it turned out, he would work directly on Watson's "revenge" computer. By the mid-1950s, Aiken was in decline and out of step with current technology. He continued to alienate Harvard administrators, students, and the agencies that could fund his projects. He never appeared to temper his disdain for IBM. Many years later, Tom Watson Jr. invited him to IBM headquarters in an attempt to mend fences. IBM flew him by company jet to Endicott and then on to IBM headquarters in a limo[32]. Several IBM executives greeted him upon arrival. As many confidential ideas were likely to come up in the meeting, they asked to sign a standard nondisclosure document. He refused, got back in the limo, and left.

# COLOSSUS, THE SECRET DIGITAL CODE BREAKER

In 1974, F. W. Winterbotham published *The Ultra Secret*[1]. Winterbotham, a decorated World War I pilot who served in the MI6 Intelligence Service during World War II, described his role in Ultra, the name given to the massive and highly successful British program to break the German war codes. The very existence of Ultra was a closely guarded secret for over a quarter century until Winterbotham's account was published.

The war produced significant advances in computing technology in the United States. There was progress in Europe as well, but much of it was shrouded in secrecy and did not surface for many years afterward. The German computer innovator Konrad Zuse toiled in anonymity in Berlin and had several of his machines destroyed by Allied bombing. His work was unknown until well after the war ended.

For the British, it was not about firing tables but about code breaking. Their dogged pursuit to break the German ciphers led to the Colossus machine, a computer that equaled or surpassed ENIAC. However, it was not until the Winterbotham book that Ultra and Colossus came in from the shadows. The code-breaking story began in 1923 when a German engineer built an encryption machine that his company promoted as a business communications tool. In use, two identical typewriter-sized machines were set up, one at the sending location and one at the receiving location. Each machine had a keyboard with twenty-six letters. As the sending operator depressed a letter on his machine, the selected key would connect via a rotor to one of twenty-six letters in lights. The illuminated light represented the selected encrypted letter. The selected light and thus the selected encrypted letter, would change with each depress of the same letter key. This was the result of the constantly turning rotor. At the receiving end, a machine with an identical setup would decipher each letter, and hence the message. Adding additional rotors within the machine would increase the potential combinations of input and encrypted letters exponentially. The designing engineers based the encoding scheme on the Vigenere cipher, the same one that Charles Babbage cracked in 1854.

Though originally designed to encrypt business communications, the German Army adopted the machine in the 1920s for field communications. They renamed the machine "Enigma." In 1929, an Enigma machine fell into the hands of Polish intelligence. They immediately set to understanding and breaking the encoding scheme. They selected twenty outstanding math students, schooled them in cryptography, and turned them loose. At the same time, the French confirmed the Germans had adopted the system but enhanced it by increasing the rotors from one to three. The Polish team cracked the new three-rotor machines. German army field units unwittingly helped the Poles with their tendency to use repetitive sequences in messages, creating clues, or what cryptographers called "cribs." It was obvious, for example, what a message sent on April 20 would contain. That was Adolf Hitler's birthday.

**Figure 19**. Four-rotor Enigma machine, capable of 1.5 quadrillion possible combinations. U.S. Government photograph.

However, the problem became one of time. The German Army changed the settings on Enigma every 24 hours. The manual process used by the Poles was not fast enough to keep pace. In 1938, the team set up a rudimentary mechanical device to mimic the action of the rotors and speed up the decoding. They initially called the device the

Cyclometer but it evolved into the Bomba[2]. That same year, however, Germany was gearing up for war and the army increased the number of rotors to five. In addition, some Enigma machines were fitted with a wiring plugboard that increased the difficulty in decryption.

With the Germans threatening to overrun both Poland and France, the Poles pulled the British into the secret in 1940. Sir Hugh Sinclair, head of MI6 intelligence, had purchased an estate known as Bletchley Park in 1938, for use in case of war. Located 50 miles north of London, it was central to the metropolis and the surrounding universities. He set up the Government Code and Cypher School and began assembling a large staff to continue the work of the Poles. Recruiters searched for individuals that might excel in cryptography. They used a variety of techniques, including selecting those who could solve the *Daily Telegraph* crossword in less than 12 minutes. By the fall of 1938, the staff had risen to 80 but had no mathematicians. That soon changed, and one mathematician in particular would rise to prominence.

Alan Turing had attended King's College at Cambridge when it was the center of the mathematical world. Max Born, Max Newman, and Arthur Eddington were on the faculty, and John von Neumann was a visiting professor. Turing wrote a paper titled "On Computable Numbers[3]" in which he defined what came to be called the "Turing machine." Max Newman helped him get the paper published by the London Mathematical Society. It was hugely influential and contained the basic definition of a computer. Upon graduating in 1934, Princeton University accepted Turing as a graduate student under the direction of Alonzo Church. Church was an acclaimed mathematician and logician. His PhD thesis on mathematical logic was also influential in computer design. Turing's office at Princeton was near von Neumann's, who became a mentor. Von Neumann was already interested in computing and offered Turing a job for the following school year. Turing decided to return to England, receiving a fellowship from King's College at Cambridge.

Shortly after returning to Cambridge, the Government Code and Cypher School recruited Turing for code breaking. They assigned him to Hut 8 at Bletchley Park[4]. The various "huts," or small buildings, constructed on the estate would eventually house about 9,000 personnel, with 80 percent women. Turing had taken a course on cryptology at Princeton, but his new focus would be using his mathematics and logic background to assess the probabilities of selected letters in the text of an intercepted message. For example, he knew that an encoded and decoded letters would never be the same. As with the Poles, he recognized that a library of cribs would be essential. The German word for "one" (*ein*) would likely be present in most messages, so it was an obvious addition to the list of cribs. A Hollerith machine team in Hut 7 catalogued and analyzed the cribs. Though there was a war on, Bletchley reserved Hut 2 for beer, tea, and recreation.

Shortly after arriving at Bletchley, Turing started working on redesigning the Polish Bomba, now rechristened the Bombe. Turing worked with fellow Cambridge mathematician Gordon Welchman and Tommy Flowers, an engineer with the British Post Office research center at Dollis Hill. Their design was electromechanical, with a prototype model named "Victory" up and running in 1940. The machine sought to duplicate the rotor positions of Enigma by constantly spinning the rotors and testing. Trying all possible combinations would take too long. The machine bypassed the letters and combinations that Turing identified as not candidates. The bombe also used the growing catalog of cribs and ran until it had the correct rotor settings. A separate group dialed the rotor settings into Enigma emulator machines located in another hut to decode the message.

Bletchley would need more than a single bombe machine. Turing, along with fellow cryptanalysts Gordon Welchman, Hugh Alexander, and Stuart Milner-Barry, wrote Winston Churchill in October 1941 about the lack of funding for their work. With a quick memo, Churchill ensured the funds flowed immediately[5]. Turing worked with Harold "Doc" Keene of the British Tabulating Machines (BTM) in the production of the machines. BTM was the company started by former U.S. Census director Robert Porter. BTM would manufacture over 200 of the Enigma Bombes. Each Bombe was seven feet high, seven feet wide, three feet deep, and weighed a ton[6]. It took the Enigma Bombes about 20 minutes to decode the rotor settings. The machines ran around the clock, operated by "Wrens" (Women's Royal Naval Service) personnel, a group that grew to more than 2,000. The Wrens and the Bombe machines were physi-

cally located far from Bletchley Park, due to the risk of German
bombing.

With the Enigma code decrypted and an infrastructure in place to
handle the huge volume of intercepted messages, a bigger challenged
loomed. The Germans transmitted Enigma messages in familiar
Morse code. In 1942, British listening stations detected a completely
different type of transmission. Bletchley technicians soon identified
the transmission as an encoded data stream, one used for teleprinters
or stock tickers. It was a binary variation of the Baudot code. The code
breakers at Bletchley correctly surmised that the new transmissions
reflected a new kind of encryption device at work.

The Germans had in fact deployed a more sophisticated machine,
designed for the highest-level communications between Hitler and his
general staff and their field commanders. The Germans engaged the
Lorentz Company, a maker of teleprinters, to develop a more secure
machine. As their teleprinters already worked in Baudot code, they
simply added a front-end encrypting mechanism consisting of 12
rotors with over 500 possible rotor settings, resulting in an unfath-
omable number of possible combinations. They included an additional complication —the insertion of a random key in each message.

The Lorenz machine and its cypher would have been virtually
unbreakable had it not been for a German operational mistake. That
occurred on August 30, 1941, when an operator sent a 4,000-char-
acter message that was garbled in transmission. The receiving operator
asked for a re-transmission. Operators were supposed to change the
Lorenz settings with each message, but this time the operator left his
settings unchanged. In addition, he shortened many of the words, not
wanting to rekey all 4,000 characters.

With the two messages side by side, John Tiltman, a British Army
intelligence officer with cryptographic skills, was able to see the
encryption key and identify that the message was using the Vernam
cypher. He passed the two intercepts on to Bill Tutte, a Canadian
code breaker. In what experts consider as one of the outstanding intellectual feats of the war, Tutte not only decoded the message but also
worked out exactly how the Lorenz machine was constructed and
operated without ever seeing the actual device[7]. The Post Office team
was able to build a Lorenz replica from Tutte's description. However,

the on-going decryption task remained. The method that Tutte used would take up to six weeks of painstaking manual analysis. This was of little use with the Germans changing the settings every 24 hours. There needed to be a faster, automated process.

A team was set up under Max Newman in Hut 11. Newman, a mathematics professor from Cambridge with an interest in computing logic, had been Turing's tutor at the school. His team included several engineers and sixteen Wrens. They dubbed Hut 11 the "Newmanry." Newman asked the Dollis Hill research laboratory to help in the design and development of the machine. Turing lobbied to have Tommy Flowers join the effort. Flowers had graduated from the University of London, joined the General Post Office in 1926, and moved to the research lab at Dollis Hill in 1930. Before the war intervened, he was exploring the use of electronic circuits for phone exchanges.

Flowers designed a unit that would compare two high-speed paper tapes, one containing the intercepted text and the other containing known settings of the Lorenz device. The paper tapes were read at high speed by optical readers, a technology that would not fully mature for another 15 years. Newman knew that the core of the new machine had to be its electronic counting circuits. From his Cambridge days, he knew Charles Wynn-Williams, a physicist who had worked with Ernest Rutherford and who had been experimenting with vacuum tube designs. Newman had Wynn-Williams, who was currently working on British radar development, transferred to Bletchley to develop a digital binary counter unit. Wynn-Williams designed a grouping of vacuum tubes with relays to count the number of matches between the two tapes[8].

Newman's team completed the machine in June 1943. They christened it the Heath Robinson, so called because of its kludge design, particularly with the two paper tape transport units that looked like large bed frames. Heath Robinson was the British equivalent of American cartoonist Rube Goldberg[9]who was famous for creating designs of impossibly complex mechanisms to accomplish incredibly mundane tasks. Though the Heath Robinson included vacuum tube circuits, it proved to be slow and unable to keep up with the intercept traffic. Flowers suspected that the problem was a combination of too many relays and too few vacuum tubes. He proposed a new machine with up to 1,600 vacuum tubes, knowing that it would be at least a thousand-fold faster. He anticipated the initial reaction to such a device would focus on tube failures. Like Presper Eckert, he surmised that the reliability problems with vacuum tubes would virtually disappear if they were simply left on. British Tabulating Machines, the existing manufacturer for Bletchley devices, was resistant and pressed to have Flowers removed from the project. Newman was also skeptical. He had previously proposed going electronic, but rejected it when it did not appear to be feasible. Flowers, a master of the understatement, indicated that his idea was "not received well.[10]" Newman dropped the proposal.

Newman asked BTM to build a new machine. BTM suggested another relay-based machine, as it was not interested or proficient in using vacuum tubes. Meanwhile, an undaunted Flowers pulled together a team of fifty at Dollis Hill and charged ahead, developing a prototype device with the 1,600 vacuum tubes[11]. He used some of his own money on the project, started in February 1943 and finished 11 months later. Now convinced of the vacuum tube approach, Newman asked Flowers to develop an operational machine. Wynn-Williams designed the vacuum tube counter ring, opting to count in binary, as it needed only five tubes. They decided against calling the machine "Heath Robinson II", but used the vastly more impressive name, Colossus.

The first Colossus performed so fast that it was dubbed the Mark I version, and MI6 commissioned ten improved machines. The upgraded Mark II had 2,400 vacuum tubes and was five times the speed of the Mark I. The machines used binary mathematics and logic. The Mark II had five parallel processing and counting units that the cryptologists programmed to discern patterns in the intercept text that would lead to the daily rotor settings of the Lorenz. The new machines retained the paper tape transport mechanism of the Heath Robinson, as the unique "bed frame" arrangement would speed the paper tapes at an incredible 53 miles per hour[12]. Turing had his probability logic and the library of cribs programmed into the machines, using a combination of patch panels and switches. Counting was essential to the process, as it tracked the number of times a specific letter appeared in the text. The exceptional digital speed of Colossus enabled this type of otherwise laborious, iterative approach. Colossus reduced the time required to determine the Lorenz settings from Bill Tutte's six weeks to just six hours. Once Colossus arrived at the day's rotor settings, the teams operating the Lorentz emulators took over and deciphered the intercepts.

**Figure 20.** Colossus No. 10 in Block H, Bletchley Park. At right are the high-speed paper tape readers (the "bedsteads" from the Heath Robinson machine). This is a Mark II Colossus with 2,400 vacuum tubes, rows of switches and patch cord panels for programming, and a printer-typewriter for output. From the collections of the National Archives of the United Kingdom.

Bletchley completed the Colossus in record time and put it into service on June 1, 1944. Colossus worked German intercepts in the lead-up to D-Day on June 6. A Colossus decrypt confirmed to Eisenhower that the Germans still expected the Allied invasion at Calais and were not moving any armored divisions to the real landing sites in Normandy[13]. The breaking of the German codes with the Enigma and Colossus machines was a very tightly held secret by the British. The code name was "Ultra," and it occupied a special place of its own well above "top secret."

The British completed Colossus several years ahead of the ENIAC. Computer historians now consider Colossus the first programmable, electronic computer, though the machine was unknown for 25 years. The Heath Robinson lived on with a new mission. The paper tape comparison function was important to the cryptographers. They upgraded Heath Robinson to Old Robinson, then to Super Robinson and finally to Mrs. Miles[14].This last version compared four paper tapes simultaneously. The team named it after a woman who just had quadruplets.

Unlike ENIAC, Colossus was not commercialized at war's end because of the Official Secrets Act. British Intelligence ordered Flowers to burn all the supporting documents and dismantle or destroy all but two machines[15]. Britain used the two left until the 1960s to decode Soviet messages. Tommy Flowers finally received clearance to write about Colossus in 1983[16]. Flowers, Turing, Newman, and all the others involved at Bletchley Park received little recognition. The government suppressed Alan Turing's paper on the subject of decryption, "The Applications of Probability to Cryptography[17],"for 70 years.

Tony Sale led a team that reconstructed the Colossus, finishing in 2008. A contest was set up pitting the rebuilt Colossus against all takers. The winner would be the first to decode three messages sent from a Lorenz machine. The rebuilt Colossus cracked the Lorenz setup in 3.5 hours. That was in line with the wartime machines. However, a software developer with a 1.4 GHz laptop took less than a minute to determine all twelve Lorenz rotor settings and win the contest. Ironically, the software developer was a German.

The estimated speed of the Mark II Colossus machine was 4,000 cycles per second, compared to 1,000 cycles per second for ENIAC. It was a very fast machine for its time. Thus, at the end of the war, there were just two digital electronic computers in operation—the ENIAC in the United States and Colossus in Britain. ENIAC got all the press, the accolades, and a good deal of the impact on computer development. The major impact of Colossus was from the machines that its key players, including Turing and Newman, went on to design.

I ask the reader for a slight digression here, though I will remain on the subject of Ultra. The British were understandably circumspect about this hugely valuable secret. They were especially concerned with bringing the Americans in on it. However, the United States ran the Allied war effort and Eisenhower and his commanders needed the Ultra intercepts. Winterbotham designed a process to distribute Ultra messages to field commanders. He decided to train a group of top-flight British and American officers in Ultra and assign them to the major Army groups. He required these liaison officers sign the British Official Secrets Act. Once deployed in the field, they were vastly outranked by the field commanders they served, in part to disguise the weighty role they were playing and the decisive secret that they held.

My father, Lt. Paul F. Shaffer, was one of those Ultra liaison officers, officially part of the American Office of Strategic Service (the OSS, the forerunner of the CIA). He reported directly to General Jacob Devers, commander of the Allies 6th Army Group. Ultra remained a secret until 1974, when Winterbotham published his memoir. And true to the type of individual that Winterbotham had sought, my father never talked about his experiences with Ultra until well after the publication of the Ultra book, and even then, he offered few details.

# BEYOND COLOSSUS

C olossus was not the only computing machine developed in Europe during the war. In Berlin, one individual working in his parents' living room designed not one, but a series of computers. His name was Konrad Zuse, and he would work in near-isolation from 1936 to 1945. Zuse attended the Technical College of Berlin, graduating in 1935. He joined Ford as an advertising designer before moving to Henschel to work as an aircraft designer. This work required heavy computation of linear equations and led him to explore automated approaches.

His first machine was the Z1, developed between 1936 and 1938[1]. Like Atanasoff, he needed to invent a memory format to store data. He devised a purely mechanical memory based on metal sheets with holes for pins. A pin in the hole was a binary "1" while the absence of a pin was a "0." It was not unlike a player piano, but on a metal roll. This format provided 384 bits of memory[2]. However, it would prove far less reliable than the capacitor drum of the ABC. Like the ABC, Zuse used an electric motor to provide a clocking of one cycle per second. He worked out the device's logic in Boolean terms. He was familiar with the Leibniz step reckoner and in fact, was able to read Leibniz's original descriptions in Latin and mimic those arithmetic techniques. Choosing binary as the basis for the machine resulted in multiplication being much simpler. Even with the usual tricks, his multiply operation still took 20 seconds. Allied bombing in December 1943 destroyed the Z1.

Zuse was drafted in 1939 but was able to continue his work. A friend, Helmut Schreyer, encouraged him to use telephone relays. Completed in 1940, the Z2 retained his unique metal sheet memory but replaced the mechanical computation with 600 relays. The Z2 was larger and took up several rooms in his parents' flat. Zuse then moved on to his next computer, the Z3, completed in May 1941. It was a fully programmable computer supporting loops and automatic processing. He dispensed with the metal strip memory and relied fully on relays. He used 600 relays for the ALU compute unit, 600 relays for logic, and 1,400 relays for memory. He got multiplication down to three seconds. Allied bombing destroyed the Z3.

Henschel Aircraft provided funding for his next computer, the Z4. He started work on the Z4 in Berlin, but increased bombing led him to move the partially finished machine to Gottingen, about 160 miles to the southeast. The Z4 continued to use relays, but he went back to his "player piano" slotted metal for memory. The Z4 was to be a commercial endeavor. He had plans to build 300 machines.

Zuse was well aware of the difficulty in programming his machines with native machine instructions. With the Z4, he decided to address this and at the same time complete a thesis for his doctorate. He proposed a mathematical programming language called Plan Calculus that was not unlike FORTRAN. In 1945, as the Allies were closing in, he secured a Wehrmacht truck and moved the Z4 to Hinterstein, a small village in Bavaria, 70 miles southwest of Munich. He stashed the machine in a barn[3].

The Allies prevented Zuse from working on the Z4 until 1949. At that time, he had it re-assembled and put to work. The Z4 assisted in the computations for the Grande Dixence Dam in Switzerland, among several other projects. As of 1951, the Z4 was the only digital computer in operation on continental Europe[4]. Thomas Watson learned of the Z4 and purchased options on any approved patents. Like Atanasoff, Zuse would have liked to work with and for IBM. However, IBM showed little interest unless he had patents. Zuse would not be able to form his own company until 1955, due to Allied restrictions on the German post-war computer industry. He released the Z22 with vacuum tubes and core memory in 1958 and the Z23 in 1961 with transistors. He would end his "Z" run with the Z43, when financial problems forced the sale of his company to Siemens.

---

## AMERICAN CRYPTOGRAPHIC PROGRAMS

Bletchley Park was not the only cryptographic center during the war. The United States had not one but two code breaking groups—the Signal Intelligence Service (SIS) in the Army and the OP-20-G department (known simply as "G") in the Navy. With Watson, it was all hands on deck for the war effort. This included sacrificing some key engineering people like Ralph Palmer and Steve Dunwell.

The Army drafted Dunwell and assigned him to the SIS. The unit had its origin in World War I when William Friedman and his wife, Elizabeth Smith Friedman, led America's code breaking efforts. Friedman formed the SIS in 1930, with a peacetime staff of three. It would grow to 10,500 during the war. Friedman, whose SIS team had already broken the Japanese Purple code using IBM punched-card machines, was aware of Dunwell's work with tabulating machines and recruited him specifically for the cryptographic center at Arlington Hall, Virginia. Dunwell's ability to customize machines, as he had done with the interface switch for Wallace Eckert at Columbia, would be highly valued at SIS. Dunwell would call this "hot rodding" the IBM equipment.

SIS developed its cryptographic methods between the wars, starting with IBM alphabetic tabulators in 1934. The Friedmans got SIS off to the right start by hiring brilliant young mathematicians to seed the team, the successful approach taken by the Poles and the British. Dunwell arrived at SIS in 1942, shocked at the old IBM and Remington equipment in use. Dunwell immediately got to work. He met with Watson in New York City to request new tabulators as well as 5,000 Lake relays for his "hot rodding.[5]" Dunwell had just been married and had one day left on his honeymoon. Watson suggested he go to Endicott and stay at his suite at the IBM Homestead[6].

The SIS installation grew to 400 tabulators, with a staff of 1,200 processing over a million punched cards daily. Dunwell modified the tabulating machines reading at only 150 cards per minute to do more than a million character comparisons a second in searching for cryptographic patterns. As Dunwell put it, "We might beat out a machine that was running 10,000, a hundred thousand times as fast" by maximizing the "sorting, collating, measuring, and matching" with plugboard programming. One could only imagine the plugboard wiring on his hot-rodded tabulators. Dunwell was generally aware of Turing and the Bletchley Bombes but did not work on Enigma, as it was the responsibility of the Navy G group.

Ralph Palmer, the engineering manager in Endicott, joined the G group in 1943. He brought along several of his engineers. Palmer graduated with a BS in electrical engineering from Union College in 1931 and joined IBM the following year. He formed his electrical team in 1937. Watson promoted him to engineering manager in 1939. By 1941, he was overseeing Halsey Dickinson and Byron Phelps in creating IBM's first vacuum-tube-based add/subtract circuit.

The Navy G team started code breaking in 1921. The headquarters location was on Nebraska Avenue in Washington, DC. This site would grow into one of the largest IBM punched-card installations, with hundreds of IBM machines churning through millions of punched cards per week[7]. Even so, it was not getting the job done, so the Navy contacted Watson directly in 1941 and asked for help. He assigned an IBM engineer to assess new ideas and pass them back to the labs.

The Navy also created a G outpost at NCR in Dayton. With NCR unable to produce cash registers during the war, it committed to a number of war projects. The Navy's G mission was to develop a more advanced code-breaking machine along the lines of the British Bombes. They assigned Palmer to the Dayton lab. He led work on magnetic memories to replace or supplant punched cards. He also experimented with electrostatic (CRT or cathode ray tube) memory.

The top priority for the U.S. Navy was German U-boat communications. By late 1941, the German Navy had increased the number of rotors to four on its Enigma machines. Bletchley Park had been unable to adapt the existing Bombes to the enhanced machine. During a blackout period lasting ten months, no U-boat messages were inter-cepted. Unfortunately, this came at a desperate time in the Atlantic battle, with German submarine wolf packs decimating Allied convoys. The pace of U-boat sinkings increased by a factor of four, with 2.6 million tons of shipping sunk in the first half of 1942, representing an estimated 500 ships.

The lack of progress by the British led the U.S. Navy to accelerate the development of an American Bombe. James Pendergrass of the "G" group had attended the Moore School lectures and pressed the Navy for a von Neumann-type computer for cryptanalysis, to process at high speed the massive volume of intercepts. The project fell to the NCR operation in Dayton. The key to the project was the role of NCR's Joseph Desch. He was an expert in vacuum-tube-based elec-tronic counting circuits and was familiar with Wynn-Williams's work in England. In fact, Desch had already submitted a patent application for his own design, one that would run up against the patent that IBM's Halsey Dickinson had filed. In addition, NCR and Desch had worked with Vannevar Bush and Samuel Caldwell on Bush's Rapid Arithmetic Machine[8]. This further improved NCR's chances of securing funding from the Navy. Desch had formed NCR's electrical laboratory in 1938 and had risen to be NCR's director of research, following in the footsteps of Charles Kettering.

Desch and his team were working without a physical Enigma machine or details on the British code-breaking approach. Bletchley had promised both but failed to deliver. Desch knew the right approach to the project was a vacuum-tube-based machine. However, his initial design required 70,000 vacuum tubes and would likely not be ready before the war ended. Instead, he proposed a massive electro-mechanical machine that had sets of four-wheel groupings that modeled 16 Enigma devices linked together. The contraption spun at high speed, lubricated by a massive oil cooling system. Alan Turing visited in December 1942 and showed the G team a shortcut that reduced the number of bombe wheels from 336 to 96[9]. He also wrote a memo to the U.S. Navy criticizing the design, which the Navy did not share with Desch and his team. Turing also spent time at Bell Labs, where he would meet with Claude Shannon for tea and talk cryptography.

The Navy pushed back on the Desch design but finally concluded it was the only viable approach given the time and urgency. They gave NCR a blank check to complete the device. The price tag quickly rose to $4 million. The NCR program would employ over 1,000 people. After many delays, NCR produced two prototypes, nicknamed Adam and Eve. Each machine was seven feet high, eight feet long, and weighed in excess of two tons. The machine ran fast and noisily until it reached the four Enigma rotor settings, and then it abruptly shut down. Even with a set of cribs, the Bombes would run an average of 18 hours to come up with the settings. They estimated the machines ran 20 times faster than Turing's Enigma Bombes and 200 times faster than the original Polish Bomba, though far slower than Colossus. NCR would produce 120 of these massive machines[10]. The Navy cryptographic team was successful in using the NCR Bombes to restore U-boat intercepts. One of the first decrypts run by the Navy in June 1943 resulted in the sinking of U-217[11].

BOMBE

Figure 21. U. S. Navy Cryptanalytic Bombe at NCR Dayton, 1944. Used successful against the 4-rotor German Naval Enigma machine. A Navy WAVE technician is setting the starting rotor positions. U. S. Government photograph.

## LOS ALAMOS AND THE MANHATTAN PROJECT

Long before he accosted the abacus salesman in Rio, Richard Feynman played an outsized role in wartime computing. Feynman graduated from MIT in 1939 with a degree in physics, and continued to Princeton for graduate study. The physics staff there included Eugene Wigner, Albert Einstein, Wolfgang Pauli, and John von Neumann. Wigner, another one of the Martians, had worked with two other Martians, Leo Szilard and Edward Teller, to urge Albert Einstein to meet with Franklin Roosevelt and describe the dire ramifications if Germany developed an atomic bomb.

After Feynman received his doctorate in 1942, the U.S. Army came calling, looking for physicists. Feynman turned down a job offer from Bell Labs and joined the Army[12]. His first project was -- no surprise – designing a calculator to generate artillery firing tables. He did not make much progress, and the project represented a huge waste of his talents. The University of Wisconsin offered a teaching job but Robert Oppenheimer intervened, recruiting Feynman for the Manhattan Project. He started in the T building at Los Alamos, the "T" standing for the "theoretical" group headed by noted physicist Hans Bethe. Bethe would wow him with the mathematics of approximation that Feynman would later use to great success with the abacus challenge in Rio.

The theoretical physicists relied heavily on computing power. As von Neumann well knew, the computations involved in simulating the explosion of an atomic bomb were daunting. Los Alamos had started with human computers, using Marchant, Monroe, and Friden desk machines. The more powerful Marchants soon became the standard. The initial group of "computers" grew to twenty, mainly wives of the scientists. The massive calculations taxed the Marchants, and they frequently needed repairs. To avoid losing broken machines for weeks sending back to the manufacturer, Feynman and fellow physicist Nicholas Metropolis began to repair the machines themselves[13].

Still, progress was painfully slow. Dana Mitchell, a physics professor from Columbia, suggested a computing setup similar to what his fellow professor Wallace Eckert had assembled at Columbia. Bethe agreed. Stanley Frankel put together the requirements and shepherded the order to IBM. The initial shipment included three 601 Multipliers, a 402 tabulator, keypunches, and sorters. IBM was not told where the machines were going, and no one would be available to set them up. IBM did provide the name of one of its top customer engineers who was already in the Army. Los Alamos ordered him to report for duty at P. O. Box 571, Santa Fe. That was the code name for Los Alamos. In the meantime, Feynman and several colleagues went ahead and installed the equipment. Feynman quickly became an expert at wiring plugboard panels. Always needing more multiplications, he worked out a method to run the three multipliers in tandem.

The 601 Multipliers did not do division, which was a requirement. IBM's drafted customer engineer suggested that IBM could develop a prototype with that function. The lab contacted John McPherson with the request, and IBM quickly produced and delivered a modified 601 Multiplier machine supporting division. Feynman, along with Nicholas Metropolis, took over the IBM computing operation. They worked with the Marchant pool to lay out the calculation steps required for the implosion problem. They then translated those steps into programming on the punched card tabulators. Feynman and his team organized a contest between the hand computers and his new IBM machines[14]. Initially, the hand computers kept pace as Feynman experimented with ways to optimize the use of the punched-card equipment. However, the machines soon far exceeded the hand computers' pace. The machines also ran 24 hours a day, seven days a week, never got tired, and never made mistakes.

Feynman would describe a "disease" caused by working with the machines at all hours, resulting in burnout[15]. Frankel succumbed to the disease and Feynman took over. He would develop a team of "boys" to develop new applications. The "boys" worked on the problem to estimate the energy released in a Trinity explosion. Los Alamos engineers soon overwhelmed Feynman's punched-card section with scientific problems. Thomas Watson became involved and he offered use of the newly developed Watson Scientific Bureau in New York to augment the Los Alamos work.

The Mark I was not the only IBM project during the war to address firing tables. Demand for ballistics calculation was insatiable, simply off the charts even in the late stages of the conflict. Bell Labs produced five models of the Stibitz design. With the new fast relays invented by Clair Lake, an IBM relay calculator was a natural project. Ben Durfee joined Lake in the machine design. Halsey Dickinson developed the multiplication process. The completed machine was compact, tabulator-size, taking up only six square feet. Inside were 2,500 of Lake's relays, with 1,500 used for processing logic and the rest organized into 28 accumulators. Dickinson adopted the now-standard multiplication algorithm pioneered by James Bryce. He enhanced it by starting the next multiplication before the current multiplication had completed. As usual, a central rotating shaft provided the overall synchronization. The machined needed two plugboards to program the 28 accumulators and up to 2,000 patch cable

connections. The relay machines were fast, computing six multiplications per second, roughly 20 times the speed of the Mark I at Harvard.

IBM delivered two machines to the Army's Ballistic Research Lab
in December 1944. This lab now had several computing sections: the
IBM punched-card group, the Bush differential analyzer, and the IBM
Relay Calculator. In addition, it was waiting on the Stibitz-designed
Bell Model V and the ENIAC. The IBM system was not terribly
successful in use, particularly compared to the Bell Model V. The
Model V was programmable, with superior logic capability, enabling it
to run problems for hours unattended. The IBM relay machines were
at least available during the war. Neither the ENIAC nor the Bell
Model V arrived until the conflict was over.

Three additional IBM Relay machines were produced, with one
installed at the Naval Ordnance Center at Dahlgren and two sent to
the Watson Scientific Computing Bureau at Columbia. Columbia
dubbed their machines "Virginia" and "Nancy."[16] Wallace Eckert
attached them to his IBM tabulator and programmed through card
input. He used the machines in 1946 to produce the "Ephemerides of
783 Minor Planets", a calculating effort that resulted in one hundred
pages of tables. The most important legacy of the IBM Relay Calculator is that it blazed the trail for the IBM 604, a vacuum-tube-based
multiplier.

# WATSON'S RESPONSE TO HARVARD: THE SSEC

After the Mark I snub by Harvard in August 1944 and even before the post-mortem investigation was completed, Watson asked McPherson and Bryce to build a "super calculator" to stick it to Harvard and Aiken. For Watson, this was personal. Watson also decided that he would funnel future IBM scientific grants to Columbia, not Harvard. Watson then turned to his Columbia University friend, Wallace Eckert, and enticed him to leave the Naval Observatory and join IBM to establish a pure science department. IBM would also fund a new laboratory to house the department. Columbia requisitioned the run-down Delta Phi fraternity house and this became the Watson Scientific Computing Lab. Watson paid the salaries of the staff.

Eckert's immediate priority was to build his research team. His first hire was Herb Grosch, the second PhD at IBM. Grosch's background was in mathematics, physics, and astronomy. He had worked at the Watson Astronomical Bureau at Columbia. During the war, he sat across the hall from Eckert's Air Almanac office at the Naval Observatory. His additional claim to fame, though not something to boast about at IBM, was being the first IBM hire with a beard. While at the Watson Lab, Los Alamos asked Grosch to provide additional calculations. He also assisted Richard Feynman with implementation of the customized 601 Multipliers. Eckert also brought in Llewellyn Thomas, a leading physicist and applied mathematician[1]. Though Thomas was not familiar with computing technology, he could state any scientific problem in mathematical language to facilitate the machine programming. John Lentz, Byron Havens, and a number of other researchers also joined Eckert's team. John von Neumann was a frequent visitor at the lab, keeping up to date on IBM's current technologies as well as running Los Alamos programs.

Watson's Harvard revenge project was underway in late 1944. In 1946, the Moore School held the ENIAC press conference, prompting a lavish and adoring article in the *New York Times*. One can picture Watson reading quotes such as "an amazing machine," "an electronic speed marvel," and "perhaps the greatest marvel of electronic ingenuity." He might have been particularly galled at "not a single moving mechanical part." The piece ended with the pronouncement, "So clever is the device that its creators have given up trying to find problems so long that they cannot be solved.[2]" Watson was in the midst of spending millions on his "Super Calculator" and now he reads these accolades.

Watson leaned on John McPherson and Frank Hamilton to accelerate the Super Calculator work. The amended goal was a machine that was faster than the Harvard Mark I and more useful to science than the ENIAC. Watson also green-lighted a second project, an electronic multiplier machine that would become the IBM 603. IBM initially developed the multiplier to be able to say they had a vacuum-tube-based, electronic machine in their sales manual.

As if Watson and IBM needed more incentive, the summer of 1946 brought the Moore School lectures. IBM was not invited. According to Herb Grosch, it may have been because Mauchly and Eckert did not want to stir up any competition from the dominant office equipment company[3]. More likely, it was a reflection that academia did not hold IBM in high regard and viewed the company as simply a successful combination of aggressive salesmen and old punched-card technology.

Watson would end up spending a reported $1 million to build his machine[4], a huge amount compared to the roughly $500,000 price tag for ENIAC and $600,000 for Mark I. Moreover, the $1 million price tag grossly understates the total spent. It was clear from the start that this was a special priority for Watson, and this resulted in significant expenditures in time and materials going unrecorded. If an expense was questioned, then all it took was something along the lines of "this is for Mr. Watson's special project" to obtain the funds. John McPherson led the project and ensured that there were no impediments.

Wallace Eckert developed the specifications for the new machine. He found an excellent resource in Rex Seeber. Having been rebuffed by Howard Aiken for his ideas on Mark I improvements, he jumped at the opportunity to "work for IBM and help them build a bigger and better computer than the Mark I.[5]" Eckert designated him the program's chief architect but he would have a lot of help. He moved to Endicott to work side-by-side with Frank Hamilton, the designer of the Mark I and now responsible for the overall development of the super machine. Hamilton was a big man and an aggressive leader. Seeber was slight and retiring. The two became great friends.

For his part, Eckert saw the machine's design as an exercise in optimizing the available technologies. Those included electro-mechanical counters (which he called "wheels") along with relays and vacuum tubes. As a scientific machine, the conversation always started with multiplication speed. Wheels took six seconds. Relays took one second. Vacuum tubes took 20 milliseconds, or 50 multiplications per second[6]. They did not want to build a device with 20,000 or more vacuum tubes, and so they gravitated to a hybrid design that married the three technologies. Eight banks of vacuum tubes provide 160 digits of fast electronic storage and system logic. The team used fast Lake relays for logic switching and as temporary data registers. In addition, they deployed paper tape loops for auxiliary tables and subroutines. Seeber and the team completed the basic design in a couple of days, as the deficiencies of the Mark I pointed the way[7].

It would be a quasi-stored program device, with both data and program instructions on the same media. Each instruction contained the location of the next instruction, which facilitated the modification of the next instruction address, and thus a logic branch within the program. The machine's official name, the Selective Sequence Electronic Calculator, or SSEC, reflected this capability. The paper tape devices used regular IBM 80-column card stock as a large, wide roll, before it was cut into cards. The result was essentially 80-column cards in a continuous loop. There were 36 of these special punched-card readers, providing roughly 400,000 bits of storage, albeit serial in nature and slow in access. IBM built a special machine called the "Prancing Stallion" to convert IBM cards into this expanded paper tape format[8]. The system emulated the von Neumann architecture but without the latest technology. Seeber designed the SSEC to treat instructions as data. It was the very powerful stored program concept but limited by the slow program loops on paper tape.

The IBM team completed the SSEC in just sixteen months. It contained 12,500 vacuum tubes and 21,400 relays. The bulky tubes were military surplus that Hamilton had tracked down, an unusual cost savings move for IBM[9]. The system's ALU (Arithmetic and Logic Unit) was a modified IBM 603 multiplier using the James Bryce-designed multiplication algorithms. In August 1947, IBM assembled and tested the equipment in Endicott and shipped it to New York City, for installation at IBM's new World Headquarters at 590 Madison Avenue. Technicians set up the SSEC behind a large street-level window, formerly a women's shoe store called the French Bootery[10]. IBM had bought out the lease. As usual, Watson insisted on extra design touches for maximum effect as he had done with the Harvard Mark I. He had the street facade of the building renovated. He wanted a clean look without masses of cabling between the machines. This resulted in one of the first computer installations that used a raised floor.

**Figure 22.** IBM Selective Sequence Electric Calculator (SSEC), installed at IBM's 590 Madison Avenue headquarters in 1948. This is the famous airbrushed picture. Reprint Courtesy of IBM Corporation ©.

As you entered the showroom, the massive card-based reader units dominated the back wall. The walls to the left and right held the electronics. All three walls were finished in modern glass panels. The center of the room held the accounting machine printers and the master console. Watson had a dedication plaque made for the SSEC. He drafted an inscription, which read, "This machine will assist the scientist in institutions of learning, in government, and in industry to explore the consequences of man's thought to the outermost reaches of time, space, and physical conditions[11]."

There was something else in the showroom: three enormous black floor-to-ceiling pillars. Shortly before the formal dedication, Watson surveyed the showroom, pointed to the pillars, and commanded their immediate removal. Of course, the pillars were there to support the 20-story building. To assuage his distress, the formal press pictures of the SSEC had the pillars airbrushed out[12].

**Figure 23**. The SSEC as Thomas J. Watson. Sr., saw it, with the huge black pillar dominating the scene. Reprint Courtesy of IBM Corporation ©.

IBM held the formal dedication in January 1948. They invited a Who's Who list of prominent scientists. James Bryce was too ill to attend the event. He died a year later. Bryce was one of ten men selected as "greatest living inventors" at the centennial of the U.S. Patent Office in 1936[13], and at that time, he was just getting started. Like the Mark I, the SSEC was noisy and one could hear it working from the street. The ENIAC was quickly ushered out of sight after its famed press conference. In contrast, the SSEC stayed right there on Madison Avenue. Though it was an expensive project, the resulting publicity proved to be a major benefit for IBM. The film noir movie *Walk East on Beacon* prominently featured the system. The press adopted the label "Super Calculator." The publicity surrounding the SSEC led the public to associate computers with IBM, at least until Mauchly and Eckert's UNIVAC appeared a few years later. For its brief time in the limelight, the SSEC was the world's best-known computer.

Eckert and Seeber hired a strong math-centric team to program the SSEC. Seeber, as director of the SSEC, brought in John Backus, Harlan Herrick, Ted Codd, and a number of others. Betsy Stewart, who had worked at the Watson Lab, was the lead operator. Backus forged a somewhat unusual path to 590 Madison Avenue. He had been expelled his freshman year from the University of Virginia for lack of attendance. He served in the Army during the war. After a false start in medical school, he ended up at Columbia, graduating with a degree in mathematics in 1949. Shortly after graduation, he visited the SSEC[14]. It was just two miles from campus. On a tour of the machine, he remarked that he was looking for a job. The tour guide escorted him upstairs to meet with Seeber. Seeber gave him an impromptu oral test on mathematics and hired him on the spot. Backus would spend three years on the SSEC before moving on to invent the Speedcoding compiler and later leading a team along with Harlan Herrick that developed the FORTRAN programming language. Ted Codd would go on to invent the relational database.

The SSEC programmers worked directly in binary machine code. As it relied partially on plugboards, it was very awkward to program. It could read an instruction from memory, from mechanical registers, and from the special 80-column paper tape, and store the result in any of those elements. It supported branch and loop operations.

Wallace Eckert made the decision on what program he wanted run first on the machine. To no one's surprise, that would be a lunar ephemeris, the positions of moon. He wanted to compute the positions for 100 years in the past and 100 years into the future at 12-hour intervals. This represented 150,000 positions. Eckert described each position as needing "11,000 additions and subtractions, 9,000 multiplications, and 2,000 table lookups" but took all of seven minutes run time[15]. This program would keep the SSEC busy off and on for the next six months.

Watson made the system available to scientists at no charge. Los

Alamos ran several nuclear fission simulations. The first was the so-called Hippo program, organized by the omnipresent von Neumann. Herman Goldstine's wife, Adele, and von Neumann's wife, Klara, both of whom had programmed the ENIAC, coded the program. The ENIAC had been too limited in memory to run Hippo. The program would run on the SSEC around the clock for days on end. Many others lined up to use the machine. In a very short time, it was fully booked. GE was the first paying customer, running an analysis of steam flow for its gas turbines.

In raw arithmetic performance, the SSEC would run 3,330 additions or 50 multiplications per second[16]. This compares to the ENIAC benchmark of 5,000 additions or 333 multiplications per second. The relatively modest multiplication performance was a result of the scientific need to multiply two 14-digit decimal numbers. In terms of real-world performance, the ENIAC was faster with some problems but needed the arduous rewiring for any reprogramming. In addition, its limited instructional memory meant that any large program—Hippo, for example—needed to be broken up into smaller programs that ran separately. The net effect was that the SSEC was far more functional. Showing a singular lack of insight, Presper Eckert called the SSEC "some big monstrosity over there that I don't think ever worked right.[17]"

The SSEC achieved one of the goals that Watson had set for it: eclipsing the ENIAC. With respect to the Harvard Mark I, the SSEC overachieved. It was 250 times faster than the Mark I and 30 times faster than Aiken's subsequent solo effort, the Mark II. The SSEC lasted in the spotlight for four years, operating from 1948 to 1952. At the end, IBM wheeled it out and replaced it with IBM's first mainframe computer, the 701 Defense Calculator.

The SSEC was very successful in a number of other ways. It reestablished IBM's reputation and working relationship with the scientific community. Though only part of the way to a modern digital computer, it captured in the public imagination the idea of a mechanical "brain," taking that mantle away from the ENIAC. The project also pushed the Watsons and IBM forcefully into electronics. In addition, quite a number IBMers, engineers and programmers trained on the machine.

# THE VON NEUMANN MACHINES

John von Neumann usually found his way to the hottest new machines in computing. After completing his influential draft report on computer architecture in 1946, he was anxious to build his own computer, to his own specifications. He would have no problem with funding. He secured commitments from the new Atomic Energy Commission, several military agencies, and RCA. His budget was a healthy $800,000, though this would balloon to roughly $2 million when a new building to house the computer was added in[1].

His bigger challenge was convincing the Institute for Advanced Study (IAS) at Princeton. The IAS was a special academic environment for contemplation and pure research. A project of this nature was quite foreign. Robert Oppenheimer was back at the institute from the Manhattan Project and supported von Neumann's project. Oswald Veblen, who had urged the Army to fund ENIAC, was also in favor. Von Neumann finally secured formal approval by threatening to leave for MIT or the University of Chicago. Both universities were anxious to have him.

Von Neumann's next challenge was assembling a team to design and build the computer, now called MANIAC (Mathematical and Numerical Integrator and Computer). He originally planned on Presper Eckert to lead the effort. This was a gross miscalculation. Both

Eckert and Mauchly were beyond irate, feeling von Neumann stole their ideas and then published them under his name. This in turn led to the loss of patent rights for inventions contained in EDVAC. In addition, both were fully booked in setting up their own commercial company based on the EDVAC design.

Von Neumann quickly moved on from Eckert's refusal. He asked Julian Bigelow, one of the key architects of the ENIAC, to take the lead. Bigelow, an MIT graduate, worked for a time at Sperry before joining IBM in 1938. He met Howard Aiken during this time in Endicott. He went back to MIT during the war and worked with digital computing pioneer Norbert Weiner on anti-aircraft fire control. Weiner recommended him highly to von Neumann. Von Neumann also added Herman Goldstine, Adele Goldstine, and Arthur Burks to the team. They enjoyed working with von Neumann and were happy to be working on another cutting-edge project. Burks would leave the effort in 1948, wanting to return to the University of Michigan. He also signed on as a part-time consultant for Burroughs, working again with Irven Travis.

The assembled team set about developing a pure "Von Neumann machine." It would become the blueprint for most early computers. The immediate challenge was memory. The ENIAC, despite having 18,000 vacuum tubes, had very little actually working memory, just 20 ten-digit decimal numbers. The von Neumann architecture would require that program instructions and data both reside in memory during execution. Burks and team estimated that they would need enough memory to store 4,000 numbers of 40 binary digits each. They stated this "exceeds the capacities required for most problems that one deals with at present by a factor of about 10." This would translate into more than 100,000 bits or flip-flops, a number vastly impractical with vacuum tubes. This led to the consideration of other hi-speed storage devices.

Von Neumann was familiar with Eckert's invention of mercury delay lines but wanted no part of the size, expense, or problems associated with them. RCA had a research lab located in Princeton. Von Neumann had met Jan Rajchman, a prolific inventor for RCA, at the Moore School lectures. He encouraged Rajchman to pursue a vacuum-tube-based memory unit that RCA called the Selectron[2].

Rajchman's invention was a cathode ray tube ten inches long and three inches in diameter that could hold 4,096 bits of information. It was a dramatically superior device compared to mercury delay lines. The target memory for the IAS computer would require just 40 Selectron tubes, providing 160,000 bits of memory. It was fast, operating at electronic speeds, not the speed of sound as was the case with the delay lines. Really, there was only one problem: Rajchman and RCA were unable to perfect the Selectron.

Bigelow heard of the Williams and Kilburn iconoscope at England's University of Manchester. The first cameras for television would use iconoscope technology. Freddie Williams and Tom Kilburn adapted it as a storage medium. Bigelow met with them in Manchester and then stopped by Cambridge to meet with Maurice Wilkes and Alan Turing, who were both working on von Neumann architecture computers. Von Neumann decided on the Williams tubes, an electrostatic memory. It was a cathode ray tube that provided 1,024 40-bit words of memory and could be randomly accessed in 30 milliseconds. The Williams tubes etched a grid pattern of dots that represented binary zeros and ones. Much like Atanasoff's condenser drum, Williams tubes periodically refreshed the grid before the dots faded.

Von Neumann publicized progress on the IAS system, as was his normal mode. Von Neumann soon discovered that he did not get along with Bigelow, and replaced him with James Pomerene (who would later join IBM)[3]. Von Neumann's interest in the project gradually declined, and Goldstine assumed leadership. John McPherson came down to look at the machine in process. McPherson was driving the SSEC program but wanted to understand what von Neumann was doing. Von Neumann had become a part-time consultant to IBM.

The Von Neumann team started work in 1946, but they did not have a final machine until 1951. There evolved the idea of the "von Neumann constant[4]," which specified how long it would be from any point in time until the IAS was finished, implying that development might take forever. In the end, the wait was worth it. The IAS was a first-class Von Neumann computer. It was very compact, the size of a "grand piano.[5]" It weighed just 1,000 pounds, compared to the 27-ton heft of the ENIAC. The binary-based Arithmetic and Logic Unit was comprised of just 1,700 vacuum tubes[6]. The machine used paper tape input at first, then added IBM card readers.

**Figure 24**. Robert Oppenheimer and John von Neumann in front of the IAS (Institute for Advanced Study) computer. Upper section housed 1,700 vacuum tubes and lower section contained banks of Williams tubes. It was a powerful von Neumann architecture system that weighed in at just 1,000 pounds.

The IAS machine was a fully stored program computer. It supported branching, looping, and subroutines. As a binary machine, it had to deal with the issue of conversion between binary and decimal. Von Neumann and Goldstine coded a simple routine of just 47 words of memory to handle the conversion, taking negligible time. Programming was still in machine code. The first program was no surprise: It was a hydrogen bomb explosion simulation.

The IAS computer was a hot machine. It performed about 16,000 additions or 1,400 multiplications per second. This was about five times the speed of the EDVAC. With this success, there was talk of a second IAS machine, but the institute was not supportive. It felt the existing machine had proven the basic technology and that additional advances should come from industry. Von Neumann's machine operated until 1958. It ended up at the Smithsonian Museum of American History. Leon Harmon, a future Artificial Intelligence (AI) researcher at Bell Labs, worked on wiring for the IAS. He met Albert Einstein, J. Robert Oppenheimer, Edward Teller, and John von Neumann while in Princeton. He would later say, "Von Neumann is the only genius I ever met.[7]"

As was his standard operating procedure, Von Neumann put all of the concepts used in the IAS computer into the public domain. He wanted to jump-start the emerging computer industry. He certainly succeeded in this, as many early computers were of the von Neumann design. In addition, many adopted the IAC (Integrator and Automatic Computer) suffix that began with ENIAC. Los Alamos produced a clone of the IAS system, also named MANIAC. Nick Metropolis, who had attended the Moore School lectures and worked on the Manhattan Project with Frankel and Feynman, did the design. There were actually three MANIACs, with MANIAC I and MANIAC II at Los Alamos, and MANIAC III at the University of Chicago[8].

The Oak Ridge National Lab built the ORACLE computer. "ORACLE" did not stand for anything, bucking the tradition. Completed in 1954, it featured a combination of vacuum tubes, transistors, and Williams tubes. The JOHNNIAC was right up front where its design came from. JOHNNIAC stood for "John von Neumann Numerical Integrator and Automatic Computer." The Rand Corporation built the machine using the Selectron vacuum tube that RCA finally perfected.

The von Neumann design proliferated overseas as well. The SILLIIAC (Sydney Version of the Illinois Automatic Computer) was the first computer in the Southern Hemisphere. The BESK machine (the Binary Electronic Sequence Calculator) was the first system in Sweden, designed by several young engineers who spent a year at Princeton learning the IAS design. Saab used the system for aircraft design. The BESM system stood for "Large Calculating Machine" in Russian[9]. Herman Goldstine had no idea how the Russian team got their design information.

As was noted earlier, Maurice Wilkes had arrived late to the Moore School lectures due to a delay in his ocean passage. He spent the five-day return voyage on the *Queen Mary* developing a plan to build an EDVAC-type computer[10]. Wilkes had received his PhD at Cambridge in the 1930s, taking some of his classes with Alan Turing. He had worked on the university's Meccano analyzer and later became the director of the school's computing laboratory. Wilkes needed funding for the computer project and found it in a most unusual place, at a British tea company. J. Lyons and Company was an institution in Britain, with over 200 tea shops, 250 Corner House restaurants, and a scattering of hotels[11]. Its leadership was struggling with administration of this vast enterprise, using countless accountants and desktop adding machines to manage operations. The company was far ahead of its time in recognizing the commercial potential of computers. John Simmons, a senior executive, sent two individuals on his staff to the United States in 1947 to assess the state of computing machines and the potential for a system that Lyons could deploy. They met with Herman Goldstine at Princeton and learned that Maurice Wilkes was building an EDVAC-type computer right in their own backyard, at Cambridge University. Upon their return, Simmons met with Wilkes and agreed to provide 3,000 British pounds in additional funding to speed up Wilkes's project in exchange for assistance in developing a computer for Lyons.

Wilkes' planned computer was the EDSAC, for "Electronic Delay Storage Automatic Calculator"[12]. His goal was a reliable system that could support a wide range of users at Cambridge. This led him to a conservative approach in the design, hewing very closely to the design of EDVAC. As the computer's name indicates, he decided to use Presper Eckert's mercury delay lines for memory, with 32 units providing 8,700 bits of fast storage. He understood the concept as he had worked in British radar development during the war. Alan Turing was not a fan of Wilkes's design, stating that it was too much like the American machine and broke no new ground technologically. Yet, this was exactly the point. The EDSAC was not fast, producing 700 additions or 185 multiplications per second. Where it excelled was in programming and ease of use.

The EDSAC featured one of the first assemblers, where English mnemonics substituted for binary instructions and the machine itself converted the program into machine code. Wilkes also proposed use of subroutines, precoded segments of programs. He felt that with a robust library of subroutines, most programs would just about code themselves. He asked David Wheeler, a mathematician, to work on this. Wheeler identified common functions and stored them in paper tape form in a cabinet. His library grew to 87 subroutines, including such functions as exponents and logarithms[13]. When a function was needed, the programmer pulled subroutine from the cabinet and copied it into the new program's paper tape. Proper addressing to and from the subroutine was maintained by a piece of code called the Wheeler Jump. Later, more sophisticated assembler versions eliminated the "jump."

Wilkes soon learned that even with subroutines and the assembler, programs still failed. He held a conference in 1949 for British computer scientists. Preparing a demo program for the conference, he designed a differential integration program of 125 instructions. He struggled to correct all the errors (Grace Hopper's "bug" term had not yet reached him). He would comment, "A good part of the remainder of my life was going to be spent in finding errors in my own programs[14]." This led the EDSAC team to develop aids to debug programs on the EDSAC, including memory dumps and instruction traces.

Wilkes began EDSAC development in 1947, and the machine was operational on May 6, 1949, running a program producing a table of squares[15]. Wilkes and his team actually completed the EDSAC before the EDVAC, an outcome caused by the disruption of Mauchly and Eckert. The EDSAC saw wide use at Cambridge. Three Nobel Prize winners made use of the machine. David Wheeler wrote up his programming work as a thesis and received the world's first PhD in computer science. Wilkes would team up with Wheeler and Stanley Gill to write *The Preparation of Programs for an Electronic Digital Computer*. Fred Brooks, a future giant in computing, would call it the most important book in the history of programming[16].

Wheeler took a leave from Cambridge to go to the University of Illinois, where a series of Von Neumann computers were being developed. The ORDVAC (Ordnance Discrete Variable Automatic Computer) went to the Army's Ballistic Research Lab at Aberdeen. ILLIAC (Illinois Automatic Computer) was a twin of ORDVAC and stayed at the University. SILLIAC (Sydney version of the Illinois Automatic Computer) landed at the University of Sydney[17].

And, what happened to the computer project at J. Lyons? Lyons assigned one of the company's engineers to work on EDSAC, providing firsthand analysis of the potential for a Lyons commercial machine. That engineer had just completed the design of a hot dog cooking machine so the computer project might have been a bit of a stretch. Wilkes recommended engineer John Pinkerton to lead the design and development[18]. Pinkerton's team developed the LEO computer, short for "Lyons Electronic Office"[19]. The first model, the LEO I, debuted in 1951, followed by the obligatory royal demo for Princess Elizabeth[20]. The LEO I used 6,000 vacuum tubes for binary logic, 64 mercury delay lines provided 72,000 bits of storage, and an array of punched-card devices from BTM (British Tabulating Machines). LEO programmers used Intercode, an assembler they devised.

J. Lyons put LEO I to work on the company's commercial applications. Each tea shop phoned in their orders. The central office would keypunch the orders and run them on the LEO. It was not long before word got around about the LEO I, and other companies inquired about having their own LEO machine. Lyons decided to take the plunge and set up LEO Computers in 1954. One prominent organization that bought LEO computers was none other than Tommy Flowers's British General Post Office (GPO). That seems somewhat ironic, where a pioneer in digital computing ends up buying its computers from the corner tea shop.

As EDSAC and LEO were under development, there was equally important computer project 300 kilometers northeast in Manchester. Max Newman, fresh from Bletchley Park's "Newmanry", led the effort at the University. He had all the knowledge of the Enigma Bombes, the Heath Robinsons, and the massive Colossus machines in his head. He had also taken several physical parts of the Colossus machine to Manchester to use as a reference. Newman had returned to Cambridge after the war but soon concluded that the school was not a good candidate for a computer project, though Wilkes would prove him wrong. Newman accepted a position as head of the mathematics department at Manchester in 1946.

Newman decided to start with a small demonstration project. He brought in some of his former Cambridge students—Tom Kilburn, Geoff Tootill, and David Rees—to staff the project. Rees had recently returned from the Moore School lectures. Newman also wanted Tommy Flowers but he was not available. He tapped Freddie Williams, the head of the university's electrical engineering department, to be the lead engineer. This was the same Williams who produced the Meccano version of the Bush differential analyzer. He also recruited Turing, who had been unsuccessful at getting funding for his own computer project called ACE (Automatic Computing Engine).

Williams and Kilburn developed the Williams tubes, the cathode ray tubes (CRT) that provided the first random-access memory. They used three standard 12-inch CRT tubes for the project. The first provided 1,024 bits of memory, the second held intermediate results, and the third contained program instructions. Newman called their machine the Small-Scale Experimental Machine (SSEM) but the name "Manchester Baby" took hold. Completed in 1948, it was considered the first electronic stored-program computer. None of the machine's core engineers—Williams, Kilburn, and Tootill—knew anything about computers coming into the project[21]. Newman provided their education. Programming the "Baby" combined binary instructions with buttons and switches. As a pilot program, the only math operation supported was subtraction. Their first test program intended to calculate the highest divisor of two to the eighteenth power (262,144). The machine ran for nearly an hour before returning the correct answer (131,072). It performed 3.5 million instructions during that time. Newman's team substantially enhanced the "Baby" and renamed it "MADAM", for "Manchester Automatic Digital Machine." The instruction set grew to 26 and they added a magnetic drum for larger data storage. The new processor featured 1,600 vacuum tubes.

Alan Turing joined the team and developed the programming system. After Bletchley Park, Turing worked at the National Physical Lab (NPL) southwest of London. This was the British equivalent of the U.S. Bureau of Standards. Turing had worked with von Neumann and was familiar with both the ENIAC and EDVAC computers. He was of course, intimately familiar with machines developed at Bletchley Park. He also had the opportunity to meet Konrad Zuse. However, he put his own ideas into ACE, the Automatic Computing Engine. His use of the word "engine" was a homage to Charles Babbage[22]. Turing worked at NPL for two years but was frustrated the lab was only interested in a test machine. The NPL's reticence on committing to a full-bore Von Neumann computer was in part due to Turing's over-engineered design. Turing was really trying to push the technology forward. He finally gave up on ACE and joined Newman at Manchester. Architects at NPL scaled back Turing's design and completed the Pilot ACE in 1950. Running at one megahertz (one million cycles per second), it was the fastest computer of its time. English Electric developed a production version of the computer called DEUCE (Digital Electronic Universal Computing Engine) and sold 33 systems[23].

Manchester's MADAM machine went operational in April 1949 as the Manchester Mark I. The British Ministry of Supply visited, was impressed, and arranged for funding for Ferranti, a defense contractor, to build a commercial version of the computer. It became the Ferranti Mark I. It featured the programming language Autocode that used algebraic expressions several years in advance of FORTRAN. Nine Mark I systems were built. William Penney, who had worked with Richard Feynman at Los Alamos, ordered one for the British nuclear program[24].

With their wartime experience and the many post-war projects, Britain was well positioned to lead the computer industry after the war. However, the government selected the wrong company to lead the charge. Ferranti, as a military contractor, had many other product lines, and had limited experience in commercial marketing and sales. The story might have been different had the government selected some other company, such as the British Tabulating Machines Company. In any case, the British failed to build on their advantage, and in a very short time, the United States and particularly IBM left them far behind.

Turing was a very eccentric individual. A long-distance runner, he would sometimes run from Cambridge to London for meetings—a distance of 40 miles[25]. He was a homosexual at a time when Britain still had draconian laws on the books for this. In 1952, Turing reported a robbery at his house and during the police investigation, he admitted to a homosexual relationship. He was charged with indecency and having pled guilty, was given a choice between prison and chemical castration. He chose the latter. Two years later, he committed suicide. He was only 41. Due to Britain's Official Secrets Act, Turing's stature and heroics during the war were not known until 1974 when Winterbotham's book came out. The British government did not issue a formal apology until 2009. Efforts to issue a pardon for his conviction languished for years, often stymied by politics. Queen Elizabeth II finally signed a full pardon in 2013. Turing's picture appeared on the fifty-pound note in 2021.

PART THREE

# THE WATSON JUNIOR YEARS

# THE RELUCTANT WATSON

When the war ended, Thomas Watson was 72 years old. He would run IBM for ten more years, but wondering who would take over his company was a central concern. Watson had married Jeanette Kittredge in 1913 before he moved from Dayton to New York. Jeannette was a debutante from a wealthy railroad family in Dayton. Watson would say that convincing her to marry him was the toughest sale he ever made. Thomas J. Watson Jr. arrived a year later, followed by Arthur and then two daughters, Helen and Jane.

Young Thomas grew up in Short Hills, New Jersey, a wealthy commuter town across the Hudson River from New York. He was a poor student, with his behavior earning him the nickname "Terrible Tommy Watson.[1]" He would eventually outgrow this and would become just "Tom" to set him apart from his namesake father, who was "Mr. Watson" or simply, "the Old Man." Watson groomed his eldest son in the company from an early age, taking him along in his travels to the far reaches of IBM. Young Tom even gave a speech at the 100% Club when he was all of 15. His speech was short, quite unlike his father's marathon orations.

He attended the Hun School, a prep school in Princeton, New Jersey, where he continued his streak of poor grades. This was in contrast to his younger brother, known as Dick, who did a bit better in school. Come time for Tom to go to college, even his father's consider-able pull could not get him into Princeton University. Watson did succeed with Brown University, but only as a favor called in and only if Tom maintained good grades.

Tom was the archetypical rich kid, with more than enough allowance to live large. That included the money to learn to fly airplanes. He flew for the first time when he was ten years old and secured his pilot license when he was a freshman at Brown. His parents bought him the hot new automobile of the 1930s, a 1934 Chrysler Airflow. His grades did not improve and Brown came very close to expelling him. Tom admitted he spent a good deal of his time carousing and flying. Still, he managed to graduate with a business degree.

Watson gave him a made-up job that took him on a paid junket around the world visiting IBM locations. This was the summer of 1937. That fall, Tom was a sales trainee in Endicott, taking the classes at the IBM Schoolhouse on North Street. The manager of the sales training program turned out to be the former headmaster of the Hun School[2]. He knew firsthand what kind of individual he was working with. As for Tom, he faced the culture and sales training program that his father had created. This included morning songs that heaped praise on IBM leaders. He came away thinking it was a borderline cult atmosphere. Instructors and classmates treated him quite differently as the founder's son and heir apparent. They elected him class president.

Tom struggled with the nuts and bolts of punched-card equip-ment. He needed a tutor in order to pass the plugboard exam and graduate from sales school. The two long winters in Endicott nearly did him in. His reward for passing sales training was a prime sales territory in the financial district of New York. He received a huge order from United States Steel in the first few weeks of January 1940 that fulfilled his quota for the full year[3]. In IBM jargon, this was a "bluebird," meaning you did not earn the business; it just flew in the window.

Tom was not particularly excited about working at IBM, even with all the benefits that his status provided. By his third year on quota, he would make a few sales calls in the morning, go flying in the afternoon, and hit the nightclubs in the evening. He may not have stuck it out at the company if the war had not intervened. With the start of hostilities, he joined the Air National Guard, coming in with over 1,000 hours of flight time[4]. He was concerned that his eyesight would disqualify him for full-time service with the Army Air Corps. But in 1940, Roosevelt mobilized the Air National Guard units and he began active service as a military pilot.

His first assignment was at Fort McClellan in Alabama followed by a posting to a squadron flying submarine search missions along the California coast. He hated his unit's commander and asked his father for help. Watson Sr. asked his longtime trusted assistant Fred Nichol to talk with General George Marshall, the Chief of Staff in the War Department. Marshall arranged a far more agreeable posting at the Command and General Staff School at Fort Leavenworth[5]. Tom had just married Olive Cawley, a *Vogue* magazine cover girl, and she was able to join him on the base. His charmed life continued unabated.

During the war, his job was to train pilots using the new Link trainer. He enjoyed the work and excelled at it. With excellent recommendations, General Follett Bradley, the head of the 1st Air Force, selected Tom as one of his pilots, along with Leland Fiegel, his chief pilot[6]. They would be piloting the B-24 Liberator, the plane that Ford Motor would end up producing at the rate of one per hour at its mammoth Willow Run plant. One of the first missions was flying General Bradley to Moscow to discuss the status of the planes that the United States was supplying to Russia. The General was also carrying a letter from Roosevelt for Stalin.

The destination of this secret mission was simply "Plainfield." The B-24 had a 3,000 mile flying range. Combined with the need to avoid war zones, this required a circuitous route to Moscow. The flight path took the plane from Washington to Puerto Rico, British Guiana, and on to Natal, Brazil. The Atlantic crossing to Accra, Ghana was the longest flight leg at 2,500 miles. They continued to Cairo and Tehran before arriving in Moscow. Given that the Russians wanted the advanced B-24 but were receiving the lesser Douglas A-20, their arrival in a new B-24 may not have been the best optics. Winston Churchill arrived for consultations with Stalin while they were there.

Tom and Leland plotted the return flight by continuing east to

Yakutsk in Siberia. This was the route of Lend-Lease planes to Russia. A steady stream of A-20, B-25, and P-40 aircraft, numbering some 8,000, flew through Yakutsk during the war. Talking to his crew at a refueling stop, Tom learned what they thought of him. All of them to a man would prefer to volunteer for combat assignments rather than continue to report to him[7]. This reflected his entitled leadership style, the eagerness to please his superiors, and the borderline nasty way he interacted with his subordinates. His volatile temper, a trait that he shared with his father, did not help. He had received his first lesson on managing people and set about earning their trust.

Tom received significant promotions, rising quickly to lieutenant colonel. As an aide to General Bradley, he accompanied the general to the White House for a meeting with Harry Hopkins, Roosevelt's key assistant. Hopkins was vetting Bradley to be ambassador to Russia; Roosevelt ended up selecting Averell Harriman instead. General Bradley opted to retire and General Junius Jones replaced him, a real "fuddy duddy" in Tom's words. He flew the general to Burma to check on the status of Claire Chennault's supply mission to the Chinese. This was the famous "Hump" route, where lumbering C-47 cargo planes had to climb to 21,000 feet to cross the Himalayas. They stopped to refuel on Palawan, an idyllic island in the Philippines[8]. Tom would later name his yacht after the island. After taking General Jones on the dangerous Hump route, he asked Chennault for a permanent Hump assignment but the Army Air Corps turned him down. By this time late in the war, he had amassed about 5,000 hours of flight time.

With the end of the war nearing, he thought seriously about a career as a commercial airline pilot. General Bradley, counseled him to go back to IBM. Before mustering out, the Army Air Corps assigned Tom to a staff position at the Pentagon. This required management training at Fort Leavenworth. He would recall the education there as superior to his IBM training. He learned how to get along with all types of people. Moreover, he learned management from a staff position, including how to execute the directives of your manager. Once he rose to an executive at IBM, he would simply reverse those guidelines.

At the war's end, the senior Watson was running a vastly larger company with a workforce nearing 50,000. Tom returned to the company in January 1946. He was no longer the fun-loving, unfocused person that he had been when he joined the Army Air Corps. He had grown and matured with his wartime service. Yet, he was still a Watson and was given another prime territory. It was also in Manhattan, in an area that included the new Watson Scientific Bureau. He quickly made his quota and attended the 100% Club in early 1947.

For Watson, one 100% Club appearance was sufficient to prove his son's readiness to move up in the company. His father promoted him to vice president, one of five, and assigned him as an assistant to Charlie Kirk, his father's key executive and a potential heir apparent. Tom realized this was another step he did not earn. Kirk was a hard-charging executive and clearly wanted the top job when the old man stepped down. Now, however, he was tasked with tutoring Junior. Tom did not even have his own desk. Rather, he had to pull up a chair next to Kirk's big desk[9].

Tom was then 32 years old. He saw Kirk as an unpolished and ruthless executive. Kirk was comfortable in his role as the bad guy. Watson would tell Kirk to "handle" someone and he would usually end up firing the person. Kirk was also blunt. He told Tom about the used cash register scam his father ran for NCR, and his subsequent felony conviction. Shockingly, this was news to Tom. In 1947, Kirk and Tom traveled to Europe to visit IBM locations. While they were in Lyon, France, Kirk suffered a heart attack and died. Tom was now second in command of the IBM Company.

In 1940 on the eve of the global conflict, IBM revenues were $45 million with income of $9 million. The workforce was 12,000. The peak year during the war was 1944, when revenues were $143 million, income was $10 million, and the workforce had grown to 21,000. The company saw 70 percent of its employees who had served during the war return to the company. Watson was torn between reducing the size of IBM in return to peacetime operations or gearing up for more growth. He saw drops in business results in 1945 and 1946, but the following year revenues climbed to $139 million and income ballooned to $24 million. After 1947, IBM growth was explosive, with IBM posting revenue of $666 million and income of $73 million in 1955. The company workforce would reach 56,000.

There was a downside to losing Charley Kirk: Tom now had to deal directly with his father. This would be a problem, as both were headstrong and hot-tempered. At IBM headquarters at 590 Madison Avenue, sometimes referred to as "Galactic Headquarters," Tom's office was on the 16th floor and Watson was on the 17th. To Tom, at least two aspects of his father's IBM had to change. As the company continued to grow, his father's one-man organizational structure with roughly forty direct reports was not sustainable. In addition, he would have to find a way to address the cult-like atmosphere that his father had built up. He had seen this in full bloom at the 1947 100% Club, where 850 sales representatives made the club (out of roughly 1,000 reps). They would see an ego almost unbounded, with the songs, tributes, adulation, and the lengthy concluding oration. For now, the two Watsons would have to find a way to work together.

IBM's slow start in electronics was not a new thing. Ben Wood had prodded Watson on the subject in 1938, alluding to the speed of light as the only limiting factor. This was still a tough sale for a hugely successful company that was 100 percent dependent on electro-mechanical technology. The wartime experience of many IBM engineers would gradually turn the tide. Ralph Palmer and Steve Dunwell were particularly influential. They briefed Watson, painting a picture of how far behind the company was and the dangers of missing this technology shift. Palmer was in charge of an electrical team in Endicott, but initially with just four engineers.

Tom pressed for the transition to electronics. Besides Palmer, he had Cuthbert Hurd and Jim Birkenstock in future programs and Wally McDowell, the Endicott lab manager, all pointing the way. Birkenstock went so far as to say that card machines were "doomed." In 1946, Tom visited the Moore School to see the ENIAC computer. He admitted that at the time he did not see the future clearly. He concluded that only very large research organizations would want such a monstrosity. Referring to the ENIAC, Tom would much later question, "Why I didn't think, Good God, that's the future of the IBM Company.[10]"

By 1949, Tom had religion on electronics. He declared that all

IBM machines would be electronic within ten years. He would have to keep up the pressure to make that come true. That included moving his father to electronics. In frustration, he would tell Watson, Sr. that all he had in Endicott was, "a bunch of monkey-wrench engineers[11]." Even by 1952, IBM electronic sales were only $9 million out of a total of $334 million, not even 3 percent[12].

Tom Watson was effectively running the company by 1950, though any major decisions would still be made by his father. Tom had a trusted individual on his executive team, his CFO Al Williams. Williams joined IBM in 1936 in sales but moved to become the company's comptroller in 1942. He served on the War Production Board during the war, the group that would manage the "Arsenal of Democracy" spending of $183 billion. Returning to IBM, he quickly rose to the top financial position in the company. In 1950, Williams did a study of R&D investment by IBM compared to several other similar companies, including RCA and GE. They were spending three percent of revenues on R&D, compared to IBM's 2 percent[13]. Tom would eventually raise the IBM's R&D rate to 9 percent.

Tom was getting a similar message from Red LaMotte. LaMotte had joined the company before it was IBM as a service technician and rose in the ranks by switching to sales. He was instrumental in the Social Security Administration win during the Depression and eventually became vice president of sales. Red did an informal analysis of the current computer projects he was seeing in the field, which came to 19. IBM was not involved in any of them. He asked Tom what IBM's plan was.

Tom's initial answer was to hire more engineers. Endicott was still IBM's main laboratory, with 300-400 engineers. His father trusted a much smaller and older group, the ones least likely to understand electronics. Tom asked Wally McDowell, the lab director, to begin hiring, focusing on electronics. McDowell sought to clarify the request, asking Tom if he meant "a few dozen." Tom replied, "perhaps a few thousand.[14]" McDowell would increase the workforces at Endicott and Poughkeepsie from 1,000 to 4,000, with nearly 2,000 of the hires being engineers[15]. McDowell drove the increase but got no points on employee relations. According to one of the engineering hires, Arthur Samuel, McDowell was "a very hard-hitting man that nobody could get along with, who bawled his employees out unmercifully, and had a reputation of being a very hard man to work with.[16]" Nevertheless, he got the job done.

In addition, Tom changed the focus of the two departments that did future product planning. As James [17]Birkenstock relates, "From 1953 to 1956, the Product Planning and Market Analysis departments shifted IBM product and systems development emphasis toward electronic data processing systems and away from electromechanical punch card accounting machine systems.[18]" Birkenstock endured a bumpy ride in becoming Watson Jr.'s key marketing lead on electronics. He worked as a golf caddie at a country club in Burlington, Iowa. Several members sponsored him at the University of Iowa. There, a chance meeting with a visiting IBM executive resulted in a job offer. He joined the company in 1935. After the standard six-week sales training program, the company sent him to the St. Louis branch office. Working on his own, he closed a new punched card installation at a local grocery chain while still a sales trainee. This marked him early for advancement. He would also benefit by having Charley Kirk as his branch manager.

Kirk tapped him to rescue a failed card installation at an Army base. Success garnered a trip to Endicott to meet with Watson Sr. and a wartime assignment to train the Army on IBM's MRU (Mobile Records Unit) truck-mounted systems. There, he teamed with another rising star, T. Vincent Learson. Birkenstock was a successful branch manager in Kansas City as the war ended. He attended a branch managers' meeting in Endicott in December 1945 where he presented a plan to expand the number of branch offices to address post-war demand. Watson Sr. was at the meeting and promoted him to General Manager of U.S. Sales on the spot. Not realizing what his new prox-imity to Watson Sr. meant, he ended up at odds with Watson on several issues with Kirk poised to either fire or demote him. He planned to resign before Watson Jr. stepped in and offered him the Future Demands position, reporting directly to him. He would now have an outsized role in IBM's rise in computing and electronics. His treatment at the hands of the two Watsons was emblematic of the changing of the guard at IBM.

## IBM'S FIRST ELECTRONIC MACHINES

The company had to start somewhere with electronics. Tom asked Ralph Palmer to expand his existing electronics team in Poughkeepsie. At the time, Palmer's small lab worked out of the manor house on the former Kenyon estate[19]. He had his engineers working on the 407 and 408 accounting machines. The first electronic project would be a vacuum-tube version of the 602 Multiplier. The first multiplier, the 601 Multiplying Punch, was the Bryce electromechanical machine announced in 1933. It could multiply two numbers in about six seconds. The 602 Multiplier, announced in 1946, was also purely electromechanical, though it used relays in place of Claire Lake's counter wheels. It performed all four arithmetic functions, not simply multiplication. It was not very reliable and IBM quickly replaced it with the 602A, which was dubbed the "602 that worked.[20]" It was also the end of the line for mechanical calculation at the company.

Halsey Dickinson had been working on electronics for IBM machines since the late 1930s. Using the counter wheels of the 601 as a guide, he designed an add/subtract device with vacuum tubes. It was a ten-tube counter ring similar Presper Eckert's ENIAC counters. He submitted a patent application for the concept in 1937. Wally McDowell asked Palmer and Byron Phelps to join Halsey to produce a workable machine. Phelps added relays for the shifting of partial products in the multiplication process. He also switched to Binary Coded Decimal (BCD) for decimal representation. BCD was a modification of the zone-digit pattern on the 80-column IBM card that debuted in 1928. BCD only required four bits per decimal digit, cutting the number of vacuum tubes required in half. A demo version was ready in 1942 when the war halted progress.

When Tom Watson returned from service, he visited the Endicott lab. Dickinson had linked a high-speed punched-card machine to a black metal box about four feet tall. Tom asked what the box was doing. One of the engineers told him it was multiplying using radio tubes. The engineers explained that it could multiply ten times faster than the punched-card machine could read. In fact, the box spent nine-tenths of its time waiting for the punched-card mechanism to catch up. Tom had failed to see any immediate opportunity in the gigantic ENIAC, but Dickinson's little electronic black box excited him. "That impressed me as though somebody had hit me on the head with a hammer,[21]" he recalled. That moment marked his awakening to electronics and its potential role in future IBM products. He told his father that IBM should produce Dickinson's device immediately.

Dickinson and Phelps went to work on a complete machine, using an overall design done by James Bryce. The resulting device was the IBM 603 Electronic Multiplier, announced in 1946. It used the Halsey circuits and the Phelps modifications, and had 300 vacuum tubes under the covers. It was actually the first mass-produced commercial electronic calculating device. Despite the achievement, there were a couple of problems. During development, Palmer expanded the multiplication function and dropped division. The machine was also hard to manufacture and maintain, as it used large, full-size vacuum tubes. IBM pulled the 603 from the market after eight months and production of just 20 units.

Ralph Palmer tasked Steve Dunwell with revamping the 603. At war's end, the Army's SIS intelligence group held Dunwell back until they were confident they could run the punched-card operation without him. He did not return to IBM until August 1946. He left the service with the rank of lieutenant colonel and received the Legion of Merit for his service. Dunwell returned with unique insights into the use of electronics, along with firsthand knowledge of the Enigma Bombes and the Colossus computers. Once on board, he started on the "602 that worked" and then the IBM 603. He met with Dickinson and discussed the "hot rodding" modifications he had made at SIS, including a machine that could read a card and punch the result into the same card. He asked SIS to send one of those machines to Poughkeepsie. Palmer added one other engineer, Jerrier Haddad, to the program. He had joined IBM in 1945 after getting his electrical engineering degree from Cornell. Despite his youth, he stepped in as overall designer.

Their goal was to add the missing arithmetic functions—add, subtract, divide, and cross foot—and build in programming capability. Addition remained at one machine cycle, with multiplication taking six cycles and division anywhere up to sixty cycles. They raised the clock speed from 35,000 cycles per second (30 KHz) to 50,000 cycles per second[22]. The performance was outstanding, with 2,000 additions, 70 multiplications, or 60 divisions each second. The transformative change was in programming. Basic functions still used the plugboard, but the team supercharged this by enabling up to sixty instruction steps to run with each punched card read[23].

To facilitate the increased function, the number of vacuum tubes increased from the 300 in the 603 to 1,250 in the new machine, the 604[24]. The 603 was a mess to build and service with 300 tubes. How could over a thousand tubes not make the situation worse? The answer came from Ralph Palmer. He designed a new tube that combined the vacuum tube, resistors, and capacitors in one self-contained package called a "pluggable unit.[25]" In addition, the pluggable unit was far smaller than the vacuum tubes on the 603. The net result was a machine with a smaller footprint that was easier to manufacture and service. However, Palmer's innovation did not end there.

Figure 25. IBM 604 pluggable unit with the compact vacuum tube on the left and the resistors and capacitors on the right. Courtesy of Norwegian Museum of Science and Technology.

Palmer meant to address a long-simmering issue with vacuum tubes. As Atanasoff and Berry had discovered years earlier, the major commercial suppliers designed their vacuum tubes for analog radio use. That was where the volumes were. They were less than ideal for digital applications. Coupled with the need for higher quality and reliability, these issues led Palmer to explore manufacturing vacuum tubes in-house. He set up a lab in that former pickle factory on the Hudson that had housed the Munitions Company for IBM during the war[26]. He felt that a focus on quality controls would yield more reliable tubes.

He hired Arthur Samuel, an international expert in vacuum tubes[27]. Samuel had gone to MIT, where he worked with Vannevar Bush and the differential analyzer. After completing his master's degree in 1928, he joined Bell Labs. He spent the next 15 years there, working for the most part on vacuum tubes while amassing many of his 57 patents. In 1946, feeling underappreciated and underpaid, he joined the faculty at the University of Illinois as an electrical engineering professor. With the unveiling of ENIAC that same year, he pressed the university for a computer. He met with Mauchly and Eckert. He made the rounds of MIT, Harvard, and Penn. He discussed computers with John von Neumann at the Institute for Advanced Study. Putting what he learned in a proposal, Samuel secured $110,000 to build a computer. This would become the ILLIAC project. Meanwhile, he decided he wanted to work closer to computers. He contacted the only person he knew at IBM, Wallace Eckert. This led to an interview with John McPherson and then a meeting with Watson. IBM paid him $20,000 salary, four times what he earned at Bell.

Working out of the pickle factory, Samuel soon realized it made no sense for IBM to get into the vacuum tube business. RCA, GE, Sylvania, and others focused on it full time. He was also tired of vacuum tubes, and well aware that the Bell Labs transistor would soon completely eclipse the technology. His rejection of Palmer's plan to manufacture vacuum tubes prompted a meeting with Watson. It did not take long before his explanation was far too technical for Watson and he asked Samuel to take it up with Tom. Samuel was surprised how little Tom really understood about electronics and computers[28]. He spent more time with him and found him to be a quick study.

IBM announced the 604 Electronic Calculating Punch in 1948. The business plan called for a modest 75 units. Over the next ten years, IBM would lease 5,600 of them[29]. That would require 1,250 tubes for each machine, or over seven million vacuum tubes, plus a significant number of replacements. Palmer no longer needed to build tubes in-house as manufacturers saw the numbers and now focused on producing vacuum tubes to IBM's specifications.

Watson was very impressed with the new machines. He frequently toured the engineering labs and plants and would on those occasions frequently stop when he saw something he liked. He would tell the managers to show his appreciation "in the usual way[30]." This meant salary increases for the key individuals involved. As for Tom, the success of the 604 further cemented the view that electronics was the future, and in that regard, IBM was ill-prepared and as it turns out, already far behind.

The 604 Multiplier was a solid first step, allowing the adaptation of IBM's "bread and butter" tabulating to scientific applications. However, this was a baby step when compared to the emerging challenges. The engineers that Ralph Palmer had worked with at Navy "G" in Dayton were off to a strong start at Engineering Research Associates (ERA) in St. Paul. The Navy wanted to keep those engineers together after the war and helped them set up the company. ERA was emerging as one of the leading computer companies in the United States. Even the British were farther ahead, working to commercialize the Manchester system. However, the direct assault by Mauchly and Eckert with the UNIVAC machine represented IBM's biggest challenge.

## NORTHROP AIRCRAFT AND THE CARD PROGRAMMED CALCULATOR

There was one large IBM account that was not looking for a computer: Northrop Aircraft. Despite the need for advanced computing, their punched-card operation just kept plugging along. In 1946, John McPherson organized a computation forum. Wallace Eckert, still a Columbia University professor at the time, described his Watson Astronomical Bureau setup with its custom control box between the accounting machine and the multiplier. Northrop decided to try a similar approach. Three individuals were instrumental in developing the concept at Northrop: William Woodbury, Greg Toben, and Rex Rice.

Woodbury had received his engineering degree at Caltech and spent World War II working at Remington Rand. He drove a taxi in 1947 before joining Northrop. His first project was a complex aerodynamics problem. He decided to explore using the existing IBM punched-card machines. Rex Rice had graduated in mechanical engineering from Stanford and worked on wing design for Douglas Aircraft. He moved to Northrop in 1946 to work in the same area. He would lead the punched-card project and be its key promoter. Greg Toben had an electrical engineering degree and spent time as a tabulator operator at a commercial company after graduation. Northrop hired him to manage their IBM machine shop. He developed an electrical engineer's understanding on the machines internals.

Their goal was to connect their IBM 405 accounting machine with the 603 multiplier in order to calculate missile trajectories. Much in the manner of the Steve Dunwell switch for Wallace Eckert at Columbia, they used long plugboard wires to connect the machines together. Critically, they changed the workflow to enable instructions and data to be punched onto cards and fully processed on the 405 accounting machine without manual intervention. The combination of card instructions and IBM 603 plugboard wiring produced a very powerful system. The punched-card information worked in concert with the plugboard to simulate a computer program, hence the eventual name given to the arrangement: "Card Programmed Calculator." Northrop nicknamed the setup "Betsy" and had it running in the spring of 1948[31]. For Northrop, Betsy represented an affordable computing system, a "poor man's ENIAC.[32]" The IBM customer engineer for Northrop had to have been involved in this. However, perhaps like Colonel Renault in Casablanca, one can imagine him saying he was "shocked, just shocked" that someone had been fooling around inside his IBM machines.

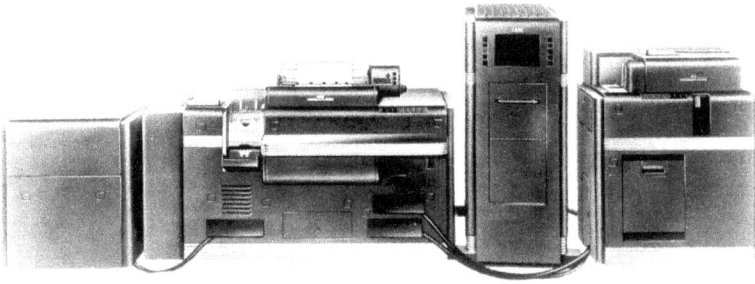

**Figure 26.** IBM Card Programmed Calculator. The "poor man's ENIAC" was based on the concept developed at Northrop Aircraft. Reprint Courtesy of IBM Corporation ©.

Northrop's director of missile guidance, George Fenn, was impressed with what his team had done with their IBM machines. He was told that IBM was enhancing both components of the Northrop system, the tabulator and the multiplier. He traveled to New York, along with Woodbury and Toben, to meet with John McPherson[33]. They wanted to upgrade the card-programmed arrangement with the new 604 Multiplier and 3402 accounting machine. They wanted IBM to build this new combination. McPherson flatly told them "no."

Fenn then produced a letter from Jack Northrop to Watson detailing the importance of the request to national security. Though Northrop was small compared to the big aircraft companies like Douglas, Convair, Lockheed, and Boeing, McPherson knew where Watson would come down on this. He told Fenn he would look into it. After they left, he pulled Herb Grosch into his office and the two sketched out a design on a flipchart that would become the IBM Card Programmed Calculator. Steve Dunwell took the sketch and developed a new design, using fewer relays and depending less on the plugboard. Dunwell passed his revised design to Jonie Dayger in Endicott to engineer and produce. Dayger would become far more famous later as the "father" of IBM's 1403 printer.

IBM flew Northrop's current machines to Endicott and transformed them into an official IBM Card Programmed Calculator (CPC). Northrop received the updated equipment six weeks later. Punched-card input worked in conjunction with the new 604 Multiplier and the 402 Accounting Machine plugboard to simulate extensive program logic. The card input supported seventeen mainly arithmetic instructions with no limitation on the length of a program. Conceptually, the CPC was similar in function to Watson's SSEC machine though far faster due to the electronics. Though not a stored program computer, it came close. It was more precisely defined as an "externally programmed automatic calculator." The advanced use of the plugboard would be a "precursor of microprogrammed control" that would become essential later with IBM's System/360 program[34].

George Fenn had demoed the Northrop system at IBM's 1948 Research Forum in Endicott[35]. Woodbury presented the combination at the 1949 Research Forum, where the formal announcement of the IBM Card Programmed Calculator was made. There were at least fifty prospects for the system before any specifications, prices, or delivery dates were completed. In the early 1950s, IBM would deliver over 700 CPC systems at the same time Remington Rand was shipping only 14 UNIVAC computers. IBM also produced high-speed versions of the CPC for the Army's Ballistic Research Lab. These were called the "Aberdeen machines."

The CPC was able to tackle very difficult applications, including engineering and scientific problems involving integration, differentiation, matrix multiplication, and simultaneous equations. Applications included wind tunnel analysis, jet engine thermodynamics, helicopter vibration analysis, carrier catapult take-off simulation, and missile trajectories. As an example, the simulation of the first 13 seconds of missile flight took 80 compute hours on the CPC, versus an estimated 3,000 hours to run the calculations manually[36]. The machine could do more. North American Aviation reportedly used its CPC at idle times to predict the results of horse races.

The aircraft industry led the charge. Boeing, Chance Vought, Consolidated Vultee, Curtiss Wright, Douglas, Fairchild Aircraft, Grumman, Lewis Flight Center, Lockheed, North American, Northrop, Republic Air, and United Aircraft all installed CPC systems. The Rand Corporation would install six or seven CPC machines before it got its JOHNNIAC, that von Neumann-architected computer[37]. Northrop would eventually have five CPC machines.

With the unexpected avalanche of CPC orders, it was soon

obvious that installation and programming would be prime customer hurdles. Watson turned to Cuthbert Hurd and his applied science department to address that. Hurd was one of IBM's early PhD hires and one with quite a resume. He had mathematics degrees at Drake and Iowa State College before getting his PhD at the University of Illinois. He did post graduate work at both Columbia and MIT before becoming a mathematics professor at Michigan State in 1936. During the war, he taught, what else, mathematics as the U.S. Coast Guard Academy. After the war, he served as dean of Allegheny College. He attended the Moore School lectures in 1946. He left academia in 1947 to become chief mathematician at the Oak Ridge National Laboratory. It was there Hurd set up and used IBM tabulating equipment, including the 602 (well, the 602A, the one that worked) and 604 Multipliers. He ran a shop similar to the CPC operation at Northrop.

Hurd hired John von Neumann as a consultant at Oak Ridge. Perhaps because of his association with von Neumann, he received an invitation to the SSEC dedication in New York City in 1948. He would have liked to run some Oak Ridge applications but the machine was already fully booked. Impressed by the SSEC and looking for a step up from punched-card management, he applied for a job with IBM. Given his stellar background, Tom Watson personally interviewed him and hired him to lead the new applied science department, providing an interface between customers and IBM research and development. He replaced McPherson as the key catalyst for new projects.

Hurd was described by a fellow IBMer, Herb Grosch, as a "natural executive," dynamic, one who hired well, delegated, was great with customers, and had excellent marketing vision. He was an "empire builder" in a good way[38]. He also worked well with Tom and with James Birkenstock. His hires would include future IBM luminaries John Backus, Fred Brooks, and Ted Codd, among many others. He was very knowledgeable in software and even taught a class in the subject for executives. As the new director of applied science, Hurd was also responsible for the SSEC installation.

Accelerating the success of the Card Programmed Calculator was one of Hurd's first challenges. He decided he needed a group of field technical reps deployed across the ten sales districts in the United

States. They would not be on quota, but assisted customers in their installation and use of the CPC. He hired young men with advanced degrees in mathematics and science and trained them on the system. Don Pendery was one outstanding example[39]. As a new applied science rep, his first assignment was to work on the CPC installation at Boeing. He was able to devise the mathematics and the needed CPC instructions to get the company's first application going. Reassigned to the Southern California district, he now worked in the center of the aircraft industry, with virtually every company taking delivery of a CPC system. He learned what each company was doing in computation and could feed additional requirements back to the Future Demands group.

The responsible sales executive, Vin Learson, was unnerved by these quasi-sales representatives and sought to have them put on quota plans, but Hurd successfully resisted[40]. Ironically, Hurd would take over Learson's position when Tom Watson promoted Vin to VP of sales. Hurd's "men" enhanced IBM's existing cadre of field technical support, the women in the System Service who supported punched-card installations. Hurd's team would become that model for future IBM regional technical specialists, or "top guns." The two groups would evolve into IBM systems engineers.

# THE RISE OF UNIVAC

Mauchly and Eckert continued to ramp up their new company, often in the face of withering criticism. Howard Aiken made fun of their plans for a commercial computer company, believing the potential market was microscopic. But, Grace Hopper did not agree. She left Harvard and went to work for Mauchly and Eckert in 1949. With start-up funding from Eckert's father, Mauchly and Eckert had formed the Eckert-Mauchly Computer Company (EMCC) in late 1947. Starting with a core group of about 36 employees, the company rapidly ballooned to 134 employees by the following year, with no orders on the books and no revenue coming in. An immediate priority became securing more cash.

The Navy Air Mission Test Center at Point Mugu, California, was the site of an early post-war computer project. The Navy wanted a system that could process telemetry data from missile launches. It put the system out for bid in 1947. Raytheon and EMCC responded. The Raytheon bid was $650,000, which it reduced to $350,000 and won the contract[1]. Robert Campbell had left Harvard and Howard Aiken's computer operations in late 1946 for Raytheon. Richard Bloch joined him there three months later. The two would lead the design team for the computer formally called the RAYDAC (Raytheon Digital Automatic Computer) but more widely known as the Hurricane[2]. Raytheon had an advantage in the bid process, as its computer was binary.

For EMCC, there seemed to be potential in the aircraft industry. Aircraft manufacturers were big users of punched-card machines for scientific computation and were always on the lookout for faster options. However, at war's end, contracts for new airplanes dried up. The U.S. Air Force was established in 1947, replacing the Army Air Corps. It needed to have an ongoing pipeline of cutting-edge strategic aircraft. Northrop Aviation was one company that could supply this need. Jack Northrop had started in the early aircraft industry as a draftsman, rose to engineer at Douglas Aircraft, and was chief engineer at Lockheed, where he designed the famous Lockheed Vega plane. He started his own eponymous company in 1939 and made significant contributions during the war. After the war, there were simply too many aircraft companies chasing too few contracts.

Mauchly visited Northrop in 1947 to discuss the use of EDVAC technology. Northrop was looking for a small navigation computer for use in an airplane. Given the size and weight of the ENIAC, it is hard to see how this conversation could have played out. The Air Force did not want to lose this small but inventive company and decided to fund the computer, called the Binary Automatic Computer, or BINAC. The negotiated price was $100,000. Critically for EMCC, the Air Force paid $80,000 of that up front[3].

BINAC would be the first in a string of EMCC contracts where the bids were far too low. Neither Mauchly nor Eckert had any real business sense. They were both academics. As president of the company, Mauchly was out of his league. Eckert had not matured much since his work at the Moore School. He would not brook any pushback on his ideas. He continued as a perfectionist, always looking to make one last change. However, now there was no Herman Goldstine around to press him to freeze the design and move on. Eckert did not give clear directions but expected the engineers to keep their heads down and execute his designs. There was little in the way of memos or documentation of the evolving plans. This was partly by design, to ensure there were no patent leaks.

In addition to raiding talent from the EDVAC project, EMCC assembled an all-star team of experienced, top flight engineers. One of those hired was Isaac Auerbach. He had gone to Harvard, where one of his professors was Howard Aiken. Before he joined EMCC, he thought it would be useful to spend some time at Aiken's Computing Center, working on the Mark I and Mark II systems. However, the minute he mentioned that he was going to work for EMCC, Aiken stopped talking with him. In Auerbach's telling, Aiken was a supreme egotist who believed he invented the modern computer and failed to get proper credit for it[4]. Furthermore, a major reason for that lack of credit was the accomplishments of Mauchly and Eckert, and their ENIAC computer. To Aiken, Auerbach had gone over to the "enemy," an unpardonable sin in his book. Aiken did not speak to him for many years afterward.

Auerbach offered an unflattering look at working with Eckert. In an early episode, he questioned Eckert about the very high clock rate of the mercury delay lines and its impact on reliability. He suggested that Eckert reduce the clock speed. Eckert declined. Auerbach was also brought in at the last minute to assist in closing new deals, including the pricing for the National Bureau of Standards bid. At $80,000, he told Eckert the figure was grossly low. Development of the UNIVAC I alone exceeded $1 million. Auerbach finally got too crosswise with Eckert and left the company[5]. He went to work for Irven Travis at Burroughs.

Patents were never far from Mauchly and Eckert's minds. An early hire was a patent lawyer. Contracts for early government customers were at their quoted prices rather than cost-plus, as was the standard. This was to ensure that EMCC retained all patent rights.

The completed BINAC used Eckert's mercury delay lines for 15,360 binary digits of main memory. Eckert opted for tape input/output in order to avoid dealing with IBM punched-card equipment. The tape decision would be a masterstroke decision. The high clock rate of the machine at four million pulses per second enabled the machine to achieve 3,500 additions or 1,000 multiplications per second. Yet, as Auerbach had tried to argue, it would come at a cost of reliability[6]. EMCC completed BINAC in 1949 at a final cost of $278,000, representing a net loss of $178,000[7].

EMCC conducted an acceptance test for BINAC in April 1949 at their facility in Philadelphia. Northrop tested a program to solve partial differential equations. The BINAC executed 500,000 additions and 200,000 multiplications in about five minutes. Jean Bartik, another former ENIAC programmer, said the BINAC ran for 43 hours without a failure. The machine was then disassembled, crated, and flown to California. Mauchly and Eckert traveled to Northrop to provide training[8].

This is where the overall concept of BINAC bordered on the bizarre. Northrop intended BINAC to be an airborne computer when fully assembled. Yet, it was clear from the start that it would be far too big and heavy to be loaded on even the largest planes, nor would it have withstood flight vibrations. Once at Northrop, EMCC left the system crated for several weeks. Finally, a Northrop engineer, not EMCC, set up the machine. The BINAC never became operational. It would operate for a few minutes, then fail. In this respect, it mirrored the early reliability problems experienced with the EDVAC. Apparently, in a move to help EMCC stay solvent, Northrop had agreed to a number of cost-saving changes, including the use of home radio vacuum tubes as the main ALU tubes. There was ample finger pointing between EMCC and Northrop. It would take Eckert over thirty years before he would admit that he had set the clock rate far too high.

Surprisingly, Northrop agreed to pay the remaining $20,000 balance of the contract, though the Air Force covered the bulk of the cost. Northrop also agreed to allow EMCC use the machine to test the mercury delay line memories in order to satisfy a contractual benchmark for one of its next customers, the National Bureau of Standards. Northrop further agreed to a public demo of the system. This was attended by James Birkenstock, Steve Dunwell, and Byron Phelps of IBM. They were able to report back to Watson Jr. on the scope of the BINAC computer as well get a sneak peek at the general plans for the upcoming UNIVAC (Universal Automatic Computer). Northrop soon moved the machine to a warehouse and never used it again. Word trickling out of this debacle did not help Eckert and Mauchly's fledging company.

Though the Air Force provided most of the funding, it seemed odd that Northrop would roll over on the BINAC as it did. Part of the explanation was politics within the company. There was, in fact, another computer project within the company at the same time. Nearby UCLA had recently produced a Bush differential analyzer, and an engineering team from Northrop made the pilgrimage to see it. This led them to work on the concept of a digital differential analyzer, something that Irven Travis had suggested to General Electric back in 1940. The system needed to solve differential equations to support real-time guidance for the Snark missile[9]. Despite the arrival of BINAC, the engineers charged ahead and built the machine -- the Magnetic Drum Digital Differential Analyzer (MADDIDA).

They wanted a formal evaluation of MADDIDA and who else would do but John von Neumann? They air shipped the machine from California to New Jersey and trucked it to Institute for Advanced Study. They set it up in a hotel room at the Princeton Inn. Von Neumann was most impressed that MADDIDA made such a trip and was up and running in 24 hours. He invited the Northrop engineers to his house for a round of his famous and very dry martinis. Von Neumann sent a letter to Jack Northrop with a positive evaluation, repeating his amazement that the machine travelled by airplane[10].

The MADDIDA and BINAC teams at Northrop collided at the highest levels of the company, with the MADDIDA team upset about the decision to bring in EMCC. Northrop never used the MADDIDA computer in the Snark missile. Political maneuverings at Northrop led the MADDIDA engineers to exit and start their own computer company. They sold six MADDIDA systems before folding the operation and selling what was left to NCR. It represented NCR's late and less than promising start in the computer business. Northrop's management apparently took the BINAC and MADDIDA debacles in stride. As a small company in an industry of giants, it continued to look for inexpensive ways to address its computational needs. Those efforts paid off with the Card Programmed Calculator.

Meanwhile, BINAC had provided EMCC a temporary lifeline. Though produced at substantial loss, it helped cash flow during critical months in 1948. At the same time, the National Bureau of Standards was taking a leading role in a number of government computer projects, including the U.S. Census. The end of the war curtailed computing projects such as firing tables, but the lure of high-speed calculating machines remained. The unveiling of the ENIAC in 1946 and the early efforts by Mauchly and EMCC heightened awareness of the new technology. All three branches of the military were looking at new computer projects.

The National Bureau of Standards (NBS) was moving forward on its own computer system. Its administrators did not have much confidence in EMCC, particularly in the company's ability to develop the machine in an acceptable time frame. The bureau decided to develop its own computer, or actually two computers, one for NBS's Eastern operations and the other for the West. NBS East built the SEAC (Standards East Automatic Computer) using the mercury acoustic delay lines. Ralph Slutz, who had worked on the IAS program, was the lead engineer. Completed in 1950, the SEAC was the first operational von Neumann computer in the United States, even beating the IAS machine.

In the West, Harry Huskey designed the SWAC (Standards Western Automatic Computer)[11]. Huskey, a professor of mathematics and electrical engineering, had been an engineer on both the ENIAC and EDVAC. He had worked with Alan Turing on the ACE computer project in Cambridge after the war. Maurice Wilkes had asked him to work with his EDSAC team but Huskey decided to join the National Bureau of Standards instead[12]. Huskey had Freddie Williams from the University of Manchester consult on the Williams tubes he planned to use on the SWAC system. Completed in 1950, the SWAC cranked out 15,000 additions or 2,600 multiplications per second. It was the fastest computer in the world until IBM's 701 arrived. The National Bureau of Standards had developed quite a computer design-and-development operation. Soon, however, political pressures would put an end to their run.

Mauchly had already been talking to the Census Bureau. The bureau was very interested but asked the National Bureau of Standards to evaluate the EMCC proposal, starting in the spring of 1946. The Census Bureau transferred $300,000 in funding for the project to the NBS, which hired George Stibitz as a consultant[13]. He left Bell Labs for private consultancy after the war and was certainly not a fan of the ENIAC or Eckert and Mauchly. Stibitz recommended that the Census Bureau look for additional bids. Raytheon submitted a proposal but at a far higher price. Stymied in searching for an alternative to EMCC, the NBS funded another study, this time with Howard Aiken, George Stibitz, John von Neumann, and Samuel Caldwell. Von Neumann's relationship with Eckert and Mauchly was beyond repair by this time. Caldwell, who had been critical of ENIAC, remained true to his analog roots. Even as digital was being firmly established after the war, he would open his lectures with, "I have come to praise the analog computer, not to bury it," a paraphrase of Shakespeare[14].

Despite generally negative input from this group, the Census Bureau went ahead with a contract with EMCC in 1948, at a price of $270,000. By this time, EMCC already had spent an estimated $400,000 in development[15]. The Census Bureau would add a second order but it would be three years before they would see a machine. The one bright spot for EMCC was that the troubled BINAC program satisfied the Census Bureau's proof-of-concept requirement, despite its total operational failure.

EMCC would need additional contracts just to break even on the UNIVAC development costs. A. C. Nielsen ordered the second UNIVAC, and Prudential came through for the third, but at price of $150,000, which was again far below cost. By 1948, Mauchly had contracts for five systems—one at Prudential and two each at AC Nielson and the Census Bureau. With the new backlog, Presper was unrestrained in engineering. Coupled with the low bids, huge cost overruns were inevitable.

The governmental market, particularly the defense industry, remained a huge opportunity for the fledgling company. Continuing to use the Bureau of Standards as an acquisition advisor, the Air Force transferred $400,000 to the NBS, and the Army Map Service sent $330,000[16]. Mauchly, Eckert, and many of their engineers had high-level security clearances due to working on ENIAC during the war. Then in 1950, Mauchly and several of the engineers at EMCC abruptly lost their clearances. There was no explanation at the time. Years later, access to FBI files determined that a meeting Mauchly had attended years before included individuals that had ties to suspected Communist organizations. This was the height of the McCarthy era, with an overwhelming focus to push all suspected Communists out of the government. This led directly to cancellation of the two military systems on order.

Because Mauchly was the lead salesman for the government and defense sectors, his security revocation severely crippled EMCC sales efforts in these areas. The ruling forced EMCC to segregate Mauchly from engineering, breaking up his successful partnership with Eckert. This had the further effect of eliminating a potential curb on Eckert's engineering excesses. It also pushed Eckert into taking on a greater management role in the company, something that he could ill afford given the many engineering challenges.

With an expanded backlog and continued heavy engineering expense, EMCC was soon again under water financially and seeking additional funding. A search led management to American Total-isator, a New Jersey racetrack operator. Henry Strauss, the founder of the company, had invented the Totalisator, an electronic tote board that displayed the status of a racetrack's horse races. Strauss immediately saw the potential of an EMCC computer. He was interested in using the machine to calculate betting odds for the company's pari-mutuel operations. He arranged a financial rescue package that included a $62,000 loan and $488,000 in exchange for 40 percent ownership of EMCC[17]. It represented a brief reprieve. However, within a year, Strauss died in a plane crash, leaving no one to champion EMCC at his company. American Totalisator wanted out of its investment and the loan repaid.

Mauchly looked for companies to buy out Totalisator. EMCC approached a number of companies, without much interest. The two most likely candidates were IBM and Remington Rand. Mauchly and Eckert met with John McPherson at the Watson Lab in New York. Wallace Eckert advised McPherson to turn them down, which he did. Mauchly and Eckert then sat down with the two Watsons. Both Watsons were concerned about the continuing financial viability of EMCC. At the same time, neither was completely convinced of the market potential for computers of UNIVAC size. In addition, Watson Sr. was skeptical about tape[18]. He had a visceral distrust of tape versus cards. He could see and relate to holes in punched cards. Finally, Watson was also concerned with the antitrust exposure to IBM if he bought the major competitor in his industry. Neither Watson was very impressed with Mauchly, but Tom would offer Eckert a significant engineering position at the Poughkeepsie lab. Eckert turned him down.

Remington Rand was their second choice. Mauchly and Eckert traveled to Florida and met with James Rand on his yacht. Rand had steered a far different path to this point than had Watson. He was unprepared for the Depression and saw his company's revenues plunge 75 percent. The government sued him in 1935 for stock manipulation[19]. He personally directed vicious strikebreaking in 1936. The Justice Department sued him in 1939 for antitrust violations. He also faced a shareholder suit the same year for his lavish spending. His company became a major military supplier during the war, which resuscitated the company's fortunes.

Rand was interested in getting into this new field, and here was a low-risk opportunity to take the plunge. He made a modest offer. It was a lifeline, but Mauchly and Eckert gave up most of their company in order to secure it. In February 1950, American Totalisator sold EMCC to Remington Rand, which became its UNIVAC division. The deal left Mauchly and Eckert with just $18,000 salaries and 5 percent of UNIVAC profits should any materialize[20]. That was unlikely, as EMCC had booked the first six orders at an average loss of $150,000 each.

At roughly the same time, James Rand doubled-down on his investment in computers by purchasing Engineering Research Associates (ERA). ERA was now one of three divisions in Remington Rand. EMCC was in Philadelphia, the newly acquired ERA was in St. Paul, and Remington remained in Norwalk, Connecticut. Rand hired Leslie Groves, the general who had managed the Manhattan Project, to run all three divisions. In addition, Rand would take advantage of General Douglas MacArthur's loss to General Dwight Eisenhower in the run-up to the 1952 presidential election and convince MacArthur to join his company as chairman of the board.

From the start, the UNIVAC and ERA teams did not get along. UNIVAC focused on technology while ERA focused on reliability and expanding commercial prospects. William Norris, the lead engineer at ERA, and Eckert at UNIVAC were constantly at odds. Eckert looked down at ERA, considering its machines to be far short of "state of the art." The third division, the original Remington Rand, was still producing tabulators in the shadow of big IBM. Despite having two sister divisions heavily steeped in electronics, this group was slow in upgrading its own technology.

Despite all the trials that Eckert and Mauchly faced, the UNIVAC program continued to move forward. However, it was quickly losing the advantage of being the first mover in a new industry. That changed dramatically with a proposal to lend CBS (the Central Broadcasting System) a UNIVAC for 1952 presidential election coverage.

UNIVAC serial #5 was bound for the Lawrence Livermore National Lab. The new lab had convinced Edward Teller to move from Los Alamos to California. He would be able to continue his work in developing the hydrogen bomb, the so-called "Super". Once he arrived, John von Neumann had advised his fellow "Martian" that his first order of business was to order a fast computer. While being tested in Philadelphia, Livermore temporarily loaned the system to CBS for analysis of the 1952 presidential election. Security remained tight, with the code name for the Livermore system was supposedly "Kidneyless,[21]" a play on "Livermore." Mauchly, still without security credentials, had Max Woodbury, a University of Pennsylvania statistician, join him at his home, and together, they drafted the programming to analyze election results as they came in.

The latest predictions prior to the election had the contest between Dwight Eisenhower and Adlai Stevenson rated a toss-up. Yet, with less than 5 percent of the vote, the UNIVAC predicted a landslide Eisenhower victory. Presper Eckert was sitting side by side with CBS anchor Walter Cronkite at the time. CBS concluded that something was wrong with their borrowed "brain" and did not air the prediction. It turns out the prediction was dead right. The UNIVAC had the results of the Electoral College as 438 to 43 for Eisenhower[22]. The actual numbers were 442 to 39. The popular vote prediction was also right on the money, at 33 million for Eisenhower. CBS had Charles Collinwood come on later in the broadcast and explain just how prescient the UNIVAC had really been. It was a game changer for UNIVAC.

Remington Rand realized very early that UNIVAC pricing was

far too low. The company raised the price first to $500,000 and then to $1,250,000. Even that high figure may not have accurately reflected the actual costs. The Census Bureau was able to force Remington Rand to accept the original low price. They installed UNIVAC #1 in March 1951. It was an enormous, room-sized installation weighing 29,000 pounds and measuring 14 feet by 8 feet and 8.5 feet high. EMCC technicians pushed the installed IBM tabulating equipment to the side. Work began in converting the enormous repository of punched cards to tape. Interviewed much later, Watson Jr. was "terrified[23]" about the Census loss. The Census Bureau would end up buying two more UNIVACs.

In 1952, the Air Force and Army Map Service installed UNIVACs. The following year, the Atomic Energy Commission installed UNIVACs at Livermore and in New York. The floodgates opened in 1954 with installations at General Electric, Metropolitan Life, Wright-Patterson Air Force Base, United States Steel at Pittsburgh and Gary, Franklin Life, Westinghouse, Sylvania, Pacific Mutual Life, and two systems at Consolidated Edison in New York.

General Electric installed UNIVAC #8 in 1954 at their Appliance Park facility in Louisville, Kentucky. GE was no slouch when it came to computing. The company had worked with Irven Travis of the Moore School on the Bush differential analyzer and sent two of its engineers to the Moore School lectures. GE had built a showcase building for the new computer and then watched it sit empty for two years due to delays by Remington Rand in production. Nevertheless, the installation was important. GE was one of the first accounts to put the UNIVAC to work on standard business tasks such as payroll and order processing. Burton Grad, who would later join IBM and lead key software initiatives, wrote a full manufacturing system for GE, including the complex use of MRP (Materials Requirement Planning). He did all of this programming in machine code[24].

Insurance companies were some of the earliest prospects for UNIVACs. Their operations were awash in millions of punched cards. Many were pleading with IBM to take action. Prudential was Herman Hollerith's first commercial customer, installing two tabulators in 1891. Edmund C. Berkeley worked for Prudential before the war. Drafted by the Navy, he ended up working for Howard Aiken on the Mark I. Back at Prudential, he pushed computerization and met with Mauchly and Eckert. Prudential had ordered a UNIVAC in 1948 but canceled the order when Remington Rand raised the price. Prudential did a review of the economics at that inflated price and determined it was not worth it. In addition, the lack of machines to convert punched cards to tape and tape to cards was a factor. The planned new applications at the company went on IBM tabulation equipment. Remington Rand had lost but at the same time, IBM had been put on notice.

Far more disastrous for IBM was the Metropolitan Life Insurance Company, another of Hollerith's early customers. Metropolitan had also tried to use punched-card machines from J. Royden Peirce before Watson Sr. bought Peirce's company and its patents. Met Life would eventually have one of IBM's largest punched-card installations, with equipment and punched-card inventories occupying three floors in its New York headquarters. It was the first insurance company to install a UNIVAC. It would order two more UNIVACs, at the new price of $1.5 million each. By the end of 1954, there were twenty UNIVAC systems in operation.

The UNIVAC I started with the EDVAC blueprint but, looking toward the commercial market, switched from binary to decimal. Eckert also used Binary Coded Decimal (BCD) to reduce memory and vacuum tube requirements. Though slimmed down from ENIAC, UNIVAC I still had 5,400 tubes and 18,000 diodes. Because of the ongoing unreliability of vacuum tubes, Eckert built duplicate error-detecting circuits to decrease the chances of data errors. Memory was Eckert's mercury delay lines, providing 12,000 digits of memory. The innovative use of the error-detecting circuits was completely offset by the utter unreliability of the mercury delay lines.

Of central importance was the decision to replace punched cards with tape. The machine's tape reel stored one million characters of data, the equivalent of 12,500 eighty-column cards. The initial printer was a typewriter unit, woefully inadequate for commercial accounts. Eckert's team added a high-speed printer at 600 lines per minute the far exceeded the 150 lines per minute speed of IBM accounting machine printers. Though marketed heavily to scientific users, UNIVAC was an excellent commercial data processing machine.

The machine implemented the von Neumann stored program architecture, which in the view of Eckert and Mauchly should have been called the Eckert-Mauchly architecture. At first, programming was only through direct machine code. Per Los Alamos physicist Robert Richtmyer, the UNIVAC machine language was superior. He would even describe it as "a marvelous thing.[25]" Richtmyer had worked on the famous Hippo program on the SSEC, so he had some credibility here.

The Census Bureau application work started early, with the first programs done by Mauchly. EMCC hired Grace Hopper and Elizabeth Holberton from the ENIAC and EDVAC programs. Hopper, who had literally written the book on programming for the Mark I at Harvard, would step beyond machine coding and blaze a new trail with assemblers and compilers. An assembler would substitute English pneumonics for binary machine instructions. One assembler statement created one machine language instruction. A compiler offered higher-level instructions that would each generate multiple machine code instructions.

The UNIVAC I was no speed demon. It was far slower than ENIAC in addition and only marginally improved with multiplication. UNIVAC II would arrive in 1958, adding core memory but retaining vacuum tubes. Addition was dramatically improved, but multiplication appeared to be a lost art, as the ratio of multiplication cycles to addition approached 100. As the UNIVAC II was firmly a commercial machine by this time, this was not a huge issue. UNIVAC was a cutting-edge electronic computer with built-in error checking, a high-speed printer, and, of course, the transformative tape storage. The biggest machine weakness at the start was the lack of full punched-card support, both for IBM 80-column cards and Remington Rand 90-column cards. However, these hardware deficiencies would pale next to the complete failure of Remington Rand sales.

# THE FIRST SUPERCOMPUTER

As the war ended, John Atanasoff had decisions to make. He had risen to director-level at the Naval Ordnance Lab (NOL) and managed a staff of one hundred in the acoustics area. With the lead taken by the BRL and ENIAC in high-speed computation, the Naval Ordnance Lab's administrators decided they also wanted to commission a large-scale computer. Atanasoff and von Neumann had met a number of times to discuss computer design. Von Neumann recommended Atanasoff to lead the project, and in late 1945, the NOL allocated $300,000 to get the project started[1].

The immediate problem was that Atanasoff already had a full-time job with the acoustics team, which would require significant time leading up to the Bikini Atoll atomic test in 1946. Atanasoff wrestled with his conflicting responsibilities. One would think that based on his groundbreaking work with the ABC machine he would jump at the opportunity to lead computer development at NOL. However, the project languished without his active direction. Perhaps he had lost the fervor he once had in Ames. Von Neumann attended many of the meetings of the NOL computer team. Eventually, he recognized the project was going nowhere and recommended cancellation. The NOL dropped the project in 1946. At the same time, von Neumann was able to use his considerable personal skills and reputation to get both the Army and Navy to fund his computer project at the Institute for Advanced Studies (IAS) at Princeton University.

Atanasoff finally completed his acoustics work in Washington, DC. He returned to Ames, Iowa, in 1948 undecided on his next step. He discovered that while he was away the University dismantled the ABC machine and discarded most of its components to make room in the physics building. Sam Legvold, a graduate student who had roomed with Clifford Berry, saved one of the capacitor drums. Many years later, Robert Stewart revealed that he was the graduate student asked to dispose of the machine because Dr. Atanasoff would not be coming back[2].

Atanasoff was being paid $10,000 a year at the Naval Ordnance Lab, a princely sum and far more than he would make returning to Iowa State as a professor[3]. The destruction of his legacy in the physics building may have helped with his career decision: He elected to remain with the Navy. In 1952, he founded Ordnance Engineering Corporation, an engineering firm whose projects included substantial work for the Navy. It would make him comfortably wealthy.

After the cancellation of the Naval Ordnance computer, the Navy turned to its Navy man at Harvard, Howard Aiken and funded the Mark II and Mark III. However, the Navy passed on Aiken's Mark IV proposal and went out to bid for a new computer for the Ordnance Lab. IBM won the contract in September 1950. It was essentially a "cost is no object" project. Watson agreed to take a $1 profit at the end. McPherson got buy-in from Tom Watson because of the leading-edge nature of the machine, to be called the Naval Ordnance Research Calculator (NORC).

To plan for NORC, Navy representatives met with the IBM contingent of John McPherson, Wallace Eckert, and a recently hired engineer, Byron Havens. According to Herb Grosch, Eckert's successful wooing of Havens was a "great prize" for IBM[4]. Havens was an utter anomaly in these rarefied technical circles. He had grown up in Pasadena and received an electrical engineering degree from Caltech. He became the head of the electrical engineering department at a university despite having no advanced degree. During the war, he worked at the MIT's Radiation Lab, alongside I. I. Rabi, the renowned physicist of the Manhattan Project who won the Nobel Prize in

Physics in 1944. Rabi highly recommended Havens, and Eckert hired him. The only chink in his substantial armor was that he was self-conscious about his middle name ("Luther")[5]. Havens would be the lead designer on NORC, with Wallace Eckert as overall manager.

The goal for the machine was ambitious. They aimed for a computer that ran at 200 times the SSEC and ten times the performance of any other contemporary computer. IBM would build the system at Columbia University's Watson Lab, now housed at a larger building on West 115th Street. The key to the new computer was a new circuit designed by Havens called the microsecond delay unit. As the name implies, it ran at one-millionth of a second, or one megahertz. It was the ultra-fast electronic equivalent of the one-second washing machine motor on the ABC or the 50-foot rotating shaft in the Harvard Mark I. The circuit took electric pulses degraded by passage through many logic gates and retimed them to pass at exactly one-microsecond intervals. With this circuit, the entire central processor of a computer, containing thousands of components, remained in synchronization, thereby providing highly reliable operation.

Work started in 1951. The team led by Havens grew to sixty. They worked on the third floor of the Watson Lab, at least in the beginning. The team soon outgrew that space, and Havens moved the operation around the corner into a larger facility, next to a Choc-Full-O-Nuts shop[6]. The NORC featured 9,800 vacuum tubes and 10,600 crystal diodes with over 200,000 total electronic components. It used both vacuum and Williams tubes for memory. The machine included one of the first input/output channels, an outboard processor aimed at offloading the central processor during the input and output functions.

Since multiplication remained the litmus test of high-speed computation, Havens' microsecond delay circuit would take the Bryce routine built into the IBM 601 Multiplier to a completely new level. As John McPherson described it:

> "The arithmetic unit featured fast multiplication using serial digit-by-digit addition and serial generation of the nine multiples of the multiplicand. A pipeline of 12-decimal adders, each of which introduced a

> microsecond of delay while adding, combined a digit from each one-digit product and output one digit of the result every microsecond. The multiplication of two 13-digit numbers required 31 microseconds.[7]"

That translated to over 30,000 multiplications per second. The machine's astounding combination of pipelining, parallelism, and fast circuitry produced multiplications at a rate that was just two times the rate of additions. By way of comparison, this ratio on most early computers was between 10 to 70 times. For the SSEC, multiplication (even with the dual partial products assist) was 67 times slower than addition.

Initially, programming the NORC was in machine code. This changed when Stan Poley created a new programming language called SOAP, for "Symbolic Optimal Assembly Program." Poley would go on to far greater fame by developing a version of SOAP for the IBM 650, the first computer to sell more than a thousand systems.

The Watson Scientific Bureau ran a number of applications before turning the computer over to the Navy. As usual, Wallace Eckert wanted to do more celestial computations. This time he was not satisfied with just the moon. He asked for the positions of the moon and all the planets in granular time segments between 1954 and 2000. This was just a ten-hour run on the NORC. A far smaller exercise that Eckert did on the SSEC in 1947 took over six months.

The Navy had a far more rudimentary test. They coded a simple program that performed two additions and a multiplication. This ran in 67 microseconds on the NORC[8]. The Navy's Mark II performed the same calculation in 83,000 microseconds. The Mark III took 7,100 microseconds. By this simple measure, the NORC was 1,250 times faster than the Mark II and 100 times faster than the Mark III. It was certainly over 200 times faster than the SSEC, as had been the goal. Gladys West, a Naval Ordnance mathematician who started her career with a Marchant calculator, became one of the NORC programmers. She was amazed, saying, "We felt the speed.[9]"

The Navy announced the NORC at a press conference in December 1954. Von Neumann was on hand, of course. Watson's old friend Ben Wood was also there. The original cost was estimated cost at $1.3 million. The final tally was $2.5 million. IBM took its $1 profit[10]. The Navy installed NORC at its Dahlgren lab in early 1955, joining the Mark II and Mark III, and an IBM Relay Calculator.

Tom Watson had green-lighted the NORC because he was interested in gaining experience at the leading edge of computer technology. It more than satisfied that goal. Elements of the NORC design went into the other scientific computers IBM was building at the time. They included Havens' microsecond delay circuit. The NORC established IBM's reputation at the frontier of computing. Possibly the first supercomputer, it was the fastest computer in the world for the next ten years. It had the fastest multiplication times ever for a vacuum-tube-based system.

# IBM'S 700 SERIES VAULTS TO #1

Tom Watson was committed to electronics and started not one but two separate projects in 1949. Poughkeepsie worked on The Tape Processing Machine (TPM) to address the UNIVAC threat. Endicott owned the Magnetic Drum Calculator (MDC) aimed at the growing number of drum-based competitors[1] like ERA.

Then, the Korean War intervened, beginning in June 1950. Watson did not hesitate to make his company available to the military. Though he had funded the Mark I and SSEC, he now wanted projects that could carry themselves commercially, even a wartime project. He asked Cuthbert Hurd, Ralph Palmer, and James Birkenstock to determine how best IBM could contribute.

Hurd created a list of sixty defense-related companies that might be a prospect for a fast computing machine. He joined Birkenstock on twenty of the visits, including Los Alamos, the Ballistic Research and Naval Ordnance labs, Wright Field, and Boeing. They came away with the conclusion that IBM could address 90 percent of the defense contractors' requirements with a single general-purpose scientific computer. This represented a huge break from IBM's history of one-off systems. Back in the office, they projected a $1 million cost to develop a prototype and $3 million cost for the entire program[2]. This was ten times the cost of the SSEC and three times the total R&D

expenditure in 1948[3]. Watson Jr. wanted more proof of the demand. Hurd and Birkenstock went back to their list of potential customers with a proposal that described a machine renting for $8,000 a month. They returned with eleven orders and ten "maybes." Despite the great potential, Hurd was surprised by the pushback he received. He faced entrenched opposition from the punched card forces who claimed that computers were too expensive, that too few would be sold, and the planned tape technology would prove unreliable. Surprisingly, even John McPherson initially lined up against the plan[4].

Work continued to build a firmer definition of the planned machine. Al Williams, the chief financial officer, ran the numbers and doubled the rental price to $16,000. The number of commitments fell to six. Hurd and Birkenstock still recommended that IBM proceed with the program and even pushed the development commitment to 18 systems. It would be an "all hands on deck" effort to complete it in the shortest possible time. Watson put the tape program on hold and many of its engineers moved to the new program. Watson named the new machine the "Defense Calculator." He continued to use the term "calculator" as he was concerned that "computer" would conjure something complex and intimidating.

Ironically, the lead executive for the Defense Calculator would be T. Vincent Learson, who had opposed the plan. He was emerging as a towering figure at IBM. This was not just for his physical size, though he was certainly was a big man, at 6' 5" and over 200 pounds. Rather, it was for his ability to get things done. The son of a Boston sea captain[5], he graduated from Harvard with a degree in mathematics in 1935. He had offers from both Remington Rand and IBM. He chose IBM in part because he recognized that IBM's electromechanical machines were superior to the purely mechanical Remington machines[6]. His choice was in spite of an attractive offer from Remington, which included a salary from day one. The offer from IBM only included compensation for room and board while in sales training, with no salary until he was on quota. In the depths of the Depression, Learson's father urged him to take the Remington offer. Instead, Learson reported to IBM Sales Preparatory Class 133 in Endicott. Upon graduating from his final sales class, he became a junior sales representative in Boston. He rose very fast, becoming a senior sales representative the following year. He loved sales and liked to call himself the "Boston peddler." He was very good at it.

In 1941, IBM management selected its two top unit record salesmen for a U.S. Army demonstration. The two selected were Learson and James Birkenstock, who was the top salesman in St. Louis. Watson promoted Learson to New York with responsibility for IBM's Army programs during the war. After the war, his ascent continued. He was branch manager in Philadelphia, followed by stints as the district sales manager in Detroit and sales executive for the Data Processing Division based in New York. He then took over the Electric Accounting Machine Division, which was responsible for developing the tape-base system.

Learson became famous for his direct, in-your-face management style. His "mere presence in a room was enough to get people's attention[7]." Tales of Learson encounters were legion. Harrison Kinney was a young, newly minted speechwriter for IBM. He drafted a speech for Learson to give at an event on the West Coast. He met Learson at the Westchester County Airport to hand-deliver the speech. He had his family in tow. Learson took a brief look at the text and forced him onto the company plane for a flight to California, leaving his family standing on the tarmac. Kinney made changes to the draft on the flight and then far more edits into that evening at Learson's hotel[8].

Chuck Branscomb was a young engineering manager in Endicott and a future senior executive with the company. His team had been working on a new sorter (the IBM 084) that would read 2,000 cards per minute. The project was running behind, and the team pushed for a slip in the schedule of three to four months. Learson called a meeting, and after some brief pleasantries he turned to Branscomb and said, "I am here today for you to tell me how we're going to ship machines in less than three months late, period.[9]" Branscomb was relatively new to IBM and certainly not fully updated on working with Vin Learson. Taking some offense at the tone, he responded, "The only way we're going to ship machines in less than ninety days late is for you to agree to ship at a lower reliability level than we think we should, period." You could hear a pin drop in the conference room. The other managers fully expected that Learson would fire Branscomb on the spot. Yet Learson agreed, and the product shipped

45 days late. Learson could be very tough, but above all, he wanted candor. He wanted the straight story and he wanted it quickly so that he could act on it.

Gene Amdahl, soon to be IBM's most prolific computer designer, says of Learson that he "listened long enough to decide, and then you almost couldn't shake him after that." You had to get to the point quickly and lead him to the right solution, the one you wanted to convey. If you meandered too much, you ran the risk that "he might arrive at his conclusion a little too soon,[10]" which could be disastrous. In another Learson first encounter story, a young executive who was familiar with Learson's reputation was giving him a presentation. Learson was slouched and expressionless. The executive stopped and exclaimed, "Sir, you just scared the life out of me.[11]" Learson laughed and waved him to continue.

Learson talked about his management philosophy in a 1982 interview[12]. Conceding that he was not a "technical man", he nevertheless concluded that he made "his reputation in IBM in managing research and engineering forces." He stressed the importance of a "good balanced mind . . . that will listen to the pros and cons of technical people when there's a choice to be made, frequently can make the right choice." He made no comment on his imposing style or his impatience with less than stellar performers.

Learson did not curb his competitiveness even in dealing with his boss. When he heard that Tom Watson would be sailing his yacht *Palawan* in the race from Newport to Bermuda, he sprang into action. He studied the race, had a leading-edge boat built, and hired the best crew. His boat, the *Thunderbird*, finished first out of 167 boats. Tom finished twenty-fourth and clearly was not happy[13].

His approach to managing people was simple: "Learson's direct. He expects people to perform" was one fellow executive's view. Bob Evans would summarize Learson's style as "gruff, tough, intolerant of bad thinking, instinctive, demanding, and wise. He assembled top-notch people for his team, grew them when they produced, and executed them when they had ample opportunity and failed[14]." Learson's own mantra was that executives should work "with a sense of urgency, a demand for excellence, and a healthy discontent for the way things are." Evans noted that Learson, "hated staffs, hated bureaucracy, hated garrulous people, and was as decisive a man as they come.[15]" As a senior executive for IBM, he kept a very small staff, not more than three or four trusted advisors. The bottom line was that whether you loved him or hated him, he got things done.

Cuthbert Hurd was familiar with the IAS computer at Princeton University and knew the team there, including Goldstine, Bigelow, and von Neumann. He had hired von Neumann at Oak Ridge, and he hired him again as an IBM consultant, with Watson's Defense Calculator as an initial project. Having worked with von Neumann extensively, Hurd was familiar with his idiosyncrasies. For openers, von Neumann liked fast cars though he was a notoriously bad driver. Hurd would have to handle the tickets that von Neumann amassed making the two-hour drive from Princeton to Poughkeepsie[16].

The Defense Calculator borrowed concepts from a number of sources. The foundation, of course, was the von Neumann architecture with the pure implementation done at the Institute for Advanced Study. The project took many elements from the TPM program. The addition of Nathaniel Rochester as lead architect brought more ideas. Rochester was an MIT graduate in electrical engineering. He had taken a course on computers taught by Bob Everett, one of five from MIT that attended the Moore School lectures. He left in 1943 to join Sylvania and worked on an advanced flight training computer called Whirlwind. Rochester joined IBM in 1948 despite having been unimpressed with IBM technology, starting with the SSEC machine.

At the start, there were discussions on whether there really needed to be separate machines for scientific and commercial applications. Von Neumann saw no reason for that distinction, assuming proper design. The proof was in the UNIVAC, which Mauchly and Eckert sold to both commercial and scientific customers. However, von Neumann's view did not win the day. The Defense Calculator would be scientific, the TPM commercial, and their many follow-on models would remain separate until the System/360 arrived in the mid-1960s. The Defense Calculator would be a binary machine with fixed-length instructions optimized for mathematical speed. The TPM tape system would be a decimal machine with variable-length data to accommodate business applications.

In December 1950, Tom Watson set up a committee to define the specifications. Detailed planning began the following month with a core team of ten. The engineering team would peak the following year at 155 engineers. They would complete the machine's design in February 1952, when Poughkeepsie shifted into high gear on the program. Ralph Palmer had overall responsibility, with Nat Rochester as lead architect. Jerrier Haddad supervised the build process. Though he had been with IBM just a short time, he would be managing a group of two hundred.

The project could not wait for completion of expanded lab facilities. Work started on the third floor of a tie factory and then moved to a deserted supermarket building[17]. The Defense Calculator team had to endure comments from the older electromechanical engineers along the lines of, "You know whose machines earn the money that pays your salary, don't you?[18]" They were talking, of course, of IBM's punched-card equipment.

Employees held Ralph Palmer in high regard. According to Roy Harper, a future IBM Fellow and one of the junior engineers at the time, Palmer was an inspiring leader. He cites one example of Palmer's management style. As the Poughkeepsie complex grew, Wally McDowell reminded Palmer that "No Smoking" signs needed to go up throughout the new buildings, just as had been the case when McDowell ran Endicott. Palmer got the job done. He mounted the signs behind each of the site's bulletin boards[19].

The completed Defense Calculator was a stored-program binary machine with the fast parallel architecture used by von Neumann at IAS. There were just 33 instructions. The binary implementation harkened to Atanasoff's add/subtract mechanism on the ABC. The engineering team wanted number accuracy of 10-12 decimal places, which converted to 35-41 binary bits. They chose 36 bits for the standard word size[20]. For addition, the system added 36-bit numbers in parallel, mirroring the same approach Atanasoff used with his 51-bit binary digits. Binary arithmetic was far faster than decimal arithmetic. The performance loss with decimal was 20 to 40 percent. In addition, binary meant a 20 percent reduction in storage space.

The engineering team expected to use the RCA Selectron vacuum tubes. When it became clear that RCA could not deliver, they switched to Williams CRT memory. There would be 72

Williams tubes, each storing up to 1,024 bits of information, providing over 2,000 words of memory. James Birkenstock negotiated patent rights for the technology from the British governmental agency NDRC (National Research Development Corporation) for $290,000. Von Neumann scoffed at this much memory, stating that any memory in excess of 1,000 words was "frivolous.[21]" The Williams memory provided fast random access. The design also included an Endicott-produced magnetic drum that provided 8,000 additional words of slower-speed storage. The logic circuits required 4,000 vacuum tubes and 13,000 diodes. Nat Rochester's engineers used Byron Luther Havens' microsecond delay circuit to synchronize system operation.

There was a greater focus on programming with the Defense Calculator. Programming began in binary machine code until the assemblers and compilers were ready. Given the 36-bit word size, two instructions would fit on one 80-column card. Nat Rochester wrote a symbolic assembly program called SO1. A rudimentary compiler called "Kompiler" (apparently no better name surfaced) translated English instructions into machine-code instructions[22]. Two customer installations, North American Aviation and Los Alamos, would also develop assemblers. There was no operating system for the machine, just a simple bootstrap program loader. It was one program at a time. The Poughkeepsie team tested a program for integrating differential equations and another to find a solution to Laplace's equations. Both ran fine and very fast. In August 1952, Cuthbert Hurd invited customers to Poughkeepsie. He encouraged them to bring test programs to run on the new machine. Los Alamos, Douglas Aircraft, Lockheed, and several others took him up on the offer.

As the planned announcement date approached, Hurd visited the twenty companies that had either ordered or expressed interest in the system. He confirmed the $15,000 monthly rental included the processor, magnetic drum, four tape units, card reader and punch, and a printer. Poughkeepsie sent the first machine to the 590 Madison Avenue showroom in late 1952, replacing the SSEC. IBM formally announced the Defense Calculator as the IBM 701 in April 1953 at an event in New York City. They invited over 150 scientists and business people, including lead speaker Robert Oppenheimer and, of course, John von Neumann. With Watson Sr. still overseeing the event, everything was first class. The *New York Times* reported on the event the next day with the lead, "Electronic Brain, '53 Model.[23]" The general press reaction concluded that IBM had finally delivered its first computer, with some news outlets calling it "IBM's UNIVAC."

Much to Watson's satisfaction, there was far more than just one Defense Calculator. Palmer's team produced nineteen 701 systems, with 18 going to customers. If there had been any doubt about the viability of large electronic computers, that ended with the 701. The Livermore Lab replaced its UNIVAC with a 701. The Los Alamos lab also took one. Eight systems went to the aircraft companies, virtually all of them except for Northrop, which would continue with its IBM Card Programmed Calculators. General Electric, General Motors, DuPont, and the Rand Corporation were the commercial customers. The balance of the machines went to government agencies—two to the U.S. Navy, and one each at the U.S. Weather Bureau and the National Security Agency (NSA). Ralph Palmer wanted to ensure that the 701 installations went smoothly. A trouble-shooting team was set up to assist customers. A young engineer named Bob Evans would make his first mark in IBM as a member of that team.

An acknowledged miss on the 701 design was the lack of floating point support. John Backus of the SSEC team in New York decided to address that. After his on the spot hire by Rex Seeber, Backus learned programming under fire, first with the SSEC and then with the 701. He led a team that programmed extensions for floating point as well as direct computation of square roots, sines, tangents, exponentials, and logarithms. They packaged the programs as Speedcode. Many purist machine language programmers looked down on Speedcode. It was true that certain Speedcode functions were slower than raw machine code, but the new high-level functions dramatically lowered coding hours.

The 701 made an appearance on television, in an early demonstration of the potential for computer-based Artificial Intelligence (AI). On February 24, 1956, Arthur Samuel, IBM's expert on vacuum tubes, pitted the 701 computer with his checkers program against a checkers master[24]. For Samuel, teaching a computer to play chess had become a lifelong avocation. It had also been the desire of Charles Babbage but he settled for Tic-Tac-Toe[25]. Samuel had read Claude

Shannon's 1950 paper "Programming a Computer to Play Chess[26]." He decided to tackle checkers, a far simpler program. He was aware of a checkers program written by Christopher Strachley[27]. Strachey used a differential analyzer to design radar vacuum tubes during the war. H worked on Turing's Pilot Ace computer at the National Physical Lab, where he wrote his first checkers program (known by the British as draughts). Constrained by memory, he talked with Alan Turing who briefed him on the Manchester Mark I. Strachey rewrote his program to take advantage of that machine's greater memory[28].

Samuel discussed his project with Nathanial Rochester and secured access to the 701 at night[29]. He was one of the earliest programmers on the machine, well before the assemblers and compilers were available. He took advantage of the 36-bit word size on the machine, mapping the bits to the 36 squares on the checker board[30]. He sought out checkers masters to comment on the design of his program, which he aimed to learn and consider all the possible moves in the game[31]. He would later describe his experience, noting ". . . a computer can be programmed so that it will learn to play a better game of checkers than can be played by the person that wrote the program. Furthermore, it can learn this in a remarkably short period of time (8 to 10 hours of machine-playing time)[32]".

Samuel's 701 checkers program won the televised matches handily. Tom Watson thought the contest would work wonders for IBM stock, which it did. It also generated very good ENIAC-style publicity for the company and represented an early milestone in AI. At the same time, IBM was somewhat jittery about portraying its computers as "brains.[33]" Samuel was not able to publish until 1959, when he released the paper, "Some Studies in Machine Learning Using the Game of Checkers.[34]" He coined the AI term "machine learning" to reflect how the computer mastered checkers through repetition and analysis.

## THE 704

The 701 program was still in process when one of Nathaniel Rochester's engineers, Gene Amdahl, was already looking at improvements[35]. Amdahl came to IBM with a sterling resume, though he never set foot in MIT or Harvard. He attended a one-room schoolhouse in South Dakota and got his undergraduate degree at South Dakota State College. He was a freshman when Japan attacked Pearl Harbor. He taught Physics as part of the Army's Specialized Training program[36] and then worked as an electronics technician with the Navy. He used the GI Bill to take graduate studies in physics at the University of Wisconsin. He was a summer intern at the Ballistic Research Lab in Aberdeen, working on both the ENIAC and EDVAC computers. Returning to Wisconsin, his faculty advisor assigned a problem to calculate nuclear forces. Amdahl spent 30 days with a desk calculator and slide rule, a very familiar story with graduate work in physics[37]. He decided to solve his own computing problem by designing the Wisconsin Integrally Synchronized Computer (WISC) as his graduate treatise. It used a magnetic drum from ERA with 6,000 bytes of memory, paper-tape input, and vacuum-tube circuitry. It had just ten instructions but those were just the ones he needed. The WISC computer could execute four instructions in parallel, an early indicator of Amdahl's computer design talent.

Amdahl graduated before his system was completed, but he sent his thesis on the design to the Ballistic Research Lab for evaluation. A copy ended up on Nat Rochester's desk, and he offered Amdahl a job at IBM in 1952. His excellent resume landed him a very high salary for a new engineering hire. Amdahl quickly found his footing in Poughkeepsie working on the Defense Calculator. He maintained a list of functions that he was unable to get into the 701, and he was ready when Rochester named him as lead designer for the 701 follow-on, the 704. He also benefited as "other experienced IBM designers were about to be committed to a joint development project with MIT to develop and produce the SAGE early warning system.[38]"

The original plan was an enhanced 701, to be called the 701A. It would feature a combination of the 701 instruction set augmented with new instructions and some limited architecture changes. This did excite Amdahl, who, like many early computer designers, always wanted to push the design boundaries. The program moved back into his wheelhouse when the project became a "clean sheet" system, the IBM 704. The new machine would no longer be compatible with the 701. At this early stage in computer design, there was no expectation of compatibility, in part because scientific installations valued new function and speed over compatible systems.

Amdahl's "fix" list for the 704 included floating-point support built into the hardware. It would be the first commercial computer to offer such support. This eliminated the need for the Speedcode program developed by the Backus team. Also on the fix list was index-ing, a set of registers added to automatically track program instruction address locations in order to return to the correct place after loops, branches, or subroutine exits. This capability provided integrated support for the "Wheeler Jump" that debuted on Cambridge's EDSAC computer[39].

The base rental for the 704 was $35,550 per month, meaning it would have been a roughly $1.5 million machine if purchase had been offered. The core rationale for the 704 was performance, and it did not disappoint. It provided four times the addition rate and doubled the multiplications per second of the 701.

There was more attention to software with the 704 program. The elimination of the need for Speedcode did not appear to bother John Backus. Rather he was looking at a project far more expansive than Speedcode. He had already discussed with Gene Amdahl how the addition of index registers would reduce the scope of developing a complete programming language and compiler. In 1953, he drafted a proposal to develop such a language for the 704. This would become FORTRAN, but it would take a few years to develop.

In the meantime, a new user group would fill in the void. In August 1955, Douglas Aircraft hosted a meeting of IBM 701 customers that were planning for their 704 arrivals. The meeting included Boeing, California Research, Curtiss-Wright, General Elec-tric, General Motors, Hughes Aircraft, Lockheed, the NSA, North American Aviation, the Rand Corporation, United Aircraft, and the Livermore and Los Alamos labs. Nat Rochester from IBM attended. The session would result in the formation of the user group SHARE.

It would play a key early role in future IBM systems and software support.

Roy Nutt of United Aircraft started the ball rolling with the Symbolic Assembly Program (SAP) he wrote for the 704. He distributed it all of the SHARE users. At a subsequent meeting, the attendees discussed the lack of an operating system on both the 701 and 704. Both systems sat idle while waiting for slow I/O operations to complete, resulting in more idle time than compute time. Customers typically used a sign-up sheet taped to the computer console to manage this incredibly expensive computing resource.

Robert Patrick at General Motors developed a basic operating system for the 704. He had started with the U.S. Air Force, running wing design calculations on a Card Programmed Calculator at Edwards Air Force Base. He moved to Convair in 1954 when the company was waiting for its 701 (#7). He traveled to 590 Madison Avenue to get 701 test times and develop programs using Backus's Speedcode, which he would call "friendly[40]." He then moved on to GM, whose Research Center was installing the last IBM 701 produced, serial number 17. It would not stay installed very long as GM would upgrade to the 704. Patrick's GM operating system produced a ten-fold improvement in jobs processed per hour[41].

At the same time, Owen Mock at North American Aviation developed a similar operating system. Both companies presented their competing proposals at SHARE. They subsequently agreed to combine their designs and split the cost. Joined by George Ryckman of GM Research, the trio developed the combined operating system known formally as GM-NAA I/O System for General Motors–North American Aviation Input-Output System[42]. Thankfully, users called it the SHARE OS, or simply SOS. The system wrote all input and output to tape for queuing. It added rudimentary job control statements (a forerunner of IBM's Job Control Language, or JCL) to define and separate each job. SOS also placed limits on programmers' direct access to registers and storage locations, a common source of system crashes.

Though Amdahl focused on the 704 design, the responsible marketing planner asked him what he thought the system's potential was. The preliminary business case expected six installations. Amdahl was insulted. He responded that at a minimum every 701 customer, a total of eighteen, would want a 704. The planner came back with an outlook of 18. Amdahl scoffed. They went back and forth until Amdahl agreed to 32. The actual number of 704 sales from 1955 to 1960 was 123[43]. Its outstanding performance propelled it to a near monopoly of large scientific computing. It was dramatically faster than either UNIVAC I or II. The new error-checking hardware led to a more reliable operation. Amdahl's 704 was the first computer where IBM was the technology leader and was not playing catch-up.

# PROGRAMMING: BEYOND MACHINE CODING

As the growth in computers accelerated in the early 1950s, an obvious bottleneck emerged—the difficulty in programming the new machines. With early computers focused on engineering and scientific problems, the mix of patch wires, switches, and even machine code was acceptable, if not preferred. Scientists in particular would accept almost any programming hurdle in order to get the most speed out of their system. Additionally, at the outset, computer memories were small, placing a premium on small and efficient programs.

The supposed "holy grail" of fast machine-language programs partially obscured the bigger picture. If you factored in programmer salaries and computer time used in debugging, the cost could be well over half of the total cost of operating the system[1]. On the commercial side, the story was different. Many new users were graduating from plugboard programming. Though wiring boards was not exactly easy, it was far more manageable than early coding languages. Plugboard programmers in punched card installations were loath to give up a skill they had labored so long and hard to master. Yet, computer technology accelerated forward and the dramatic improvement in memories, circuit speeds, and hardware costs resulted in a glaring light shining at the time and cost of programming.

The first baby steps were the subroutine libraries. Proposed by von

Neumann and Goldstine, Maurice Wilkes and David Wheeler imple-
mented them at Cambridge. Assemblers, where simple English terms
translated one-for-one to machine instructions, came next. Kathleen
Booth created the first assembler on the ARC (Automatic Relay
Computer)[2]. She and her husband Andrew made the pilgrimage to
Princeton and met with John von Neumann before returning to
England to work on the ARC. Wheeler wrote a rudimentary assem-
bler with one-letter mnemonics in 1949 for the EDSAC computer.

In the United States, Grace Hopper led the early progress. After
joining Mauchly and Eckert at EMCC in 1949, she pursued her
objective of creating an easier interface between programmer and
machine. By 1952, she had created the first quasi-compiler, called A-
o. It converted program statements into machine code while also
linking in subroutines. A-o was followed by A-1, A-2, and A-3. The
Remington Rand marketing department got involved at this point and
came up with a more exciting name for the A-3 compiler: ARITH-
MATIC. An enhanced version supporting algebraic statements was
developed and initially listed as AT-3, but again marketing interceded
and it became MATH-MATIC[3]. The progression culminated in what
Hopper had intended to do all along, the far more accessible and
general purpose B-o compiler. Marketing quickly rechristened it as
FLOW-MATIC.

## FORTRAN

Meanwhile back in New York City, John Backus set his sights on a far
more extensive algebraic language. When the IBM 704 with hardware
floating point eliminated the need for his Speedcode program, Backus
received funding for a full-fledged compiler for scientific computing,
released as FORTRAN (Formula Translation). At the time he started
the project, there were several other similar efforts underway. Besides
Hopper's MATH-MATIC, compilers called DUAL and SHACO
were in use at Los Alamos. Cambridge's David Wheeler had taken a
position at the University of Illinois and developed a similar compiler
for the ILLIAC. Much closer to Backus's concept was an algebraic

compiler at MIT. Halcombe Laning and Neal Zierler designed the compiler for MIT's Whirlwind machine. The Laning-Zierler compiler was better known simply as "George" (for reasons left to other histories).

Both MATH-MATIC and the Laning-Zierler compiler had the same issue—slow run times. Backus was an admirer of Hopper's A-2 compiler but not of its relative inefficiency and awkward syntax. In the summer of 1954, Backus saw a demo of the Laning-Zierler compiler firsthand at MIT, running on Whirlwind, and confirmed that despite executing on one of the fastest computers in the world, it ran slow.

This was the general view of what was called "automatic programming" at the time. Existing programmers felt that such compilers would slow down performance by "a factor of five or ten.[4]" Backus aimed to create a programming language that was fast. By fast, he meant that its execution speed would approach that of hand-coded programs in machine language. Backus felt that if FORTRAN was 50 percent slower, the language would be rejected.

With Cuthbert Hurd's funding in hand, he pulled together a team that included Irving Ziller, Harlan Herrick, Lois Haibt, and Robert A. Nelson[5]. Ziller had recently joined IBM and had been working on plugboards. Herrick, a mathematics graduate from Iowa State, had joined IBM in 1945 after teaching mathematics at Yale for eight years. He already had five years of programming experience on the SSEC and the 701 when Backus tapped him. He was initially disdainful of the project as it was not "real" programming in his mind. He enjoyed working in binary, dealing with the machines' registers, and getting close to the computer's "metal." Lois Haibt (nee Mitchell), a young math graduate from Vassar spent summers working at Bell Labs. She joined IBM as the company offered a salary ($5,100) twice what the Bell position paid[6]. Robert Nelson was a brand new employee assigned to do the technical typing for the team. He would later develop the core concept of VM, or the "Virtual Machine," and become an IBM Fellow.

This group developed a "marketing" document summarizing the planned compiler and toured the country talking with customers that had ordered the 704. The standard feedback was that they would not be able to get near the speed of hand-coded machine-language programs. Roy Nutt was one of the few customers that provide positive feedback. As a direct result, United Aircraft loaned him to the team. He was a great choice but somewhat unusual for this kind of project. He was a known as a "bit chaser[7]," a term used for someone who looked for machine code techniques to accomplish some task in the absolute minimum number of instructions, thus the lowest number of bits used.

The Backus team worked on the 19th floor of the building next to the 590 Madison Avenue headquarters. This top floor only had backstairs access. Their offices were next to the engine room. Machine time on a 704 was usually only available at night. They often rented rooms at the nearby Langdon Hotel to catch some sleep during the day before computer sessions at night. Development started with Hurd overseeing the project and progressed over several years through managers Charles DeCarlo and John McPherson.

The team split the compiler into six execution phases, with an owner assigned to each phase. The compiler (1) read the entire source and analyzed it; (2) then optimized DO statements and arrays; (3) consolidated the information in phases 1 and 2; (4) produced efficient instructions based on statistical analysis of instruction frequency; (5) transformed the flow into registers; and (6) assembled final program. The algebraic language and syntax were the easy part. Designing an efficient compiler was the challenge. This task would fall to Lois Haibt[8]. She would use the Monte Carlo method of frequency analysis to select the most efficient instructions. A very simple FORTRAN program illustrates the language:

```
WRITE (*,*) "Enter Fahrenheit temperature"
READ (*,*) FAREN
CELSIUS = (FAREN−32) / 1.8
WRITE (*,*) "Celsius is", CELSIUS
STOP
END
```

When executed, the program would prompt for entry of a Fahrenheit temperature, calculate the conversion, and print the result. The same program in assembler would require about 60 lines of very tedious code.

FORTRAN initially worked only with numeric data. It would soon need to accommodate alphabetic information as well. The Backus team accomplished this change by designating a prefix ahead of any alphabetic string. They called the prefix the Hollerith field, in honor of Herman Hollerith's place in IBM history. The FORTRAN encoding "3HABC" signified a three-position alphabetic field with the value "ABC."

It was not an easy development project. Backus's team of twelve labored two and a half years to produce a compiler containing 20,000 instructions. Once they completed the initial coding, extensive testing began. The new language debuted at the Western Joint Computer Conference in Los Angeles in 1957. IBM asked customers to suggest problems to test the new compiler. The local aircraft companies suggested computing airflow characteristics in jet wing design. The resulting program amply demonstrated the value of FORTRAN, both in reduced development time and in comparable execution performance to machine-coded programs.

By 1955, the compiler was in good enough shape to distribute to IBM 704 customers, of which there were 66 at the time. The binary compiler came in a box with 2,000 eighty-column cards[9]. It was nicknamed the "Tome." Two accompanying FORTRAN publications enhanced the reception and use by customers. Grace Mitchell wrote the "Programmer's Primer," and David Sayre developed the cogent 51-page FORTRAN Programming Reference manual. It became a model for future programming guides. One early FORTRAN programmer at GE estimated that the new language made programming at least five times faster. It soon became the mainstream scientific programming language. Cecil E. Leith called FORTRAN the "mother tongue of scientific computing." Kenneth Thompson, who would create the UNIX operating system at Bell Labs, would observe that, "95 percent of the people who programmed in the early years would never have done it without FORTRAN. It was a massive step."[10] It remains the primary language for supercomputers. Backus offered a clue about his motivation for the project in a 1979 interview with *Think Magazine*, saying, "Much of my work has come from being lazy. I didn't like writing programs." More and more machine language programmers came over to his way of thinking.

----

## COBOL

In 1959, a cross-industry effort embarked on creating a commercial equivalent to FORTRAN[11]. At the time, most languages other than FORTRAN were vendor-specific and focused on scientific applications. The Department of Defense (DOD) agreed to sponsor the program with the goal of a fully portable commercial language across vendor platforms. IBM had already developed a commercial equivalent of FORTRAN called COMTRAN (Commercial Translator), and Grace Hopper had followed up her symbolic assemblers with FLOW-MATIC.

The DOD brought together a somewhat unwieldy group with six computer manufacturers (IBM, Burroughs, Honeywell, RCA, Sperry Rand, and Sylvania) and three government agencies. The National Bureau of Standards chaired the group. At the outset, the two existing commercial languages vied to be the basis of the new language— FLOW-MATIC from UNIVAC and COMTRAN from IBM. The first meeting had 41 people in attendance, including Hopper and Betty Holberton (one of the original ENIAC programmers). There was a strong anti-IBM bias in the group, resulting in more elements coming from FLOW-MATIC than COMTRAN. The central theme of Hopper's FLOW-MATIC was its extensive self-documenting English statements. It was so verbose as to be bordering on the undecipherable.

This large group struggled to make progress. A core team of six (two from IBM, two from RCA, and two from Sylvania) met separately and developed the specifications. A late entrant in the language "sweepstakes" was Roy Nutt, who had left United Aircraft and along with Robert Patrick co-founded Computer Sciences Corporation. Nutt developed a programming language for Honeywell called FACT (Fully Automatic Commercial Translator) and suggested it to the group. Reaction was initially favorable, but its lack of portability forced the group to retreat to their existing design. Hopper was a prime mover, providing much of the new language's style, and traveling the country to sell it. The result was COBOL (Common Business Oriented Language). Its success was assured given its portability, the mandate of government agencies, and the dearth of good business languages. IBM invested in both COMTRAN and COBOL for a period of two years and then standardized with COBOL.

# THE IBM TAPE MACHINE – BETTER
# LATE THAN NEVER

M auchly and Eckert had launched a two-pronged attack on IBM's computing industry hegemony. Their new tape units would eliminate the need for massive numbers of punched cards. In addition, the UNIVAC I stored program computer had the potential to completely replace IBM's electromechanical punched-card machines. IBM had been monitoring this threat since 1946 and the very public unveiling of ENIAC. IBM's largest customers were signaling as early as 1948 that IBM needed to come out with a tape-based system. IBM knew that it needed to respond to the UNIVAC threat, but the company was addicted to the high profits and near monopoly of punched-card equipment. The old engineering guard was still highly represented in upper management, with a staunch supporter in Watson Sr. at the helm.

Ralph Palmer had moved from Endicott to Poughkeepsie to start the new lab there. He assembled a new team with heavy experience in electronics. The temporary lab at the Kenyon House estate had 23 rooms that were adapted as offices, conference rooms, and work areas[1]. Palmer added two more key engineers, Werner Buchholz and Jerrier Haddad. Buchholz attended school in England in 1940 while his parents remained in Germany. They died in the Holocaust. He immigrated to Canada and then on to the United States, receiving an electrical engineering degree at Caltech before joining IBM. Jerrier Haddad had worked with Palmer and Dunwell on the 604 Multiplier.

When Nat Rochester arrived in Poughkeepsie, he was stunned to find that there was not a single stored-program computer project in process[2]. He immediately started a series of lectures on the subject. His next priority was to create a prototype to help in the design and testing of stored-program concepts. Working with Buchholz and Morton Astrahan, two 604 Multiplier frames were combined. Williams tube CRT units and a magnetic drum built in Endicott provided memory. They called the contraption the Test Assembly[3]. Though it was not pretty to look at, it hewed closely to von Neumann's architecture. The finished Test Assembly first ran in 1950.

Watson had sidelined the TPM (Tape Processing Machine) with the total focus on his Defense Calculator. However, IBM already had twenty letters of intent for such a tape-based system, to rent for $8,000 a month[4]. The demand was there and UNIVAC was surging ahead, with fifteen systems already installed. As it was now a full-blown crisis, Tom Watson tapped Learson, his most able executive, to restart the Tape Processing Machine program. Steve Dunwell had become frustrated with management support for the program, and after a meeting with Tom Watson, he took a branch office technical job in Washington, DC. That lasted for just a year before he returned to Poughkeepsie and stepped back into the newly recharged tape-based program, which the team now dubbed "TPM II."

The delay on TPM actually proved beneficial. The experience with the 701 led to a complete redesign. Though the 701 and TPM II were very different in target customers, much of the engineering in the 701 also applied to the commercial system. That included the stored-program design, Williams tube memories, and pluggable electronics. The commercial-oriented internal architecture, designed by Rochester and Buchholz, was substantially different. It would be a decimal, variable-word-length machine, with computational speed a distant secondary consideration.

With the Tape Processing Machine, the first order of business had to be a tape drive. The initial drive developed read the data serially on one recording track bit by bit. It was far too slow when stacked up against the UNIVAC tape system. Engineering switched to a seven-track tape encoded in Binary Coded Decimal, but speed remained a problem. A key innovation designed by engineer James Weidenhammer, the vacuum system, solved the issue. Weidenhammer's vacuum unit would provide physical buffering of the tape during high-speed starts and stops. It allowed the tape drive to run at far higher rates (15,000 characters per second) and enabled IBM to leapfrog UNIVAC[5]. The tape media was plastic, an advance over the steel-based tape media of the UNIVAC. The 1,200-foot tape reels released with the 701 gave way to a 2,400-foot reel of half-inch tape that would become the industry standard.

Aside from the physical tape machinery, the core issue with tape was the fixed sequence of the data on the tape. Many commercial applications needed to sort or merge records before processing. Punched card processing solved this issue with sorters and collators. The Poughkeepsie team heard that UNIVAC had developed an efficient tape sort program. In addition, John von Neumann suggested a sort-merge technique in 1945, one that he and Herman Goldstine further refined in 1948. The technique was termed "divide-and-conquer." The program split the tape file into two and one by one, records from each file were compared. If they were out of order, the records were swapped. This routine repeated until the source file was in proper sort sequence[6].

In developing the 700 series of computers, it appeared that IBM had skipped the number "703." In actuality, IBM reserved the number for a dedicated tape-sorting system. It would feature a small processor with up to twenty tape drives attached. This would result in super-charging the von Neumann sort technique by splitting the input file into many more segments and running all the comparisons in parallel. According to Gene Amdahl, Poughkeepsie actually built three such systems and ended up selling them to the U.S. Treasury Department[7].

IBM announced the TPM in September 1953 as the IBM 702. It presented the new system to 100% Club members and the gathered press at the Waldorf Astoria Hotel in New York City. A television hook-up connected with the Poughkeepsie lab, IBM's first use of this media. The 702 received fifty orders almost immediately. It would be 18 months before the first delivery. The schedule did not work for

Metropolitan Life, which ordered two more UNIVACs. As production ramped up, Learson's challenge changed from production to damage control. Early product testing revealed reliability issues with the Williams tube CRT memory. The tubes periodically lost data bits. Williams tube replacement became a near daily chore for the IBM Customer Engineer[8]. This was a show-stopping issue, as the 702 architecture did not include data error checking, as did the UNIVAC. Learson ordered a production halt after the first fourteen systems. The same issue would also affect all of the shipped 701 systems, though occasional data loss was less of an issue with scientific systems. Learson pressed the lab for an alternative memory. One can only imagine how Learson "pressed."

Williams tubes had met the general criteria for main computer memories by providing fast, random access to instructions and data. It was a giant step up from Atanasoff's ABC drum and superior to the mercury delay lines Presper Eckert used in the EDVAC and UNIVAC. However, IBM concluded that the reliability of Williams tubes could not be substantially improved. Some other technology had to come to the fore.

Wallace Eckert had looked into using ferrite material to store bits of data beginning in 1946. Llewellyn Thomas on Eckert's staff continued the work, meeting with Eckert and Palmer in 1948 to share progress. In the meantime, an engineer in Howard Aiken's Harvard lab had the same idea. An Wang conceived of tiny rings of ferrite, magnetized to represent bits of data. The direction of the magnetic flow, either clockwise or counter-clockwise, signified whether the bit was a "1" or a" o." Three wires passed through each ferrite core ring— the X and Y drive wires that magnetized the ring in one direction or the other, and a sense wire to read the value. Wang found his tiny, donut-shaped ferrite cores at General Ceramics. Wang described his concept at a Harvard symposium in 1949 and then patented them as magnetic cores. Howard Aiken used them in his Mark IV.

Less than two miles away, the wartime Whirlwind computing project at MIT would also end up with magnetic cores[9]. In 1944, the U.S. Navy was looking for a more advanced flight trainer. The goal was a computer to support a mock cockpit, providing a simulation of real flight. The program, termed the Airplane Stability and Control

Analyzer, was a high priority for the Navy and resulted in initial funding of $875,000 in 1945, a sum that would escalate to an annual budget of $1 million.

MIT selected Jay Forrester, a graduate student, to lead the project. Jay's father, Marmaduke Forrester (thankfully known as "Duke") homesteaded in Nebraska. Young Jay attended a one-room schoolhouse before continuing on to the University of Nebraska. He then moved to Boston and MIT, where he would remain for the next 76 years. During World War II, he worked on an analog servo-mechanism team at MIT that the Navy planned to use to improve the speed and tracking of their anti-aircraft guns. Forrester became well versed in real-time analog controls. Perry Crawford was one of Forrester's fellow graduate students at the time. Crawford reviewed Vannevar Bush's Rapid Arithmetic Machine and wrote his dissertation on switching the Bush analog concept to digital. He promoted the concept in lectures at the Moore School and urged a similar approach with Forrester. Forrester was soon convinced enough to adopt a digital approach for the flight simulator[10].

At the same time, Forrester changed the design he had in mind from a general-purpose machine to a real-time machine. This added significant cost and time to the project. In addition, Raytheon, the prime contractor, added extensive development processes that added to the cost. The program was renamed Whirlwind, with a recalibrated price tag of $3 million, an enormously high figure for a computer at the time. The final tab would be $8 million, and it would take eight years to complete. In the end, the Navy would not end up with a flight trainer[11].

As Forrester redesigned Whirlwind, he became increasingly dissatisfied with Williams tube memories. They had an average tube life of one month, were expensive, and still did not provide the speed he was looking for. He explored other options. He heard about the magnetic cores designed by An Wang at Harvard. Now an MIT professor, Forrester asked one of his students, William Papian, to study the magnetic characteristics of the material for his master's thesis. Forrester then proceeded to develop a two-dimensional array using the ferrite rings. His design represented a significant improvement over the Wang design. Forrester asked another graduate student,

Ken Olsen, to build a prototype computer using the new cores, which proved the concept was solid. Forrester retrofitted Whirlwind with the new memory. It doubled the speed over the Williams tubes and more importantly, was completely reliable. Maurice Wilkes met with Forrester at MIT and saw the new memory. By this time, Wilkes was planning the EDSAC 2 computer for Cambridge. He immediately dropped mercury delay lines and electrostatic CRT memories in favor of the Forrester core.[12]

**Figure 27.** Core Memory Module – A 32 x 32 core memory plane
storing 1024 bits (or 128 bytes) of data.

Ralph Palmer continued to press IBM's core memory efforts at the Poughkeepsie lab. He asked Arthur Samuel to turn his attention from vacuum tubes to core memories. Samuel recommended the addition of

Mike Haynes, who had worked on the ILLIAC computer at the University of Illinois. Haynes had written his PhD thesis on magnetic memories[13]. He also discovered the small ferrite cores at General Ceramics, just as Wang and Forrester had. Joined by Erich Bloch, who would soon become IBM's lead engineer on memory technologies and packaging, the two developed an experimental core matrix and placed it in a 405 accounting machine. Upon getting reliable function, they enhanced the design to three-dimensional cores in 1954.

Unfortunately, IBM's progress came too late, as both Wang and Forrester had already secured patent rights. IBM paid An Wang $500,000 in 1955 for his patents[14]. Wang would use this windfall to form his own eponymous company, Wang Labs. Jay Forrester sued IBM for patent infringement. He sought a two-cent royalty for every core bit that IBM produced. Given that IBM might require several trillions of the bits, IBM would be looking at a payment of as much as billion dollars. James Birkenstock negotiated the dispute for IBM, which did not conclude until 1964. IBM provided a one-time payment of $13 million[15], a nominal amount compared to what the company would have owed with Forrester's formula[16].

Once core memory emerged as the answer, Learson did not hesitate. He set off a crash program to rework the 702 with the new technology. He also made the call to replace all Williams tube memories in the installed 701 systems. Learson's decision to switch IBM to core memories led to another problem: the high cost of producing the cores. Fabrication of the finished core planes required the stringing of three wires through each tiny ferrite core. The manual process to produce a single 4,096-bit core plane took 25 hours.

Ralph Palmer kicked off an effort to improve the manufacturing program. His engineers designed a device to automate core stringing. The time for stringing each core plane collapsed from 25 hours to 12 minutes. However, IBM's business cases and prices were already set based on the manual process. The automated process resulted in substantial profits for IBM. The dramatic reduction on the cost per core bit illustrated the improved efficiency. At the outset in 1953, the cost was 33 cents per core. With the new process, the cost went to 5 cents per core in 1954, and just 2 cents by 1964[17].

A total of fourteen 702 systems were produced. One went into the showroom at 590 Madison Avenue. The others went to commercial customers, including Bank of America, Chrysler, Ford, Commonwealth Edison, General Electric, Monsanto, and Prudential. Prudential had been Herman Hollerith's first tabulator customer and was one of the companies that spurned the huge price increase on the UNIVAC. It had patiently waited for the IBM tape system.

Though Learson's quick action averted disaster with the Williams tubes, a far greater concern emerged. The 702 simply did not measure up to the UNIVAC. It was marginally faster in raw computing speed, and the IBM tape drives were clearly superior. However, the UNIVAC outperformed the 702 in most commercial applications. It overlapped input, processing, and output, resulting in far more efficient total throughput times. In contrast, the 702 took each step serially, without the ability to buffer input or output. When those slow operations were active, the processor sat idle. The UNIVAC also had Eckert's built-in error-checking circuits, ensuring data integrity. The 702 lacked such a capability. Lastly, the UNIVAC was easier to program. This reflected the work of Grace Hopper and several of the women ENIAC programmers on the UNIVAC team.

Learson halted marketing of the 702 after the first 14 systems shipped. He moved rapidly to develop a replacement model, the 705, which focused on the identified weaknesses. Engineering added input and output buffering and error-checking circuits. They increased computation performance, resulting in a fourfold advantage in addition and threefold advantage in multiplication over the UNIVAC. The 705 had an additional advantage, the physical packaging. Each system component fit into an elevator, making installation far easier.

It did not hurt 705 sales when Hollywood featured the computer prominently in *Desk Set*, a 1957 movie starring Spencer Tracy and Katherine Hepburn. In an early example of product placement, the 705 appeared as EMERAC, the "Electronic-Magnetic Memory and Research Arithmetic Calculator", fondly nicknamed "Emmy[18]." Hepburn's character was Bunny Watson, possibly a reference to the "Old Man." IBM marketing outdid itself, dressing up the 705 with a wall-sized panel of blinking lights that far outshined the simple glowing Ping-Pong balls of the ENIAC.

The quick replacement of the 702 with the 705 received a

welcome but unexpected assist from Remington Rand. IBM did not expect the competition to stand still, but that is exactly what happened. With UNIVAC reps screaming for a response to the IBM 705, Presper Eckert was busy working on an advanced computer for the Livermore Lab called the LARC. For nearly a year at this critical time, he refused to divert any of his engineers to work on an enhanced UNIVAC model[19]. The UNIVAC II would not arrive until 1958, and by then it would be too late. IBM orders for large systems exceeded UNIVAC for the first time in 1955. IBM shipped the first 705 in 1956 to a historic IBM customer, the Social Security Administration. By 1959, the IBM 705 was the dominant large commercial computer.

On balance, Mauchly and Eckert, with Rand financing, had made enough of the right moves early on to capture the lead in large computers. They installed twenty UNIVAC systems by 1954. Remington Rand would produce a total of 46 systems[20]. However, a series of blunders left the door wide open for IBM. Above all, James Rand did not take advantage of his market-leading position, repeating the mistakes of the 1930s when IBM surged ahead in office equipment. He did not fully understand computers, especially what was required to sell them. He had a small sales force, split between punched card and computer sales. He had no plans to enlarge his sales staff to respond to a much larger market opportunity. He also did not use his punched-card salesmen to sell computers. In fact, if a Remington Rand computer salesman displaced a Remington Rand punched-card installation, the card salesman had to pay back commissions.

Rand also did not learn from the Watson sales philosophy, which instilled solution selling from the very start. If there was ever a need for a marketing representative with solution skills, it was during these early days of the computer. Leslie Groves also stepped in to help IBM. At the exact moment that UNIVAC was poised to control the new computer market, with the heightened visibility from the CBS election coverage and a significant investment in advertising, Groves decided to cut UNIVAC production. This sent a message to the sales force that despite the hype, the UNIVAC division was just not that important[21]. This was a sale force already seriously outgunned by IBM, not adequately trained in the new technology, and operating with conflicting sales incentives. A UNIVAC executive would conclude, "It doesn't do much good to build a better mousetrap if the other guy selling mousetraps has five times as many salesmen[22]." That was not the end of the missteps. In these critical early years, Rand would sell only purchase, as leasing required financing—and that required cash. IBM computers surged past UNIVAC. Though they were superior machines, they did not have to be, not with IBM sales and service such an important factor in customer decisions.

Nonetheless, Watson Jr. was magnanimous in his company's rise in computers. In 1953, he organized a luncheon at IBM headquarters in New York to recognize the pioneers of the new computer age. He invited Mauchly and Eckert, James Madden (an executive with Metropolitan Life), and Ralph Palmer, Nate Rochester, and James Birkenstock from IBM. The list did not include Howard Aiken[23].

# IBM 650: "THE MODEL T OF COMPUTERS"

C uthbert Hurd had led the charge on the Defense Calculator. At the same time, he was attentive to what he was hearing from the field. Sales representatives wanted a more affordable machine. They were seeing the first in a wave of small drum-based computers and had nothing in their IBM sales bag to take them on. Working with his Applied Science team, Hurd proposed a midsize machine that would rent in the range of $3,000 to $4,000 per month[1]. The Endicott lab was the logical place for such a system as Poughkeepsie was more than busy with its 700-class large computers.

The concept for the smaller machine coalesced around the latest storage device, the magnetic drum. The lab christened the program the Magnetic Drum Calculator, or MDC. Endicott warmed to the program as the additional function could extend the life of its punched-card machines. It would also build on the capabilities of the 604 Multiplier and the Card Programmed Calculator. Initial work started in 1948. Steve Dunwell in Future Demands did the original planning work. Though he had built his reputation as a card-machine "hot-rodder", he knew that a stored-program computer was the right path forward. Once Endicott got involved, Frank Hamilton, the lead designer on the Mark I and SSEC, took over. Tom Jr. reviewed the

MDC plan in 1949 and rejected it. The Korean War and the Watson Sr. Defense Calculator put the program on hold.

There was also confusion as to what the MDC really was. Corporate viewed it as simply an upgrade for 604 accounts, not fully realizing that it was a completely different animal altogether, a full-fledged computer. This contention and the resulting delays eliminated any chance that the MDC would be IBM's first commercial computer. That honor went to the 702. A further source of the confusion was the magnetic drum itself. It provided 10,000 digits of relatively high-speed memory, on a par with the EDVAC computer. However, Endicott engineers struggled with its design.

Slow progress on the MDC also revealed the schism between the Endicott lab and Palmer's new upstart lab in Poughkeepsie. Poughkeepsie was full speed ahead with electronics. Endicott, with its mechanical old guard, was not ready to let go. It was proud of the lab's heritage. With the Mark I and SSEC, it had one foot in the old IBM and the other in the new. Frank Hamilton symbolized this ambivalence as the lab director. It was not long before the lab rivalry escalated. There was a review of the MDC plan in 1950, but Poughkeepsie engineers performed it. The reason, of course, was that most of the expertise on electronics and stored programs was in Poughkeepsie. Nevertheless, it was a sore point for Endicott. In addition, Steve Dunwell had originally championed the MDC, but with the program now in competition with his TPM II project for funding, he sided with Poughkeepsie's tape machine.

The fierce opposition given his MDC proposal likely stunned Hurd. Many in Endicott believed that only large corporations, scientific institutions, the government could afford a $4,000 monthly rental price tag. The companies that Endicott sold punched-card equipment to would not be candidates for Hurd's machine. On the other hand, key executives like Learson had doubts about Endicott's capability to handle the project. Wally McDowell lobbied for the program but could not convince sales executive Red LaMotte to sign up for it. He finally escalated to Tom Watson. Tom resolved the impasse in a day[2], approving the MDC project in November 1952.

## THE WOODEN WHEEL STRADDLES COMPUTING WORLDS

There was one slight hiccup in Endicott's march to the MDC: a rival machine designed in Poughkeepsie called the Wooden Wheel. Some 5,000 604 Multipliers and nearly 700 Card Programmed Calculator (CPC) systems would be sold. However, there was a yawning gap between those solutions and the 705, which rented for over $20,000 a month and required huge expenditures in maintenance and programming.

A key driver of the CPC at Northrop, William Woodbury, was behind the Wooden Wheel[3]. After the completion of the CPC project, Woodbury moved to Princeton for graduate study. He arrived in the fall of 1950 when the IAS project was still in development. He was able to discuss computer design with Goldstine, Bigelow, and von Neumann. He found Goldstine too much of a mathematician for his tastes, Bigelow difficult to deal with, and von Neumann surprisingly uninterested in actual programming. Woodbury decided to use the concept of an enhanced Card Programmed Calculator as his dissertation project.

He left Princeton before graduating and met with John McPherson in New York about possible work for IBM. McPherson passed him to Wally McDowell, who brought him onboard as a two-day-a-month consultant. It soon turned into a full-time position. Woodbury thought that IBM had sold perhaps as many as one hundred CPC systems. He now discovered that the actual number was over 700, with zero of the proceeds coming to him or to Northrop. He argued with James Birkenstock on this but got nowhere. He concluded that his hiring by IBM was in part payback for the CPC.

Woodbury brought his enhanced CPC design to Poughkeepsie. In early 1951, he presented his concept to Ralph Palmer and received $100,000 in funding to develop it. He was able to form a strong core team for the project. His former colleague at Northrop, Greg Toben, was already working for IBM in the lab. Toben enlisted a fellow engineer, Truman Wheelock, to join the project. Wheelock was an IBM Customer Engineer in Chicago before moving to Poughkeepsie. He had been working on the Palmer's Test Assembly computer simulator. A fourth engineer, Roy Harper, joined the team. He had been a star technical specialist in the Midwest. IBM named him an "IBM Knight", an advanced field engineer able to work on virtually any project. He worked in Bryce's department at 590 Madison Avenue during development of the 603 Calculator, alongside Byron Phelps and Halsey Dickinson.

Woodbury's concept was a low-end machine costing about the same as two 604 Multipliers. It was a hybrid of stored-program and plugboard programming. Given Woodbury's affinity for the plugboard, his design was more like a plugboard machine with a small stored-program assist. Woodbury provided the mathematics, Wheelock was the electronics expert, and Harper did the internal logic. According to Woodbury, Harper was all about electronic circuits. He speculated that Harper "laid awake nights and thought about it.[4]" Greg Toben would put all the various electrical pieces together.

Woodbury envisioned the use of CRT memories in his design, an idea reinforced by their use in the IAS computer, the 701, and the 702. However, he found CRT memories used by IBM too expensive and tracked down cheaper ones outside the company, a move that did not endear him to Ralph Palmer. Woodbury's prototype machine was the "X795", where the "X" meant it was a contract program, and not yet a formal IBM product. However, Poughkeepsie engineers knew the machine as the "Wooden Wheel", a combination of the names of Toben, Wheelock, and Woodbury. Woodbury, noting the contribution of Roy Harper, said that "Wooden Wheel Harp" would have been more accurate[5].

The completed machine had 100 program steps expandable to 240, vacuum-tube-logic circuitry, and a very large plugboard. Main memory consisted of Woodbury's bargain-basement CRT units and a small core memory of 1,000 words, the most that was affordable. The hybrid nature of the machine made it far more difficult to program than either a standalone plugboard machine or a stored-program computer.

The lab delivered the prototype Wooden Wheel (X795) to Northrop for evaluation in late 1953. Toben moved to California to train Northrop on its use and then moved on to the new IBM lab in San Jose. Woodbury came out for the training, working with Rex Rice, who was still there. Woodbury took advantage of the field assignment to enjoy the good life on nearby Manhattan Beach. He later mused that he should have sent the machine to Princeton to fulfill his doctorate dissertation. An updated Wooden Wheel, designated the X797, would be production model. Erich Bloch replaced Woodbury's CRT memory with core memory. Poughkeepsie produced four of these systems and sent them to Northrop, Stanford University, and UCLA for further evaluation. The target planned monthly rental was $2,500[6].

There remained one big problem. IBM was not going to produce both the Wooden Wheel and the MDC. They were just too similar. The two machines would compete for the production go-ahead. Woodbury was optimistic about his chances. He felt the Wooden Wheel was four times as fast as the MDC. It was probably 50-100 percent faster and could be even more dominant for a given application. Woodbury also examined the Electrodata Datatron 201 drum computer and concluded that it was a better drum-based design than the MDC system. Woodbury did not stop there. He pressed the contention that the Wooden Wheel could even outperform the high-end 701 on selected problems. In his telling, the Wooden Wheel would perform six additions and one division and be done. The 701 would have many nonproductive cycles to organize the program, prime the registers, perform the calculations, and finally post the results. He estimated that upwards of 90 percent of the 701 processing time was nonproductive. Woodbury discussed his contentions with von Neumann, Hurd, and Haddad. While von Neumann was receptive to his argument, Hurd thought that the Wooden Wheel was simply a toy in comparison with the 701. It was also a very complex toy. Each application required a substantial and difficult programming effort, one that was four times the scope for the same application on the MDC[7].

However, the biggest problem was the price. Its combination of memories, plugboard, and stored program made for a costlier machine. The assigned $7,000 to $9,000 monthly rental was double the price planned for the MDC. However, all of these issues were likely secondary to the lab factor. Endicott had already spent several million dollars on MDC development. The program was marked for cancellation as many as five times, but Frank Hamilton kept reviving it. The idea that Endicott would lose the program so near the finish line to a very unusual machine produced by Poughkeepsie was simply not acceptable. It is likely that the winner was never in doubt and that IBM marketing was simply using the threat of losing its prized program to force Endicott to hurry up and finish its machine. Hurd and Birkenstock were also impatient to get a product into the hot magnetic drum market. Endicott committed the MDC to readiness within ten months.

Meanwhile, in Poughkeepsie, Ralph Palmer had his plate full with 701 and 702 development and could only commit to bringing the Wooden Wheel to market in less than three years. Endicott did not come close to its ten-month commitment, as drum patents and drum issues interceded. In the end, the MDC got the nod and Endicott won this battle with Poughkeepsie. As for Woodbury, he was sore about the CPC and now he was doubly sore about the demise of his Wooden Wheel.

The magnetic drum was central to the MDC. IBM had magnetic drum patents that it had acquired from Gustav Tauschek in 1932. His invention was a rotating drum that could store 500,000 bits[8]. A standard magnetic drum was a large metal cylinder divided into concentric rows or tracks coated with a ferromagnetic material. In most designs, there were read/write heads for each of the rows. To read or write specific data from a track, the paired head had to wait on the rotation of the drum until it reached the target data. This delay before data reads or writes was the drum's rotational latency. The drum design was an update of Atanasoff's 1937 ABC concept, which used capacitors instead of magnetic material.

During World War II, the Navy G cryptographic engineers in Dayton developed a magnetic drum.. After the Navy helped Bill Norris set up Engineering Research Associates, the new company secured a number of drum-based computer projects with the Navy. With its drum-based computers, ERA was actually the leading computer maker by unit sales in the 1950-1952 timeframe. Other companies jumped in with drum computers, including Librascope with its LGP-10, Bendix with the G-15, and Consolidated Engineering with the Datatron 201.

Consolidated Engineering was the company that Clifford Berry joined in 1942. He worked in spectrometry and developed an analog computer to analyze the simultaneous linear equations produced by the company's mass spectrometer. Consolidated formed a new company called Electrodata to market Berry's computer, called the Electrodata 103⁹. He published papers in 1945 and 1946 on using digital computers to perform such linear analysis. He urged the company to produce a general-purpose version of his spectrographic computer, which it did. It debuted as the Electrodata 201, with a vacuum-tube processor and magnetic drum memory. It was more widely known as the Datatron 201. Harry Husky, who was responsible for the Bureau of Standards SWAC computer, designed the Datatron based on the approach suggested by Berry. Burroughs would buy the Electrodata division in 1956 in order to enter the computer business. The Electrodata 205 became the Burroughs 205.

Endicott struggled with its magnetic drum. The alternative was simply acquiring a drum from a vendor. Steve Dunwell agreed and Jim Birkenstock negotiated an agreement with ERA to supply their drums while protecting IBM patents. In fact, Birkenstock recommended to Watson Jr. that IBM simply acquire ERA, an option Tom rejected because of antitrust concerns[10]. Bill Norris of ERA considered the agreement as too one-sided for IBM but still signed.

The Endicott lab continued work on its drum, focusing on performance. Most existing magnetic drums rotated at 3,400 RPM, providing an average access to data of 8.5 milliseconds. This was much slower than Eckert's mercury delay lines (at 0.5 milliseconds) but on a par with Williams CRT tube memories (10-12 milliseconds). Other drum manufacturers enhanced access time by installing a second read-write head on each track. This design, called a "revolver" drum, required complex programming to take advantage of the second head. In addition, IBM was wary of employing the technique due to patent concerns[11].

Endicott engineers took a decidedly different approach. To improve performance, they moved to a much smaller-circumference drum that raised the rotational speed to 12,500 RPM and cut the access times in half[12]. This also cut memory capacity in half, from 4,000 words to 2,000 words. Engineers made a further design change that could cut the access time to near zero. They added the address of the next instruction into the current instruction's record. If the program placed the next instruction in proximity to the current instruction, the MDC could read that instruction immediately, without the wait time for the head to move or the drum to rotate. However, this approach required some programming to work.

With the drum issue resolved, the MDC took shape as a general-purpose, stored-program computer with as much as 35,000 characters of memory, and central processing logic handled by 3,000 vacuum tubes. To keep the number of vacuum tubes low, the design used a variation of decimal called the bi-quinary format. This was in essence an electronic version of the standard two-and-five-bead abacus, with a combination of two bits and five bits used in conjunction to encode one decimal digit. Though not known at the time, Bletchley Park's Colossus machine deployed the same decimal arrangement.

The MDC had an instruction set of 44 operation codes, compared to several hundred on the 700 series machines, but each instruction was robust. Arithmetic performance was modest, at 600 additions or 77 multiplications per second, but these numbers were more than adequate for the target commercial customer set. Endicott had a lab model of the MDC up and running by early 1954.

When Cuthbert Hurd originally proposed the MDC concept, he envisioned selling at least 50 systems. The formal business plan was far more ambitious, calling for 250 systems[13]. IBM announced the Magnetic Drum Computer in July 1953 as the IBM 650, joining the IBM Multipliers in the 600-series section. It was available for just a quarter of the price of the IBM 701.

**Figure 28**. IBM 650, called the "Model T" of the computer age. The first computer system to sell more than a thousand units. Reprint Courtesy of IBM Corporation ©.

IBM delivered the first 650 to John Hancock Mutual Life in Boston. The central unit was $3,250 per month, a reader/punch added $550, and an upgrade to core memory was available for $1,500. A typical configuration went for $3,800 plus maintenance. A number of other installations followed this first one. In fact, there were thousands of others. The 650 was the first computer to ship in the thousands, with some 2,000 systems sold between 1954 and 1962. It soon gained the accolade, "Model T of computing.[14]" For IBM and certainly for Endicott, it was the perfect machine as it blended the punched card with computer programming and electronic processing.

The 650 jumped into the drum computer fray and proceeded to blow by all of the competition. The quality of the machine plus IBM's reputation, sales, programming ease, reliability, and service carried the day. Because of its success, Endicott produced a series of enhancements, including high-speed core memory, tape drives, a printer at 150 lines per minute, and the San Jose Lab's new invention, the disk drive.

With the 650 and the 700 series, IBM took the computer industry by storm. Tom Watson found himself on the cover of *Time* magazine in March 1955[15]. In the cover-story interview, Tom said he expected IBM to be a $1 billion company by 1960. IBM would reach that goal in 1957. Frank Hamilton had shepherded the 650 to the market, adding to a resume that already included the Mark I and the SSEC. In 1963, Tom Watson made him the first IBM Fellow, the highest honor within IBM. A lifetime appointment, it included the funding and freedom to take on any project. It also came with such benefits as automatic attendance each year, along with one's spouse, to the premier IBM recognition event, the Golden Circle.

IBM found success with the 650 in a combination of its price and programming. The new machine supported machine code programming, but IBM was highly unlikely to find many customers that wanted to do that. The Watson Scientific Lab was an early recipient of two 650 systems. Stan Poley, who had written the optimizing SOAP program for the NORC computer, would do the same for the 650. His Symbolic Optimizing Assembly Program not only provided for conversion of English mnemonics to machine code but also copied in selected subroutines. Virtually all 650 installations used his SOAP program. Poley's compiler also took advantage of the 650's unique drum design and placed instructions and data at the best drum locations to reduce access delay.

It turned out his SOAP program was simply the opening salvo in an avalanche of assemblers and compilers for the 650. A team at Carnegie Institute of Technology created Internal Translator (IT), a FORTRAN-like compiler. It read in source algebraic statements and punched a SOAP source deck. The SOAP compiler converted it into an executable program. Many more compilers followed, including ADES II, BACAIC, BELL, CASE SOAP III, DRUCO I, DYANA, EASE II, ESCAPE, FAST, FLAIR, GAT, KISS, MAC, MITILAC, MYSTIC, OMNICODE, RELATIVE, SIR, SPIT, and SPUR. There was even the SPACE program, which stood for "Simplified Programming Anyone Can Enjoy." Most of these compilers used that two-step process in conjunction with Poley's SOAP program.

Regardless of which tool you selected, the execution process was similar. You keypunched your source code into 80-column cards and placed the selected compiler deck ahead of your source program cards. The compiler would punch out a machine-language card deck or, if using SOAP, that program would punch the execution cards. Finally, you would add your data cards at the end of the execution deck and run your program.

Don Knuth and the Case Institute of Technology provided an early 650 success story. Knuth described the IBM 650's arrival on campus. The student newspaper called the newly acquired machine the "IBM 650 UNIVAC[16]." Once installed, Knuth started programming. He began first with SOAP and moved on to SOAP II when it became available. He then decided to develop his own assembler, which he brazenly called SOAP III. In 1958, he tried to cram as many features as possible into his drum's 2,000 characters of memory. The university purchased core memory the following year, and Knuth wrote Super SOAP to take advantage of it. There followed HAND SOAP, which used SOAP assembly but allowed manual or "hand" placement of the drum locations. Knuth continued to graduate school at Caltech, where he developed an ALGOL compiler on the university's Burroughs 205 drum computer, which was really the Berry-inspired Electrodata 205.

Tom Watson recognized early on that a lack of programmers would limit the expansion of IBM in the computer market. He provided a 650 to MIT and then offered IBM 650 systems to colleges at substantial discounts. The once very select fraternity of computer programmers grew rapidly. Because of the sheer numbers of 650 installations, most of the world's programmers at the time were 650 programmers.

# THE FIRST DISK DRIVE, AKA "THE BOLOGNA SLICER"

In the early 1930s, IBM was still working on Ben Wood's original request for a test-scoring machine. A high school teacher in Michigan had invented a scoring device based on Wood's multiple-choice form. His device used a high-resistance sensor that detected the lead from the scoring pencil. He sent Wood a description of his invention and received an invitation to Endicott to review his approach with IBM engineers in 1934. IBM bought the rights to his design, which the company developed into the IBM 805 Test Scoring Machine. Announced in 1938, it remained on the market until 1963. IBM also hired the schoolteacher. His name was Reynold Johnson[1]. He spent the next 18 years in the Endicott lab, working on test scoring, time clocks, keypunches, and printers. He amassed over 50 patents, "some of them fairly good,[2]" as he would say.

In January 1952, Wally McDowell, IBM's Director of Engineering, asked Johnson to start a new lab in San Jose. IBM had trouble attracting engineering talent to the Hudson Valley from the West Coast. The new lab would flip that dynamic. There was already an IBM card plant in San Jose, producing 16 billion 80-column cards a year, still a very big business for the company[3]. The area had a critical mass of large IBM accounts, including most of the nation's aircraft companies. Douglas Aircraft was coming to the end of a major contract and a number of their engineers would be available. There was another benefit of the West Coast location of Johnson's lab. It was then a 12-hour plane trip from the East Coast. This would reduce the oversight from IBM Corporate and give him space to develop his plans.

Johnson rented a simple commercial building at 99 Notre Dame Street in downtown San Jose and started hiring[4]. He was limited to 50 people at first. Reynold asked Lou Stevens, a fellow engineer who was working on the Defense Calculator in Poughkeepsie, to join him at the new lab and help recruit new hires. Stevens had graduated from the University of California at Berkeley, so the entreaty to work on the West Coast was not a hard sell. Stevens placed ads in the local papers for engineers, scientists, and technicians. Reynold and Stevens were under strict orders not to recruit East Coast IBMers. The lab was up and running with 30 engineers that July. Johnson had 60 on board by the end of the year. The big question was, "what would they be working on?" Johnson had the latitude to figure that out.

At the time, it seemed that all the needed mass storage options were in place, with punched cards, tape, and magnetic drums. The issue was whether there was a breakthrough technology that did not presently exist or a current technology that needed perfection. Reynold formed a team of his engineers to study how current punched-card customers were using data in order to identify where a new technology might help. His team received local help from a marketing representative out of the San Francisco office (Ed Perkins) who would take Reynold's engineers to Bay Area customers that were struggling with punched cards processing. Reynold's team explored magnetic drums, tape loops, magnetic plates, and rotating magnetic disks. All were variants of a medium doped with a ferromagnetic coating. The magnetic orientation of a specific ferrous particle defined whether the bit was a "1" or "0", as was the case with core memory.

During World War II, German engineers developed a primitive tape recording machine called the Magnetophon. Singer Bing Crosby asked a U.S. company (Ampex) to develop the technology so that his live performances could be recorded. Johnson's team noted that tape technology had progressed and the tapes themselves were inexpensive. However, the fact that tape was sequential was a deal-breaker.

In acquiring Gustav Tauschek's company and his services, IBM would also acquire his drum patents, and Endicott had developed the magnetic drum as the main memory for the IBM 650 computer. The drum approach was simple in design, and it afforded random access. However, the density of the data was low, a costly a read/write head was required for each track, and the drum's latency reduced access times. In addition, the expense of a drum device climbed rapidly as it grew in capacity. An enhanced drum seemed like a dead end.

The final area Johnson's team evaluated was a rotating magnetic disk. Jacob Rabinow, an engineer working at the National Bureau of Standards, had designed a multiple-platter array with a read/write head that moved from one disk to the next[5]. Each disk of his proposed device would hold 500,000 bits of data. Ralph Palmer and his Navy G team in Dayton had attempted a similar rotating disk mechanism for use in code breaking. They intended to store massive volumes of intercepted messages across two spindles of disks and then perform comparisons to look for intercept patterns or anomalies. It would be a high-speed upgrade of the dual paper-tape mechanisms on the Heath Robinson at Bletchley Park. The Palmer machine, called the Oscillograph Full Selector, would use a group of oscilloscopes to read the intercept characters. The concept of a random-access rotating disk looked promising, but was it feasible—and, if so, at what price?

Johnson decided in early 1953 to run with the magnetic disk as the primary focus of the new lab. The stated goal was a large disk that provided access to any data in less than one second. The key challenge was the access time, as the disk mechanism would have to seek the target concentric track, wait for the data to circle around, and then read the data. In contrast, the read/write head on the drum was fixed on each track. The only issue was the rotational latency.

Johnson's team realized that they needed expertise in magnetic recording. Lou Stevens had just the person in mind[6]. Al Hoagland had been a fellow electrical engineering graduate student at Berkeley. The university had received funding from the Office of Naval Research (ONR) to build a small computer called the CALDIC (California Digital Computer). Hoagland worked on the drum memory for the machine. His PhD thesis was on digital magnetic recording. Hoagland met with Stevens and another former graduate student who had also worked on CALDIC, a new hire named John Haanstra. They encouraged Al to join the disk project in San Jose. A dinner in San Francisco with Reynold sealed an arrangement. Hoagland would finish his doctorate and then join the Berkeley faculty, but in the interim, he would work part-time as a consultant to the San Jose lab.

About this time, IBM received a Request for Proposal (RFP) from the U.S. Air Force Supply Depot in Ohio for an instant-access inventory system. The request went to thirty companies. The RFP (Request for Proposal) described the existing system, a classic punched-card flow that used tub files. The tub files contained the master punched cards, such as the customer and inventory records. An inventory transaction required the operator to merge the transaction card with the corresponding master card from the tub file. They did this by hand or through machine collating.

Johnson's lab viewed this as the perfect application for its new random-access disk drive, which would replace the tub file. However, a complete response to the RFP required both a magnetic disk drive, which did not exist, and a small computer built around it, which also did not exist. Rey Johnson was sufficiently ambitious to embrace the entire concept, asking Bill Goddard to develop the RFP response. IBM did not win the business, but the RFP response ("A Proposal for a Rapid Random Access File") became the roadmap for the lab's mission. Johnson asked Stevens to take over responsibility for the program. Rey was the visionary but Lou Stevens was the aggressive manager who could make it happen.

The task looked good on paper, but the proposal asked more questions than it answered: What kind of disks, how to coat the disks with magnetic material, how to rotate them, and at what speed? What would the read/write heads look like, and how could they scan over the rotating disk without crashing into the surface?

The engineering team decided on aluminum disk platters that were 24 inches in diameter and would spin at 1,200 RPM. The engineers had to increase the thickness of the disks until they would rotate without any wobble. They coated the disks with a ferrous material, an oxide paint similar to the one used on the Golden Gate Bridge. An engineer applied paint with a Dixie cup filtered through a ladies' nylon stocking while the disk was spinning. The disk was then oven baked[7].

With the new disks successfully coated and properly spinning, the engineers turned to the problem of flying read/write head over the disk surface. Their solution was to pump air—a whole lot of it—into the path of the recording head. A dedicated air compressor was added to maintain constant head flight height. The final technical hurdle was the servomechanism. This unit had to move the read/write head back and forth across the concentric data tracks on the platter to the target track. There was a slight complication. In order to have the target capacity of five million characters of data space, fifty of the 24-inch platters would be required. Finally, the drive needed to meet Johnson's goal of access within a second.

There was also one major nontechnical issue and that was the IBM board of directors. Still led by Watson, the board wanted to cancel the project because of its potential negative impact on the punched-card business. Rey Johnson and his senior engineers presented to the IBM board at 590 Madison Avenue in 1953. Watson Sr. introduced them with comment, "These are the gentlemen that are going to replace the punched card.[8]" Watson was successful in having the program officially axed. After the meeting, Tom Watson took a dejected Johnson aside and told him to proceed, regardless of what the board had said. Tom saw clearly how the magnetic disk could revolutionize the industry, and he wanted IBM to take the lead. Though Tom was fully on board, IBM marketing remained to be convinced. Field sales had a hard time conceiving of a customer's need for five million characters of storage, let alone future plans to double that amount.

Johnson agreed to keep working on the program but told Tom that there was no plant to build the new disk file once his team finished the design. Tom arranged to fund a new plant in San Jose and signed Berkeley architect John Bolles (who would later design Candlestick Park) to create a design for the San Jose manufacturing facility on Cottle Road. His design featured dramatic floor-to-ceiling windows. The centerpiece was the main building (Building 25), which looked out on a broad reflecting pool.

Work on the new drive, sometimes referred as the "bologna

slicer,[9]" progressed, and Rey Johnson was soon able to make a note in his calendar, "First data written and read back successfully." Ralph Palmer would have recognized the look of the completed machine. It was very similar to his wartime design in Dayton, right down to spinning fifty magnetic disks. The capacity of five million characters came straight out of the Air Force RFP, where there were 50,000 master records of 100 characters each in the tub file. The drive capacity was the equivalent of 62,500 punched cards. There were 100 disk surfaces on 50 platters, each with 100 concentric recording tracks, with each track having 500 data characters. The lab produced fourteen prototypes and shipped them to customers for evaluation and feedback.

IBM announced the disk machine, formally the IBM 350, in 1956. Reynold Johnson would later comment, "This is the product that spawned the magnetic disk storage industry—an industry that has since come to generate annual revenue of 23 billion dollars[10]." Though "disk" is the accepted term today, the new technology began life as DASD, for Direct Access Storage Device.

**Figure 29.** IBM 350 Disk Drive (1957). The first disk drive was known fondly as the "bologna slicer." Developed by Reynold Johnson and his engineering team at the San Jose Lab. Reprint Courtesy of IBM Corporation ©.

An interesting side story is relevant to IBM's disk drive announcement. In 1956, UNIVAC was in a position to bring a magnetic disk to market. Remington Rand's ERA division in St. Paul worked on the concept. Eckert and Mauchly lobbied hard to stick with the magnetic drum, likely to avoid depending on ERA for any component. The UNIVAC drum system was composed of ten drums but together, they held only 1.8 million characters of data. There would be no ERA disk project. It appeared that IBM chose right and UNIVAC chose wrong.

The bigger picture was the new horizons of commercial applications that opened up. As the work on the Air Force RFP had amply demonstrated, the alternative to mass storage for most companies at the time was punched-card collation or tub files. The fast, affordable, random access of the magnetic disk was the final key piece to move to real-time applications, and on to on-line interactive systems. Watson was correct in stating that the drive was going to replace the punched card, though it would take some time.

As work progressed on the disk drive, Johnson marshaled a separate engineering team to design and produce a computing system built around the drive to automate the tub file process. This would become the Random Access Method of Accounting and Control, or RAMAC. He selected John Haanstra as chief engineer for the project. Haanstra started with IBM in Poughkeepsie after graduating from Berkeley and jumped at the chance to return to California. He was a very capable engineer and an even more capable executive, clearly bound for greater things. He was also a big and imposing individual prone to pounding on tables to convey his point[11].

RAMAC would include the new disk drive, a card device, console, printer, and a processing unit. The goal was a complete entry computer at a target rental of $2,000 per month. The weak link would be the printer. William Woodbury, of the CPC and Wooden Wheel, ended up working for Rey Johnson in San Jose. He volunteered to develop a printer, not fully understanding that he was competing with Larry Wilson, a master of IBM's electromechanical machines. Woodbury's effort was short-lived, and he ended up in contention with management and was soon on the street.

The completed RAMAC was a curious blend of technologies. It was another decimal computer with a small core memory of 100 characters augmented by 3,200 characters on a magnetic drum. It used 604 Multiplier vacuum tubes for circuits, becoming one of the last IBM tube machines. RAMAC also retained the plugboard, so programming involved a combination of machine code and plugboard wiring.

Programming for random access to the new disk presented the next challenge. Hans Peter Luhn of the Poughkeepsie lab led a team that developed the concept of indexed data files[12]. His team allocated two areas on the disk surface for each physical file, one for the actual data and the other for an index that contained the logical key (i.e., item master number) and the actual address on the disk drive for the data. A program request for a specific item master record would search the index, find the matching key and the location, and then seek the actual record. This index structure cost roughly 20 percent of the drive's capacity but dramatically decreased the overall access time.

The RAMAC system debuted in September 1956. The complete system rented for $3,200, missing the target of $2,000 by a substantial margin. Despite this, IBM would sell over 1,000 of the RAMAC systems. The first customer was Chrysler's MOPAR parts division. Hewing to the lab's target environment, the Chrysler application was an inventory processing application that previously used a large 80-column card tub file.

# WATSON JR. TAKES OVER

Watson continued to have the final word at his company nearly until his death in June 1956. Tom had been running the company for a number of years, but his father did not want to let go. As Tom relates, "He wanted to make me head of IBM, but he didn't like sharing the limelight.[1]" Tom had to make a number of changes to help him run the company day to day. He relied heavily on his CFO, Al Williams, and his lead sales executive, Red LaMotte.

The "Old Man" had an impressive run. From a troubled and humble beginning, he shepherded his company to great heights. He led the rise from the modest $4.1 million revenue of C-T-R to $892 million, and took income from $1.3 million to $87 million. He paid dividends every year except 1912 and 1914. He never had a losing year. A $1,000 invested in C-T-R stock in 1914 would have grown in value to $148,000 in 1952, the largest capital gain of any stock during the period. IBM revenue grew from 1929 until 1956 by an average of 14 percent per year[2]. The company became the 37th largest manufacturer in the United States. The company's workforce rose from 1,346 to 73,000[3]. Aside from those gaudy numbers, Watson built the modern transnational company, with a vibrant and enduring company culture. Despite his many faults, he was clearly on a very short list of the best businessmen of the 20th century. He would probably trade that accolade for simply "world's greatest salesman" which is what the New York Times called him in its obituary[4].

**Figure 30.** Thomas J. Watson, Sr., better known within IBM as the "Old Man" during the time when he was struggling with the decision to turn over the company to Tom Jr. Reprint Courtesy of IBM Corporation ©.

At the time of his father's death, Tom was 41, lean and trim at 6' 3." He was also prematurely gray, adding a certain gravitas to his new position at the head of IBM. Early on, Al Williams asked him about compensation. Williams was surprised and somewhat annoyed when Tom said he wanted the same deal as his father[5]. He felt that Tom had yet to earn such a lucrative arrangement. However, Tom was adamant, and Williams reluctantly agreed. Tom's earnings would be a quarter of 1 percent of annual profits after dividends. With 1956 income at $87 million, that came to $217,500. Given where young Tom Watson would take the IBM Company, the arrangement was a bargain.

Tom's most immediate priority was IBM's management structure. He realized how top heavy and inefficient it was, with everything going through one person. Tom convened a meeting of his top 100 executives in Williamsburg, Virginia, with the objective of creating a more efficient corporate line and staff organization. Though both groups would own results, any disagreement between the two groups would bubble up to the Corporate Management Committee (CMC), which he would chair. The resulting organization chart featured six standalone product divisions, World Trade, and a separate corporate staff group. He added an additional feature called the non-concur process as a check against any decision moving too fast. Virtually anyone up and down the line could raise an objection (termed a "non-concur") at any point and force the issue up one level in the management chain.

Even with the new organization, issues still bubbled up to the top. Tom adopted the management philosophy of tackling issues with dispatch. He would say, ". . . solve it, solve it quickly, solve it right or wrong." If it turned out wrong, you could go back and adjust, now armed with the knowledge of what did not work. His management style evoked the style of his father. He had frequent fits of rage. Bob Evans recalled making presentations to Tom and the CMC, calling it "heart attack alley.[6]" Kevin Maney would quote a number of people who worked with Tom as saying he was "not a nice man.[7]" Like his father, he liked to manage by competition, coupled with extra compensation for the winners. He rarely used question marks in his memos, as he was never asking; he was telling. He had a short attention span. He wanted any issue that he had to deal with boiled down to a typewritten recommendation on one 3" by 5" card.

Despite this, Tom was uniformly praised for his vision, intelligence, and leadership. Bob Evans would say, "Of all the executives I have met across the world, Tom Watson Jr. was the most electrifying and most powerful leader.[8]" Tom Watson progressed from that day in Russia where he had a near revolt of his B-24 crew. He made one management adjustment that would directly affect him: He set the policy of the IBM CEO retiring at age sixty. Tom traveled frequently to IBM branch offices during this early period. He found they were still stuck in the cult-like atmosphere that his father had encouraged. There were the songs of praise, slogans, conservative dress, and ever-present pictures of their leader. Tom was committed to changing all of this. He agreed to soft collars on dress shirts, eventually going as far as allowing blue button-down shirts. He would later allow liquor at IBM functions.

He also wanted to shed the stodgy image that IBM projected. He described walking down Fifth Avenue in New York and stopping at an Olivetti branch[9]. The Italian manufacturer of typewriters and small computers ensured that the look and feel of its showrooms matched the products on display—smart, colorful, modern, and exciting. He asked the general manager of IBM Holland about Olivetti, and he sent Tom a portfolio of equally smart, colorful, and modern images from the company's product materials. With those in hand, Tom asked his team where he could find an industrial designer who would be equal to the task of reinventing IBM's image. He was directed to Eliot Noyes, who Tom actually knew from the war when Noyes was head of the Army Air Force glider program. Noyes, known as "El Noyes," had been the curator of industrial design at the New York Museum of Modern Art.

Noyes first took on IBM's Madison Avenue showroom, mixing an all-white floor with brilliant red walls. He also redesigned the main IBM lobby around the corner, which was looking like it remained stuck in the 1920s. He moved on to the look of the next IBM computer to occupy IBM showroom, the 702. Its redesign was cleaner and completely modern. The contrast with the previous occupant, the 701 Defense Calculator, was dramatic. Circling back to Olivetti design, Tom asked Noyes to redesign the IBM Selectric Typewriter. Noyes took his cues from the Italian manufacturer, with bright colors and curves to complement the typewriter's unique feature, the golf-ball typing element. One final design assignment was a revamp of the IBM logo. Tom enlisted Paul Rand, a noted graphic designer, to produce the 13-bar logo in 1956 and then an 8-bar version in 1972.

Tom had far more substantive projects in the queue. He would commission over 150 plants, laboratories, and new offices from 1956 to 1971[10]. Noyes cautioned him to find the best designer for each project and not rely on a single individual. Tom had selected local architect John Bolles for the San Jose plant and Eero Saarinen for the plant and lab at Rochester. He would go back to Saarinen to design the new IBM Research building at Yorktown Heights. Tom's mantra was ". . . you ought to be able to look at an IBM factory, at an IBM product, even at an IBM curtain, and say, it's IBM.[11]" He also took on IBM education, an essential element of the IBM of his father. Tom looked at General Electric's education programs and found IBM behind in both customer engineering and management training.

Though IBM's business was growing at a very brisk rate, there was one issue casting a pall over the company. The Justice Department had filed suit against IBM in 1952, accusing the company of monopolizing the punched-card industry, with upwards of a 90 percent market share[12]. The reaction from Watson Sr. was predictable. He would fight it to the end. In his mind, this was pure and simple punishment for building a successful business, with great products, outstanding sales, and satisfied customers. His singular focus on punched cards through the years was his strength. If others could not match his dogged pursuit of the technology in order to compete, that was not his problem.

It was supremely disingenuous on his part, given the tactics he deployed when he was in sales and the sordid scheme he ran for NCR. In the beginning, Tom sided with his father in fighting the suit. They sought to redefine the "industry" as far bigger than punched-card machines, including everything from ledger cards to adding machines. The Watsons also argued that the company had considerable opportunities to engage in anti-competitive behavior, which they avoided. They would cite turning down Mauchly and Eckert when they offered the UNIVAC business or neglecting to invest in a new technology that would become Xerox. However, the real story behind Xerox varied substantially from how IBM portrayed to the Justice Department lawyers. James Birkenstock, who was handling patent issues for the company, met with a patent attorney in 1947 representing a client who had an "electro-photographic printing machine." Birkenstock mistakenly took the opportunity to Watson Sr. who did not understand the concept and quickly said "no." Birkenstock relayed the answer to the patent attorney. Birkenstock would consider his handling of this opportunity a massive "fumble." He should have sought approval from Watson, Jr.[13] His "fumble" was the company that became Xerox.

The problem for IBM was the Justice Department track record in antitrust suits. It rarely filed unless it felt it could win, and it did in fact win in at least 90 percent of its cases. Watson had been there already, with the 1936 antitrust decision on IBM's monopoly of punched cards. It was a bitter pill then and one that he had not forgotten. Yet, after several years without a resolution of the current action, Tom came to the realization that opposing the lawsuit was a mistake. He felt it would sap the company financially and take the focus away from competing in the new field of electronics at a critical time. IBM had already fought off a lawsuit from Sperry Rand in 1955 that contended that IBM's 90% share of the tabulator market was clearly a monopoly[14]. Tom started a campaign to change his father's mind. When he pressed the issue directly, it usually ended with both tempers flaring. He enlisted trusted friends of his father to move him gently in the right direction, with little success. By the end of 1955, Tom was essentially in control of the company and started working with the Justice Department lawyers on a settlement. IBM hammered out an agreement and Tom signed it in January 1956.

The agreement was not perfect, but it allowed IBM to move on. IBM agreed to offer its machines for purchase. With typical five-year product cycles, this opened the door for leasing companies. IBM also agreed to sell its software licenses and patents at a reasonable price. This was the beginning of the plug-compatible market, which would bedevil the company for years to come. IBM would also provide servicing information to third-party maintenance companies. Finally, the company would refrain from monopolistic sales practices such as unhooking competitor's orders, disparagement of the competition, and the marketing of unannounced products. The company produced a booklet called *IBM Business Conduct Guidelines* that required reading and certification by everyone in IBM sales. Finally, the agreement required IBM to place its services business into an independent subsidiary called the Service Bureau Corporation, to run at arm's length.

## RUSSIAN BOMBERS AND SAGE

Jay Forrester's Whirlwind project for a Navy flight trainer ended in 1949 when the Navy bailed out of the program, due to ballooning costs and lack of progress. That same year, the Russians exploded their first atomic bomb. With the Soviets possessing the bomb and the planes to deliver it, the Air Force convened a board to explore air defense options. George Valley, an MIT physics professor, was on the board. He decided to look for a computer fast enough for the task and discovered that it was right there on the MIT campus—the Whirlwind. The Air Force decided the MIT computer was promising and agreed to chip in $500,000 in funding for 1951. MIT reimagined Whirlwind as a strategic air defense system called Whirlwind II or by its more formal name, the Semi-Automatic Ground Environment (SAGE). With the mission change, the Air Force took over all funding and brought Canada into the project. MIT's president, James Killian, was tentative about taking on such an effort. He wanted to get the university back to academics. The prospect that the program could make MIT and the Boston area a center for the electronics industry finally won him over. He agreed to create the Lincoln Lab to run the SAGE program. Jay Forrester's earlier decision to design a real-time computer was now prescient, as was his switch to core memory. Valley and Forrester proceeded to develop a detailed description of SAGE, completed in 1953. Their design added one other important element to the plan: the transistor.

Vacuum tubes were clearly a dead end—an expensive, power-consuming technology with reliability concerns. Furthermore, they dictated the elephantine size of early computers. In the late 1940s, engineers at Bell Labs in Murray Hill, New Jersey, explored some form of electronics for telephone switching that would supplant vacuum tubes. Three researchers -- Walter Brattain, John Bardeen, and William Shockley – studied the problem. Brattain was a gifted experimenter in materials science, Bardeen was a quantum physicist, and Shockley, the leader of their team, was an expert in solid-state physics. Bardeen and Brattain had been graduate students at the University of Wisconsin with John Atanasoff.

They looked at the materials options. There were good conductors like copper and poor conductors like sulfur. In between, there were semiconductors like silicon and germanium. With this later group, you could improve their conductivity by contaminating them with elements like arsenic or boron. The three researchers created a test device with three layers of silicon where a small charge in the voltage applied to the middle layer would produce a large increase or amplification of the current through the entire device. This was essentially the same electrical reaction produced by a vacuum tube. Bardeen and Brattain created a prototype in 1947 that worked. Bell Labs moved immediately to patent it. Bell named the new electronic device the transistor. Shockley was upset at his exclusion from the patent application and barred the two men from additional research. Shockley proceeded to develop an improved transistor. Bardeen and Brattain complained but ended up leaving Bell Labs. All three would share the Nobel Prize in Physics in 1956. Bardeen would go on to become the only person to win the Nobel Prize in Physics twice. He won his second Nobel in 1972 for work on superconductivity.

The transistor was transformative in a number of ways. The most obvious characteristic was its size. It was roughly one-fiftieth the dimensions of the vacuum tube. Then, there was the power consumption. The transistor used one-millionth of a watt, compared to a typical vacuum tube that drew a full watt or more. This advantage directly translated in to far less heat, heat that would require some method of dissipation. The final and greatest advantage of the transistor would emerge later. Its innate structure would enable the further miniaturization of transistors through etching on silicon.

Back on SAGE, by 1955 Jay Forrester had built a small prototype machine using transistors called the TX-0 (Transistorized Experimental Zero). It used 3,600 transistors, supplied by Philco at $80 each[15]. The total cost of $288,000 seemed to be a deal breaker for use of transistors on SAGE, as its computers would use far more. Transistors were not quite ready for prime time.

Whirlwind II was a prototype machine. Though it was an expensive program, its costs would pale compared to the deployment of the full system. IBM won the contract to develop a pilot implementation called the Cape Cod system, completed in 1956. The successful test of the concept set the stage for the development of the complete system. The Air Force plan called for 23 SAGE locations in North America, each containing two massive digital computers. Doing the math, Tom Watson realized that the full contract for SAGE would be as impactful to IBM as the SSA contract was in the 1930s. Remington Rand had the same idea and very much wanted the business. Rand representatives met with the MIT team on James Rand's yacht along with their new board chairman, General Douglas MacArthur. Leslie Groves hosted the meeting.

John McPherson led the IBM response[16]. He got Tom involved immediately. Tom would later comment, "I worked harder to win that contract than I worked on any other sale.[17]" He hosted Jay Forrester at the IBM plants and introduced him to the key engineers. Three potential vendors were being evaluated—IBM, Remington Rand, and Raytheon. Forrester concluded that IBM had the production facilities, the engineering expertise, and most importantly, worked smoothly as a team. He was concerned about the infighting between the Remington Rand divisions. He ranked the bids as IBM #1, Remington #2, and Raytheon #3. IBM got the contract. As big as the deal was, it was only for the SAGE hardware. The programming contract was won by the Rand Corporation, to be supervised by MIT's newly created Lincoln Lab. Robert Crago, one of IBM's SAGE engineers, observed, "We couldn't imagine where we could absorb 2,000 SAGE programmers at IBM when this job would be over.[18]"

The IBM code name was "Project High," so named after High Street in Poughkeepsie, the site of the necktie factory in town where work began[19]. Tom immediately assembled a team of 300 engineers and scientists. The IBM contingent worked with Forrester and the Lincoln Lab in designing the SAGE computer. The project soaked up resources, with engineers and technicians taken off key projects for the high profile and high-risk endeavor. IBM also built a new plant and lab at Kingston, twenty miles up the Hudson River from Poughkeepsie, to support the effort.

The Whirlwind II prototype featured 16 parallel math units for high performance, using 5,000 vacuum tubes just for this purpose. Forrester had grown concerned when he did the math on vacuum tube failures, with existing tubes having an average life span of just 500 hours. The real-time, online production SAGE computers would have 55,000 vacuum tubes and needed to run around the clock. Forrester studied the problem and concluded that silicon contamination in the nickel metal cathode was the culprit. Working with Sylvania, silicon-free cathodes dramatically increased average tube life. Nat Rochester was an engineer at Sylvania at the time and used the improved vacuum tubes in the SAGE arithmetic and logic unit (ALU) design.

The SAGE computer's formal name was the AN/FSQ-7, which stood for "Army-Navy Fixed Special eQuipment." This indicates that all the good names for computers were taken, and whoever coined the acronym forgot that the Air Force now funded the project. Each SAGE center would have two AN/FSQ-7 systems, each weighing in at 250 tons and taking up an acre of floor space[20]. There were 50 displays with light pens attached to these duplexed systems. With core memory and the 16 parallel ALU computing circuits, the SAGE computers were shockingly fast, at 50,000 additions and 33,000 multiplications per second[21]. The two "A" and "B" computers provided for redundancy and real-time maintenance, an innovation that would later be termed fault-tolerance.

Each one of the 23 SAGE centers consisted of a four-story block-house built to withstand significant blasts, even nuclear explosions. The SAGE centers received data on incoming threats from a large network of radar stations. The first site was operational in New Jersey in 1958. The last site went online at North Bay, Ontario, in 1963. SAGE was designed to detect enemy bombers, but by 1963, intercontinental ballistic missiles (ICBMs) made the system virtually obsolete. Still, the system formed the backbone of NORAD (North American Aerospace Defense Command) until the 1980s when the Air Force replaced the SAGE systems with faster systems from Hughes Aerospace. Two AN/FSQ-7 systems remain—or at least parts of them do, one at the Boston Computer Museum and the other at the Mountain View Computer Museum.

The total SAGE project cost was about $10 billion, representing three times the cost of the Manhattan Project. Tom Watson was right in seeing the importance of winning SAGE. The contract to develop and manufacture 56 SAGE computers was worth in excess of $500 million[22]. That represented 4 percent of total IBM revenues and 80 percent of IBM stored-program revenues for the years SAGE was under development. At its peak, IBM had 7,000 employees working on SAGE. Tom created a new division—Federal Systems—to run the project and others like it. The first two SAGE prototypes required 589,000 ferrite cores[23]. There was also the added benefit of winning the contract from Remington Rand. SAGE would have been a much-needed lifeline for the UNIVAC division where Presper Eckert was wasting his limited engineering resources on the Livermore LARC program.

Perhaps of greater value to IBM than the financial impact was the technology. It accelerated IBM's adoption of magnetic core memory. Other SAGE innovations such as fault-tolerant real-time processing, duplexed systems, printed circuit boards, and interactive CRT devices with light pens would find their way into future IBM systems. Most immediately, the SAGE experience went directly into the 701 replacement, the IBM 704.

# AMERICAN AIRLINES AND THE SABRE SYSTEM

In the summer of 1951, Blair Smith, an IBM marketing rep, was traveling from California for a pre-announce class on the 702 computer in Poughkeepsie[1]. He was sitting in the back row of an American Airlines DC-6 flight to New York City. It was supposedly the safest seat on the plane. An older man sat down in the seat next to him. He needed a shave and a change of clothes. The stewardess stopped and addressed him as "Mr. Smith." Blair remarked that they shared the same last name. The older Smith was very talkative and soon had ferreted out Blair's entire life story. He discovered that Blair was a sales rep for IBM and had sold a 701 to Douglas Aircraft. Blair mentioned that he had in fact just made a sales call on Donald Douglas, with Tom Watson Jr. joining him. The older Smith then told him that Tom Watson was a good friend of his. It was at this point in the conversation that Blair Smith realized that he was talking to the CEO of American Airlines. Cyrus Rowlett "C.R." Smith had been head of the airline since 1934. It turned out that C.R. would leave his New York office at the spur of the moment to check out flight operations on the West Coast. He would hop on the ten-hour propeller-driven plane trip without bags or even a toothbrush.

A true salesman, Blair Smith launched into a description of the upcoming IBM 702 computer and, having some experience in airline reservations, described how the new computer might solve the airline's reservation problems. C.R. asked him to visit the American Airlines reservation center at LaGuardia Airport and follow up with him on what he thought, including what was wrong with it and what IBM might be able to do. Blair Smith was able to discuss the chance meeting with Tom Watson at the 702 class, where he received the go-ahead. Within 30 days, Smith sent a letter to C.R., copying Tom, outlining the possibilities of a new online reservations system and suggesting a joint development effort.

Blair Smith was uniquely qualified to take advantage of the meeting with the American Airlines CEO. He had started on IBM tabulator equipment at the Agricultural Adjustment Administration (AAA) in 1934 then moved to Boeing, where he integrated the 601 Multiplier into the company's card operation. Wiring plugboards became his forte. He began working at California Shipbuilding in 1942, a start-up that would build over 400 Liberty ships during the war. In 1948, he moved to Western Airlines as its punched-card manager.

In 1950, IBM hired him as a sales representative. While still at Western Airlines, he ordered a machine late in the year that helped the local IBM branch reach 100 percent. The branch manager returned the favor by hiring him. The manager also broke the news that Smith would have to shave his moustache. Smith skipped all the preliminary punched-card classes and went straight to Sales School in Endicott, where they were still singing IBM songs. He was on quota in September 1950 and made a full year's quota in three months, in part by selling 701 systems to the RAND Corporation and Douglas Aircraft. He also took over the Northrop account. He was actually there the day they delivered the Wooden Wheel and removed the BINAC computer. Given his territory in the center of the aircraft industry, he started the Digital Computer Association user group to allow interchange among his 701 accounts. It was very unusual for someone like him to have spent 16 years in technical positions and switch to sales. However, the knowledge and experience just super-charged him in sales situations. IBM promoted him to headquarters as manager of market analysis.

Blair Smith did not forget about his conversation with C.R. Smith.

He went to LaGuardia and assessed its current reservation operations. His work at Western Airlines had already partially informed him of what he would find. C.R. had taken over the airline when it had just 85 planes. He said early on, "Any employee who can't see a day when we will have a thousand planes had better look for a job somewhere else[2]." American's reservation system began as a manual operation, using large posting boards in main airport locations. Agents needed to exchange messages to track the seat count of each flight, resulting in a process that could take up to 45 minutes. Charles Amman at American set up a system in 1944 with the Teleregister Company that electrically connected the flight boards at American's main office with similar boards at the outlying airports. This system, called the Reservisor, would only signal if a flight was below 75 percent booked, a safe level to allow a remote agent to proceed unilaterally with ticketing. All subsequent tasks, including queries of alternative flights, remained manual.

By 1952, even this rudimentary system was breaking down. This was at the time when magnetic drum computers were making the news. Amman worked with Teleregister to migrate the airline's existing system to run on a drum-based computer. Teleregister installed the new system, the Magnetronic Reservisor, at La Guardia about the time that the two Smiths were having their in-flight conversation. The new Reservisor had a single drum containing the number of open seats on each American flight. It turned on a light on the agent's desk device to indicate availability. Though this new system was a step forward, it still required the involvement of several different agents, took as long as three hours to complete a single booking, and was prone to errors. Meanwhile, C.R.'s vision of 1,000 American Airlines planes was well on its way, along with an impending fleet change to far faster Boeing 707 jet. Call volumes would rise to 83,000 per day[3].

It took some time for IBM and American Airlines to define a workable approach. They reached a preliminary agreement in 1954 to produce a new system that would maintain 40,000 reservations and handle the 83,000 calls per day, while providing access to reservation records in less than three seconds. American planned to deploy the system to 100 company locations and over 1,100 travel agents via leased lines[4]. The project got going in 1957, acquiring the name SABRE, for Semi-Automated Business Research Environment. Tom Watson personally selected Perry Crawford to lead the design. As you may recall, Crawford had written his MIT doctorate thesis on digital logic, influencing both Eckert's ENIAC counter design and Jay Forrester's switch on Whirlwind from analog to digital. Crawford joined IBM in 1952.

With the American Airline project, progress was slow despite pressure from the IBM sales team. Robert Head, one of Crawford's senior engineers who had joined IBM from GE, commented, "At IBM, make no mistake, the salesman was king.[5]" The sales team pushed numerous changes to the project scope, resulting in modifications that would end up comprising 90 percent of the completed system and slowed development.

In the end, the extended timeline would work in the project's favor. IBM was in the midst of the SAGE program and the SABRE team could incorporate many of its innovations to support rapid, real-time function. The project would also benefit from San Jose's ongoing work on the disk drive. The original 350 disk drive was too limited in capacity and speed for reservations. The San Jose lab was working on the next-generation drive, the Advanced Disk Facility. The new drive would fit nicely into the SABRE timetable. There was also the changeover from vacuum tubes to transistors. IBM would have to work through the 709, the replacement for the 704, and the massive gamble that was the Stretch project before arriving at the right computer for SABRE. That would be the transistorized 7090 that arrived in 1960.

With the drawn-out timeline for SABRE, the idea of a prototype or experimental system made sense. Plans immediately coalesced around the new IBM 650—small, powerful, easy to program, and available. IBMer Bill Elmore built an impressive demo system. It required fully 25,000 lines of code[6], which hinted at the scope of the full SABRE programming. IBM would need to program the far more extensive application in assembler to ensure adequate performance. In addition, IBM would develop a custom operating system called ACP (Airline Control Program) from scratch. IBM applied upwards of 200 developers to the task of writing over a million lines of code[7].

Work proceeded from 1955 to 1957 without a signed agreement from American. At that point, Learson stepped in and asked Blair Smith to evaluate and determine if this was a "go" or "no go." Blair determined that he needed another "chance" meeting with C.R. Smith to secure a commitment. He hung around the coffee shop at American Airlines headquarters until he bumped into C.R.[8] He told C.R. that he was having trouble getting a decision from American's marketing and sales VPs, who remained unconvinced of the savings identified for the new system. With intervention by C.R., American and IBM soon signed a contract for $28.5 million and the program was back on track.

## THE ADVANCED DISK FILE

In the San Jose lab, Rey Johnson began development of a new disk drive even before the launch of RAMAC. John Haanstra took the lead in assessing the market need and developing the specifications for the Advanced Disk File (ADF). The goal was to increase access and density by factors of ten. A key ancillary goal was the elimination of the huge amounts of compressed air used in the original 305 disk drive. With the success of RAMAC, Rey Johnson secured funding and was able to move his team to the now-completed plant and lab on Cottle Road in San Jose.

Work got underway in earnest in 1957. Alan Shugart was named engineering manager. He had joined IBM in 1951 as a punched-card customer engineer and worked on RAMAC. Al Hoagland and Jack Harker were among the many engineers on the program[9]. Harker had graduated from Berkeley High in 1946 and enlisted in the Navy's radio electronics program. He left the Navy as an electronic technician's mate second class and enrolled at the University of California Berkeley. The school was crowded with returning servicemen, so he transferred to Swarthmore, where he graduated in 1950. He returned to Berkeley for graduate studies. In 1952, he spotted the small ad that Lou Stevens had placed in the *Daily Cal* newspaper looking for engineers. He was the lab's eighteenth employee.

Harker and Hoagland were essential in the new disk drive's transformative design, which pioneered several major enhancements over the original 305 drive. They added separate read/write heads for each recording surface, arranged in a comb-like unit that would snake in and out of the stack of disk platters. This unit would rest in the middle of the disks in order to achieve the fastest access times when activated.

The engineering team redesigned the read/write heads to be aerodynamic. They would skim over the disk surface naturally without the need for forced air. The new heads would also "fly" lower, slipping down to 250 micro-inches, compared to the flying height of the 305 drive at 800 micro-inches[10]. This increased the recording density of the platter by a factor of 13. The disk rotation speed increased from 1,200 RPM to 1,800 RPM, resulting in faster access times. One element that did not change was the preparation of the disks. Technicians still poured the ferrite oxide magnetic slurry over rotating aluminum platters, which then cured in an oven. The team did dispense with the Dixie cups and nylon stockings, though.

The resulting drive, the IBM 1301, provided up to 28 million characters per unit and 500 million characters across 16 units. With the dramatic increase in capacity and performance, Alan Shugart had to manage pressure from the SABRE program team, which was anxious to get the new drives. IBM promoted Haanstra in the middle of the project to an executive position at Corporate.

Engineers working on this Advanced Disk File program wrote technical articles for the *IBM Journal of Research and Development* in 1959. The engineers also presented at the Joint Computer Conference. This published information, which should have been IBM trade secrets, led three companies—Bryant, Computer Products, and Telex —to make competing, cloned disk drives. This would grow into an immense and ongoing problem for IBM and the San Jose disk operation.

At the center of SABRE was the custom agent set, a fully interactive display station. This consisted of an IBM Selectric printer-keyboard and an attached console panel. Once the caller identified the desired flight leg cities, the agent would select a schedule card for that route that identified all of the flights, both on American and other airlines. The console displayed the availability of each flight with panel lights. The agent reversed the schedule card to select the return flight. The agent could then complete the booking by typing the passenger information on the Selectric keyboard.

The project, which had started in 1954, went into trials in 1963 and full operation in 1964. The SABRE system ran on two transistor-based 7090 computers, with one doing online reservations and the other processing batch jobs but ready to take over. The new Advanced Disk File provided 50 million characters of storage per unit and 800 million characters total. By 1965, the SABRE system was handling 7,500 reservations an hour in addition to handling 30,000 fare quotations, 40,000 passenger confirmations, and 40,000 inter-airline requests[11]. IBM announced the SABRE system as the IBM 9000 Series Airline Reservation System and began marketing to the other airlines.

The SABRE program experienced severe cost overruns, with the final development tally running to at least $40 million. There were no repercussions as the investment firmly established IBM as a leader in real-time commercial applications. American, Pan Am, Delta, and Eastern would sign up for the IBM system. United went with UNIVAC, and TWA selected Burroughs, but IBM eventually won both of them back.

American Airlines and the SABRE system would have an unusual relationship over the years. The airline would spin off SABRE as a separate company in 2000 and later leave the reservation system altogether, only to return. U.S. Air sued the SABRE holding company in 2011 for favoritism on flight displays. American Airlines would acquire USAir in 2011 and decided to continue the suit against the very organization that it had founded. The issue was decided in American Airlines' favor in March 2022[12].

# STRETCHING COMPUTER POWER

E dward Teller, one of the Hungarian "Martians," relocated from Los Alamos to the new Lawrence Livermore National Lab at Berkeley to continue work on the hydrogen bomb. The establishment of a puppet state in Hungary by the Soviets in 1949 only served to increase his dedication to the effort. Heeding John von Neumann's advice, Teller ordered and installed a fast computer, the UNIVAC #5, which CBS temporarily requisitioned for the 1952 presidential election. The UNIVAC's limited memory soon became a problem in modeling nuclear fission implosions. Programmers needed to split their programs into smaller segments to run on the machine. In 1951, the lab ordered an IBM 701. It was significantly faster than the UNIVAC, easier to program, but not as reliable. It was upgraded to two 704 systems in 1954. They provided twice the performance of the 701, with more memory, floating point support, and far greater reliability.

Yet despite this substantial progress in computing power, Livermore was already looking for an even faster machine in 1955. The lab invited proposals for this faster system, called the Livermore Advanced Research Computer, or LARC for short. Cuthbert Hurd led a team, along with Steve Dunwell and Gene Amdahl, to meet with Teller and the lab engineers. Amdahl developed a preliminary design for a machine that would be five times the stated requirement, delivered in 42 months at a cost of $3.5 million[1]. The lab received a competing bid from UNIVAC that met the specifications and would be ready in only 29 months for $2.85 million. Before the lab made their decision, Ralph Palmer withdrew the IBM bid. He had determined that IBM was on the cusp of a technology change from vacuum tubes to transistors and the proposed IBM machine would be obsolete even before it arrived at the lab.

Remington Rand won the LARC contract by default. Perhaps disappointed at the time, this would turn out to be a very good deal for IBM. UNIVAC would spend the next five years developing the LARC, not the 29 months it had promised. Though the LARC was a fully transistorized system, Palmer was right in his prediction that it would be obsolete on day one. Furthermore, the LARC system used limited and expensive core memory, was difficult to program, and was extremely bulky. The lab had to construct a new building just to house it. It did achieve the stated goal of 125,000 floating-point multiplications per second, roughly 300 times the speed of the UNIVAC I[2]. It would be the world's fastest computer for less than a year, in 1960[3].

More importantly for Remington Rand, LARC was another financial disaster. Bid at the price of $2.85 million, it cost at least $19 million to develop, if not more[4]. The UNIVAC division produced only two LARC machines. Remington Rand merged with Sperry in 1955, and Harry Vickers replaced James Rand as CEO. Vickers was under the impression that he was now at the helm of a leading computer company. He was also unfamiliar with the estimating skills of J. Presper Eckert. For IBM, the major benefit of losing the bid was having Eckert fiddle away his division's engineering resources on LARC instead of the mainline UNIVAC business. IBM would cruise by UNIVAC with the 700 series and never look back.

Despite the loss at Livermore, IBM was not willing to walk away from supercomputers. With Tom Watson's support, Ralph Palmer asked Steve Dunwell to lead an effort to develop an IBM machine of breakthrough speed. There were a number of different approaches to consider, including an enhanced NORC or a modified SAGE design. Engineers in the Poughkeepsie lab were in the midst of exploring a concept called the "Datatron." Dunwell and Werner Buchholz, later joined by Gene Amdahl, started working on the project in 1954. They discussed their evolving design over coffee meetings and documented in it in a series of updates called the Datatron memos[5]. There would be 63 of these memos spanning the period from 1955 to 1959. As the urgency of the project heated up, Palmer asked his key engineers to forgo vacation plans for the summer of 1954 to pull together a formal specification. He had Dunwell, Amdahl, Buchholz, and several other engineers in these sessions. What emerged was a machine that would potentially have performance that was 100 times that of the existing 704. Palmer and Cuthbert Hurd agreed that the Datatron concept had the most potential.

Hurd went back on the road to test interest in such a super-fast system. He went to Los Alamos, the Rand Corporation, and the National Security Agency. In April 1955, he took the team back to the Livermore Lab, where Amdahl pitched the concept at 40-50 times the performance of the 704. That September, with more confidence in the emerging plan, Dunwell and Amdahl visited Los Alamos with the revised Datatron plan, now estimated at 100 times the 704 performance. Both Livermore and Los Alamos would sign letters of intent. Hurd also received commitments from the National Security Agency and the RAND Corporation. At the 1956 Eastern Joint Computer Conference, Dunwell presented the Datatron concept while Presper Eckert presented the LARC[6].

Hurd's commitments were sufficient to secure support from Learson and Tom Watson. Ralph Palmer put together an impressive team for the project, adding new talent to the core of Dunwell, Amdahl, and Buchholz. James Pomerene had joined von Neumann's IAS project and by 1946 was its chief engineer. IBM hired him in 1956. Ted Codd was an Englishman who had degrees in mathematics and chemistry from Oxford and flew for the Royal Air Force during the war. He was one of Rex Seeber's hires on the SSEC, working as a mathematical programmer. Erich Bloch was born in Germany. He was attending college in Zurich when his parents perished in the Holocaust. He emigrated to the United States and studied electrical engineering at the University of Buffalo, then joined IBM in 1952. He would become an expert in core memories and electronics packaging. Harwood Kolsky was a professional hire from the Los Alamos lab.

Uniquely qualified, he had a physics degree from the University of Kansas, worked with Dunwell in the Army SIS during the war, and got his physics doctorate from Harvard in 1950. He started working at Los Alamos the same year. Dunwell convinced him to move to IBM.

Dunwell found a couple of additional key members by hitting the universities. John Cocke had just graduated with a PhD in mathematics from Duke University. He would later become far more famous as the father of RISC computing. Dunwell enlisted Fred Brooks at Harvard. When he was thirteen, Brooks had read about the Harvard Mark I in *Time* magazine. He followed an undergraduate degree in physics at Duke with dual studies in mathematics and physics at Harvard. He was anxious to get into the computer field, but his faculty advisor said, "You're too late. You can't get in on the ground floor." However, "you can catch the first landing![7]" He started in the Aiken Computer Center, with the Mark I on one side of the room and the Mark IV on the other. Howard Aiken was his faculty advisor and Ken Iverson was his graduate instructor. Brooks described Aiken as "domineering." Though known as "the boss", he worked hard with his graduate students. He met with them at the end of each weekday for coffee. Iverson stayed at Harvard for five years but left when denied tenure. He joined IBM in 1960 in Research and invented the APL programming language. The company named him an IBM Fellow in 1970 and he won the prestigious Alan Turing Award in 1979[8].

Brooks kept his computing focus during summer breaks. He spent his first summer in Endicott, working at IBM on punched-card equipment. When the factory shut down for two weeks in the midst of the summer, he went to IBM school and learned to program the 701, which was still in development. Since no assembler was yet available, Brooks wrote his programs in machine code. The next summer he worked for Bell Labs. The third summer found him at North American Aviation, programming a missile tracking application on the company's new 701 system. He also wired a plugboard application. The last summer he worked for Marathon Oil Company in Ohio, developing a 40-state payroll application on Marathon's 650 computer. This work became the basis for his graduate thesis. While at Harvard, Aiken enabled him to meet John Mauchly, Presper Eckert, and Konrad Zuse. He programmed the Mark IV, which had no compiler or assembler, just a relay box to encode machine instructions.

Steve Dunwell hired Brooks, as well as his wife, who would work on transistor design under Werner Buchholz[9]. While at Harvard, Dunwell also hired Gerrit Blaauw, another graduate student supervised by Aiken. Blaauw had grown up in the Netherlands and won a Watson scholarship to study in the United States.

This high-powered engineering team faced the challenge of designing a machine with 100 times the performance of the 704 at a time that no individual component technology would come anywhere close to that objective. Using the latest transistor circuitry, core memory, microcode, and instruction design would still leave you significantly short of the target. Core memory was now six times faster. Leading-edge transistors were ten times faster. The design team believed that they could combine those improvements in hardware speeds with architecture changes and software tricks to bridge the gap to 100-fold performance. The idea was to "stretch" current technology in this manner rather than wait years for normal technological progress. Given this, they changed the name of the program from Datatron to Stretch. From the beginning, there remained a difference of opinion as to what was possible. Fred Brooks estimated that performance could be as high as 200-fold times the 704 while Gene Amdahl was far more conservative at 40 to 50 times that of the 704.

That Stretch would be a fully transistorized machine was a given. The Soviet Sputnik launch in 1957 had created a frenzy to upgrade U.S. technology. The Air Force had declared that it would only consider computer proposals based on transistors. IBM had been experimenting with transistors to replace vacuum tubes for many years, but the economics remained a reach. In 1954, Gordon Teal of Texas Instruments improved the transistor by substituting silicon for germanium as the substrate material. Silicon avoided the "leak" of bit on or off signals that characterized germanium. Silicon could also function at a greater temperature range. To publicize its improved and less costly transistor, Texas Instruments produced a transistor radio the size of a deck of cards that sold for $49.95. Tom Watson bought 100 of the radios and passed them out to his executive team[10]. His not so subtle message was that IBM risked being late to transistors if they did not move faster. He followed this ploy with a formal mandate. Wally McDowell defined the objective in 1958, stating that all future IBM machines would be based on solid-state electronics, which meant transistors. IBMers knew it as the "solid state in '58" memo[11].

Stretch would have transistors—lots of them. The design called for each machine to have 210,000 transistors[12]. The higher density and increased reliability of transistors permitted such a dramatic escalation in numbers. In addition, Ralph Palmer's decision to withdraw from the LARC proved correct, as the transistors of the time were not well suited to the high-speed switching circuits found in IBM computers. The drift transistor arrived soon thereafter and addressed Palmer's concerns. Given the huge numbers of drift transistors needed for Stretch and later IBM systems, Palmer decided to build an automated manufacturing process to produce them economically and profitably. The resulting automated line produced 3,600 transistors per hour[13]. The price of transistors had been $80 at the outset. The automation cut the price to $1.50[14]. Palmer then decided to move the transistor line to Texas Instruments and focus on the packaging of the transistors onto the processor cards that would go into each system. In just five years, IBM would produce its ten-millionth transistor, and still purchased more from outside sources.

Palmer would tap Joseph Logue to combine the transistors into a package that would go into the Stretch machine. Logue had been a professor of electrical engineering at Cornell University. He joined IBM in 1952 and designed the vacuum-tube pluggable modules for the IBM 650. He also had a critical role in upgrading IBM's electronics skills. He conducted classes, called Logue's "College of Digital Knowledge,[15]" which focused on transistors. He had one of his classes transistorize the 604 Calculator, replacing the 1,400 vacuum tubes with 2,200 transistors. The need to combine a large number of transistors with additional electronic components led to a packaging format called the Standard Modular System (SMS). SMS used 2.5-inch by 4.5-inch boards that held three to six transistors along with resistors, capacitors, and diodes on one side and connectors on the other. The completed SMS modules shipped in bubble wrap, one of the first uses of this new material[16].

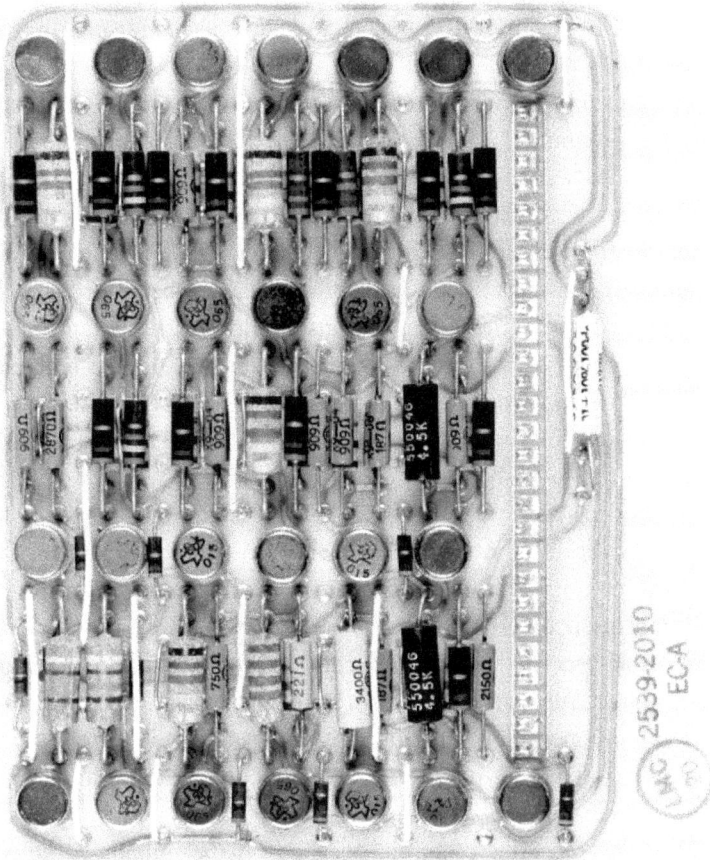

**Figure 31.** Stretch SMS (Standard Modular System) logic card. Each system had 4,025 double cards and 18,747 single cards for a total of 169,000 transistors plus myriad diodes and resistors. The transistors are the silver circles in the image. Reprint Courtesy of IBM Corporation ©.

The design team's focus was "stretch" performance. There was a dramatic difference in speed between the new transistors that comprised the system's arithmetic and logic processor and the speed of core memory. The transistors were cycling at 300 nanoseconds while core memory was seven times slower. The obvious solution was to figure out ways to feed the central processor more work. The team proceeded in three major areas: (1) early preparation of future instructions, (2) processing multiple instructions in the processor in parallel and, (3) buffering input and output to keep the processor running at full speed.

The first initiative involved the staging of multiple instructions from the program[17]. However, because of loops and branches, a program did not always proceed serially through the program. Stretch would proceed in parallel with current processing to identify and load the next instruction. This could be the next physical instruction in the program or a guess as to what instruction was next based on prior flow. Both the next and "guess" instructions would be fetched and queued up for the processor, with up to 11 instructions in some form of execution. This queuing would be called pipelining. A problem arose when the guess next instruction was incorrect. To address this, a holding area called the look-ahead buffer was created that would include the most recent instructions processed. If the system determined an out-of-order condition, it would back out the instructions from the look-ahead buffer until it reached the point of divergence from the real program sequence. The look-ahead buffers would take up 16 percent of the transistors in the machine. Instructions fetched in the correct sequence were run in the processor, four at a time. Stretch could attain a peak rate of one million instructions per second when the predictive instruction fetch and parallel execution combined with the faster speeds of the hardware.

The last area of "stretching" focused on the imbalance between the ponderously slow speed of input and output and the speed of the processor. The team designed a buffer processor called the Exchange that independently managed I/O units such as card readers and printers, leaving the central processor free to keep running. This innovation would later become the familiar channel of IBM mainframes.

Stretch was, of course, a binary machine. It would convert back and forth from decimal and binary as needed. A key change was the expansion of decimal notation from six bits to eight bits. Werner Buchholz would become famous for coining the term "byte" to describe this collection of bits[18].

Multiplication was of central importance with Stretch. With binary, there was no need for the single-digit multiplication tables used in decimal machines. The Stretch team opted to use octal (base eight) representation, with octal representing a shorthand for binary. This approach reduced a 48-bit binary multiply to sixteen carry and save additions followed by one normal addition. As a scientific machine, Stretch needed to support floating-point arithmetic. For this purpose, the team created a separate processor unit comprised of tens of thousands of transistors. This resulted in a floating-point addition in 1.2 microseconds and a multiplication in just 2.4 microseconds[19]. The ratio of multiplication to addition speed, which was 67-to-1 for the SSEC, was now 2-to-1 in Stretch. For raw multiplication performance, Stretch was 80 times faster than the 704.

A final architected change for throughput in a real-world setting was the ability to execute multiple programs at once, or multiprogramming. Stretch pioneered an interrupt mechanism that managed the switch from servicing one program to servicing another.

Gene Amdahl felt that Nat Rochester had designated him as the lead architect on Stretch, or at least that was his perception. Based on his lead roles with both the 704 and the LARC proposal, he developed very specific ideas on Stretch architecture. Dunwell wanted to explore how the Stretch design might be used by commercial accounts. This had been, in fact, the original plan for the 709 system. It would combine the planned 705, a commercial system, with the 704, a scientific system, to create a system for both environments. Amdahl argued that this would impede the laser focus on maximizing speed, and thus affect scientific value. He was able to win that battle. Once the 705 became a separate program, Dunwell invited Gene's input on its architecture. He ended up not using any of Amdahl's ideas, much to Gene's irritation.

The debate on a commercial focus for Stretch continued through the summer of 1955. Amdahl would not win this battle. In the end, Nat Rochester told him that Dunwell was in charge of the Stretch project and would make such decisions. Amdahl went back to his office and drafted his resignation letter[20]. However, before submitting it, he called his brother in California, who worked for Ramo Wooldridge, a missile contractor. He wanted to know if he had a job there if he left IBM. With his already outsized reputation, the response was of course he did. He submitted the letter. In the heat of the moment, his immediate goals were to leave IBM and get back to

California. He seemed to forget about the part about what he would be doing. His job at the missile contractor was administrative management of government contracts, not computer design.

At the time of Amdahl's exit, Dunwell was 41 years old and had 21 years of experience in the computing field, with a lengthy resume of success. Amdahl was 32, and though he had shined in three major development projects, he had been with IBM for just three years and had limited experience in the management of large projects. After Amdahl left, a "Three in One" group met to pursue the commercial angle[21]. The concept was to split Stretch into three sections—the base system, a separate processor called Sigma tuned for scientific performance, and the Harvest extension for the NSA. Poughkeepsie adopted the plan in 1957 but dropped it a year later.

Amdahl was missed, but Palmer's Stretch team had so much talent that the program did not miss a beat. Taking advantage of Harwood Kolsky's former employment at Los Alamos, a joint planning group was set up between the mathematicians at Los Alamos and IBM to consult on the evolving design. They held ten meetings before the design was final in 1957. By this time, the core Stretch team in Poughkeepsie had grown to over 300. As potential costs soared, the marketing planners got involved. To produce a business plan that would not be under water, the planners cut the number of transistors 210,000 to 169,000. They set the price for Stretch at $13.5 million. The overriding issue with the program remained the performance promise. It was critical to track the team's progress to the 100-fold target. Kolsky and John Cocke built a simulator to gauge performance.

The completed Stretch machine became the IBM 7030. It was a fully transistorized computer with roughly a million bits of core memory. That was six times the memory of the 704. At the time, this was an opulent amount of memory. The electronics were oil cooled. The machine itself was 15 feet long and 5 feet deep. Harvest added another 20 feet in length. IBM delivered the system in ten boxes, each the size of a refrigerator, to facilitate transport in elevators. The console for Stretch was a modified IBM Selectric typewriter. Software support was minimal. There was an assembler—Stretch Assembly Program (STRAP), a rudimentary operating system with a small resident supervisor to support multiprogramming, and the FORTRAN compiler, which had just been announced.

**Figure 32.** The Stretch computer (IBM 7030) with project director Steve Dunwell. The system "stretched" computer performance and pioneered many aspects of modern digital technology. Reprint Courtesy of IBM Corporation ©.

Cuthbert Hurd received early commitments from Los Alamos, the Rand Corporation, the National Security Agency, and the Livermore Lab. Even though the price escalated during development, eleven customers signed up at the new $13.5 million price. Eight customers took delivery, with the Los Alamos and Livermore Labs installing in 1961. Those installations were followed by the Atomic Weapons Research Lab in the United Kingdom, the U.S. Weather Bureau, the Naval Weapons Lab at Dahlgren, Virginia, the French Atomic Energy Commission, the NSA, and the MITRE Corporation (a commercial subsidiary of MIT). Those installations comprised eight systems. Poughkeepsie build a ninth Stretch, which IBM retained.

The NSA Stretch was a custom installation that had its own

unique history. In 1952, the Army and Navy intelligence groups combined to form one organization focused on cryptography, the NSA. James Pendergrass, who had attended the Moore School lectures in the summer of 1946, pushed almost immediately for an EDVAC-type computer to do code-breaking work. The Navy spent millions of dollars with Engineering Research Associates (ERA), receiving the Atlas I and Atlas II computers. However, those computers were wholly ill suited for cryptographic work.

The Tape Processing Machine (TPM) would have been an excellent fit but it languished as IBM focused on the 701 Defense Calculator. Though the 701 was not perfect, the newly formed NSA went ahead and installed it, followed by its replacement, the 704. In 1955, the TPM finally arrived as the IBM 702. The NSA took delivery and would take five of the follow-on system, the 705.

John McPherson had chaired an advisory group that worked with the NSA. He leaned heavily on Steve Dunwell to evaluate how the Stretch computer might fit. Based on Dunwell's intimate experience with the process of decrypting coded messages, Dunwell realized that adapting Stretch would be a significant extension to the scope of the program. It would require two separate Stretch processors, one standard unit at the front end and the Harvest extension at the back end[22]. It would also require a radically new tape technology and a new programming language geared to textual analysis. Most importantly for IBM, it would require significant NSA participation in its development. IBM would be liable for the two Stretch-class processors and the development of the fast tape technology and programming. Quoted $3.5 million at the start, the NSA funded about $2 million in research. The Stretch system price would rise to $13.5 million and the Harvest unit and new tape drives added $3 million, which then rose to $5.5 million. On top of all that, the agency would spend more than $4 million programming the system[23].

Dunwell supervised the project, in addition to his overall management of Stretch. James Pomerene was the lead engineer. The Harvest "outboard" processor had to handle huge amounts of data. It would require 500,000 transistors just by itself. IBM called the new fast tape technology "Tractor." Just like the ambitious goal for Stretch, the Tractor objective was aggressive, a hundred-fold increase in tape speed. It was a cartridge tape system where each cartridge held 90 million characters of data, 15 times the capacity of standard IBM tapes. The Tractor system could hold up to 480 of these cartridges, providing 43 billion characters of data that were immediately accessible[24]. Tractor performance was many orders of magnitude faster that the Heath Robinson paper-tape readers that provided textual input to Colossus.

While Stretch had a huge and complex instruction set, it was dwarfed by the 735 instructions built into the Harvest processor. The instructions fell into categories that included data analysis, table lookups, and frequency computation. They were all in the service of the text analysis involved in the cryptographic mission. IBM's Frances Allen designed the custom compiler Alpha[25] that she optimized for cryptographic work. The NSA later extended the Alpha language, calling it Beta.

The NSA used Stretch to scour massive amounts of text looking for certain characters or sequences. It really was a modern, high-speed equivalent of Colossus. The NSA estimated that Stretch was 50-200 times faster than any other cryptographic system. It was operational from 1962 to 1976. A contributing factor to its retirement was deterioration of the Tractor tape units. IBM was not interested in developing a replacement.

IBM CEO Frank Cary attended the Stretch retirement celebration in 1976 with Jim Pomerene and Bob Evans. The NSA people in attendance had forecast that the system would have a life span of perhaps four years, but it kept plugging away for 14 years. Cary was visibly distressed when he heard this. On the flight back to New York, he ruminated that IBM's engineers had to back off building something so long lasting[26]. In hindsight, the NSA realized that the high total cost of the Stretch-Harvest program was a rounding error compared to the cost of the SAGE system. Frances Allen would become the first woman honored with the prestigious Turing Award, given only for the highest contributions to computer progress. She was also the first woman named an IBM Fellow.

The one and only litmus test of success for the Stretch program was the 100-fold performance promise. The simulator that John Cocke and Harwood Kolsky engineered was unable to provide definitive performance ratings during development given the complexity and interplay of all of the "stretching" innovations. The reduction in the number of transistors by the market planners certainly had an impact. The new drift transistors also performed slower than their specified speed due to the long connection lines between the packaging frames. The cycle speed of the core memory was also slower than predicted. There were other specific issues. The team could not improve division to the same degree as multiplication. The serial processing required on variable length instructions turned out to be slower than planned for. The architected interrupts for multiprogramming had the secondary effect of clearing the look-ahead buffers, which the system had to pause and rebuild.

By 1958, preliminary speed estimates were coming in far lower than planned. There were wide variances in specific comparisons with the benchmark 704. The advanced features of Stretch kicked in variably depending on the program. Programs that were memory constrained on the 704 would meet or even exceed the target. However, across a great many programs, the team was seeing results that were only 35 to 50 times the 704 performance. Even as late as December 1959, the simulation tests still came up at 60-100 times the 704. This provided some confidence that the target could still be met[27]. Bob Raney commented on his linear scalar program that ran in ten minutes on the 704. When executed on Stretch, the computer halted immediately. Raney described the ensuing troubleshooting: "After a half an hour or so of debugging and poking around the console, we suddenly realized the program had simply completed successfully, and within seconds"[28]. The estimate of Stretch's raw speed came in at one million instructions per second. However, the final consensus was far more sobering. It appeared that across a standard set of customer applications, Stretch was only reaching half the performance target.

Though Remington Rand's LARC computer missed its delivery promise by a wide margin and ran far over budget, it still met its performance target. This was not the case with Stretch, and it would soon become a monumental problem for IBM. In a Stretch Users' Group meeting at Los Alamos in the summer of 1961, there was little in the way of complaints about performance but rather the sense that

IBM knew it had come up short and was now working on Stretch II. There was, in fact, some tentative work done near the end of Stretch on future options that might be pursued under the name SuperStretch[29].

Those plans became academic once the Livermore Lab pressed the issue of the performance and argued for a lower price. Tom Watson engaged a consultant to analyze the performance shortfall. The resulting report concluded that Stretch performance indeed missed the goal by roughly 35%. Tom Watson reduced the price from $13.5 million to $8.5 million. Watson made the price cut for all Stretch customers, not just Livermore. This assured that Stretch would lose money and there would be no Stretch II. Tom announced the reduction at the Western Joint Computer Conference, noting, "Though we had the world's fastest and most capable computer, the specifications were not met." IBM's loss was roughly $20 million[30], which was unfortunate but a mere pittance in the company's finances. The resulting view that Stretch was an utter failure did not credit the program with the enormous impact it would have on other IBM products.

Tom was irate about the Stretch loss, but he was even more concerned about the damage done to IBM's reputation and the failure of the company's organizations and management structure to prevent the fiasco. He asked Learson to conduct a full review. The entire research and engineering community came under scrutiny. Learson's review, completed in June 1961, led to a further separation of research and development. Watson moved the IBM Research personnel that had worked in concert with engineers at Poughkeepsie to the new Research Division location at Yorktown Heights, also called "Mole-town"[31] as none of the offices had windows.

Though there was no Stretch II, there was a modest improvement program. Design changes in the branch prediction function and look-ahead buffering significantly reduced multiplication times. This positive result pointed to a more enlightened approach that Tom Watson could have taken. He could have committed to improve Stretch performance over time rather than cutting the price and walking away. In his 1990 book, Tom recognized his impulsive mistake and granted that his overreaction on Stretch chilled any work on supercomputers in IBM for many years. It also left the door wide open for Control Data to own that market.

The brunt of the repercussions fell on Steve Dunwell. Tom stripped him of his leadership role in Poughkeepsie and shunted him to a "penalty box." Within IBM, this meant a demotion to research, corporate staff, quality programs, or a sales branch in some far-off location. Dunwell's penalty box was a basement office at Yorktown Heights, working completely out of the mainstream on computer-aided instruction.

Despite the performance shortfall and the embarrassment to IBM, Stretch was the fastest computer in the world when it arrived in 1961, and it would remain so until 1964, when the Control Data's 6600 model would take over the crown. At the time of the Stretch debut, the ERA Atlas computer was the only computer that even came close. Stretch would perform 400,000 multiplications per second, with Atlas coming in second at 200,000, and Presper Eckert's LARC was at 125,000. The CDC 6600, designed by legendary computer architect Seymour Cray, would reach three times the performance of the Stretch.

More than that, Stretch pioneered many supercomputer and general computer innovations. The list of firsts included instruction prediction with the look-ahead buffer, pipelining, the eight-bit byte, the Exchange I/O processor, multiprogramming with program interrupts, and octal-based multiplication. Mike Flynn called Stretch "a bridge too far.[32]" He would lead the design of IBM's next supercomputer, the System/360 Model 91. Interestingly, many called his Model 91 a "cleaned up Stretch."

As for Steve Dunwell, he spent a long time in his "penalty box." On April 7, 1964, the day of the System/360 announcement, he drafted a letter to Tom Watson that said in part:

> "The new System/360 is in many respects the image of Stretch. It is important to me that you know this, for I hope that in time you will look upon the Stretch contribution to our technical heritage as an excellent bargain.[33]"

In 1966, Watson spoke at an awards dinner and took pains to single out Dunwell for his Stretch work. He went on to announce Dunwell's selection as an IBM Fellow. Dunwell was 62 years old at the time and had been with IBM for 41 years. He decided to take early retirement. Along with his wife, he renovated and operated the Poughkeepsie Opera House. Harwood Kolsky had a very high opinion of Dunwell. He viewed him as a superb manager of a large complex project and as a visionary, who had the audacity to throw caution to the wind and reach for a hundredfold increase in technology in one gigantic leap. He said of Dunwell, "His real genius was that he saw where IBM should be five or ten years hence and was able to put together a huge project over the endless objections of everybody.[34]" The exciting project attracted the best young computing minds. It was no accident that quite a few other IBM Fellows came from the Stretch project, including Harwood Kolsky, Nate Rochester, John Cocke, James Pomerene, Robert Henle, Ted Codd, John Backus, and yes, even Gene Amdahl[35].

## STRETCH BECOMES THE 7090

IBM won the contract for the Ballistic Missile Early Warning System (BMEWS)[36] to detect incoming ICBMs. The plan required delivery of a 709 system to get started but also specified replacement with the follow-on machine for the 709. The new 7090 was the immediate beneficiary of the Stretch program and was not be burdened by its costs. The 709 was successful but tube-based. The 7090 would be transistor-based. Bob Evans, for one, could not wait to move to transistors and led the 7090 project.

Robert O. Evans, known as "Bob" or "Boe," grew up in a small Nebraska town and majored in electrical engineering at Iowa State University in Ames. He joined IBM in 1951 as a junior engineer working on the 701[37]. During that program's rollout, he set up the nationwide troubleshooting and support group for customers. Jerrier Haddad noticed and promoted him as his Administrative Assistant (AA). He then tapped Evans to run IBM's Navy operations, which included the development of the NORC computer. Early on, Evans saw how engineering could rework the Stretch transistor design to produce a transistor version of the 709. With the threat of competitive transistor machines looming, he pursued some back channels to get the field to say they needed such a system. Evans was a strong barrel-chested man used to getting his way, yet universally admired by those who worked for him.

Work actually began on the 7090 program in the midst of the 709 development. It incorporated many of the Stretch innovations, including silicon-based drift transistors, fast memory, and disk. It also used the new electronics packaging, the Standard Modular System (SMS). At first, it was simply the 709-T program. There was some tension between the 7090 and Stretch groups, as both were looking to create new benchmarks of computer performance. IBM announced the 7090 just a year after the 709 debut. It actually reached the market before Stretch shipped. It provided six times the performance of the 709. The substantial gain in speed was the result of the combination of 50,000 drift transistors replacing vacuum tubes, and new, fast core memory. The 7090 was dramatically smaller in footprint than either the 704 or 709, while compatible with each. It was a classic IBM mainframe, with a classic mainframe price tag: $3 million to purchase and $65,500 per month to rent. The Naval Ordnance Lab installed two 7090 systems to augment the processing of its NORC computer. The BMEWS contract resulted in four 7090 systems.[38] Over five hundred 7090 systems were sold, representing a total of $1.5 billion in revenue[39].

# IBM IN THE 1950S

I BM continued to do well with Tom at the helm. Between 1956, the year he assumed command, and 1964 when the System/360 revolution began, IBM revenues grew from $696 million to $3.2 billion, an average gain per year in excess of 18 percent. Income increased from $73 million to $431 million, an average jump per year of 22 percent. The IBM workforce increased from 56,000 to 150,000. Tom and Al Williams stopped taking options in 1958, with Tom commenting, "we don't want to look like pigs[1]." By 1966, Tom's profit sharing was down to a quarter of 1 percent. That would still translate into $1.3 million on income of $526 million.

Meanwhile, IBM's competitors were struggling. As IBM surged to the industry lead, the trail of competitors left behind was known as the BUNCH, standing for Burroughs, UNIVAC, NCR, Control Data, and Honeywell. This group became known as the Seven Dwarfs when RCA and General Electric entered the field. Burroughs had started when William Burroughs, a bank clerk, invented an adding machine in 1880, patented it in 1888, and started American Arithmometer. The company entered computers in 1956 with the purchase of Electrodata, a division of Clifford Berry's Consolidated Engineering Corporation. The Datatron 205 drum computer became the Burroughs 205[2]. The company's computer business grew with both low-end bookkeeping machines and high-end mainframes. Their land-mark design was the B-5000, arriving in 1962. It was transistor-based with virtual memory and other cutting edge technology features. Robert Barton designed the machine. He got his start in computing working for Cuthbert Hurd in the IBM Applied Science group, then followed in the footsteps of Irven Travis and Robert Campbell in joining Burroughs[3]. Burroughs would merge with Sperry in 1986 to become Unisys, a name that came out of an employee contest.

Bill Norris of ERA sparred with Presper Eckert at UNIVAC and the head office of Remington Rand in Norwalk[4]. Taking over responsibility for both the ERA and UNIVAC divisions in 1955 did not substantially improve matters. By 1957, he had tired of the political infighting within the company and led a number of engineers to exit ERA and form Control Data. The new company focused at first on its magnetic drum-based programs. Seymour Cray, a junior engineer, was a late volunteer to join the new company. Cray designed the transistor-based 1604 and set CDC on its path to supercomputer dominance.

In 1955, Honeywell participated in a joint venture with Raytheon to develop a computer. The result was the Datamatic 100. It was a behemoth, checking in at 35 tons and sporting 3,600 vacuum tubes and 500 transistors. It competed with the IBM 705 and had a faster tape drive, one that could read data going forward or backward. Honeywell would go on to buy GE's computer operations in 1964, getting GE's Bull subsidiary in France with the deal.

General Electric had early experience with computing, with the Bush differential analyzer. The company had one of the first commercial computer installations with UNIVAC #8 at Appliance Park in Louisville. It made a big splash in the industry by winning a $50 million contract for computerization at the Bank of America[5]. The company collaborated with MIT and Bell Labs on Multics, the precursor of UNIX. GE would work with Dartmouth College to develop time-sharing, carving out an early and successful niche.

RCA had blazed a glorious trail in radio and television. Founder David Sarnoff saw an equally bright future in computers and oversaw the company's entry into the industry in the mid-1950s. RCA could have been a major factor in the early days of computers with its Selectron vacuum-based memory tube, but struggled to perfect it, allowing core memory to take over. The company became a subcontractor on the SAGE program. RCA's first computer, the Bizmac, was very SAGE-like, with 30,000 vacuum tubes and 60,000 diodes. Its first transistorized computer, the RCA 501, was successful and applied RCA's vast experience in electronics packaging with the compact logic cards used in the system. It also took a page from the Watson playbook and had the industrial designer of its televisions and phonographs, John Vassos, design the modern look of the machine. Sarnoff died in 1970, and the company essentially walked away from computers the following year.

In 1951, NCR entered the industry by acquisition when it bought Computer Research Corporation. This was the company started by the MADDIDA engineers at Northrop Aircraft. In 1957, NCR teamed with GE to build its first computer, the NCR 304. ATT, which was NCR's largest account, bought the company in 1991. NCR reemerged as an independent company in 1997 and left the computer business the following year. It moved its headquarters from the iconic "Cash" location in Dayton to Georgia in 2009.

IBM and UNIVAC were neck and neck in market share in 1954 with 46 systems installed. A year later, IBM had 86 computers installed, with 190 more on order[6]. IBM exploded in share with the addition of medium-size computers like the 650, where IBM had 369 systems installed (seven times all the others) and 920 more on order (four times all the others). UNIVAC did not have an offering in this space. Both James Rand and Harry Vickers failed to have the vision, focus, and sales and marketing to maintain their early lead[7]. Though they had the opportunity to dominate the fledging computer market, they were unwilling to make the kind of investment required to compete with IBM.

# THE SPERRY TRIAL: WHO INVENTED THE COMPUTER

I n 1966, Richard K. Richards published the book *Electronic Digital Systems*[1]. At the front of the book was this statement:

> "The ancestry of all electronic digital systems appears to be traceable to a computer which will here be called the Atanasoff-Berry Computer. This computer was built during the period from about 1939-1942."

This was quite a statement, as the world already knew the first "brain" was the ENIAC and the inventors were John Mauchly and J. Presper Eckert. Richards had trained as a theoretical physicist and graduated from Iowa State College in 1943, a year after John Atanasoff and Clifford Berry left for war service. After Richards returned from the war, he joined IBM and spent seven years developing digital computer concepts. He wrote a technical book on his IBM work (*Arithmetic Operations in Digital Computers*)[2] in 1955. Coming from Iowa State and writing about digital computers led him to learn about the small computer project at Ames. He discovered a story in the *Des Moines Tribune* on the computer from January 15, 1941. Intrigued, he contacted Atanasoff, who was busy running his very successful engineering company. Atanasoff suggested that he talk to Berry.

Richards tracked down Clifford in Pasadena and started corresponding. Berry answered his questions and eventually sent him an extended letter with a detailed description of the computer.

In 1953, Robert Noll of IBM's patent office talked with his counterpart at Consolidated Engineering Corporation (CEC). CEC was concerned that the ENIAC patents might compromise the drum-based computer it was launching[3]. This led to a dialogue with Clifford Berry and the story of the ABC computer at Iowa State. When Berry brought up the visit of John Mauchly, IBM now understood the possibility that Mauchly derived the ENIAC from the Atanasoff-Berry Computer. IBM's patent attorney, A.J. Etienne, contacted Atanasoff[4]. In June 1954, IBM lawyers met with Atanasoff and asked for his help in overturning the ENIAC patents. IBM provided copies of the ENIAC patents, which Atanasoff had never seen. Atanasoff agreed to work with IBM. However, within a few months, the company soon stopped communicating.

Sperry Rand sued IBM in December 1955, accusing the company of monopolizing the punched-card market with a 90 percent share. IBM countersued Sperry six months later, citing infringement on 35 IBM patents. During this time, IBM apparently decided that it would be too difficult to press the case that Sperry's ENIAC patents were invalid. Instead, the company began negotiations with Sperry, and in 1956, agreed to pay $10 million for the ENIAC patent rights[5]. James Birkenstock negotiated the agreement for IBM[6].

With IBM royalties in hand, Sperry now targeted the other companies in the computer industry. It demanded upwards of $150 million from the group and viewed their patents as potentially worth a billion dollars[7]. Bell Labs was one of the Sperry targets. They refused to pay. In 1962, Sperry Rand sued. Judge Archie Dawson upheld the validity of the ENIAC patents. Bell was unable to prove any derivation with the ENIAC. The ruling served to embolden Sperry Rand and chill the others.

Sperry Rand filed a lawsuit against Honeywell and Control Data in 1967 in Washington, DC, citing patent infringement. It demanded royalties of up to 2 percent of the selling price of their computers. Honeywell's general counsel, Henry Hanson, was an Iowa State classmate of Richard K. Richards and came into possession of his book. In

April of that year, Allen Kirkpatrick, a patent attorney for Control Data, visited Atanasoff. He had also read the Richards book and asked for help in the Sperry lawsuit. Honeywell attorney George Call also approached Atanasoff with the same request. Atanasoff agreed to work the case, and Honeywell and Control Data decided to join forces. Atanasoff was more open to committing his time now as the effort appeared more likely to succeed.

The day that George Call visited Atanasoff coincided with a nearby computer science forum featuring John Mauchly as the guest speaker. Call decided to attend. The moderator introduced Mauchly as the co-inventor of the ENIAC, but went on to mention the Richards book and its conclusion that ENIAC was not the first computer[8]. Mauchly, shocked by the introduction, went to great lengths to defend his pioneering role and minimize any influence from Atanasoff or his machine. For George Call, this was quite a gift as it provided a preview of how Mauchly would testify if the case went to court[9]. With the information from Richards's book and the cooperation of Atanasoff, Honeywell and Control Data countersued Sperry Rand for monopolistic practices and fraud. The trial would be in Minneapolis with Judge Earl Larson presiding.

Henry Halladay was Honeywell's lead courtroom attorney. An outstanding trial lawyer, he was the beneficiary of a vast trove of documentation on the case collected and cataloged on a computer. Atanasoff had maintained extensive personal files, including all of his correspondence with Mauchly. After Mauchly gave his testimony, Halladay cross-examined, refuting each contention made by Mauchly. In most cases, he used Mauchly's own words from the letters with Atanasoff. In the end, he forced Mauchly to admit he spent five full days in Ames evaluating the ABC, that he received extensive schooling on the machine from Clifford Berry, and that he had unfettered access to the 35-page ABC manual. He also admitted the many additional conversations he had with Atanasoff when they were both working at the Naval Ordnance Lab. Mauchly's letters praising the ABC invention and citing its influence in his own work were the most damaging. The correspondence also made clear that Atanasoff was in the midst of patenting his digital computer and was only sharing information out of professional courtesy.

Once he agreed to participate in the lawsuits, Atanasoff moved to document the ABC project. He retrieved the famous 35-page ABC manual and additional design drawings. He contended that anyone could take the manual and designs, and build their own ABC system. And, that is exactly what he did. He commissioned his company's electronics expert build a replica of the ABC at his facility in Maryland[10]. He and his wife, Alice, tracked down 1940s-era vacuum tubes. He pulled all his correspondence with John Mauchly and Berry's master's thesis on the ABC. He also secured a deposition from his first wife, Lura, who corroborated Mauchly's intense interest in the computer while in Ames and the suspicious time he spent alone with the 35-page "green book". Atanasoff's depositions and court testimony ran to 1,252 pages[11]. His testimony was complete and persuasive, whereas Mauchly appeared unconvincing and evasive.

The most explosive revelation during the trial was the existence of the secret licensing pact made between IBM and Sperry Rand made in 1956. At the time, the two companies controlled 95 percent of the computer industry. The Justice Department was aware of the deal but said nothing. IBM had substantial prior knowledge of the ABC that questioned the validity of the ENIAC patents. Atanasoff had also written to IBM in 1940, offering to share his invention. That same year an IBM engineer visited Ames to review the system and Atanasoff would then meet with IBM's director of Market Research. A decade later, IBM was again aware of Atanasoff and the ABC, leading to discussions aimed at overturning the ENIAC patents. Yet, IBM decided to secure the best possible licensing deal with Sperry Rand and keep quiet. There was one further wrinkle: IBM had countersued Sperry Rand to contest the patents. With the licensing agreement in place, IBM elected to go through the motions of the lawsuit with sham representation aimed at losing its own case or getting a quick dismissal[12]. This kept the licensing arrangement secret and deprived the industry of the knowledge that IBM had gathered, which could have led to an early invalidation of the ENIAC patents.

The trial ran for five years and generated a transcript that ran over 20,000 pages[13]. Judge Larson posted his 246-page written decision on October 19, 1972. That was the day before the "Saturday Night Massacre," when Richard Nixon fired Watergate special prosecutor

Archibald Cox and both his attorney general and assistant attorney general resigned. That news overshadowed the trial result.

Judge Larson ruled the ENIAC patents were invalid. He cited the failure to apply for the patent within a year of first use. Edward Teller had testified for Sperry Rand. However, it was his use of the ENIAC in November 1945 for hydrogen bomb calculations that constituted this first use. Mauchly did not submit the patent application until June 1947, more than a year later[14]. In addition, Larson cited the derivation of key ENIAC patent elements from the ABC computer. It does appear this result was Judge Larson's goal from the start, as the ENIAC patents represented an antitrust danger to the entire computer industry. He also cited IBM's actions with Sperry as restraint of trade but noted that the statute of limitations had expired.

Finally, Larson recognized John Atanasoff and Clifford Berry as the true inventors of the digital computer. His statement read, "Eckert and Mauchly did not themselves invent the automatic electronic computer, but instead derived that subject matter from one Dr. John Vincent Atanasoff.[15]" Atanasoff's testimony was convincing. Judge Larson could have accused Mauchly of lying under oath. Judge Larson went on to state that ENIAC was the result of many inventors besides Mauchly and Eckert, including Thomas Sharpless and Frank Mural (the master programmer), Arthur Burks (the multiplier and system logic), Robert Shaw (function tables), Jack Davis (accumulators), Jeffrey Chu (the divider), Joseph Chedaker (electric circuits), and Harry Huskey (the reader and printer)[16].

The decision was not the end of the controversy over who invented the digital computer. John Mauchly and Presper Eckert continued to lobby against the ruling. Despite his testimony to the contrary under oath, Mauchly described the ABC as "a pile of junk that wouldn't do anything.[17]" Computer forums continued to introduce him as the "co-inventor of the first digital computer[18]" or "the co-inventor of the modern computer as we know it." He also tried to refute the testimony of John Brainerd, the dean of the Moore School, who argued that ENIAC was the product of a large team rather than the sole invention of Mauchly and Eckert. Mauchly continued to insist that ENIAC was a stored-program machine because the "instructions" resided in the Master Programmer. With his actions and public comments after the trial, it is easy to understand why his testimony was unsuccessful. He died in 1980, but his widow continued the fight, writing *Who Invented the Electronic General-Purpose Digital Computer* in 1984.

Presper Eckert remained at Sperry. He had basked in the ENIAC glow for nearly 30 years and received the National Medal of Science in 1969 from President Lyndon Johnson. The trial tarnished all that. In a 1991 speech in Japan, he commented, "The work by Dr. Atanasoff in Iowa was, in my opinion, a joke. He never really got anything to work[19]." Arthur and Alice Burks decided to rebut the public relations campaigns of Mauchly and Eckert, authoring an extensive technical analysis of the ABC and ENIAC. Arthur Burks was the principal architect of ENIAC. Alice Burks started as a "human computer" at the Moore School and became one of the ENIAC programmers. After the book-length analysis of Atanasoff and the ABC machine, the Burks summarized by noting the ABC had, " vacuum tubes to compute, condensers to hold numbers, and relays to control the activities, together with rotating drums and card-punching equipment, all working at coordinated speeds. You can hardly believe this has all been accomplished by one professor, one graduate student, a physics shop, and a few hourly workers – for less than $7,000."[20]

Though it occurred very late in his life, Atanasoff finally received proper recognition. The University of Florida and University of Wisconsin awarded him honorary degrees. The city of Ames organized a parade down its main street. Iowa State built an official ABC replica. It had no trouble finding the $350,000 in funding to construct it. The Cosmos Club in Washington, where Herman Hollerith had once roamed the halls, had previously inducted Atanasoff based on his mathematics and physics credentials. His invention of the digital computer was unknown at the time but was now recognized[21]. One of the Honeywell trial lawyers ran into Dr. Blagovest Sendov, a Bulgarian mathematician with the Bulgarian Academy of Science[22]. Sendov recognized "Atanasoff" as a Bulgarian name. He was elated that a fellow Bulgarian was now the official inventor of the computer. John and Alice Atanasoff visited Bulgaria at Sendov's invitation where they were grandly feted.

IBM knew in 1954 that the ENIAC patents rested on a shaky

foundation. One can assume that the decision to pay Sperry Rand $10 million in 1956 was directed by Tom Watson and agreed to by Watson Sr. True to his training by John Patterson at NCR, Watson Sr. always focused on amassing patent rights to protect the business.

Events could have turned out dramatically better for IBM had the company chosen the Henry Ford approach. You could simply substitute George Selden for Mauchly and Eckert to understand Ford's challenge. Selden was a patent attorney who conceived of a four-wheeled, powered horseless carriage in 1879. The gasoline engine had been invented seven years earlier, and the idea of adding the new engine to a carriage was not particularly novel. Nevertheless, Selden patented the concept and then used his considerable patent skills to keep it alive over the years. He never actually built anything. He sold his patent to an electric car company in 1899 but retained a percentage for himself. The car company went bust but realized it owned something far more valuable, the patent for the automobile. It formed an industry cartel called the American Licensed Automobile Manufacturers (ALAM) and required any automobile company to pay royalties on each car it sold. Ford Motor emerged as the leading car company. Its early market share rivaled IBM's 1950s market share. ALAM demanded that Henry Ford start paying royalties but he refused. ALAM sued in 1909 and won. Ford still refused to pay and appealed. Two years later, a second judge ruled in his favor and the cartel-like structure of the automobile industry fell apart[23].

Unfortunately, IBM made a different choice in 1956. The patent payment to Sperry Rand was monopolistic in nature and not any different from the payments that ALAM required of car manufacturers. Judge Larson's ruling in 1973 could have included severe penalties for IBM or fostered antitrust actions by the Justice Department. It was not IBM's finest hour.

# THE WORLDWIDE ACCOUNTING
# MACHINE

T hough IBM had 90 percent market share of the punched card devices in the United States, it was less dominant in Europe, where the market split was closer to 50-50. IBM competed with the Powers subsidiary, known as Powers-Samas. There was also a new kid on the block. A Norwegian engineer at Storebrand Insurance in Oslo gained experience on their Hollerith machines. He concluded that he could do better. His name was Fredrik Bull[1]. Storebrand sent him to the United States to study the IBM equipment in more detail. The company provided $10,000 for Bull to build a better tabulator. He designed a technically superior machine while sidestepping the Hollerith patents. He patented his own design in 1919. He was successful in the Nordic countries and on the verge of a breakthrough in Europe when he died at just age 42 in 1925. His patents ended up with a Swiss company. They hired Bull's chief engineer, who insisted on relocating to Paris. The new company became Groupe Bull.

Groupe Bull prospered under the radar until 1953 when it introduced the Bull Gamma 3, which married an electronic calculator with 400 vacuum tubes and core memory to its Gamma series of punched-card equipment. It was plugboard-driven with up to 64 programming steps. Critically, it was far more powerful than the IBM 604 Multiplier and integrated in a complete system like the Card Programmed

Calculator but at a much lower rental price. The new Gamma system also addressed a new and important application requirement, the need to recalculate interest on a daily basis in French banks. The IBM 604 performed this function before the Gamma 3 arrived. By 1955, Bull had wrested away 50 percent of the French punched card market.

The Bull challenge was a serious one for IBM operations in Europe, which was becoming an ever-larger part of the IBM Company. Hollerith had started international operations early with a focus on country censuses. Watson Sr. would frequently note that the United States had only 6 percent of the world's population and was serious about expanding the company's reach. He made this clear in 1924 by adopting the new name "International Business Machines." He had the maxim "World Peace through World Trade" emblazoned on the headquarters building.

By 1949, IBM operated in 78 countries with $50 million in sales. In the early years, the international operations were known as the Foreign Department. The British Tabulating Machines Company (BTM) was the licensee in England. Former U.S. Census director Robert Porter agreed to pay a royalty of 25% on IBM equipment sales. BTM played a major role in World War II, including its work with the Enigma and Colossus machines. Like IBM, the company faced major decisions after the war with its tabulator business and the emerging market for electronics and computers. However, BTM remained tied to IBM's onerous 25 percent royalty rate, limiting its ability to invest in the new technologies. Sir Raleigh Phillpots, the BTM chairman, met with Watson Sr. and Tom in 1949 to discuss a renegotiation of the royalty. Watson was not interested, and the two parties decided to discard the agreement and go their separate ways[2].

Watson formed the World Trade Corporation later that year and appointed Arthur "Dick" Watson as vice president of the new subsidiary. Dick was Tom's younger brother by five years. Like Tom, he was not a great student, but did get into Yale, where he excelled in foreign languages. During summers, he worked in the machine shop in Endicott. After his work of the Mobile Records Units (MRU) during the war, he re-joined IBM in 1947. Watson saw World Trade as a possible landing spot for Dick and took him on a tour of IBM's

European operations in 1948, a trip made in the middle of Dick's honeymoon[3]. Dick would rise to president of World Trade by 1954.

World Trade markets were different from the U.S. market, with far more medium-sized businesses and very limited funding of large computers by governments. Groupe Bull took full advantage of this difference. Dick Watson pushed for greater autonomy and, in particular, an increased role in new products that provided a better fit for his markets. IBM Research was in New York City and the major labs were nearby in Endicott and Poughkeepsie. In contrast, the establishment of labs and a research mission were just getting underway in Europe. The French and German labs were the largest, with 300 to 400 employees at each[4]. The British lab had perhaps 30-40 people. The Dutch lab was small, and the Nordic lab was still in the planning stages. As was the case in Endicott, many of the engineers had expertise in electromechanical devices, not electronics. These emerging labs faced an uphill battle. Watson instilled the concept of lab competition but the European labs were at a distinct disadvantage. They would begin a new project only to have one of the U.S. labs hear about it and try to outdo them. Typically, the final product ended up more tuned to the U.S. market.

For Dick Watson, the success of the Gamma 3 machine pointed to the sweet spot that IBM needed to address. Punched-card shops could afford to move up to the more functional environment of the Gamma 3, but larger stored-program systems were completely out of reach. The IBM 650, with monthly rentals in the range of $4,000, helped, but it was at the top end of what was affordable. Customers could implement a punched-card installation for about $2,500 per month. Groupe Bull feasted on the gap between IBM's punched-card systems and their entry stored-program computers.

In 1954, Dick Watson pushed his father and brother to step up to the issue. IBM France president Baron Christian de Waldner visited Watson to press the case[5]. The elder Watson liked to have aristocrats, such as the Baron, in his European leadership. It somehow burnished his own standing. Dick asked Waldner to pull together a project to address the Gamma challenge. They called the effort the "Worldwide Accounting Machine", or WWAM for short. Principal work started in the French and German labs. The German design team hosted a meeting in the summer of 1955 with engineers and planners from both France and the United States. Two of the key French engineers—Maurice Papo and Gene Estrems—would spend a year in Poughkeepsie developing the WWAM design. Their focus was mainly how to apply electronics to IBM's current offerings in order to compete directly with Bull.

The French, German, and Poughkeepsie labs competed on the WWAM concept with the French design winning. It featured stored-program support like the IBM 650, but high cost estimates forced the designers to scale back. They reverted to a plugboard-centric system with core memory. In order to reduce expensive core needs, the French engineers came up with the word mark concept that enabled variable-length instructions. The completed design looked similar to both the Bull Gamma 3 and the rejected Wooden Wheel.

If given the go-ahead, the French lab would be responsible for the central WWAM machine while the German lab would engineer a new printer. The German printer design emulated the stick printer that San Jose had introduced on the RAMAC system[6]. It was an eight-sided vertical "stick," with 64 characters, eight on each face. The stick would both rotate and move up and down to select each character, and then move horizontally along the platen for the next print character. The printer ran at 3,000 characters per minute, or roughly 30-50 lines per minute on the page. However, the design proved unworkable, as it required far too much oscillation to achieve reliability goals.

Poughkeepsie did a detailed review of the WWAM plan. Jacques Maisonrouge owned the WWAM requirements for World Trade and spent considerable time in Poughkeepsie. He had joined IBM in 1948 and worked on the SSEC, writing machine language programs. He also worked on the 604 Multiplier before returning to France and transferring into sales, a move that would eventually lead him to succeed Dick Watson as head of World Trade. Ralph Mork had the lead role on WWAM in the Poughkeepsie lab. An MIT graduate, he had joined IBM in 1949 in Endicott. He worked on SAGE before the promotion to lead executive for Electronic Accounting Machines. It was an unusual move as he now had a very critical mission but with a staff of just one—himself. Mork began his new role by asking whether there was an alternative to the path for Europe such as the Card

Programmed Calculator. In fact, World Trade had created a modified version of the CPC for commercial applications, adding more relay memory, but it was too slow to compete with the Gamma 3.

Meanwhile, the French team had a WWAM prototype running in the Paris lab by 1957. The original target rental was $1,250 per month, but once U.S. market planners finished analyzing it, the cost estimate came in at $4,500 rental. The Paris team whittled it down $3,100, but that still was not enough. And, there was another factor involved beside pricing. Vin Learson was skeptical that the French and German labs were up to such an important project. It would be the first major program for World Trade and would require significant multi-lab coordination. Tom Watson agreed with his concern. Towards the end of 1957, Learson sent Mork to Europe as director of engineering and looked for a replacement that could wrest control of the WWAM program.

He tapped Chuck Branscomb, that young 084 Sorter engineer who had the audacity to challenge him on the slipped sorter schedule[7]. Branscomb was nearly finished with a degree in mechanical engineering when the Army drafted him and sent him to Korea. He was assigned to the 107th Mobile Records Unit, where he learned plugboards and tabulation equipment. He ran an application with 36,000 cards that represented each of the GIs stationed in the country. After mustering out of the Army, he took advantage of the GI Bill and completed his master's at North Carolina State in 1950. His service experience with IBM punched-card machines led him to an interview in Endicott that same year. He hired on as a junior engineer with the master's grade salary of $75 a week. Though sporting a lofty degree, he still went through standard engineer training, now running for eight months. He went to work for Larry Wilson, an engineer who was already a legend in the company for his punched-card engineering. Wilson had gone to the University of Illinois and worked on punched-card equipment there. He joined IBM in 1939 and worked at the Census Bureau during the war as head of their tabulation group. He developed into a whiz with card equipment. By 1954, Branscomb had risen to engineering manager and Wilson was now working for him.

Branscomb started his new assignment as Mork's replacement

with a formal assessment of the WWAM program. He went to Paris to attend a WWAM review chaired by Maisonrouge. He asked Wilson to join him on the trip. After presentations on the processor design by the French team and the stick printer by the Germans, he came to an important conclusion. In his mind, the era of the plugboard was over, and its inclusion in WWAM was its fatal flaw. He was leaning toward canceling the program and asked Wilson for his view. Wilson, the master mechanical engineer, had this mantra about engineering design: "When you get the right design, it's gonna be simple, it's gonna cost less to build, it's gonna be more reliable, and that's what you want.[8]" Though Wilson advised him to "do what you think is right," it was his mantra that confirmed Branscomb's decision. The WWAM design was simply not the right design. He recommended cancellation to Learson. This would be a politically charged decision. He knew that Endicott Lab manager Jim Troy and Ralph Palmer were both supportive, and he was confident that Learson and Tom Watson would agree. However, it would be a bitter pill for Dick Watson and World Trade.

Learson agreed with Branscomb's recommendation on WWAM but still viewed a solution to World Trade's Gamma market exposure as his top priority. He tasked Branscomb and Endicott to come up with a new approach. To soften the blow, Learson asked Poughkeepsie to produce a stopgap machine. Ralph Palmer had his team develop a fully transistorized multiplier. They completed the IBM 608 in 1957 with 3,000 transistors under the covers[9]. Ralph Mork, now on the other side of the Atlantic with World Trade, agreed to take the new calculator and build it into a temporary offering. Max Paley, a Poughkeepsie engineer who had designed the central logic of the 604 Calculator, would spend a year in Europe with the program. He added core memory and paired the revised machine with a card reader/punch and tabulator. The resulting combination was not very different from the Gamma 3. World Trade offered it at a competitive price and sold nearly 150 units, taking the pressure off the lack of a WWAM machine.

# THE 1401: THE CHEVROLET OF COMPUTERS

Meanwhile, back in Endicott, Chuck Branscomb needed a permanent WWAM solution. Per the Larry Wilson's directive, it had to be one that was simple, would cost less, and be more reliable. He would turn to Francis Underwood for the answer. Underwood had followed a slightly different path before reaching this critical juncture at IBM[1]. He grew up in Binghamton, the town next door to Endicott. He went to a technical high school, one with a heavy emphasis on mathematics and physics as well as machine shop work. During the war, he went through the Navy's V5 flight program, first at Hobart College and then at Cornell. The war ended before his air squadron went into service. He joined IBM in 1947. Without a doctorate or a master's, or even a college degree, he started in the machine shop, working on the 604 Calculator. He applied for a position in engineering, was accepted, and worked on the 402 tabulator redesign. Soon, he was an instructor for new engineers. Still interested in new horizons, Underwood took a class on circuit design given by MIT graduate Fred Fosse, who had learned binary logic from Claude Shannon. Underwood started working on logic design and ended up teaching computer architecture to his fellow engineers. He earned a promotion to the Advanced Systems Development Department. That was where Ralph Mork found him.

Mork had him working on other projects while WWAM was going through its rounds of revision and review. Now, Branscomb wanted a fresh pair of eyes of the project and told Underwood to look for a different path forward. The objectives remained the same—create a punched-card replacement system to challenge Bull and hit the target rental price of $2,500.

Underwood refocused on the central issue for the WWAM rejection: the cost overrun due to the price of core memory. The French team had set core memory for the WWAM at 1,900 bytes. What if that was reduced to 1,400 bytes, a figure he selected purely by what needed to be to be affordable. Core memory cost at the time was roughly $10 of rental for each 100 bytes[2]. Underwood validated his memory decision by testing it with the French interest calculating application. Underwood coded the function in machine code within his 1,400-byte allocation. In addition, he embraced the French team's word mark concept of variable-length instructions. It neatly solved the issue of optimum instruction length. There could be a mix of short and long instructions, and the net effect was a further reduction in core memory.

Next, Underwood looked at the WWAM plugboard. With its central role in punched card machinery, he was not surprised it remained in the design. It was second nature to engineers who had grown up with the device. This was also the approach of the Bull Gamma 3. Given his experience in computer architecture, Underwood toyed with the obvious alternative of going all the way to a stored-program computer. However, this would surely be too expensive. Or, would it? To Underwood, it was obvious that the WWAM was a compromised design, with one foot in unit record technology and the other in stored program. Did it have to be?

Getting under the machine covers of the French WWAM prototype, he made a critical discovery. Nearly half of the circuitry inside the machine was there to control the interface between the plugboard and the logic circuits. That had to be a significant cost. Eliminate the plugboard and its hubs, and all the electronics supporting them would evaporate, along with their cost. He was now nearly all the way to a stored-program machine.

Underwood decided to start over with a clean sheet stored-program design. Branscomb concurred, noting the difficulty in setting up plugboards as another nail in the WWAM coffin. Based on his own experience with plugboards, he related that "for each application (i.e. payroll, billing, inventory control) you need at least three plugboards— one each for the reproducer, the collator, and the accounting machine." He knew the first one was relatively simple but the other two were complex. The WWAM design was a kluge and installations would be needlessly complex, inhibiting the program's success. Karl Ganzhorn, who was at the German lab and later would become its director, would comment on the use of the plugboard, ". . . for WWAM, this concept was never even questioned.[3]" It was indeed ingrained.

Branscomb endorsed the new concept and secured the commit-ment of Bob Evans, who was now director of engineering in Endicott. Underwood plunged ahead. His placeholder name for the new program was the "Stored Program Magnetic Core Memory Variable Instruction Length Word Length Accounting Calculating Machine." However, since this was 1958 and barely a year after the Russian Sputnik launch, he changed the name to SPACE, for "Stored Program Accounting and Calculating Equipment[4]. By 1959, Branscomb had a 40-person development team working on SPACE.

Branscomb and Underwood had one additional hurdle -- to produce a business case that would receive approval from the bean counters, something that had eluded the WWAM team. For this, Branscomb would rely on Sheldon Jacobs. Jacobs started with IBM in Chicago as a sales representative and moved to Endicott as a product planner in 1957. With his sales background, he decided to take to the road, visiting punched-card customers to understand what they needed and how much they were spending. With a couple of tabula-tors and a calculator, their monthly rental ran from $2,000 to $2,500. He described the planned replacement system with its quantum leap in functionality and got very favorable feedback, giving him confi-dence to work up an aggressive business case.

He set the entry rental at $2,500. IBM had a 10-to-1 ratio of product cost to rental, meaning that the entry machine would have to cost out at $25,000 or less[5]. The planned cost was close. He ended up $50 over but still got the plan approved. Jacobs forecasted 2,200 systems in the United States and 4,700 worldwide. Those were astounding numbers for such a large system. Tom Watson stopped by Underwood's office and asked if IBM could really sell that many and Underwood of course said yes. That turned out to be the one and only time he met Tom face to face, and fortunately, he was ready with the right answer.

The business plan also needed the field to approve. Ralph Palmer leaned on Don Spaulding, a marketing guy that Palmer had brought into Poughkeepsie to represent field requirements. Spaulding was now Vin Learson's key marketing aide, and one of only six in Learson's very small staff. Palmer asked Spaulding to push the field to say the new machine was needed. It was an easy request for Spaulding.

The new WWAM, now SPACE, would need a printer. The German stick printer did not appear promising. As if by serendipity, an alternative emerged. Endicott was working on a new printer. It was significantly faster than current IBM printers, employing a unique new technology. However, there was a problem. It was destined for the big IBM mainframes. Branscomb was not in the least deterred and recognized it immediately as a key linchpin in the SPACE program.

IBM always seemed to lag when it came to printers. The company was caught flat-footed in 1914 when Powers added a printer to its tabulator. Even with that threat met, IBM tabulators remained weak on printing. The current top-of-the-line tabulator, the 407, topped out at a mere 150 lines per minute. In contrast, the UNIVAC I printer was running at 600 lines per minute in 1952. Most of the existing printing mechanisms were in fact vertical, like the German stick design. Earlier IBM tabulators had type bars that moved up and down. The more recent 407 had a type wheel that rotated vertically relative to the continuous form paper. The common thread through all of them was the wavy line of poor print quality.

Fred Demer, an engineer on the print team, had been experimenting with new approaches to reaching higher print speeds. This overstates the case. There are in fact two approaches to printing—vertically or horizontally. He decided to give the latter a try. He assembled a continuous metal chain containing slugs for each character in the print set. The chain loop would rotate horizontally at high speed while a set of hammers would fire on the required character at the exact moment the chain was at the designated position on the print line. Demer built a prototype with just a few slugs and hammers and was able to crank it up to over 1,000 lines per minute[6]. Something else emerged when he ran the tests. The print quality was outstanding, with no wavy lines. There were slight imperfections in the horizontal line where the hammer fired and slightly missed the correct spot on the paper. However, the human eye did not react to these misses as it did with vertical mistakes. The eye was used to proportional spacing and so the print line looked perfect.

**Figure 33.** IBM 1403 Print Chain, the secret to the printer's success, designed by engineer Fred Demer. Reprint Courtesy of IBM Corporation ©.

With the Demer prototype showing promise, the program had the green light to proceed. The team worked under the direction of Jonie Dayger, a future IBM Fellow who had earlier developed the CPC redesign under Steve Dunwell and then managed the Wooden Wheel prototype for Northrop. The printer work was not without setbacks. A near-crisis ensued as the high-speed print chains would break during testing.

The speed target for the new line printer was 600 lines per minute. A print chain with slugs for 48 characters was the standard— with 26 alphabetic characters, the 10 numbers, and 12 special characters. The print chain moved at 17 feet per second. Before the printer could advance to the next print line, it needed to select and print each character on the line. When a desired character slug reached its target position, a transistor would fire an electro-magnetic driven hammer against the chain, sending the character slug into the print ribbon and the image onto the paper. Multiple print hammers printed simultaneously, resulting in a fast-moving symphony that combined with the rapidly moving chain to complete each line.

The innovation did not stop with the print line process. The next print line could be the next one down the page or somewhere on the next page. The paper carriage mechanism had a high-speed advance that would accelerate the paper to 80 inches per second without ripping the form out of the guide sprockets. The printer handled multi-part continuous forms at 600 lines per minute with excellent print quality. It was four times the speed of most competitive printers[7].

The fact remained that the Dayger's team designed the printer for 700 and 7000 mainframe customers. The resulting business case forecast was relatively small given the limited inventory of large systems. Branscomb and Jacobs saw the synergy of the printer with the SPACE program. A punched-card customer would potentially move up to SPACE just to get a very fast printer. A 650 customer could install SPACE with the new printer for the cost of another 650 with a slow printer. Some large accounts could choose to install SPACE as an outboard processor to run the printer instead of managing the printing on the mainframe. The bottom-line message was that adding the new printer to SPACE would help both programs and ramp up the printer numbers.

The new printer (to be released as the 1403) was not the only timing windfall for SPACE. The new system would use transistors in great numbers. The SMS packaging of transistors used in Stretch and the 7090 paved the way for SPACE. Branscomb and Jacobs could piggyback the SPACE volumes on top of the SMS plan. Each SPACE machine would use 10,600 transistors and 13,200 diodes on 2,300 SMS cards[8]. Other elements of the system also fell into place. Larry Wilson's team in Endicott designed the 1402 combination reader and punch. In addition, San Jose would finish the 1311 ADF (Advanced Disk File) in 1961, with two megabytes of data on removable disk packs.

Francis Underwood created a simple instruction set for SPACE tuned for business applications. A base assembler called SPS (Symbolic Programming System) was included as well as the FORTRAN and COBOL compilers. The software team provided an advanced assembler (Autocoder). It was far more functional than SPS but took up more memory. Autocoder would become the most popular programming language, especially as the 1400 series expanded over time into larger and larger systems. However, given the initial focus of the system at the low end, the key programming additions were FARGO and RPG.

FARGO was a dramatically different programming language. Developed by an IBM Systems Engineer at a customer location, FARGO stood for "Fourteen-o-one [IBM 1401] Automatic Report Generation Operation". FARGO excelled in smaller 1401 accounts that replaced existing punched-card equipment. In a sense, FARGO was a tabulator emulator. It simulated the punched-card cycle—read a card, do some processing, and produce some output. This was also the typical flow of business applications.

FARGO programming[9] involved the filling of a series of specification sheets to define the job. The input sheet defined the fields in the transaction or master file. The calculation sheet covered all the needed computations. The output sheet defined the headings, detail print lines, totals, and file output or updates. With the sheets completed, the programmer keypunched them into 80-column cards and placed them behind the FARGO deck, ahead of the application cards. The entire deck went in to the card reader. There was no compilation, just load and go. For someone familiar with setting up jobs with plugboards, FARGO made perfect sense. For someone new to computers, it was an easy language to master.

There was an early concern that FARGO would be a temporary program and customers that invested in it might be left behind. However, early favorable reports on FARGO led IBM to develop a formal replacement for the language called Report Program Generator, or RPG. It maintained the general approach and philosophy of FARGO while significantly extending its programming functionality. Barbara Wood and Bernie Silkowitz developed the first version of RPG with some assistance from Wilf Hey (who would later become famous for coining the phrase "garbage in, garbage out"[10]). Wood was a field systems engineer with a number of punched-card installations. She viewed programming from the plugboard perspective, as did her customers.

The RPG compiler arrived in 1961, relegating FARGO to the sidelines. The difference in developing applications in RPG versus languages like COBOL and FORTRAN was dramatic. Those languages were procedural. This meant you started with a blank freeform coding sheet and developed all of the processing from scratch. With RPG, all of the code required to set up and run the full transaction processing cycle, including group totals, was built into the compiler. The programmer simply had to define the input fields, calculations, and output, and RPG would take care of the rest.

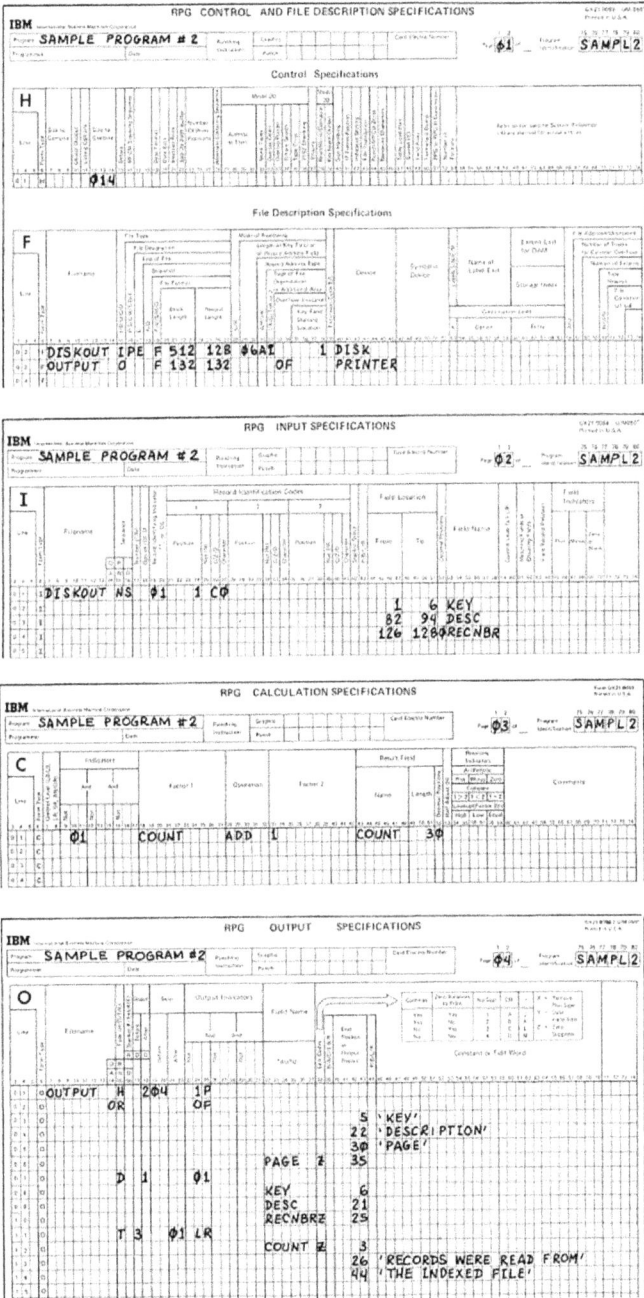

**Figure 34.** Specification sheets for a simple RPG program. The F sheet defined the input disk file (DISKOUT) and the printer file (OUTPUT).The I sheet defined the fields within the disk file. The C

sheet did the calculations, in this case adding to a total count field.
And, the O sheet laid out the report.

For those customers who did not want to do any programming, no matter how easy, IBM also developed application suites for selected industries and bundled the software with the system. SHARE would take customer applications and distribute them to members.

The WWAM replacement system that Chuck Branscomb, Fran Underwood, Fred Demer, and many others worked on was coming together. Bob Evans became director of engineering in Endicott in 1958 and described the 20-hour days required to push the project forward and get it announce-ready. John Haanstra had risen to assistant general manager of the General Products Division (GPD) and had overall responsibility for the launch. SPACE became the 1401 on October 5, 1959, via closed-circuit television to over 50,000 customers, prospects, and IBMers in 102 cities. What they saw was a sleek, modern computer at a very attractive price point. Though designed for commercial applications, the system was still capable of 3,200 additions or 420 multiplications per second. It provided seven times the performance of the 650. Time-Life installed the first 1401 to print magazine address labels. They converted 40 million punched cards to tape in the process[11].

**Figure 35.** IBM 1401 System (1959), designed by Francis
Underwood with a key assist by the 1403 printer (on right). The first
computer to sell more than 10,000 systems. Reprint Courtesy of
IBM Corporation ©.

IBM received 5,200 orders in the first five weeks following the announcement[12]. Haanstra immediately moved to double the planned first-year production to 1,700. The entry system remained at the target $2,500 monthly rental, but the typical system rented for $6,500, still a fraction of the cost of a 700- or 7000-class system. Sheldon Jacobs's forecast was 5,000. The 1401 blew by that number, eventually selling close to 20,000 systems across all of the 1400 series models. Much like the 650 before it, the press called it the "Model T of the computer industry." The 650 should probably remain the "Model T" and the 1401 can be viewed as the "Chevrolet" of computers[13]. By 1962, IBM 1401 revenue exceeded accounting machine revenue for the first time[14]. IBM had the corporate jet rechristened with the number "1401." By 1965, IBM 1400 systems comprised roughly half of the world's 26,000 computers. The 1401 converted a sizeable number of plugboard programmers into computer programmers.

The original impetus for the 1401 was the increasing competition in Europe from the Gamma 3. Bull had continued to enhance the Gamma line to stay ahead, culminating with its Gamma 60, a transistorized machine with core memory and advanced multithreaded processing. However, Groupe Bull designed and priced it to compete with IBM's 700-series machines. Europe had few large companies that could afford that class of computer. The Gamma 60 arrived in 1960 in the midst of the 1401 wave and sold only 20 systems.

With the runaway success of the 1401, IBM asked several key players in its development, including Branscomb and Underwood, to tour the IBM labs in Europe. They were told to take their wives along, and then cautioned not to spend too much time at the labs. Watson Sr. would have appreciated this kind of recognition.

The 1401 success was the result of many astute decisions and a whole lot of luck. The table was set with the work that the European labs did on WWAM. This provided a starting point for Fran Underwood to reimagine their approach with a computer design that incorporated several of their key innovations, including variable word lengths with the word mark. The 1401 program was also able to capitalize on the SMS circuit board business case and the pent-up demand across unit record and 650 accounts. Then, there was Fred Demer's printer. The 1403 printer was a major factor in 1401 volume sales, and the printer would go on to sell 23,000 units.

Despite the great numbers, there were a few acknowledged misses with the 1401. The sleek modern look of the system suffered with large connecting cables between the units. Branscomb called it an "eyesore[15]." He corrected this mistake with the next IBM small system —the System/3. The much bigger miss was at the low end. Virtually no one ordered the entry machine at $2,500. The average system rented for a healthy $6,500. The 1401 missed most of the low-end market. That would have to wait for the System/3 and its follow-on systems. Finally, Fran Underwood missed on memory needs, as so many others before him had. He surmised that most 1401 applications would not need more than 4K of core memory and set the instruction address size accordingly at three digits, providing what seemed at the time substantial future growth to 16K. That did not turn out to be the case.

The 1401 years were the glory years for Endicott. The next 1400 series model was already in the works before the 1401 announcement. Bob Evans met with customers who pressed him on the value of compatibility. Don Spaulding visited Endicott during planning and warned against making it incompatible with the 1401. Evans asked Richard Case to work on the new design. Case expanded the address size from three digits to five digits, resulting in an increase of the maximum core memory from 16K to 80K.

The San Jose Lab had been busy designing the 310, a planned follow-on to the very successful 305 RAMAC. Larry Wilson, the electromechanical wizard, had moved to the lab to work on the 310. William Woodbury of Wooden Wheel fame would join him. Sheldon Jacobs arrived in 1960 as the product planner. Ralph Palmer had the program transferred to Bob Evans in Endicott, who decided to cancel it but retain its centerpiece, the new 1311 ADF disk drive with its removable disk packs. The system that would emerge was an entry model, the 1410[16]. Announced in 1962, it was somewhat less successful than the other models of the family. It had the random-access technology, which was relatively new, but no tape support, which remained a key requirement. The 1410 turned out to be mostly compatible. You did need to reboot the system in 1401 mode to run 1401 programs. The use of the term "reboot" here is only for modern clarity. The correct term at the time would be "IPL," for Initial Program Load. Customers performed the IPL by pressing the big blue button found on the console of most IBM processor units. An interesting change with the 1410 was the inclusion of two early and rudimentary operating systems. Endicott released them as PRS-108 for tape-oriented operations and PRS-155 for disk operations, two less than imaginative names

A larger 1400 model, the 1460, arrived in 1963, providing twice the performance of the original 1401. World Trade started planning for an entry model at an even lower price point than the 1410. However, the System/360 program eliminated the need for their planned 1430 system. At the same time, a key driver of the 1400 series, the 1403 printer, was not sitting still. Endicott speeded up the printer to 1,100 lines per minute and then again, to 2,000 lines per minute. The dramatically increased speeds further validated Fred Demer's big bet on a horizontal rotating chain.

# A VERITABLE MAZE OF COMPUTER
# MODELS

W hile Fran Underwood was a one-man skunkworks trying to come up with a way to salvage the World Wide Accounting Machine program, the Endicott and Poughkeepsie labs were busy with far more critical matters. They needed to satisfy Tom Watson's "solid state in '58" decree and provide a growth path for IBM's two major commercial vacuum-tube-based product lines, the 650 and the 705. Both labs were working on replacement designs. The arrival of the RCA 501 in early 1958 added to the urgency. The 501 was fully transistorized, with advanced pluggable circuit boards. Ralph Palmer wanted to have one plan going forward, despite the vast differences between the two IBM systems. A typical 705 system rented for $17,000 while the average 650 was $3,000-$4,000 per month. Both labs coveted the huge gap between the two systems. Palmer decided to pit the two labs against each other, long a familiar tactic of Watson.

Rex Rice, formerly of Northrop Aircraft and the Card Programmed Calculator, joined IBM in 1955 in Poughkeepsie. He worked on the 608 and then turned his attention to a complete system that would fit between the 608 and the 705. His concept became the 750. Given his background at Northrop, it started out as a plugboard machine. Rice maintained, "eight to twenty-eight words of high-speed memory is all that one needs" as long as you have plugboard programming to handle the rest[1]. Gerrit Blaauw disagreed and switched the design to a stored-program computer.

Meanwhile, in Endicott, the lab worked on a replacement for the wildly successful 650, with an eye towards growing into that midrange gap. Like the Gamma 3, the planned system, dubbed the 660, combined core memory, transistors, and a magnetic drum. To remain compatible with the 650, the new system needed to include the odd instruction format that stored the drum address of the next instruction. In addition, 650 instructions were fixed-length decimal, unlike the 1401 that used the word mark to support variable instruction lengths. This difference translated into 40 percent more cores, or about a $1,000 increase in monthly rental.

Evans called the competition between the two labs "vigorous," likely an understatement. Dal Allan, who worked at Endicott, commented on the antipathy between the two labs: "Poughkeepsie viewed itself as the brains and those were idiots at Endicott.[2]" This is where it started getting messy. Palmer asked the Applied Science group to evaluate the two designs. It recommended Endicott's 660 design. Bob Bemer and a team in the Programming group also evaluated the two proposals. They recommended the Poughkeepsie 750 to their line executive, Roger Bury, who apparently got it backwards when he reported up the chain of command, declaring Endicott's 660 the winner[3]. In any case, Palmer elected to combine the two designs and award the program to Endicott, with the composite machine designated as the IBM 7070. At the time, Palmer was concerned about Poughkeepsie's bandwidth, as the lab was working on both Stretch and 7090 systems at the time. Evans, a big admirer of Palmer, would later say this decision was Palmer's biggest mistake.

Though they had apparently won the design competition, it soon became evident that Endicott was not sufficiently experienced with large stored-program designs or the demands of manufacturing large systems. They were in over their heads. Then, in May 1959, Palmer's already flawed game plan simply blew up. Tom Watson initiated a reorganization that cleaved IBM manufacturing and development into two new divisions. The Data Systems Division (DSD), working primarily operating out of Poughkeepsie, would handle large systems that rented for more than $10,000. The General Products Division (GPD), based in Endicott, Rochester, and San Jose, would work systems under $10,000. The Data Processing Division in White Plains would remain the sales and service arm for both.

This change meant that Endicott's 7070 program, slotted just above that $10,000 bar, would have to move to Poughkeepsie. The Poughkeepsie lab had apparently lost the design contest and now had to build its competitor's system. Poughkeepsie was not in the least amused. Not willing to accept defeat, Gerrit Blaauw redesigned the 7070 specification into a full stored-program system called the 70AB. Steve Dunwell argued that the 70AB would be twice the performance of the 7070 and less expensive. Vin Learson stepped in and compounded Palmer's mistake by demanding that Poughkeepsie build the 7070 first and worry about a better machine later. He tapped the very able Max Paley to lead the effort. Bob Evans, whose team in Endicott had produced the design, would later admit that the 7070 was a "dog.[4]"

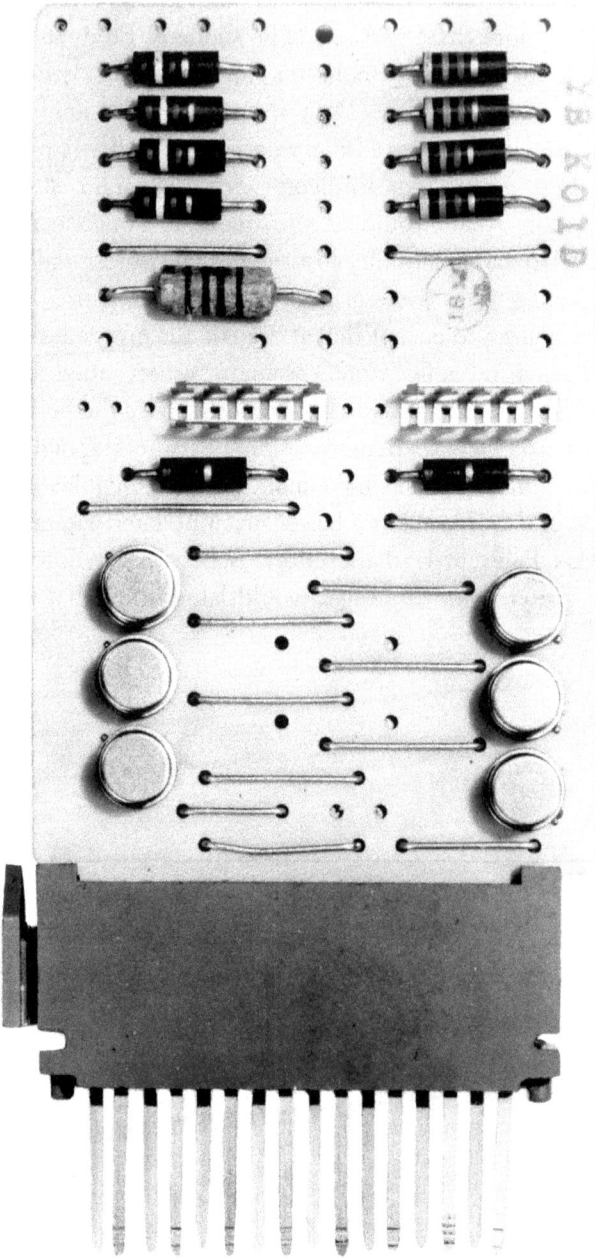

**Figure 36**. Simpler SMS (Standard Modular System) logic card for the 7070 computer, the star-crossed "winner" of the competition between Endicott and Poughkeepsie. Reprint Courtesy of IBM

Corporation ©.

In the end, Poughkeepsie produced the 7070 as Learson wanted and announced it in 1958. An IBM 650 customer would not recognize the computer that emerged. It was a monster weighing in at well over 20,000 pounds, sporting 30,000 transistors and 22,000 diodes attached to 14,000 SMS cards[5]. The configuration could even get larger by adding the huge 350 disk drives from San Jose. In addition, the 7070 was not compatible with either the 705 or the 650. For 705 customers, the instruction set was different, requiring recompilation among many other changes. It was a significant regression in technology, sporting the now outmoded magnetic drum. IBM 705 customers spurned the new system. For 650 accounts, the story was similar. Though IBM portrayed the machine as a transistorized and compatible 650, it clearly was not. The magnetic drum was included but the drum-addressing scheme, central to 650 applications, was gone. Moreover, 7070 required a simulator to run 650 programs. This is what IBM had done to its most successful computer line.

The other and overriding factor was price. The average purchase was $1 million, and the monthly rental was $24,000. The world's best-selling computer should have rated an attractive successor. The 7070 clearly was not it. A small silver lining was the use of SMS circuit boards in the 7070. It enhanced the 1401 business case, which provided a much better growth path, though the 650 customer would have to wait a year for Underwood's SPACE machine. It was ironic that the lowly World Trade WWAM concept morphed into the wildly successful 1401 program in Endicott and then provided a growth path to Endicott's pride and joy.

Despite its many shortcomings, the 7070 ended up with 525 customers, though many of them were underwhelmed. Many 705 accounts were unwilling to move, resulting in a crash program to develop a fully compatible replacement. That became the 7080 system, renting for $48,000 a month. IBM offered those customers that bought the 7070 upgradable models with the 7072 and 7074, providing up to 20 times the performance. Poughkeepsie would also get some of the 1401 magic. Evans asked Learson to move the planned high-end version of the 1400 series to Poughkeepsie. The lab would merge 1401 architecture with 7000 series mainframe hardware and announce the 7010 in 1962. This series of fiascos would have a silver lining. It would lead to IBM's greatest computer.

# THE "BET THE COMPANY" GAMBLE

I BM's center of gravity remained in New York, with the two major development labs in Endicott and Poughkeepsie. World Trade continued to lobby for a greater product role. Research had moved from New York City to the impressive new facility 40 miles north in Yorktown Heights. Tom considered moving IBM headquarters to this site but opted in 1961 to create a new IBM campus in Armonk, New York.

Following Watson's "solid state by '58" mandate, the company had completely shifted its product line to transistors. Poughkeepsie's 700 series became the 7000 series while Endicott replaced the venerable 650 with the 1401. However, as early as 1959, there was a sense that despite its rapid growth, the company had reached a plateau. In 1960, IBM grew only 9 percent, while the computer industry was expanding at a higher rate. IBM's revenues were twice the total of the next eight computer companies combined and its market share topped 75 percent[1]. Still, Tom Watson found this sag in growth unacceptable. IBM revenues increased nearly tenfold during the 1950s. However, the company had up to 80 percent gross margins, so there was plenty of room for others to make money if IBM were to stumble. Tom wanted to start planning for the next generation of IBM systems.

IBM's current strategy certainly influenced Tom's ideas. He

wanted to get ahead of the technology curve and make a big bet to secure the company's future. He was willing to jettison the present incremental approach that enhanced each system with upgrades as the technology evolved. Rather, he wanted to replace all of IBM's existing systems at once with a brand new family of machines, with all of them embracing the latest technologies. He asked Al Williams to become IBM president in 1961 and assume responsibility for financing such a "bet your company[2]" program. He asked Learson to be his point man for the effort. The code name for this massive investment was "NPL," for New Product Line, an understated name for what would be a vast and very expensive enterprise.

NPL aimed to address the fundamental flaw in IBM's product programs. Each system used a different architecture, which translated to duplicate development and engineering costs. Separate models for scientific and commercial customers represented a needless extra expense. Varying peripheral interfaces required development to produce separate input/output machines for each system, leaving fewer resources and funding to match competitor's enhancements. The same situation existed for programming. Each IBM computer line needed to develop unique system programs. The common thread here was the lack of compatibility. With a common family of systems, IBM could focus hardware and software investment against a single architecture. Evans had seen the folly of incompatibility with the 7070 debacle. His boss Don Spaulding came to a similar conclusion, a position reinforced by the benefits of a 1410 that was compatible with the original 1401.

**Figure 37.** Bob O. "Bo" Evans, the key executive who tamed the Endicott-Poughkeepsie rivalry and managed the development of System/360. Reprint Courtesy of IBM Corporation ©.

In the beginning, the constant search for faster and faster calculation resulted in different architectures, technologies, and computer models. However, as the bulk of the computer market switched from scientific to commercial, the value of raw speed declined precipitously. The computer architects were still on the job, and architects just want to architect. As Tom Watson outlined the NPL strategy to launch IBM into the future, both U.S. labs were in the midst of gearing up to enhance their existing computers. Poughkeepsie would replace the 7000 series. Endicott wanted to extend the 1401. The labs in Europe had their own plans. The British lab pressed for approval of its SCAMP computer while the Germans worked on the System 3000 system that would fit just below the 1401.

In the wake of the 7070 and the failed 70AB plan, Fred Brooks became the planning manager for the Poughkeepsie lab. Not one to shrink from failure; he worked with Gerrit Blaauw to expand the 70AB concept into a more comprehensive plan to replace the entire 7000 family of systems, to be called the 8000 series. Brooks and Blaauw designed five processors with a sevenfold performance span that began with a small commercial system (8103), small scientific (8104), midsize commercial (8106), midsize scientific (8108), and large scientific (8112)[3]. Each machine was unique and incompatible with the others. They also ignored Steve Dunwell's press for combining commercial and scientific into one system. Work had already progressed on the 8106 midsize model.

Brooks visited Bob Evans in Endicott seeking his support for the 8000-series. He wanted Evans to build the entry models in Endicott. Evans, with the hugely successful 650 and 1400 series computers, had absolutely no interest. Brooks' entry models would compete directly with his systems. The $10,000 dividing line between the labs should have made it clear to Brooks that this was not his territory. Evans was not amused by the request, nor was John Haanstra. Relations between the labs were further chilled.

Brooks and his team charged ahead and presented the 8000 plan to Learson and the executive team in January 1961. Brooks thought the meeting went well with the exception of Learson, who got more agitated as the presentation progressed[4]. Spaulding had impressed on Learson the folly of investing in incompatible systems. In addition, Brooks' plan featured older technology, using SMS logic cards instead of Solid Logic Technology (SLT), the newest and fastest circuit packaging. The plan also completely ignored the 1401 success and its substantial install base. Brooks realized too late that the only person in the room that mattered was Learson, and Learson was not happy. Despite this, Poughkeepsie went ahead and built a prototype machine and Brooks pressed for an announcement of the new line in late 1961 or early 1962.

Learson did not waste any time in addressing the 8000 issue. He was smart, decisive, and could be ruthless when required. In this case, he intended to use his special technique he called "abrasive interaction.[5]" He would switch the management of opposing groups to force heated dialog and produce a resolution. He moved Evans, the engineering manager of the 1400 program in Endicott, and inserted him above Fred Brooks in Poughkeepsie. Evans met with Learson and Spaulding to receive his marching orders. He was asked to evaluate the 8000 plan, then as Learson put it, "If it's right, build it and if it's not right, do what's right.[6]"

That night, Learson moved Brooks' existing boss, Max Ferrer, to the new lab in Boulder, Colorado, and replaced him with Evans. Evans moved to Poughkeepsie as the new director of the 8000 program, reporting to Poughkeepsie lab director, Charlie DeCarlo. Evans invited Brooks to a local fish place to discuss resolving the issues surrounding the 8000. However, Brooks and his team were unwilling to engage. They "circled the wagons" and avoided any cooperation with Evans. The obstruction went all the way up to DeCarlo, with Brooks and DeCarlo working behind the scenes to expedite the 8106 and plan for its announcement. DeCarlo appeared as clueless about Learson's management style as was Brooks. Once Learson heard about the lack of cooperation, he relieved DeCarlo and replaced him with Jerrier Haddad, Evans' boss from Endicott. Haddad ensured cooperation. By May 1961, Evans had seen enough and recommended cancellation of the entire 8000 program. He based his decision on use of outdated SMS logic technology, the limited plan for peripherals, and the seeming random collection of models with no central animating philosophy. His recommendation to Learson appeared to kill not only the 8000 but also the planned 1401 follow-on, and pave the way for Watson's NPL project.

With the 8000's demise, Evans was persona non grata in Pough-keepsie. He had to move fast to avoid chaos and get buy-in and focus for the new direction. He had won the battle, but he knew he savaged Brooks and the entire 8000 team in the process. Broken fences needed mending. He took the senior Poughkeepsie team to the Gideon Putnam Hotel in Saratoga Springs to work out how they would release "warm-overs," interim 7000 machines to hold customers until NPL models were ready[7]. There would be midlife kickers to the high-end 7090, with the 7094 and 7094 II models. The lower-cost 7040 and 7044 were scaled-down versions of the 7090 though, of course, not compatible. The 7044 was actually very similar to the SCAMP machine that IBM's Hursley lab in England was working on. John Fairclough took SCAMP to Poughkeepsie and benchmarked it against the 7044. It did very well, but once more, World Trade came out on the losing end. Learson axed the program[8]. However, Dick Watson was able to leverage the loss of SCAMP by securing a significant World Trade role in the NPL program.

As for Fred Brooks, he expected to reassigned, the typical "reward" for being on the losing side. Yet, much to his surprise, Evans tapped him to lead the entire NPL program. Brooks' reaction: "You could have knocked me over with a feather when he asked me to take the crown jewels." Jerry Haddad's counsel helped. He told Brooks that he "never knew anybody that regretted working for Bob Evans[9]." Brooks would learn to agree with that. He would later rate Evans alongside Tom Watson Jr. and Howard Aiken as his greatest bosses. Steve Dunwell had chosen well when he plucked Brooks from Harvard. Brooks would prove to be very much like Dunwell, able to see the future clearly and convince people and whole organizations to follow his vision. Being a good salesman certainly helped. When that failed, he knew Evans, known as the "bulldozer[10]," would have his back.

Evans did have one caveat for Brooks. He needed to accept Gene Amdahl as the lead designer. This was the Gene Amdahl who left IBM in the middle of the Stretch project. IBM realized that though Amdahl had a volatile personality, he was a brilliant computer archi-tect. IBM had been working since 1960 on getting him to return to IBM. Manny Piore, the head of IBM Research, traveled to California and took Amdahl and his wife out to dinner. Promising a central role in a new supercomputer project, he convinced Amdahl to move back East. Amdahl would say that the environment at IBM Research was less bureaucratic than his earlier days at Poughkeepsie. It helped that his new position was significantly higher in the organization. Piore wanted him on Project X, aimed at regaining the high end mojo lost by the failure of Stretch. The assignment did not last long. Evans met with him and encouraged him to join the NPL project. He invited him to a June 1961 presentation that discussed the issues of compatibility and outlined the plan for upward and downward compatibility among the NPL models. Evans would say of Amdahl, "He could visualize what happens internally in a computer", was "a brilliant architect", and he had "yet to see his peer.[11]" Amdahl joined NPL on the condition that the product family included a supercomputer design at the high end.

It was clear why Vin Learson, with his aversion to any staff, selected Don Spaulding as his main and nearly sole direct report. Spaulding was always a step ahead, apprising Learson of the right course of action. Spaulding knew the cancellation of the 8000 series was not the end of it. He needed all of the stakeholders to be invested in the NPL plan and working cooperatively to devise a blueprint for this "bet the company" program. He suggested that Learson set up a task force to define the NPL plan. Learson agreed, and said, "Look, I'm gonna get one product. The whole spread of the computers . . . in one family.[12]" Spaulding took the cue and named the planning group name "SPREAD", for "Systems, Programming, Review, Engineering, and Development". The group's simple mission would be to discard all of IBM's current systems and replace them with a broad, new compatible line. This led others to label "SPREAD" as "Spaulding's Plan to Re-organize Each and All Divisions.[13]"

Though Learson cancelled the 8000 series, Spaulding was not convinced that John Haanstra, the VP of Engineering for the General Products Division, would agree to walk away from his 1401 plans. Learson opted once again to force the issue with another "abrasive interaction." He named Haanstra chair of the SPREAD team. Haanstra's backing of NPL was essential, as the low-end volumes currently represented by the 1401 were critical to the overall business case. Learson expected that as the chairman, Haanstra had to work in concert with the rest of the SPREAD team.

The group had 13 members. Besides Haanstra as chair and Evans as vice chair, there was Brooks and Amdahl from the U.S. labs, John Fairclough from World Trade, Bill Heising from Programming, Martin Kelly from San Jose, Jerry Svigals from Competitive Analysis, Cy Rosen from the Data Processing Division, Herbert Hellerman from Research, Doug Newton from the Advanced System Development Division, Bruce Oldfield and Joel Aron from Federal Systems, and Walter Johnson from the corporate staff[14].

Even at the outset, Haanstra felt that the group was too mainframe oriented. As chair, Haanstra felt that he could more than adequately represent Endicott's smaller system views. This seemed to work as designed until late 1961 when Watson promoted Haanstra to president of the General Products Division. He quickly reverted to his prior role as the single-minded promoter of the 1401. He set up a team in California working on his division's special follow-on, the 1401 Super. Evans, Fairclough, and Brooks remained to push the group forward. Tom Watson and Learson pressed the team to reach a conclusion by the end of the year. With deadlines for a report approaching, the SPREAD group moved to the Sheraton New Englander motel in Cos Cob, Connecticut in November to focus their energies[15]. They completed the work within eight weeks and issued an 80-page report.

The SPREAD group was not unanimous in their view of the New Product Line. Haanstra, of course, was working behind the scenes on the Super, a directly competitive product to NPL. Brooks and Amdahl had argued against moving forward with the project. John Backus of FORTRAN fame thought it too risky, saying there were, "too many eggs in one basket.[16]" Others questioned how it could possibly handle the conflicting needs of commercial and scientific users, and small and large accounts. Evans said that despite all the contention, the stars of the meetings were Fred Brooks and John Fairclough. The decision on the NPL program would be left to the executive team.

The December 1961 SPREAD report described five new models (Models 30 through 70) that would represent a fifty-fold performance range from top to bottom, a range that would increase to two hundredfold with the release of several additional models. This new product line would replace all of IBM's existing computers, including the 701-704-709-7044-7090-7094 scientific series, the 702-705-7080 commercial series, the 650-7070-7074 midrange series, and Endicott's 1401-1410-1440-1460-7010 series.

The new family would be equally adept at both scientific and commercial applications. They would share the same instruction set. The models would be upwardly compatible, allowing customers to grow through the range. Solid Logic Technology (SLT) would comprise the internal circuitry, as the group felt that reliable and affordable integrated circuits were still years away. All models would use the eight-bit byte, as the expansion of character sets would exhaust the 64 characters available in the six-bit byte. All models would support the same input/output devices based on a common, standard interface, the I/O channel pioneered by Werner Buchholtz on Stretch. IBM would announce 44 new peripheral devices with the line. Separating I/O from the processors in this manner meant a new processor could be swapped in over a weekend.

For the first time, software would be a central focus. IBM would develop a new operating system to run across all of the models of the line. It would support all of the major programming languages along with a new high-level language called PL/1 (Programming Language 1). PL/1 would handle both scientific and business applications, a combination of FORTRAN and COBOL.

Marketing planners assumed a 20 percent annual growth rate for the IBM install base because of the new line. In 1961, that install base was 29 million points (a point represented one dollar of monthly rent). The plan forecast growth of that base to 151 million points by 1970, a fivefold increase and an astounding 20% annual compounded growth rate[17].

In January 1962, the IBM executive team reviewed the SPREAD report and the business plan for NPL at the Yorktown Heights Research Center. John Haanstra made the presentation to demonstrate that, at least at the time, he was supportive of the plan. There were some reservations expressed and comments such as "too ambitious." However, with the awareness that Watson and Learson were determined to move forward, the executives reached agreement. Learson closed the session by simply saying, "Do it.[18]"

To recap, the "do it" he intoned involved obsoleting all of IBM's existing computers (seven major lines) and replacing them with a broad new line that was incompatible with all of them. This audacious gamble was likened to General Motors announcing that it was dropping all of its car brands and replacing them with a new mega-brand of compatible automobiles called GM-1, GM-2, and so forth. However, this comparison falls somewhat short. GM was dominant in the car market but never had more than a 50 percent share of the market, and that was back in the early 1950s. In contrast, IBM was abandoning its entire product line and starting from scratch at a time when it owned at least 70 percent of the computer business.

# THE NEW PRODUCT LINE GETS UNDERWAY

D espite all the troubles in getting the NPL to executive approval, the planning work would be the easiest part. The project required the coordinated efforts of all the IBM labs around the world. Thousands of programmers would write millions of lines of new software. IBM would build a number of new manufacturing plants. At the same time, the first system would not ship for five years, so the labs had parallel efforts in place to refresh existing hardware and software. IBM had 116,000 employees in 1961, a number that would double to support the program. In addition, Al Williams had to find the money to pay for it all. Learson told Bob Evans to make the new line, "economical as hell, simple to operate, and the best on the market.[1]" That was a clear objective but it would not be a simple task.

Learson had overall responsibility. With his promotion to VP of Development for the Data Systems Division in 1962, Bob Evans was in operational charge of the program, with a worldwide team of 3,500 as the project got underway. Fred Brooks had the lead in defining the plan. He had his favorite designer, Gerrit Blaauw, as well as Gene Amdahl, on the design team. Jerrier Haddad replaced Ralph Palmer as the Poughkeepsie lab director. IBM named Ralph Palmer an IBM Fellow and he turned his attention to other pursuits.

The first order of business was to tackle the basic design. At issue was the speed/cost conundrum, how to create a common architecture that would achieve low cost for entry models while being fast enough for the high-end models. This was on a par with combining two wildly dissimilar systems, the 1401 and the 7090. For large scientific applications, you would need large memories and a large instruction address size. This would saddle the smaller systems with ten extra bits in each address that would never be used.

The eight-bit byte was part of the Stretch design. With NPL, there would many fights on the issue before a resolution was reached. Going to the eight-bit byte meant that engineers had to rework every I/O device, affecting 44 new I/O devices and 144 total development items (machines and features). Brooks and Amdahl had almost came to blows over the issue, with Amdahl arguing to stay with the current six-bit byte. They battled back and forth. Brooks threatened to leave. Evans sided with Amdahl, as he believed that he was more critical to the project. Brooks resigned. DSD president Bill McWhirter stepped in, met with Brooks, and convinced him to stay. He also authorized a raise. It was Evans's role to inform Brooks of the pay bump, and he made it clear that it was not his idea. Brooks was not in the least amused. Evans settled in 1962 in favor of the eight-bit byte. He would say later that it was the best decision he ever made[2]. The eight-bit byte, formerly known as EDCDIC (Extended Binary Coded Decimal Interchange Code), was derived from the Hollerith codes.

The instruction address size was the next issue to resolve. The high-end models with large core memories would need a large address space. Virtual addressing, if adopted, would require an even larger address space. The low-end models could get by with a 16-bit address space but would be burdened by anything larger. Ten weeks into the design process with no easy answer in sight, Amdahl suggested a design competition. Brooks agreed, and set up thirteen teams of designers to work the problem. The exercise actually fostered team-work among the architects, to Brooks' great satisfaction. The teams led by Amdahl and Gerrit Blaauw both solved the address sizing problem with an additive register. Blaauw was tapping into the design he had developed for the ill-fated 8000 series. Amdahl brought his ideas from Project X. Brooks was the sole evaluator, and he selected the design of the Amdahl team. Address sizes of 32-bit and 48-bit were considered, but cost considerations resulted in a compromise at 24-bit. It was the right decision at the time but would cause problems for IBM mainframes for many years to come.

Brooks codified 8-bit byte and 24-bit address space decisions in a document called "System/360 Principles of Operation." This became the roadmap for the project[3]. The design team completed a first draft in late 1962 and the final version in the fall of 1963, when specifications were frozen. At that time, they considered the document crisp and "short" at only 570 pages[4].

On the hardware side, a group led by Eric Bloch started evaluating the options for the logic technology in 1961. The 7000 and 1400 series machines used SMS (Standard Modular System) logic, the simple transistor modules used in Stretch. Besides representing an older technology, SMS was plagued by the same issue as IBM's computer lines —compatibility. To support all of IBM product lines, the logic designers created over 2,500 variants. On the competitive front, RCA had released its cube logic module with the 501 system which packed in 100 times the electronic components of SMS[5].

IBM Research developed a replacement for SMS (under code name COMPACT for "Cost-Oriented Machine Program in Advanced Computer Technology")[6], called Solid Logic Technology, or SLT. Engineering treated wafers of silicon the size of a half dollar with a photosensitive solution. Ultraviolet light then etched transistor and diode elements on the wafer. A machine sliced the wafer into small chips, each 28-thousandth of an inch square containing single components, which were in turn mounted on the SLT ceramic module. Each module was a half-inch square and contained between six and 36 components (transistors, diodes, resistors). Each one of these modules was the equivalent of an entire SMS card. A final step populated the SLT modules on circuit boards. The objective of SLT was four times the performance of SMS at half the cost.

Eric Bloch and his logic team considered another option. A giant technological leap occurred in 1959 with the invention of the integrated circuit, or microchip. The transistors, diodes, and resistors were no longer discrete components but were all etched on silicon. Texas Instruments and Fairchild Semiconductor devised the integrated circuit at virtually the same time. William Shockley had left Bell Labs after the invention of the transistor, in part because Bell listed his name below those of Bardeen and Brattain as inventors. He formed Shockley Semiconductor in Mountain View, California, to be close to his aging mother. He found it difficult to convince Bell Labs engineers to join him on the West Coast as the East was still viewed as the hub of advanced technology. Silicon Valley was not Silicon Valley yet. Even so, Shockley had no problem finding top engineers locally. However, just a year into the new company, his authoritarian management style led to an uprising. Robert Noyce, Gordon Moore, and six other engineers left to form Fairchild Semiconductor. They each chipped in $500 then received additional funding from Sherman M. Fairchild, the son of George Fairchild of C-T-R fame. Shockley was beyond irate and dubbed the departing engineers the "traitorous eight.["]

Gordon Moore, working with Jean Hoerni, developed a process to etch circuitry on silicon, eliminating the need for discrete components altogether. At Texas Instruments, a long-time collaborator with IBM on memories, Jack Kilby came to a very similar insight, putting transistors and resistors on silicon with the components connected by gold wires. It was the etching design by Noyce and Hoerni that drove the new industry and helped shift the center of technological invention to California.

Since working on SMS packaging, Joseph Logue, the former Cornell electrical engineering professor, turned his attention to the new integrated circuits. In early 1964, Logue moved to the Components Division to work full-time in this area. He was surprised at the lack of work done by the very group that should have been leading the microchip effort. In fact, he encountered pushback as the Components team felt his plans threatened their SLT program. Though shaken, Logue continued to develop a plan for a phased implementation of integrated circuits. His first phase would be Monolithic Systems Technology (MST) with roughly 100 circuits per chip. The second phase would be New Generation Technology (NGT), featuring significantly higher component density encased in a ceramic module.

Magnetic core memory continued to be dominant as no replacement technology had emerged. At the same time, magnetic core continued to be fantastically profitable for IBM based on its automation of its manufacturing. The company had reduced price per core to a third of a penny, from 33 cents in 1953. However, IBM had reached the limit of wringing cost out of the process, and further work would be more expensive than simply moving to integrated circuits. Moe Every led the search for a solid-state replacement for core memory. When he failed to make meaningful progress, Every was relieved of his role in late 1963 and replaced by Erich Bloch. Bloch decided to refocus the effort on enhancing core speeds. He targeted a range of core memory speeds that would support the wide 50-fold performance range of NPL models. He began with core memory with a 2.5 microsecond cycle time, then used smaller physical cores and shorter wiring to reduce the cycle time first to one microsecond and then further to 0.75 microsecond. He then expanded the core performance range by expanding the width of data transfers. The low-end Model 30 would use the slowest core speed and transfer one byte at a time while the high-end models would use the fastest core and transfer eight bytes per cycle.

Bloch's team considered one other memory element, called Large Capacity Store (LCS). It would later be known as "cache." This was an intermediate-speed memory between magnetic core and disk access. It was planned to be core-based but at far lower cost per bit. Fred Brooks lobbied Erich Bloch on LCS, but no viable formula was found and the concept would have to wait until later.

The NPL plan would have five models at announce. Evans had tapped Max Paley to be the overall engineering manager for the five models, working for Fred Brooks. The Model 30 would be the entry model, assigned to Endicott. The Model 40 program went to the Hursley lab, a result of Dick Watson's lobbying after the WWAM cancellation. Fritz Trapnell was the director of the Hursley lab and had pressed for this assignment. The Model 30 and Model 40 would account for half of the units in the entire program[8]. The Poughkeepsie lab would own the Models 50, 60, and 70, with Peter Fagg in the lead. He worked on the 1410 and was the engineering manager for the 7010, 7040, and 7044.

An interesting conjunction of IBM programs occurred in 1962. Bob Evans and Eric Bloch were meeting with John Fairclough at the

Hursley lab when Evans received a call from the IBM team at the Atomic Weapons Research Center in Aldermarston, just 30 miles north. The Center was installing a Stretch machine and a series of problems had stymied progress. Evans proceeded to "fill the sky with airplanes" as the IBM phrase went in order to resolve the issue. It was fitting that Stretch would so intrude on important System/360 deliberations.

# THE SOFTWARE CHALLENGE

Though the NPL hardware design posed issues in balancing small versus large models and commercial versus scientific applications, software for the NPL would prove to be far more diffi-cult. This was a reflection of the ambitious scope of the software plan and the dearth of system software in prior IBM systems. NPL came at a time not so distant from the days when there was no software. A programmer wrote in machine code and loaded it in the computer one job at a time. The new line would jump from this very limited base to a comprehensive portfolio of software products. The most important program was the operating system, to be called OS/360.

Bob Evans could easily measure the scale of the challenge by lines of code. The IBM 650 had 10,000 lines of code. The 1401 jumped an order of magnitude to 100,000 lines, consisting mainly of compilers and a few basic utilities. The initial estimates for OS/360 came in at a million lines, while the final tally was closer to ten million lines[1]. Besides the core operating system, each of the 44 new peripherals devices required programming. In addition, the software system needed to support decimal and binary numbering, variable length fields, and floating-point calculation. Brooks reiterated the overarching NPL imperative programs would be both upward and downward compatible. Evans and his team would marshal thousands of programmers, in the United States and Europe, for the task. And, many of those programmers were already busy working on software for the existing IBM platforms.

There would be little to build on from the outset. The software family tree for OS/360 was sparse. IBM customers had handled most system programming, highlighted by the SHARE Operating System (SOS) completed in 1959. IBM finally got into the act in 1961, releasing the IBSYS OS for the 7090 and 7094 systems, patterned after the SHARE system. At the other end of the product line, the Endicott lab the rudimentary control programs for the 1401 (PRS-108 for tape systems and PRS-155 for disk systems).

Fred Brooks recognized the importance of software to the success of NPL. He recommended that Evans select a strong leader to oversee the effort. Evans' did not have to look too far[2]. Carl Reynolds reported to Evans as systems and planning manager. Like Brooks, Reynolds went to Harvard but took a detour with the Marines before graduating with a master's in physics from Brown University. He did a stint at the Naval Ordnance Lab in Washington, DC, before joining IBM in 1954 in the Applied Science Department. Brooks would continue to drive the hardware and Reynolds would lead software. Brooks needed immediate help in synchronizing the hardware plan with software requirements. Reynolds assigned George Grover, his overall software architect, to the task.

The initial programming teams were spread across several locations. The large system division had programmers in the Time-Life Building in New York City and at the Poughkeepsie lab. There were only 30 programmers in Poughkeepsie, working out of Building 705. That building also housed the lab's cavernous data center, jammed full with Poughkeepsie's machines —the 7070 and 7080 for commercial accounts, and 7040, 7044, 7090, 7094, and 7094 II for scientific accounts. All of those incompatible systems in one room was a testament to the wisdom of NPL. With a massive influx of additional programmers, Building 705 would soon be full. The lab hastily constructed a companion building (706) next door, with a raised-floor data center matching that in 705. It would soon overflow with all of the NPL machines. However, at the very beginning, there was no physical NPL hardware to develop and test on. The early core soft-

ware team built a basic assembler and a simple OS called Basic
Programming Support (BPS). The developers used the Stretch system
in Building 701 for testing.

Endicott had moved its programmers from New York to Endicott
so that they could work side by side with hardware engineering. James
Frame, the team's manager, favored this arrangement. He was a liberal
arts major from St. Johns College who programmed the input-output
control system (IOCS) for the star-crossed 7070 before joining
management.

It was clear from the start that the one sweeping operating system
would not work, but the thought prevailed that a single operating
framework with multiple versions or flavors could accommodate the
broad spectrum of application types and hardware models. The baseline framework would evolve from the prior designs such as IBSYS
and the 1400 series programs. The core operating system would
comprise four major components—the supervisor, scheduler,
input/output control system (IOCS), and the suite of compilers and
utilities.

Initial progress was slow. Richard Case wrote a memo in October
1962 on the challenges of developing the software for NPL. He criti-
cized the lack of progress, stating, "Those guys are really fouled up
and getting nowhere.[3]" Brooks received a copy of the memo and
confronted Case, challenging him to join the software team. Case
agreed and became Grover's immediate manager. In early 1963,
Grover conducted planning sessions at the Anchor Inn in Poughkeepsie with staff from the three labs (Endicott, Poughkeepsie, and
Hursley) in attendance. Case brought along flipchart paper to capture
their ideas.

The discussion identified four operating system (OS) variations,
which Case labeled I, II, III, and IV across the top of the flipcharts.
Someone suggested the name "Four Romans" and the name stuck.
Romans I and II would be low-end operating systems assigned to
Endicott. Roman III was a midrange OS assigned to Hursley, and
Roman IV was the high-end OS assigned to Poughkeepsie. Romans I
and II would run in 4K and 8K of memory respectively. Roman III
would require 32K and Roman IV needed 128K. Those initial
memory estimates doubled very early in development. Programs on

the first three OS platforms had to run on any larger model or operating system variant. With the Romans sketch of the project complete, Reynolds asked Case to work up an estimate for the total cost of development. At such an early state, he projected the tab at $100 million. Management reduced his figure to $33 million. The final tally would be closer to $500 million[4].

Scott Locken in charge of the OS/360 design. Locken was an electrical engineering graduate of the University of Denver, served in the Army in World War II, and joined IBM in 1950 as a service technician. He led development of the 1410 and 7010 operating systems. Multiprogramming was a requirement and an early complication for the OS/360 design. Locken could turn to two prior multiprogramming models, the American Airlines SABRE system and the SAGE program. Martin Belsky was one of the architects for SAGE[5]. He had joined IBM in 1954 as a programmer for IBM's Service Bureau Corporation (SBC), a subsidiary that Watson Sr. created in 1932 to offer rental time on punched-card machines. As a newcomer to SBC, Belsky had to do operator shifts on the equipment until the next newcomer was hired. Coincidentally, that next newcomer was George Grover. Belsky, along with a small team, was assigned to assist in the multiprogramming design.

The Input/Output Control System (IOCS) was a key element of the new operating system. With so many new peripheral devices planned for NPL, it was simply not possible to integrate that many machines without a common interface. The 702 had a standard interface, and the 709 had expanded this and introduced a rudimentary channel. With Stretch, the Exchange architecture provided a far more comprehensive blueprint for handling input/output requests and freeing the central processor. The decision to move to an eight-bit byte already mandated changes to all input/output devices, so programming could code the common interface at the same time. In most cases, the I/O channel would be an additional processor. This further increased the cost and complexity of the lower models.

By the end of 1963, Fred Brooks had the hardware plan in good shape and logged into the Principles of Operation master document. He volunteered to work on software, reporting to Reynolds. Brooks had an initial $60 million budget that would grow as he ramped up the number of programmers. He wanted the majority of the programmers in Poughkeepsie and decided to move the 110 programmers still working at the Time-Life Building in New York City. A slight complication arose. Many of these programmers were Orthodox Jews who refused to move unless there was a synagogue in Poughkeepsie. Brooks helped organized a new synagogue[6].

**Figure 38.** Fred Brooks, who would lead both hardware and software development. Reprint Courtesy of IBM Corporation ©.

In contrast to the hardware architecture and engineering on the physical machines, the software plan was a mess. Brooks organized a retreat to assess the current design and develop a working plan. He also rearranged his immediate team, tapping Martin Belsky as overall architect, with Scott Locken owning the OS versions and Dick Case responsible for the compilers and utilities.

Trouble began almost immediately with the Endicott lab. To start with, Endicott was resentful that Poughkeepsie was leading the NPL programming project. This was simply a continuation of the long-time friction between the two labs. James Frame, the Endicott software manager, had decided very early on that the concept of a single OS that was upward and downward compatible was just not feasible, at least not for the critical Endicott model, the entry Model 30.

To move 1401 systems to the Model 30, Endicott needed an OS that would fit it into 8K of memory and support tape-only and small disk configurations. Endicott was on the hook for Romans I and II but would need to incorporate significant additional function for compatibility with Romans III and IV. That function would likely go unused on the entry models. For Roman II, the minimum memory now stood at 16K, with the overhead target for the resident operating system code at 6K. That would leave only a space of 10K for application programs. Endicott also wanted an entry model with only 8K, which obviously would not work when the base OS code needed 10K by itself.

Frame compounded the memory issue by refusing participation in NPL planning sessions. His programming team was heads-down working on two new 1401 models, the lower-cost 1440 and high-performance 1460. The situation dramatically worsened in November 1963 with Honeywell's announcement of the H-200 computer. It came with a software package called Liberator that enabled the H-200 to run IBM 1401 programs without conversion. Honeywell wanted a piece of Endicott's core and very successful 1401 business.

Haanstra and Endicott had been concerned about the NPL plan and its impact on the 1401 program from the very beginning. The Endicott lab built a prototype 1440 with the new SLT logic circuitry to demonstrate how competitors could easily leapfrog 1401 performance. In addition, simulation testing led Haanstra to believe that 1400 series programs would run slower on NPL machines. These results prompted Haanstra to start the 1401S "Super" program, aimed at a much faster 1401 machine. The Honeywell H-200 confirmed his thinking, as the competitive machine offered a fivefold performance edge over existing 1401 systems at a lower price, along with the Liberator software to ease the switch. Haanstra's team then benchmarked a prototype 1401 Super on native NPL hardware, resulting in a six-fold performance improvement over the 1401.

In January 1964, Haanstra called a multi-day meeting of group executives in White Plains, New York. He presented his secretly developed plan for the 1401S to replace the Model 30 at the low end of the new line. He described the encouraging results of his "Super" tests on NPL hardware. Haanstra also presented the plan to Tom Watson, who purportedly described the revelation as "the finest fiftieth birthday present a man could wish for.[7]" Haanstra then doubled down and scheduled the 1401S announcement for February 1964, two months prior to the target date for the NPL announce.

In the meantime, Frame redirected his developers to work on Haanstra's new machine. He became evasive and uncooperative with the NPL team. For Reynolds and Brooks, this was simply untenable. Besides the negative impact to the overall NPL business case, Frame and his team were also on the hook for critical components needed across all the versions of OS/360. Reynolds escalated the issue and Learson replaced Frame with Earl Wheeler. Frame said that his removal was part of "stamping out the last vestiges of Haanstraism.[8]"

Brooks would have help with his Haanstra problem from an unlikely source, the Hursley lab. The lab had been established in 1958 seventy miles south of London, located at first in a Queen Anne mansion set on a 100-acre estate. Vickers Supermarine used the property during the war to develop the famous Spitfire fighter plane. John Fairclough was leading Hursley projects. He had served in the Royal Air Force before earning an electrical engineering degree at the University of Manchester in 1954. He was there during the Max Newman era, with the Manchester Baby, the Mark I, and Ferranti machines. He worked for Ferranti for three years before joining IBM. He moved to the United States to work on the Stretch before returning to Hursley, where the staff had grown to nearly 400.

Fairclough started with the SCAMP (Scientific Computer and Modulator Processor) design. Expensive electronics and a weak sales forecast sank the business case, and World Trade cancelled it in 1961. He persisted. SCAMP II was a larger system with a revised business case. It also failed to hit the profit bar. Not to be denied, he proposed SCAMP III with a volume outlook based on purely commercial accounts. He had a viable business case but his revamped system completed directly with Poughkeepsie's mid-priced 7044. Fairclough's team ran benchmarks to show it was a better machine. Nevertheless, Learson killed it[9].

However, there was a silver lining. The SCAMP machines used microcode. The concept of microcode originated with Maurice Wilkes. He had formulated his ideas on "microprogramming" in 1951 after returning from the Moore School lectures on the *Queen Mary*[10]. He envisioned a new fast kind of memory he termed "control store" that would be a better option for certain critical repetitive functions. He first implemented the concept on the EDSAC 2 computer. David Wheeler returned from the University of Illinois and designed an instruction set for this specialized read-only memory[11]. Fairclough soon realized that his "control store" could also emulate a computer's entire instruction set. He could achieve this emulation embedding the full instruction set in high-speed vacuum tubes or transistors. This was, of course, prohibitively expensive. However, since this control store would be read-only, he could devise a new simpler yet fast alternative medium for the task, something in between disk and core in speed.

Fairclough's team developed such a memory for SCAMP, called TROS (Transformer Read Only Store)[12]. TROS was a module consisting of a "sandwich" stack of 128 Mylar sheets. Each sheet contained 256 etched copper wires fed through 60 electromagnets called transformers. A wiring path that was active was a "1" bit while a non-conducting path was a" o." He created the "o" bits by simply punching a hole through the designated wire. A complete module of 128 sheets could hold 1,920 bytes. Since the TROS module was constructed in this manner, any changes to the embedded instructions required physical alteration of the Mylar wiring or more likely, replacement of the entire module.

Fairclough's SCAMP was not the first commercial use of microcode. That distinction belongs to the LEO III, the Lyons teashop computer. The Leo III arrived in 1961[13], a transistorized system with core memory, a multitasking operating system , and its own business programming language (CLEO, for Clear Language for Expressing Orders) that was similar to COBOL. LEO III used microcode to provide "master" control over task processing. Orders were brisk at the outset, and its largest customer, the General Post Office (now British Telecom) continued to order machines.

Though the LEO III was an outstanding small business computer, it suffered from insufficient marketing, low margins, and, most critically, a lack of funds to grow the system and effectively compete against IBM. On top of that, the parent company was dealing with declining revenues from its teashops and restaurants. Lyons merged the LEO division with English Electric in 1963. In 1968, the British government orchestrated the consolidation of the remaining UK computer companies into International Computers Limited (ICL). The LEO had reached the end of the line, though not before the Post Office issued an emergency request for five more LEOs. The Post Office would switch off the last LEO in 1981.

Fairclough's microcode concept would come to the rescue of a potentially beleaguered NPL program. If released, Haanstra's Super would cut the total NPL forecast in half. It would also compete for SLT modules, which would become the limiting factor on NPL hardware deliveries. It was time for Learson to step in. Apprised of the benefits of Fairclough's microcode approach, he asked Fred Brooks to run a contest between the Fairclough emulation method and Haanstra's simulation. On the very night of Haanstra's reveal of the 1401S in January 1964, Fred Brooks flew three engineers (William Harms, Gerald Ottoway, and William Wright) to Endicott to work with the lead architect on the Model 30 (William Hanf). They were to convert the relatively simple 1401 instruction set to run in microcode. Brooks brought them coffee while they worked through the night[14]. The Model 30 with 1401 instructions running in microcode performed at four times the speed of the standard 1401, or roughly 80 percent of Haanstra's Super performance. Crucially, the microcode approach would also allow 1401 customers to install the Model 30 and take their time in moving to native NPL mode. Brooks took the morning flight to White Plains and presented the results.

Learson's view was that 80 percent was close enough. He checked with Jim Dezell, a system engineering manager in the Eastern Region. Dezell had already been helpful on the Liberator issue. He arranged a visit to a Honeywell customer and they verified that the conversion software did in fact work. He agreed with Learson that the emulation results were more than adequate[15]. Learson also reviewed the results with John Opel, head of the DPD sales organization, who withdrew the division's support for the 1401S. On the third day, the executive team formally agreed to the emulation plan. Haanstra did not attend the meeting but sent his engineering manager. Tom Watson was able to see the benchmark results and get the full story on the impact of a separate 1401S machine on the NPL program. Learson commented on the decision, "It is obvious that in 1967 the 1401 will be dead as a Dodo bird. Let's stop fighting this.[16]" IBM cancelled the 1401S program in February. The following month Watson placed Haanstra in the "penalty box" and replaced him as GPD president with Charley Frizzell.

The use of microcode rippled throughout the NPL product line. It provided 1401 emulation on the Model 20 and added 1440 and 1460 support on the Model 30. The Model 40 added 1410 and 7010 emulation while the Model 50 added 7070, 7072, and 7074 support. The Models 65 and 85 had microcode emulation for the 709, 7040, 7070, 7080, and 7090 machines. Emulation could even outperform native performance in many situations. The higher-performance Model 65 would run 7090 emulation faster than the native 7090[17].

# DEEPENING TROUBLES

By 1963, the situation in the field was getting dire. The NPL program looked to be a long, hard, difficult slog to the finish. Most work had stopped on follow-on machines while competitors continued to roll out systems that challenged the ones IBM had in the field. The year was a down year for IBM, with revenues only growing by 7 percent.

As the project drew on, Al Williams became more and more concerned with IBM's cash position. Even over and above the huge outlays for NPL development, he was missing a substantial amount of cash. Tom Watson asked John Opel to dig into the problem. Opel was not yet 40 years old but had been on a steep upward career path. He had fought in the Pacific in World War II and used the GI Bill to continue his education, culminating with an MBA at the University of Chicago. He planned to stay in academia and update the prevailing textbook on economics or alternatively, return to Missouri and take over his father's hardware store. A fishing friend who happened to work for IBM offered him a job in sales. He accepted, dropping his other plans. He began by selling unit record equipment. During the course of his rise in the company, Tom Watson saw him perform in a sales training class in Endicott and tapped him as his AA, the ultimate springboard for further advancement.

Opel surveyed the plants but had little to show for the effort. He decided to visit each plant and force a complete accounting. He soon discovered the missing cash in work-in-process inventories. The lengthy NPL development cycle had "marooned" significant factory production short of release and sale. Opel was able to identify roughly $150 million in work-in-process goods tied up in the factories[1].

When Fred Brooks took over software responsibility, no firm target date had been set for the NPL announcement and availability of the various OS versions. The software teams were not working toward a deadline. Instead of heads-down coding to meet a committed date, the programming teams engaged in on-going design debates. At the same time, the software architecture was in flux, with essential elements such as multiprogramming still incomplete. For Carl Reynolds, this was a problem. He did not want to commit to a date until he knew the scope of the project and his large team sized the effort. Yet, the pressure continued to mount. In April 1963, he estimated OS delivery at December 1965 for Romans I, II, and III, and 1966 for Romans IV. Hardware development was moving much faster, and looking at an April 1964 announcement for the new line. In September 1963, Reynolds reluctantly agreed to the announce date despite the continued turmoil in software development. Many in his organization viewed the commitment as wildly unrealistic, and that would indeed prove to the case[2].

In mid-1963, Brooks had moved responsibility for the OS architecture design from Scott Locken to Martin Belsky. Locken continued to press Brooks on this issue. He contended that if Brooks had left the architecture with him, the software would still be late but it would be right. In May 1964, Brooks finally agreed and moved the design back to Locken's team. He gave Belsky's group an oversight role as a compromise. This lasted about six months before he gave Locken full ownership and assigned Belsky to the high-end OS, Roman IV.

George Grover's early memory estimates had been 2K for Roman I, 8K for Roman II, 32K for Roman III, and 128K for Roman IV. Scarcely three months after this projection, he had to double the memory requirements[3]. The architects presumed that the team could contain the resident OS program to 16K, leaving room for at least one application program. Simulator tests in October 1964 demonstrated that this might physically work but would not be very functional. In addition, it required a disk drive to swap segments of the OS in and out of the main memory. It was becoming clear that the plan to have one operating system was doomed.

And then, Grover and Locken announced the decision to increase minimum memory for OS/360 to 32K. This was a complete nonstarter for Endicott. John Gibson, who was Haanstra's group executive, allocated $1 million for the Endicott lab to develop separate OS programs for the low-end disk models. On the last day of 1964, Evans and his team officially discarded the idea of four versions of OS/360 to cover all models. OS/360 would have the three Roman versions for the larger models, and Endicott would be on the hook for four operating systems to cover the space below. Those four programs were Basic Programming Support (BPS), Batch Operating System (BOS), Tape Operating System (TOS), and Disk Operating System (DOS). The first two OS versions would only require 8K, and the others would run in 16K. Earl Wheeler, who had replaced Frame in September 1964[4], was yet another alumnus of Union College, following in the footsteps of Ralph Palmer, Halsey Dickinson, Byron Phelps, and George Stibitz. He graduated with an electrical engineering degree and joined IBM in 1955. Drafted by the Air Force, he programmed on the 650.

Wheeler's team, unchained from the OS/360 design, could lean on the prior operating systems developed in Endicott. They completed the BPS version first as it was critical to early development and test activities. The other three OS programs borrowed heavily from programs developed for the 1410 and 7010 machines. BPS and BOS were simply transitional in nature. Wheeler dropped the tape version (TOS) when disk prices came down. In the end, DOS was the only long-term survivor. His expanded OS mission led Wheeler to enlist the help of the labs in San Jose and Rochester. San Jose signed up for the low-end assembler and the RPG compiler. Rochester agreed to take over software maintenance for the 1401 series, an assignment that would prove valuable to the future of the lab.

Brooks assumed management of OS/360 in February 1964. He subsequently decided to accept an offer to become a computer science professor at the University of North Carolina. He discussed this with

Tom Watson in the summer, planning to start at UNC in September. When his time came to leave, OS/360 was still in chaos. Tom wanted Brooks to stay one more year and worked hard to get him to say "yes." He offered to send someone to teach Brooks' classes and agreed to have Brooks spend one week a month in Chapel Hill. In addition, he said that IBM would be there when the university needed a new computer, a promise that would result in a $900,000 computer system to anchor the university's computation center. The subsequent growth of computing talent in the area would also lead Tom to build a new plant and laboratory at Research Triangle Park in nearby Raleigh.

With Brooks poised to start his one week per month, Bob Evans recommended a replacement to Carl Reynolds - Frederick "Fritz" Trapnell[5]. Trapnell grew up in San Diego and served with the Marines during the Korean War. His father was a famous aviator and Navy admiral. Upon his return from Korea, Fritz graduated from Caltech, where he worked on the university's first computer, the Clifford Berry-inspired Electrodata 205. He also worked on the LGP-30 (Librascope General Purpose), created from a design by Caltech alumnus and Los Alamos tabulator designer Stan Frankel. Frankel returned to Caltech after his security clearance was revoked. Both systems were drum machines that competed with the IBM 650. Hal Martin, a fellow Caltech graduate, recruited Trapnell. Martin joined IBM in 1951 as an engineer in Endicott and was now working for Ralph Mork at World Trade headquarters in Paris.

Early on, Trapnell had a personal situation arise and asked to resign from IBM. Not wanting to lose him, Martin offered Fritz a posi-tion working for him in Paris. Trapnell rose fast and took over as the Hursley lab's director in 1961, reporting to Byron Havens. John Fair-clough worked for him, and they collaborated on the SCAMP machines. In early 1964, Trapnell moved to White Plains as the state-side World Trade executive. There, he wrote a memo decrying the slow progress of software development. Brooks challenged him to join the team, as he had done earlier with Richard Case. Trapnell at first demurred, saying he was happy at World Trade. Brooks asked him to read his memo again. Brooks was the consummate salesman and usually got his way. Trapnell signed on, starting at Poughkeepsie in March 1965.

The defection of Endicott from the OS/360 plan resulted in three Roman variants remaining[6]. Roman II became the Single Sequential Scheduler (SSS), supporting the execution of one program at a time without multiprogramming. It would later be renamed the Primary Control Program (PCP) and run in 32K of main memory. Roman III changed to the Multiple Sequential Scheduler (MSS), providing for multiprogramming with a fixed number of partitions and program tasks. IBM would later adopt the common name that was used for it— MFT, for "Multiprogramming with a Fixed number of Tasks." MFT required 128K of memory but needed 256K to run well. Roman IV became the Virtual Memory System (VMS). This high-end operating system provided multiple variable-sized program spaces. The VMS name soon changed to MVT (Multiprogramming with a Variable number of Tasks). All three OS/360 variants shared the same set of APIs, Job Control Language, database access methods, print spooling, and input-output control system.

By the time Trapnell entered the fray, the OS/360 program had acquired a nickname, the "Big Oz." The road leading to Poughkeepsie, New York Route 9, was now "the yellow brick road." There were at least a thousand programmers working on the project, with over 500 in Poughkeepsie alone. Trapnell's first order of business was to assess where the overall effort stood in relation to the promised schedule. Endicott had already delivered BPS, BOS, and TOS. Earl Wheeler told Carl Reynolds that DOS would slip from the plan date of December 1965 to March 1966. Reynolds and Trapnell reviewed the 40-plus software products in development and laid out a master schedule that would deliver them across twelve separate releases.

Setting this revised schedule was the easy part. They still needed to present it to IBM Corporate for approval. Reynolds and Trapnell arrived at the meeting to find Vin Learson in the back of the room. With a good amount of trepidation, Trapnell presented the revised VMS schedule. Learson could be tough, but above all, he valued competence and complete transparency. Because Trapnell and Reynolds had done their homework, Learson accepted the delay. After the meeting closed, John Opel sat down with them. He stressed the bright side of the schedule slip. Customers would be paying the higher rents of their current systems for longer until they installed the lower-rent System/360 machines. After this commitment, Poughkeepsie did not miss any schedules.

After the announcement in April 1964, the top priority became PCP (Romans II) , as early customers (mostly Model 40s) needed something to get started on. PCP was relatively simple compared to MFT (III) and VMS (IV), but it still represented 440,000 lines of code, a huge program at the time. Carl Reynolds canceled MFT and scaled back VMS. The system architects designed VMS to have completely flexible memory usage, unbound by partitions. This became impossible, as the original memory structure did not plan for it. As a result, the programmers of the compilers and utilities did not constrain their programs from modifying areas of main memory. This would not be resolved until the System/370 arrived. Trapnell replaced VMS with the simpler MVT (Multiple Variable Tasks) which had fixed memory partitions. This change did not sit well with Larry Cohn, the chief architect of VMS. He resigned from the company[7]. The field pushed back on the cancellation of MFT, Poughkeepsie reached a compromise where they would only deliver a base variant of MFT. Fortunately, lead developer Tom McCallister had continued work on MFT in his spare time, enabling a release of MFT in 1967. MVT, the ultimate OS/360 variant, shipped over two years late, included in the last of the twelve drops in the schedule devised by Reynolds and Trapnell.

Fritz Trapnell would have help in his high-pressure role with Big Oz. It would come from someone who had completed an analysis of the System/360 programming effort and thoroughly trashed it. His name was Watts Humphrey[8]. Humphrey earned a degree in physics at the University of Chicago, where Nobel Prize winner Enrico Fermi was one of his professors. He then walked across campus to complete an MBA at the Booth School of Business. Humphrey joined Sylvania in 1953, where he led a team to produce the slyly named MOBIDIC (Mobile Digital Computer), outbidding IBM, RCA, and Philco. It was a 704-class machine patterned after the SAGE systems and funded by the Army. They wanted a replacement for the Mobile Records Unit (MRU) punched card configurations used for battlefield accounting. The MOBIDIC computer would reside in a large trailer and provide real-time information. Sylvania was not interested in taking the concept commercial, so Humphrey resigned and applied to IBM in 1959. With his stellar background, Tom Watson did the interview in his office on the 16th floor at 590 Madison Avenue.

In 1964, Learson asked Watts to manage a bid for a Federal Aviation Administration (FAA) air traffic control system. Learson instructed two division presidents that for the purposes of this effort, they would be reporting to Watts. Only someone like Learson could pull this off. Humphrey proposed modified Model 50 machines. IBM won the business, representing $100 million in revenue. It was one of the largest contracts in IBM's history. Learson soon tapped Watts again, this time to lead a frantic effort to develop System/360 time-sharing, a major miss from the announcement.

In late 1965, Watts wrote his white paper on OS/360 programming. He did not pull any punches, describing the current effort as thousands of programmers busy coding but not working off any real, committed plan. Learson read the paper and challenged him to fix the issues by promoting him to Director of Programming. It seems that the prior Director of Programming had been cheating on his wife with his IBM secretary and someone filed an Open Door complaint. Tom Watson handled the issue directly and fired the director, opening the door for Humphrey, whose stellar work on the FAA bid and the time-sharing program had proven his worth.

Humphrey arrived shortly after Brooks left for North Carolina. Reynolds and Trapnell had already revised the OS/360 plan, but the work continued to be challenging and the new schedule was seriously exposed. Humphrey started with yet another complete review. The result was "some thirty-on technical capabilities were decommitted[9], the IBM jargon for cancellation. Humphrey had not been in his new assignment for long before Learson called a meeting with him and his 30 senior managers. Learson kicked off the meeting in his inimitable style by launching into a tirade about the slip in the OS/360 schedule, complaining that the hardware was starting to ship and the software was still not ready. As was Learson's standard policy, he demanded an answer and a committed date, and he wanted it in two weeks. Humphrey said it would take 60 days to assess and come up with realistic schedule, to which Learson responded, "Okay, God dammit.[10]"

Though he came late to the party, Humphrey would stay to the end. The simple Endicott OS programs, BOS and TOS, had shipped on schedule in February 1965. PCP, the most pressing OS/360 version at the start, arrived in March 1966. Endicott's DOS slipped a couple of months to June 1966. The reinstated OS/360 MFT would ship in October 1966, about ten months late. The flagship OS/360 MVT would not be ready until August 1967, almost a year and a half after its original ship date. The compilers (assemblers, FORTRAN, COBOL, PL/1, and RPG) all shipped in late 1965.

Memory space remained a problem. The 32K minimum was set early in the program and based on the cost of core memory and the perceived willingness of customers to pay it. At the time, the architects viewed it as an outlandish amount of memory. This arbitrary restriction resulted in substantial additional work for the programmers. Gene Lindstrom, working in San Jose, had a common experience with the memory cap. He pleaded for more space for his assembler program. Fritz Trapnell had allocated 14.5K for him to work with. At this size, he would have to split the program into two programs. He did the work but told Trapnell that the best he was able to do was one program at 14.5K and the other at 15K. Trapnell reiterated the limit was 14.5K. Lindstrom went back to work and split the assembler into three programs. In hindsight, the memory limit was simply an IBM made-up issue. Customers quickly realized that they needed more memory to make OS/360 work effectively. Most customers had no problem upgrading their memory, usually to at least 128K. This also provided a nice bump in IBM revenue.

One additional software objective of NPL was the new common PL/1 programming language, a combination of FORTRAN and COBOL. The code name was APOLLO, supposedly for "All Procedure-Oriented Languages Lumped into One." Carl Reynolds pulled together a team to decide on the needed features. Called the 3x3 Committee, it had three members from the user group GUIDE, three from SHARE, and three from IBM. The SHARE members were especially concerned that IBM would withdraw support for FORTRAN once the new language was available.

Reynolds assigned the project to the Hursley lab. Hursley had lost Roman III with the restructuring of the Four Romans, so it had the resources to tackle this. The early versions of PL/1 had both function and performance issues. Fred Brooks was critical of the design, saying it was too feature-filled, what he called "gargoyles and all.[11]" In the end, the basic premise was its undoing. COBOL users shied away from what they perceived as a scientific language, and FORTRAN users decried the interminable syntax and verbosity of COBOL. Though it was supposed to be the new mainframe programming standard, it found far greater acceptance in minicomputer and PC environments, with adoptees not burdened by any preconceptions. Despite this, there remained a window of opportunity for the language. As was the case with RPG, other computer manufacturers would have fallen into place if it appeared that IBM was going to press PL/1 as a standard. However, IBM failed to push it forward aggressively and soon abandoned the effort altogether. PL/1 did become widely used internally within IBM. Reynolds wanted all OS/360 code written in a PL/1 variant called PL/S.

The experience of developing the software for System/360 was painful. Designs changed, schedules slipped, and costs escalated. The early business case estimate was $40 million, a figure marked down from Richard Case's $100 million estimate. Tom Watson addressed the issue while speaking to customers in 1966, "We are investing nearly as much in System/360 programming as we are in the entire development of the System/360 hardware. A few months ago, I was informed that the bill for 1966 was going to be $40 million. I asked Vin Learson what he thought the cost would be for 1966, and he said $50 million. Twenty-four hours later, I asked Watts Humphrey, who is in charge of programming. He said, 'It's going to be $60 million.[12]'" At this point, Tom stopped asking, for fear the number would further escalate. It did. The final tally was close to $500 million.

Chuck Branscomb recalled a meeting he had with Tom Watson and Fred Brooks, well after Brooks had left for North Carolina. Tom asked about the cost and delays of the OS/360 programming. Brooks responded:

> "I had two groups of people that had a different view of what we should do. One group said; 'If you get the architects out of the way, we'll meet the schedule.' The other group said: 'Finish the architecture whatever you

do; otherwise this thing is going to be a mess for a long time!' The schedule pressure was really hard on us, and so I picked the first group that said they can meet the schedule. [13]"

Brooks told Tom that his decision probably cost IBM $100 million. Learson had a more sanguine view, commenting, "we did what Charles Kettering, an engineering genius and president of the GM Research Division always advised against: we put a delivery date on something yet to be invented.[14]" Fritz Trapnell was even more critical. He believed the original concept of one common operating system across the entire line was never feasible. The effort to force that plan probably set the program back a year[15]. Fitting OS/360 into 32K when memory cost was no longer an impediment added countless programming hours to the already mammoth project. The System/360 started with an outlook of one million lines of code and then ballooned to ten million, two orders of magnitude greater than the 1401. At the end, IBM had two major operating systems, DOS and OS/360. Both would live on for many decades.

Tom Watson understood hardware development far more than programming. The tremendous escalation in cost and the extensive slips in schedule had endangered the entire NPL program. He asked Brooks, who had managed both hardware and software development, "What's the difference?" Brooks told him he would need to think it over. Brooks already had some opinions on the subject. OS/360 was only IBM's second major operating system project. It was both leading edge and incredibly ambitious. Multiprogramming presented a far more complex environment than simple batch. There was a bit of the "second system effect" where the architect attempts to stuff in all the features that fell out of the first project, an outcome called "creeping featurism" or "feature shock." Then there was the lack of a final and frozen design before mass programming began. Once the design was complete, a single chief architect should have had the final say on any mid-course changes. Brooks also realized his mistake in switching the design role from Scott Locken to Martin Belsky.

The composition of the programming team was another issue. Experienced coders tackled the early programs. The expanding scope required more and more programmers, many just out of college without much coding on their resumes. Brooks recognized the difference between good programmers and not-so-good programmers. Good programmers were dramatically more productive, producing perhaps five to ten times the output. His experience led Brooks to write a book about programming, called *The Mythical Man-Month*[16]. The central conclusion, embodied in the title, argued that adding resources to a late project would only make it later. He observed facetiously, "The bearing of a child takes nine months, no matter how many women are assigned.[17]" The OS/360 project was as large, complex, and diverse in lab locations as had ever been attempted. The addition of a single new person required the existing development team to slow down and integrate that person into the project. That person is thus not an incremental man-month of new production but a brake on overall progress. Brooks dedicated his 1975 book to Tom Watson and Bob Evans. It came to be viewed as the "bible of software engineering", so called because like the bible, "everybody quotes it, some people read it, and few people go by it.[18]"

The experience of OS/360 was a time of crisis, and the unrelenting pressure took its toll. Carl Reynolds was Exhibit A for burnout. He left IBM in January 1966 after three years at the helm of the pressure cooker. He was a great manager, "cool under fire,[19]" and a big loss for IBM. Fred Brooks wanted out of fourth-level management and a return to technical matters, so he decamped to the University of North Carolina. George Grover, the original OS/360 architect, moved on to IBM Research. Larry Cohn, the VMS program manager, resigned when Reynolds rebuffed his design. His replacement, Larry Foster, stepped down after only six months. Jim Frame survived his banishment and resurfaced as head of the new software lab in Santa Teresa, California. Martin Belsky left the OS/360 project before the end. Fritz Trapnell left IBM in June 1967, just after the final operating system (MVT) had shipped. He admitted to being burned out, as so many others were. He was young and not yet steeled to this kind of unrelenting pressure. He would later join Gene Amdahl in Amdahl's new company building IBM mainframe clones.

# THE SYSTEM/360 ARRIVES

With the mammoth NPL effort still underway, the field grew increasingly starved of new products, even with the temporizing machines. There was growing pressure to announce the new line. The original plan called for a phased release over a 15-month period. Competition on several fronts forced the issue. RCA had announced its broad Spectra 70 line, GE released the 400 and 600 series systems, and Control Data debuted the high-end CDC 3600. The announce plan changed. IBM would announce the entire line at the same time.

The target date was April 7, 1964—a Tuesday. Apparently, Tom Watson set the day, as he sailed on weekends and was not always in the office early on Mondays. John Opel, as head of the DPD marketing division, managed the ambitious announcement. IBM conducted 200 reporters by train from Grand Central Station in New York City to the Poughkeepsie plant. The company held simultaneous rollout events in 165 cities across the United States and World Trade, with total attendance estimated at over 100,000[1]. Those who turned out witnessed the coming-out party for third-generation computing technology comprised of 6 new processor models and over 150 total products. Tom Watson opened the session, with Bob Evans and Fred Brooks providing the details.

The program known for several years as the New Product Line became the System/360. Many of those working on NPL did not learn of the actual name until announce day. System/360 was chosen to reflect a single family of systems covering the complete circle of computing needs, from small to large, and from commercial to scientific. The six new processors spanned a 50-fold performance range, which would grow to a 200-to-1 range with later models. The entry Model 30 processed 33,000 additions per second while the high-end model calculated 750,000 additions per second. Aimed at replacing the 1401, the Model 30 was six times the performance at 50 percent less cost. Though the Model 30 was the entry system, it still was comparatively large, taking up the space of a two-car garage[2].

**Figure 39.** IBM System/360 (1964), a watershed event in computer history and the audacious "bet the company" effort by Watson. Jr. Reprint Courtesy of IBM Corporation ©.

None of the new models were compatible with existing IBM computer lines, aside from the extensive built-in emulation capability.

The new models (30, 40, 50, 60, 62, and 70) ranged in monthly rentals from $2,700 to $115,000, with the purchase range from $133,000 to $5.5 million. They all featured SLT memories and microcoded control stores. Brooks and his hardware architects achieved performance gradations with a combination of CPU processor, core speeds, and the width of internal data paths. The CPU cycle time would range from 54 nanoseconds on the later high-end Model 95 down to 900 nanoseconds for the later entry Model 25. IBM soon replaced the announced high-end computer, the Model 70, with the faster Model 75. It provided ten times the performance at 40 percent less cost than the 7090 system it replaced.

As if the 360 announcement was not big enough, the World's Fair in New York City opened two weeks later. Tom Watson emulated his father's 1939 celebration with a big, brash splash at the exhibition[3]. The IBM pavilion covered over an acre at Flushing Meadow. A giant ball 90 feet high dominated the site. The ball was actually a Selectric typewriter ball with the letters "IBM" imprinted 4,000 times across its surface. Visitors entered the ball-shaped IBM pavilion through a "grove" of 45 sheet-metal trees festooned with 14,000 Plexiglas leaves. The centerpiece to the 54,000-square-foot pavilion was the main theater located within the Selectric sphere. Five hundred spectators took their positions in a seating area called the "people's wall." A master of ceremonies clad in a tuxedo descended on a tiny pedestal from high above to welcome each group[4]. A 12-minute multimedia presentation flashed simultaneously on 15 large screens showing highlights of the new Information Age[5]. Ten million visitors would file through the IBM pavilion during the course of the fair.

Eero Saarinen did the overall design of the IBM pavilion, one of his last commissions for the company. Charles and Ray-Bernice Eames, famous for their modern chair designs, did the interior work. Tom Watson did not spare any expense to put IBM's best foot forward. As his father had done, Tom orchestrated an IBM Day. There were 12,000 IBMers on hand with former general and president Dwight D. Eisenhower as the principal speaker. He spoke fondly of his relationship with Watson Sr. while stressing his total lack of understanding about computers[6]. Watson, Sr. had been instrumental in the Eisenhower taking the job of Columbia President in 1948[7].

In the initial weeks after the announcement, the slow pace of orders was concerning. The sheer breadth of what IBM unveiled took time to digest. By the end of the first month, IBM had over a 100,000 orders, though all but perhaps 1,100 of those were positional orders, not firm orders. By July, the volume of orders, and especially firm orders, had accelerated dramatically, with 2,200 firm orders and many thousands more in the backlog. The fear of too few orders gave way to too many. In addition, the average order was far larger than the marketing planners projected.

The hoopla on announcement day and the continuing presence at the New York World's Fair masked the challenges ahead. IBM was only three years into the effort, and the last System/360 model would not ship until June 1971. The first shipment, a Model 30, would go to McDonnell Aircraft in June 1965[8]. The initial excitement quickly faded as the reality of the work ahead became clear. Many thousands of additional workers needed to be hired and five new plants remained to be built. Watson's team also had a slew of technology issues to be resolved.

The month following the announcement, Tom Watson made a dramatic change in the leadership of the program. He put his brother Dick in charge of all development and moved Vin Learson to lead sales efforts. Dick had done a great job in growing World Trade to a business nearly equal in size to IBM in the United States. He was only 45 years old, five years Tom's junior. Tom felt that he should provide an opportunity and path to the top job at IBM for Dick. This would require more experience at the U.S. corporate level. He felt a logical place for that growth would be as head of all research and development, the spot held by Learson. He had concluded, perhaps naively, that Learson had put the System/360 program on a solid footing and Learson's skills could better be used to drive sales. Learson was not at all happy about the change. He had shepherded the project since its inception and looked forward to enjoying the fruits of his great labor. He commented in the usual Learson way, "I'm bringing the marvelous 360 to the marketplace and you bring the goddamn brother over to run my project.[9]"

Both Dick Watson and Learson would be peer senior vice presidents reporting to Tom. He could step back and let them manage the day-to-day issues. Before he put the change in place, he added one caveat to both of them: He did not want the lack of any customer function to endanger sales. He asked Learson to identify them and Dick to marshal the development resources to deliver them. This was the normal relationship between sales and development, but it would prove to be problematic.

The acceleration of System/360 orders presented the first crisis for Dick. Each system would need SLT modules and core memory. Ralph Palmer had pressed Tom to have IBM produce the SLT modules. Emanuel ("Manny") Piore, the head of Research and Al Williams agreed. This approach provided more control, more profit, and protected IBM's intellectual property[10]. However, SLT production was significantly more complex than the core memory or transistor manufacturing processes. IBM spent $100 million to construct the SLT plant at East Fishkill, 15 miles southeast of Poughkeepsie. Tom Watson had to explain to the IBM board of directors that while regular plants came in at $40 per square foot, the SLT facility would cost $150 per square foot[11]. There remained a need for some outside suppliers. The parts that connected the SLT modules to the circuit boards, called "tabs", were soon in very short supply. Tom Wise, a journalist for *Fortune* magazine, noted that, "IBM representatives suddenly began appearing at tab plants late in the evening or early in the morning, with suitcases. They would pack all the tabs they could and fly back to Endicott to keep the production line moving.[12]"

Production started at East Fishkill in 1963 under Dr. Bernard Slade, with John Gibson in overall charge of the Components Division. Slade developed the first transistors used in radio and television at RCA before joining IBM. The first year of SLT production produced 500,000 SLT modules. The runaway pace of orders required production for 1964 to scale up to 6.1 million. Gibson asked Slade to increase the volume to 54 million the following year. He committed to 28 million, representing a more than a fourfold increase. Gibson asked him to reconsider. Slade worked with his production chief and came back with a revision to 36 million. That was not acceptable, and Gibson replaced Slade with Ed Garvey, who accepted the higher number[13]. The actual production for 1965 was 36 million, exactly what Slade had forecast. The volume requirement continued to escalate, with a target of 143 million SLT units in 1966 and upwards of 190 million the following year[14]. Slade also indicated that too high a rate of production would result in widespread defects and reliability problems. The astronomical increases in volume made his prediction a certainly. Fully a quarter of the SLT cards produced failed quality tests and had to be discarded.

John Gibson, recently promoted to IBM group vice president in charge of all of the development divisions, took the fall. Tom Watson replaced him with John Haanstra. Haanstra was in the "penalty box" for the continued championing of the 1401S, which bordered on insubordination. In the days of Watson Sr., Charley Kirk would have simply fired Haanstra. Tom took a more enlightened approach, sending the individual to a lateral or downward position, one that was quite often career ending at the higher executive levels. That was not the case for Haanstra, as his executive talents were held in high regard. As for Gibson, he also made a quick recovery from his banishment, returning to Components Division in 1965, where Jerry Haddad said he belonged[15]. Haddad observed, "components are the man's life." On the positive side of the ledger, the unit cost of SLT modules plummeted to just 40 cents as production passed 100 million annually. Learson would note, "IBM had become, in a very short time, the largest component manufacturer in the world."

SLT production was not the only issue facing Dick Watson. Core memory was a close second. After the System /360 announcement, orders poured in with configurations that included significantly more memory than had been predicted. The resulting challenge was a repeat of the SLT issue -- how to scale up production to meet the demand. There had been on-going innovation in the machines that produced core memory. The devices used hollow needles to guide the wires through the magnetic cores. The production units graduated from 16-station devices to 32 stations, and eventually to 48 stations. Poughkeepsie did the design work for core manufacturing and the Kingston plant did the production. Core production was nearly one billion in 1963. The volume increased dramatically, peaking at 20 billion cores in 1970[16].

The Poughkeepsie site continued to offload parts of its many missions in order to address the crisis at hand. It moved the tape mission to Boulder, Colorado, a university town that provided ample access to talent. The plant constructed in 1965 north of the city grew into a sprawling campus with 2.5 million square feet of space across 25 buildings. The site was asked to do core production in addition to tape drives. This was still not enough resources for the core demand. The plant general manager in Kingston had previously worked in Japan and thought that additional production could be arranged using hand wiring in Japan. He would move significant core volumes to Japan and Taiwan.

By the latter part of 1964, System/360 manufacturing and development was in disarray. Tom Watson decided to step back in and take a hands-on role. Dick Watson and Vin Learson were not amused with the renewed oversight, which required daily progress meetings with Tom. Vin and Dick did not get along. Both knew they were in a competition for the top job. Learson had taken advantage of his sales role and flooded Dick with requests for additional function. Dick was out of his league in managing a project of this scope. Their personal relationship deteriorated as the crises in manufacturing multiplied. The company was looking at significant delays in committed ship dates. By late October 1964, something had to give. Tom was grim, torn between the fate of his company and the relationship with his younger brother. There was no easy way out. The IBM Company was drifting toward the abyss. The day before the announcement of shipment delays, he relieved Dick of overall responsibility and put Learson back in charge. Tom asked his brother to run the corporate staff and World Trade, a position clearly out of the mainstream of the company. Dick complained that Tom left him with the "crumbs,[17]" but there was no turning back. Dick was completely demoralized and retired in 1970. He had described his older brother as "not only my boss but my best friend.[18]"

In fairness, it was a no-win proposition from the beginning for Dick Watson. Learson had 20 years of experience managing IBM plants and labs. Dick got very little of that in World Trade. Their personalities were radically different. Learson was aggressive, decisive, and sometimes brutal in his single-minded focus on the task in front of him. Dick had grown World Trade with a completely different skill set, one not built for the current challenge. Tom realized this belatedly. He should not have asked Dick to leave World Trade in the first place, and certainly not to drop him into the most challenging project in IBM history, a project that had been Learson's "baby" all along.

Learson hit the ground running. Knowing he could not dig out of the problems by himself, he enlisted four senior executives to ride herd over the System/360 program. He chose them for their ability to get things done, regardless of their prior missteps. Three of the four had prior stints in the IBM penalty box. He put John Gibson back into the Components Division to manage SLT production. Clarence Frizzell would oversee the Endicott and Rochester plants, Hank Cooley would be responsible for Poughkeepsie and Kingston manufacturing, and John Haanstra would oversee the Boulder and San Jose plants. Dubbed the "Four Horsemen", they would have near-dictatorial power over manufacturing and development in their assigned spheres. Watson had only recently promoted Haanstra to lead a "mega-division" called the System Development Division (SDD). It consolidated all three manufacturing and development divisions and had 14,000 employees working in eight major plants and labs across the United States and Europe. Haanstra vacated his office at SDD, thinking he would be back in a few months. Learson named Chuck Branscomb to take over. Branscomb expected that he would sit in for Haanstra for just a short period. He would be wrong.

Despite the manufacturing challenges, it soon became clear that the System/360 would be a runaway success. Six hundred systems shipped in 1965, with 7,700 installed by the end of 1966[19]. Production reached 1,000 systems per month. From 1965 to 1969, IBM shipped 33,000 systems. The Model 30 and Model 40 represented half of those numbers; at least until the entry Model 20 began shipping in mid-1966. The Model 30, the essential Endicott machine that rented for an average of $8,000 per month, would bring in $1 billion in revenue by 1972[20]. The business plan for the midrange models (40, 50, 60, and 70) was 3,000 systems. The actual number, just in the United States, was over 5,000.

The impact to IBM's bottom line was dramatic. In the five-year period after the System/360 announcement, IBM revenue grew from $3.2 billion to $7.5 billion[21]. Income was up from $431 million to $1 billion. The IBM workforce increased from 127,000 in 1962 to

271,000 in 1971. Besides the staffing in the manufacturing plants, a key segment of that increase was in field systems engineers, a team that would prove critical in helping customers install their systems. In 1964, the worldwide computer inventory was an estimated $10 billion, with IBM representing 70 percent. By 1969, the total inventory had grown to $33 billion, with the IBM share inching up to 72 percent[22].

There was also a financial impact for some of the players behind the success. Tom Watson liked to reward his top performers in grand style. He went to Poughkeepsie and requested the time cards for Fred Brooks, Gene Amdahl, Jerrier Haddad, and Bob Evans. Commenting that the salaries looked a little light for a team that successfully brought the System/360 to market, he tacked on an extra $10,000 a year to each one.

After the System/360 announcement, Brooks told his team, "Tonight the lights will be on in the other guys' offices.[23]" This was absolutely true. IBM had set completely new standards for computers, in technology, price/performance, and compatibility. The traditional BUNCH companies struggled mightily. RCA already had a compatible product line, with the four Spectrum 70 models. It added compatibility with System/360 and upped the ante by moving to monolithic integrated circuits, a debate that was still raging within IBM. Those changes proved to be insufficient. RCA sold its computer business to Sperry Rand in 1971. Sperry would later merge with Burroughs in 1986 to form Unisys. General Electric had the compatible 600 series as well as the significant head start in timesharing. It would announce the "Compatible 200" series in an attempt to go head-to-head with IBM. For IBM, the competitive environment would start to shift away from the BUNCH. The new battlegrounds would be minicomputers, supercomputers, and the plug-compatible manufacturers.

The total estimated cost for the NPL program was well over $5 billion. Tom Watson's own breakdown for the project was $5.3 billion in total cost, with 70,000 new hires, and $4.5 billion to build five new plants[24]. Tom Wise was granted access to the company in 1966 for an inside profile on the System/360 program. His filed two major stories in *Fortune*, "IBM's $5,000,000,000 Gamble" and "The Rocky Road to the Marketplace." He quoted Bob Evans as saying the project was a

"You bet your company" project. Wise illustrated the sheer bravura of this undertaking by saying, "It was roughly as though General Motors had decided to scrap its existing makes and models and offer in their place one new line of cars, covering the entire spectrum of demand, with a radically redesigned engine and an exotic fuel." Roger Smith of GM actually attempted a bet of somewhat similar scope in 1982 with his GM10 program to replace all of General Motors midsize makes. Smith's program cost $7 billion ($23 billion in 2024 dollars) and did not end well. The press called it the "biggest catastrophe in American industrial history.[25]"

Andy Grove of Intel would comment, "The 360 casted a shadow that lasted decades.[26]" James Collins, in his book *Good to Great*, rated the top three business developments of all time[27]. They were the Model T, the Boeing 707, and the System/360. Of course, one must remember that the IBM gamble was audacious but not foolhardy. There were many compelling reasons to make the huge investment. Maintaining competitive systems across so many lines was killing IBM. Competitors were locking into specific segments, making it tougher to have a winning offering in every niche. This was especially true in the area of input/output devices, where the number of machines and attachments had mushroomed for IBM. The alternative to the NPL investment was the cost of maintaining all those incompatible systems and peripherals.

In addition, customers were increasingly savvy and put up with only so many conversions before rejecting the concept of incompatible upgrades and moving on to a different vendor. Finally, software was ascendant, and even IBM could not afford to build and maintain a huge portfolio of incompatible programs across multiple systems. Watson Jr. deservedly basked in the gamble's overwhelming success. He had initiated the plan and made the decisions to insure it got to the finish line. He would comment on the System/360 in his biography, "It was the biggest, riskiest decision I ever made, and I agonized about it for weeks, but deep down I believed there was nothing IBM couldn't do." At the same time, he would recognize the role that Learson had played: "Vin was the father of the new line of machines.[28]" Yet when asked if he would do it again, he said never. He had lived through

1965, that grim year where he nearly lost his company and his brother.

Though the press generally celebrated the success of System/360, they found one area to criticize – Solid Logic Technology. They derided SLT as sheer timidity on IBM's part. The company did not agree. Though the day of monolithic circuits would soon come, the performance of the SLT modules proved their worth and validated the decision. SLT modules would achieve 1,000 times the reliability of vacuum tubes. The average Model 30 experienced one failure every three years[29]. The cost of SLT was about half that of the new monolithics. The standardization with SLT across all System/360 models provided enormous economies of scale. Despite problems ramping up production of SLT amid the crushing wave of orders, the problems were resolved and implemented in the same low-cost automatic fashion so successful with core memory manufacturing. Tom Watson would later comment on the SLT decision as "the most fortunate decision we ever made."[30]

The software architecture has endured. It has defined the structure of large computer systems for decades. As groundbreaking as the design was, IBM was still able to make it work across the entire line. Most of the analysis of System/360 software centered on the astronomical cost overruns and delivery delays. Certainly, in hindsight, there would have been many changes in how the team approached Big Oz. That included spending so much time trying to cram OS/360 into 32K of memory and charging into coding before a comprehensive design was complete. Yet those issues fade into the background as one looks into the ongoing impact and dominance of the two resulting operating systems, OS/360 and DOS. If there is a miss in this area, it is that IBM would spend many years and many dollars supporting both operating systems, a violation of the original NPL promise.

**Figure 40**. Tom Watson, Jr. and Vin Learson at the console of the System/360. An iconic photograph of the two leaning on the result of their $5 billion investment. Reprint Courtesy of IBM Corporation ©.

# A MISS AT THE HIGH END

Despite the overall success, such a large, comprehensive program could not escape without some misses, and some of them turned out to be spectacular misses. The list includes the high end, the low end, timesharing, and the rise of plug compatibles.

The high-end miss was already apparent well before the System/360 announcement. Fred Brooks had finalized his hardware plan in late 1962. In August 1963, Control Data announced the CDC 6600. The Seymour Cray-designed system packed two and a half times the performance of Stretch. It ran at two MIPS (millions of instructions per processor second[1]), faster than IBM's fastest computer at the time, the 7094, which was itself nearly three times faster than the 7090.

Tom Watson was irate at the failure to match Control Data, much as he had been with the performance shortfall of Stretch. This led to his now-famous "janitor" memo commenting on Control Data's announcement of an even faster high-end model.

> "Last week Control Data had a press conference during which they officially announced their 6600 system. I understand that in the laboratory developing this system

here are only 34 people, including the janitor. Of these, fourteen are engineers and four are programmers, and only one has a PhD, a relatively minor programmer. Contrasting this modest effort with our own vast development activities, I fail to understand why we have lost our industry leadership position by letting someone else offer the world's most powerful computer.[2]"

Watson's ire centered on the prestige factor of having the world's fastest computers and to a lesser extent, the significant revenue exposure. There were other key reasons—the ability to attract talent to the company and "technological fallout[3]" that would come with the development of high-end architecture and its flow of innovations to other programs. This was the heritage of the Stretch and would be a hallmark of the program that the high-end miss inspired: the Model 90 series.

Seymour Cray led the small crew at Control Data that Watson described in his memo. Cray would be a force in supercomputers for years to come. He was born in Chippewa Falls, Wisconsin, and was a radioman during World War II. He graduated from the University of Minnesota with a BS in electrical engineering in 1949 and a master's in mathematics in 1951. Upon graduation, one of his instructors suggested that he simply walk down University Avenue to the Navy spinoff, ERA. They hired him and put him to work on the Atlas II computer for the Navy. Marketed by ERA and UNIVAC as the 1103, it used Williams tubes and magnetic drums and competed with the IBM 701. Cray went on to design the ERA 1103A, which replaced the Williams tubes with core memory and competed with the IBM 704.

Cray joined the move to Control Data in 1957 when Bill Norris and a cadre of engineers left ERA and formed their own company. Cray designed the CDC 1604, a 48-bit transistorized system that competed with the IBM 7090. Cray did not like working on commercial computers, finding them boring. Much like Gene Amdahl, he was happiest designing the newest, fastest machine. He lived by his performance guideline that doubling the price required delivering four times the performance.

Cray was an insular and, at times, odd man. He found the corporate environment at Control Data in St. Paul too distracting. He decided to build a complete lab on land next to his house in Chippewa Falls. Given his stature in the company, Norris and his management team were not about to object. He worked with a very small team, as anything larger would complicate the design process. His hobby was digging tunnels under his house. It was his way of clearing his mind when struggling with a difficult problem, much like John Atanasoff's famous drive. Cray would say that "elves" in the tunnels provided him with the solution. With the success of his high-speed creations, the press dubbed him the "Wizard of Chippewa Falls."

As much as Bill Norris hated the politics of dealing with corporate Remington Rand, Cray took it to a completely new level. When asked for a five-year plan, his response was, "Five year plan—build the biggest computer in the world. One year goal, one-fifth of the above.[4]" Fred Brooks would comment about Cray, "The genius of his design flowed from his total personal mastery over the whole design, ranging from architecture to circuits, packaging, and cooling, and his consequent freedom in making trades across all design domains. He took the time to do designs he could master, even though he used and supervised a small team.[5]" That small team included Jim Thornton, who was the unsung hero behind the company's emerging line of supercomputers.

Working in his new lab, Cray and his team designed the successor to the 1604, the CDC 3000 series, with the 3600 model as the top end. CDC sold the lower 24-bit models as commercial systems while the 48-bit upper models went to scientific environments. The CDC 3600 sold for over $1 million. The CDC 6600, the source of Tom Watson's ire, was the world's fastest computer from 1964 to 1969. Considered revolutionary, it borrowed many techniques pioneered by Stretch, including I/O channels, pipelining, and overlapped processing. With multiple processors, it could execute up to ten instructions in parallel. Cray kept the instruction set small, using RISC (Reduced Instruction Set Computing) techniques well before John Cocke of IBM developed the concept. Cray switched from discrete transistors to etched silicon transistors designed by Fairchild Semiconductor, resulting in faster machine cycle times.

Always looking for every edge in performance, he focused on the path that his electrons would have to travel as instructions were processed. Grace Hopper had paved the way. She carried a length of wire to her presentations that was exactly 11.8 inches long. It represented the distance that light traveled in one nanosecond[6] (one billionth of a second), demonstrating that light was very fast but not infinite. With this in mind, Cray reduced the length of his electrical paths wherever possible. This resulted in a very compact, almost circular design to his systems. With Cray, there was less focus on achieving processing speed through mathematical algorithms. Multiplications required at least ten machine cycles, though they were very fast cycles.

Cray's software was a marked contrast to his hardware innovations. The Chippewa Operating System (COS) was unremarkable and, in many situations, unreliable. This was an impediment in commercial installations, but, of course, Cray did not care much for such environments. His bottom line was speed, and he always overachieved. Control Data would sell over one hundred of the CDC 6600 machines at an average price of $2.4 million, many coming at the expense of IBM[7].

As IBM struggled to match the CDC 3600 and 6600 systems, Cray moved on. He lost interest in the 6600 even before it went into production. He was already working on the 7600. Targeted at ten times the performance of the 6600, it would be on a par with the much later high-end System/360 Model 195. For this next turn of the performance crank, the increased density of the electronics generated escalating heat. Cray hired a refrigeration expert from appliance maker Amana who designed a Freon-based cooling system to keep temperatures in check. Fred Brooks would note that both Cray and Gene Amdahl recognized refrigeration as the emerging center stage in supercomputer design[8]. The 7600 would be the fastest computer in the world from 1969 to 1975.

IBM lost the Los Alamos lab in 1965 when it installed a CDC 6600 running at four to six times the performance of its Stretch machine. The lab installed four CDC 7600 systems. Cray's next project, the CDC 8600, was his first major failure. The rising cost of funding Cray's development work led Bill Norris to hedge his bets.

He asked Jim Thornton, Cray's design partner, to start a second super-computer project in parallel. Two major projects were not sustainable. This, combined with greater intrusion by corporate into Cray's work, led to a parting of the ways in 1976. Cray's CDC8600 never saw the light of day, and Thornton's machine, the STAR-100, resulted in only five installations[9].

Staying at his home, Cray had no trouble getting funding for his new company, Cray Research, given his outsized reputation. His first machine, the Cray-1, arrived in 1976 and was a great success. Selling for upwards of $8 million, Cray sold 80 systems, an astounding number for such a large system. It was ten times the performance of any existing supercomputer and was the first Cray-designed system to use integrated circuits. Continuing Cray's focus on short electrical pathways, the Cray-1 was again circular. His supercomputer looked more like seating in a cocktail lounge.

Cray went on to design the Cray-2, based in part based on his failed CDC 8600 design. The Cray-2 was roughly 12 times the speed of the Cray-1 and could process a billion instructions per second. The longest wired connection in the Cray-1 was four feet. For the Cray-2, he had it down to 16 inches[10]. With the increased compactness, even more refrigeration was required. This resulted in a closed-loop system with a liquid called Fluorinert developed by 3M. Pumped in under pressure, it provided the needed heat dissipation but was highly toxic.

By this time, Cray's company had grown to over 1,000 employees. Seymour Cray's hermitic approach to design had not changed and threatened the continuing success of the company. As Bill Norris at Control Data had done, the company set up a separate design team under Steve Chen, resulting in the Cray-XP computer, which outsold the Cray-2. The follow-on Cray-3 was a failure, and Cray Research went bankrupt in 1995. Cray would die at age 71 in a car accident the following year, the result of a reckless driver on Interstate 25[11].

## PROJECT X

The loss of high-end leadership to CDC was galling to Watson, but he had no one to blame but himself. He had contributed to the utter debacle by his rage at the Stretch failure, the subsequent investigation, and most damaging, the banishment of Steve Dunwell, a well-respected and accomplished engineer and leader. It sent a clear message that anyone that worked on high-end computers was risking their career.

With the System/360, the planned high-end machine, the Model 70, paled in performance next to the CDC 6600. The CDC super-computer ran two to three times faster than the IBM system. The high-end game plan had to change to salvage any chance of success. Though Tom Watson had chilled high-end work in the wake of Stretch, IBM Research had pressed on with two follow-on projects. They now had a renewed sense of urgency. Project X was near-term and targeted to be ten to 30 times the Stretch performance. Longer term, Project Y was planned to be 100 times Stretch's performance. Until Watson's "janitor memo," these two programs had little resources compared to the vast spending on the NPL. Project X would be the immediate focus.

Emanuel Piore had a "Super Stretch" in mind when he convinced Gene Amdahl to return to IBM in 1960 and work on Project X. The Livermore Lab received the "Watson discount" on its Stretch machine. Tom was aware that the lab used that money to purchase their Control Data machine. The same month of his janitor memo, IBM marketing told Learson that a high-end machine with ten times 7090 performance had a market of 53 systems[12]. Harwood Kolsky wrote Tom about the CDC 6600 and 7600 supercomputers, stating the obvious, that IBM was not even remotely competitive. Tom asked Piore to develop a supercomputer plan, one that would maintain performance leadership for three to four years after announcement.

Gene Amdahl formed a team with James Pomerene to tackle the challenge but Bob Evans tapped him to work with Fred Brooks on NPL architecture. This left Project X shorthanded, a situation less than ideal when one was competing with Seymour Cray. Piore trans-ferred Project X to Poughkeepsie and it became the Model 90 series program. Bob Evans took the heat for the failure at the high end. Charlie DeCarlo, who Learson had banished during Evans' 8000 Series evaluation, was given the job of fixing the high end, a move that Evans described as a "slap in the face[13]."

Faster logic technology was central to the high-end program. Though SLT had been the right decision for most of the System/360 project line, it would not be fast enough for the Model 90 models. Tom Watson wanted new technologies evaluated and laid this challenge on his brother Dick and on John Haanstra in his rehabilitated role as SDD president. Joseph Logue had already defined the progression from SLT, with Monolithic Systems Technology (MST) and New Generation Technology (NGT) the two follow-on phases. Jerrier Haddad in commissioned a study in late 1965 that NGT would not be viable until 1969 at the earliest. The crisis at the high end could not wait, and the Component division deployed several interim improvements to SLT. They would release SLD (Solid Logic Dense), Unit Logic Device (ULD), and Advanced Solid Logic Technology (ASLT) before finally arriving at Logue's MST with the System/360 Model 85 in 1968[14]

IBM announced the first salvo of the Model 90 series in August 1964 as the Model 92. Designed to compete against the upcoming CDC 7600, Gene Amdahl previewed the system at the Fall Computer Conference that year, describing a machine with a cycle time of 75 nanoseconds using the new ASLT circuit boards. The machine implemented multiplication in esteemed tradition of James Bryce. It produced multiplication in just three instruction cycles, using six accumulators to store the partial results before the final summing[15]. This turned out to be a hollow victory, as the CDC 6800 debuted at four times CDC 6600 performance. Though the field took custom orders for the Model 92, Poughkeepsie never shipped the system and replaced it with the Model 91. CDC would sue IBM in 1968, contending the Model 92 was a "paper machine" that simply sought to stall decisions on CDC computers.

Mike Flynn led the Model 91 program. He joined IBM in 1955 and spent nine months in Endicott training, learning logic and circuit design. He learned programming on the 650. Watson Sr. was still around then and visited Flynn's training class. They all stood and sang one of the songs of praise to Watson's leadership. All except Flynn, who said he would only stand for God and Country. IBM transferred him to Poughkeepsie, which turned out to be fortuitous. He worked for Fred Brooks, with Evans as his second line. He would call Evans the "heart and soul" of the System/360[16].

Announced in 1966 and first shipped in 1967, the Model 91 featured many of the Stretch innovations, including pipelining and out-of-order instruction execution. It also used the advanced floating-point multiplication process of the Model 92. With 60 nanosecond cycle time and an overall speed of 16.6 million instructions per second, the system produced 5.5 million floating-point multiplications per second. This represented roughly fifty times the performance of the IBM 7090 mainframe. It used ASLT logic modules, though manufacturing problems at Texas Instruments resulted in a nine-month delay to October 1967. This was two years after the CDC 6600 hit the market.

The Model 91 did not fare well, even with discounts on its $5 million price tag. Fifteen systems were built with four remaining internal to IBM[17]. IBM reportedly lost money on each machine. In Manhattan, two Model 91 systems were installed within a couple of blocks on one another[18]. NASA's Goddard Institute for Space Studies' machine went into Columbia University's Armstrong Hall at corner of Broadway and West 112th Street. Tom's Restaurant occupied the ground floor below the Institute's computer room. The restaurant, or rather the outside façade of the restaurant, later became very famous as the diner used on the television series Seinfeld that ran from 1989 to 1998. A few blocks away, Columbia University installed its Model 91 in a new underground computing center. UCLA used its Model 91 to support the fledging ARPANET, the precursor of the Internet. Princeton implemented a Model 91 for astrophysics research. And, Stanford installed a Model 91 at SLAC (Stanford Linear Accelerator Center) for atomic research[19]. It is interesting to note that the SLAC team did an extensive comparison of the IBM Model 91 with the CDC 6600. It stressed computational speed, ease of programming, and total cost. The University opted for the Model 91.

In the same period, Control Data produced 94 CDC 6600 and CDC 6800 systems, and 121 shipments of other models in the 6000 series[20]. The Livermore Lab would be completely lost to Seymour Cray, as it installed the CDC 3600, CDC 6600, CDC 7600, and the Cray-1. Despite all the ambitious plans and programs, Charlie DeCarlo's high end strategies failed to put a dent in Control Data's super-computer dominance. In addition, his phantom Model 92 resulted in a very expensive lawsuit for the company.

IBM was more successful with the machine just below its high-end series, the Model 85. It would become the fastest System/360 put in general production and would rack up a number of firsts. It was the first System/360 to feature Monolithic System Technology. It would also feature the first appearance of cache memory. Carl Conti and Don Gibson created the concept as a memory buffer, to hold instructions in process. They first called it the "muffer," but thankfully changed to "cache" before it got out of the lab[21]. The Model 85 was also the first IBM system with a CRT display as the main console, replacing the IBM Selectric typewriter. The Model 85 was modestly successful, with a production of 30 systems. However, its design paved the way for the System/370 machines that followed.

While Project X moved forward with its Model 90 series machines—some real, some just imagined— IBM Research continued work on the longer-term supercomputer program, Project Y. The goal was a system with 100 times Stretch performance. Progress was slow until the announcement of the CDC 6600 and Watson's "go for broke" memo.[22] Chuck Branscomb clarified the mission as, "We don't care about compatibility and all that, build the fastest machine.[23]"

Branscomb wanted an aggressive person to lead the charge and tapped Max Paley. Paley was the right choice. He had worked on the 604 Multiplier, the Wooden Wheel, the 608 Multiplier and the 628 WWAM program in World Trade before managing the manufacturing of the ill-fated 7070 in Endicott and the Stretch machines in Kingston. He ended up working for Fred Brooks on System/360. Paley had Jack Bertram, James Pomerene, Harwood Kolsky, and many on his Project Y team but no janitor. He would manage the program out of a new Research center near San Jose.

His team would use the Stretch superscalar approach combined with a ten-nanosecond cycle speed to meet the stated performance goal. Setting the 7090 performance was at one, the CDC 6600 rated at 50, the paper Model 92 at 110, and the planned Y machine at 2,500. For this project, "superscalar" meant that up to 50 instructions would be in some stage of execution at the same time. The design included expanded use of out-of-order instruction execution, multi-threading, instruction prediction, and elements of what would later become RISC technology. The design would also use the "muffer" (excuse me, cache memory).

There would be a major complication in the game plan. IBM named Gene Amdahl an IBM Fellow in 1965. He elected to move back to California and work on Project Y, now rechristened the ACS-1 ("Advanced Computer System"). However, Amdahl quickly conflicted with Paley's team and they excluded him from discussions. He worked on his own design. In early 1967, Amdahl proposed that ACS-1 be System/360 compatible despite Branscomb's admonition on compatibility. His ACS-360 design improved the business case but severely compromised performance. Amdahl relied on help in Pough-keepsie to blunt the performance issue. Each time the ACS-1 team achieved a new performance breakthrough, he would ask Pough-keepsie to tweak a System/360 machine to have ACS-360 match it.

In May 1968, Bob Evans proposed a competition between the two designs. Amdahl's ACS-360 somehow emerged as the winner. There was a complete mutiny. Half of the design team left. Paley resigned as lab director and left IBM to work as an independent consultant. John Cocke moved back to Research at Yorktown Heights[24]. This left Amdahl as lab director. Branscomb had pointedly selected Paley to run the program as he deemed Amdahl to have poor managerial skills, a characterization that Amdahl resented but one that was clearly on the mark.

Now in control of the program, Amdahl lobbied for a complete line of ACS-360 machines, with three separate high-end models. IBM corporate was only willing to fund one. Data Processing Division President Frank Cary wanted to build a few machines simply to get supercomputer bragging rights. He would not put ACS-360 into general production. In September 1970, Amdahl resigned. The ever-petulant Amdahl left IBM for the second time, with ideas to create an IBM mainframe clone. Bob Evans would comment that Amdahl was upset about many things at the time and "would have quit the next day had the hamburgers in the cafeteria been cold.[25]" Chuck Branscomb, as SDD president, had the chore to present the cancellation recommendation to Watson. Tom was initially angry but soon calmed down and accepted it.

# A MISS WITH PLUG COMPATIBLES

Amdahl's departure to create his own System/360-compatible mainframe was concerning, but it would not be the only plug-compatible challenge IBM faced. The decision to design common interfaces for all input/output devices created an even greater opportunity for competitors. Before System/360, the market for IBM peripherals, which varied by system, was not big enough for competitors to target. The emergence of plug-compatible competitors was not necessarily a miss, but a direct result of the basic System/360 concept.

The success of the new line dramatically increased IBM volumes across the product line. It also meant that IBM, at least for a time, could command higher prices. This in turn provided more even pricing room for plug-compatible manufacturers. Their pathway was all the easier as IBM had already done the invention, development, testing, and marketing, and would still provide ongoing support. They simply needed to copy an IBM device and slot it at a 15 to 20 percent discount to IBM prices. The emergence of leasing companies as a result of Tom's 1956 consent decree substantially improved the plug compatible's edge. They could bundle both the IBM and competitive machines into a multi-year lease. Finally and most troubling, IBM made the job even easier. IBM engineers felt compelled to publish technical details on their machines or attend industry conferences and give presentations on their invention, giving away the company's valu-able trade secrets.

Even during NPL development, Fred Brooks worried about the common I/O interfaces and the threat of standalone competitors[1]. As early as 1961, Bob Evans assembled a task force to study the threat. The computer market already supported a number of companies providing peripherals for computers. In the 1950s and 1960s, IBM competitors like the "BUNCH" routinely used these OEM (Original Equipment Manufacturers) to design and produce I/O devices for their systems in order to avoid developing their own. Control Data, for one, moved aggressively into this area, starting with magnetic drums and moving on to disks, tapes, printers, terminals, and card equipment. Telex was first out the chute with a plug-compatible machine, a tape drive.

San Jose's new series of disk drives provided the most inviting target. The 1301 Advanced Disk Facility (ADF) debuted in 1961. The 1311 disk, which introduced the first removable disk packs, followed a year later. Led by engineering manager Bill Carlson, the 1311 drive reduced the disk platter diameter from 24 inches to 14 inches and combined six disks with ten recording surfaces into the disk pack. The System/360 announcement added the 2311, which increased storage capacity from two million to 7.25 million bytes. The year 1965 saw the release of the 2314, which had four times the storage of the 2311. Customers labeled it the "pizza oven" due to its boxy drawer design. Those last two disk drives, the 2311 and 2314, featured the new I/O interface, enabling attachment to any System/360 model. They presented the most enticing target for competitors.

In 1952, RAMAC engineers Noyes and Dickinson published *Engineering Design of a Magnetic-Disk Random Access Memory* describing IBM's disk technology in detail. The paper was widely read[2]. In 1956, as RAMAC was starting to ship, San Jose engineers published additional technical articles in the *IBM Journal of Research and Development*. Al Hoagland documented the head design while Noyes and Dickinson provided details on platter construction and drive mechanicals. Ken Haughton, Russ Bruner, and Al Hoagland produced a technical paper on the Advanced Disk Facility (ADF) drive. The disk engineers also made presentations at the Western Computer Conference that year[3].

By this time, the key players in San Jose had changed. Rey Johnson left the lab in 1956, and Lou Stevens had replaced him as director. Several years later, Vic Witt replaced Stevens. Witt earned his reputation leading development of the tape dive for the IBM 701, but he was completely new to the disk area. Gerry Harries headed the 2311 and 2314 disk programs, reporting to Witt. Like Witt, he was not a disk person. However, his management style was even more concerning. With pressures to meet design and production targets, he drove his engineers hard. The harsh work environment under Witt and Harries was unrelenting. There were constant 48-hour work weeks alternating between high pressure and spates of monotony, coupled with a lack of technician support for the engineers[4].

In early 1968, twelve engineers[5] in the lab quit to form a new plug-compatible disk drive company. The group was labeled the "Dirty Dozen," reflecting a sense that they had betrayed IBM. The mass departure seemed well thought out. The new company, Information Storage Systems, or ISS, would be developing plug-compatible 2311 and 2314 drives. The group that left included just the right mix of engineers to make that happen. Despite this, the new company continued to raid the San Jose lab for additional people. ISS also arranged for Telex to be its exclusive sales outlet, letting the new company focus on developing and producing the drives. All of the advantages that IBM had created for plug-compatible manufacturers to prosper were apparently not enough.

The resignation of so many engineers at once brought a swift reaction from IBM. Chuck Branscomb came out to assess how this could have happened. He found that the seeds of discord had been present for quite a while but had gone unaddressed. This was a period when the whole company was under the stress of the System/360 program. A key difference was the location. Southern California had become a hotbed of technology companies. For engineers, there was little risk in leaving for a startup company, as the demand for engineers was far greater than the supply. This was not the case with IBM labs like Endicott, Poughkeepsie, and Rochester, where leaving IBM would be far riskier. At the same time, the potential rewards in higher pay and stock options were incredibly attractive.

The lab, and in particular the new Merlin program (the future 3330 disk drive), was put in disarray. One member of the Dirty Dozen was actually the product manager for the Merlin program. He left with the comment, "you're never going to make it work.[6]" The sense that this drive would take a magician to build resulted in the "Merlin" code name. In January 1968, a month after the Dirty Dozen departed, Alan Shugart returned as head of disk drive operations. He had previously led the RAMAC and ADF programs before leaving for several years to work on a high-capacity storage device for the NSA. His new mission was to restore order. Shugart was a well-respected and popular leader[7], and he was able to get the lab back on track quickly. Jack Harker took over the Merlin program. His team solved the central issue of the drive's design, creating a new read/write actuator that could handle the tighter spacing between recording tracks.

Alan Shugart restored some degree of normalcy to the disk drive organization, but his own discontent was growing. Despite his central role in the Dirty Dozen defection, Vic Witt was neither reprimanded nor demoted. Instead, IBM promoted him to run both the San Jose and Boulder labs. Shugart and Witt were polar opposites. Witt was buttoned-down and all business whereas Shugart was personable and fun loving. Ken Haughton noted that Shugart was an outstanding manager and attracted people wanting to work for him[8]. Now, Shugart was stuck working for Witt. He also sensed that he was no longer a rising star in IBM like John Haanstra. It is clear that IBM should have recognized him as one.

In 1969, Shugart received a promotion to corporate staff in New York. He quickly discovered it was not an upward career move and within two weeks, he resigned from IBM and returned to California. Memorex, the peripheral manufacturer, recognized his value and immediately hired him as a vice president, with the mission to develop a clone of the floppy disk drive his team had developed in the San Jose lab. The floppy drive was covered with patents, but Shugart was intimate with the particulars and knew his way around them. However, he could not develop the Memorex clone by himself.

Shugart parked himself at the Paddock Bar across the street from the San Jose lab with stacks of employment contracts and stock options. Somewhere between 50 and 200 IBM engineers, technicians, and planners accepted his offer to join him at Memorex[9]. The sheer scope of his raid on IBM indicates the level of his anger towards the company. He apparently took glee at taking IBM's best disk people. IBM lost in at least three ways with Shugart's departure. First, the company lost a talented engineer and charismatic manager. Second, he took a multitude of IBM engineers with him. Third, he normalized the practice of competitors using any means, including buying IBM engineers, to build clone machines.

Memorex adopted raiding IBM as a core business strategy. Between 1968 and 1971, the company hired about 600 IBMers. Jack Harker, who would become the lab director in 1972, estimated the initial Shugart "take" at about 50 engineers and felt the impact was less than one would expect[10]. He had a steady influx of new talent coming into the lab as well as engineers who moved from the recently shuttered Menlo Park lab.

Later that year, Bob Evans replaced Branscomb as SDD president, and subsequently he replaced Witt with Jack Kuehler as head of both disk and tape operations[11]. Kuehler had joined the San Jose lab in 1958 after graduating from Santa Clara University with degrees in both mechanical and electrical engineering. He had risen to lab director at IBM Raleigh when Evans tapped him to restore order in San Jose. Kuehler would have substantial "headroom" within IBM, even considered at one point for the CEO position.

IBM named Vic Witt an IBM Fellow in 1970, shortly after the Merlin drive announcement. He had run the San Jose lab and its disk programs for a decade, a period that saw leaps in disk technology and the new floppy disk drive. His time at the helm, though, came at great cost. Both the Dirty Dozen and the Shugart exodus occurred during his watch and reflected his adversarial management style. Alan Shugart led Memorex's diskette drive program, though in the smaller 5.25-inch format, not the eight-inch format invented in San Jose. He did not last long at Memorex, leaving in 1973 to started Shugart Associates. He had no problem obtaining funding for the new venture. His goal was to build a complete ecosystem around the diskette drive to replace punched-card equipment. Xerox acquired his company. In

1974, after two years with limited progress and no product, the new board fired him.

Shugart and his business partner Finis Conner saw the growing demand for personal computer hard disk drives. They formed Shugart Technology to develop a hard drive based on the same form factor as the diskette drive. Xerox threatened a lawsuit, as it owned the name Shugart Associates. Sporting a new name, Shugart released the new drive as the Seagate Technology ST-506, arriving in 1979. It offered 15 times the capacity of the diskette drive with lightning speed at just three times the price. It caught the personal computer wave, and the company prospered. When Shugart resigned in 1998, Seagate Technology was the largest disk drive company in the world.

Back at the San Jose lab, work continued on Merlin. Companies like Memorex, Telex, and the Dirty Dozen ISS were feasting on IBM's 2311 and 2314 drives but aware that the party would not last. They looked for ways to clone Merlin. Jack Clemens led Merlin after Shugart left. Clemens himself would leave to join to Telex. He had been disappointed when not asked to leave with the Dirty Dozen[12]. He was not sure of his reaction, but the lure of big salaries and stock options was attractive. He left in 1970 just before Merlin shipped. At Telex, he helped develop its 3330 clone. He recruited more engineers from IBM. He moved to Memorex in 1973 and led that company's clone 3330 project. He left Memorex and consulted with Control Data on its 3330. It is hard to imagine one individual, aside from Gene Amdahl, who damaged IBM in such a grandiose fashion.

Meanwhile at ISS, the "Dirty Dozen" start-up, work continued on IBM cloning. Dal Allan was a software developer who worked on System/360 DOS at Endicott and moved to the San Jose lab in 1969. He left in 1970 to join ISS. He described the ongoing program at ISS of hiring more engineers from the San Jose lab, targeting those who had the specific detailed inside information that ISS needed. He was unabashed in describing his own efforts to pump IBM engineers at GUIDE and SHARE meetings for their highly classified information[13]. He would also network with engineers at other Plug-Compatible Manufacturers (PCMs), as they were all in the game of feeding off IBM. None of this seemed to trigger any regrets.

ISS would succeed in hurting IBM and the San Jose lab in another way. In 1973, the UNIVAC division of Sperry Rand purchased ISS. The ISS team developed a 3330 clone for the UNIVAC 90 Series. Thus, former engineers from IBM San Jose, now at ISS, corrected UNIVAC's earlier mistake of investing in drum storage rather than disk. This unraveling of the San Jose lab in the service of competitors was certainly something that Rey Johnson would not understand. The lure of big money and the easy ability to secure it was apparently too much to resist for many.

Not everyone left. Jack Harker stayed, and as lab director, he focused his team on reasserting technological leadership. Despite the treachery and subsequent emergence of PCM competition, IBM was still doing very well with disk drives. With Merlin (IBM 3330), Harker was running the most profitable program in IBM. However, there was no resting on laurels. His team produced a double-density version of the 3330. The total number of 3330 drives sold by IBM exceeded 100,000.[14].

The 3330 was a mainframe drive. IBM's midrange systems needed a smaller, more affordable disk. Harker turned to Ken Haughton to lead the effort. Haughton had obtained a mechanical engineering degree from Berkeley and a master's from Iowa State before joining the San Jose lab in 1957, just when RAMAC was shipping[15]. He helped design the flying heads on the Advanced Disk Facility before IBM afforded him the opportunity to get his doctorate at Berkeley. Returning to the lab, he started on the midrange program, geared for the entry System/370 Model 115 and IBM Rochester's System/3.

Haughton first wanted the answer to a basic question: Should the new drive be fixed, or removable? He visited a number of customers and IBM branches, and the consensus was that the drive should be removable. This was at odds with actual customer use. Most customers never changed their removable disk packs. Nevertheless, he went along with the consensus. However, he decided to adopt the key advantage of fixed disk drives—a sealed environment. The read/write heads and assembly would be put into the removable disk pack rather than be part of the drive. This added more expense and led to research on lower-cost heads. The plan called for a drive of 30 megabytes across two disk platters. His engineers dubbed the arrangement the

"30-30" configuration, which led Haughton to name the project Winchester after the Winchester 30/30 rifle.

The read/write heads within the Winchester disk pack would make a substantial mark in disk history. The new read/write head was a smaller, low-mass head, reducing the existing head design from 300 grams down to only 18 grams. The new head would fly just 18 micro-inches above the disk surface to allow for far greater track density. The RAMAC heads had flown at 800 micro-inches and the Merlin heads came down to 50 micro-inches. Winchester's other key innovation was in head take-offs and landings. Engineering reserved an area of the disk for these functions and coated the surface with a secret patented lubricant. The heads always took off and landed in that area[16]. On top of the technical changes, the new head design came in at a quarter of the cost of the previous head designs. Winchester heads became the industry standard, and nearly all subsequent disk drives used them. The resulting midrange 3340 drive arrived in 1973 and represented a striking upgrade when compared to the original RAMAC drive. The bit density increased from 100 per inch to over 5,000 per inch, resulting in a recording density increase from 2,000 bits per square inch to 7.8 million bits per square inch. Capacity increased from five megabytes to seventy megabytes.

Haughton could not ignore the mainframe market. His Madrid program used Winchester technology to produce the IBM 3350. Released in 1975 for System/370 systems, it featured sealed disk packs like the 3340 and cranked the storage per pack up to 317 megabytes. For a time, it enabled IBM to stay a step ahead in disk technology. However, both Memorex and StorageTek soon had 3350 clones on the market. The obvious next move would have been a double-density 3350, but Harker and his team had other plans. They were already hard at work on the 3380. To no one's surprise, PCM competitors soon had double-density 3350 drives on the market, but not without problems. There was a period of one to two years where they did well in the absence of IBM competition. Ken Haughton was never a candidate for the Dirty Dozen or the Shugart exodus. He would move up to lab director in 1980 before retiring from IBM to become dean of Santa Clara University.

Denis Mee was part of Jack Harker's team in San Jose. Educated in England with an advanced degree in magnetic materials, he joined CBS Labs in 1957 and worked on stereo tape recording, which would lead to the consumer audiotape cassette. He joined IBM Research in 1962 with a focus on advanced magnetic materials. At Yorktown Heights, he marveled at the wonderful material research capabilities there. He was like a kid in a candy store[17]. Mee worked on thin film research, a technology that borrowed from the semiconductor photolithographic process to create far more compact magnetic devices. He moved to San Jose to work on the application of thin film to disk drive read/write heads.

Mee's first project was the IBM 3370, the first thin film disk drive. It amply demonstrated the transformative value of the technology. The thin film head was substantially smaller, lighter, lower in cost, and far greater in bit density. It also reduced the flying height above the disk surface by 30 percent. The experience with the 3370 program led to a quantum leap for large disks with the 3380 drive in 1980. This mainframe drive would stay in production for the next 15 years, an unheard-of time span in the disk industry. The 3380 with thin film heads reduced the individual bit spacing on the tracks to 305 nanometers. The bit spacing on the original RAMAC disk drive was 1,270,000 nanometers while the bit spacing on the 3340 Winchester drive was just over 5,000 nanometers. This decreased spacing increased density to 12 million bits per square inch, compared to the RAMAC density of 2,000 bits per square inch, a 6,000-fold jump. The 3380 matched the 3330 in success, selling well over 100,000 units. Mee would become an IBM Fellow in 1983.

The San Jose lab was not the only IBM lab to struggle with System/360 PCM competition. A similar story unfolded in Boulder about the time Shugart was writing contracts at the Paddock Bar. Boulder assumed the tape drive mission in 1965. At the time, it was tough to convince engineers to leave Poughkeepsie for Boulder. One engineer who did make the move was Jesse Aweida[18]. He had been born in Palestine and moved with his family to the United States. IBM hired him in 1958 and he worked on the transistorized 608 Multiplier before moving to tape. Once in Boulder, Aweida's mandate was the development of an industry-leading tape drive. His second line manager was none other than Jim Weidenhammer, the engineer famous for developing IBM's first tape drive with its vacuum-based buffering mechanism.

Aweida's new tape drive was the IBM 2420, announced in 1968. Aweida assisted on the business case, which projected selling perhaps 5,000 tape units at $82,000 each. IBM actually sold 45,000[19]. Aweida did a back-of-the-envelope estimate of what revenue he could expect if he created a compatible tape drive at a lower price point that would skim off perhaps 10 percent of those volumes. Startled at the result, he put together a business plan and enlisted Juan Rodriguez and two other engineers to leave IBM and form a new company. He found a financier who knew of the Dirty Dozen affair in San Jose and offered $250,000 in start-up funds in exchange for a 40 percent stake. The new company, Storage Technology Corporation (StorageTek) began operations in 1969. Aweida signed up many engineers from the Boulder lab. He had a 2420 clone drive on the market the following year.

The Boulder lab emulated the San Jose response to tape cloning. It decided to charge ahead in technology. Boulder engineers would develop the 3420 reel-to-reel tape drive, followed by 3480 cartridge tape drive that owed some of its design elements to the NSA Tractor tape design. IBM moved its tape mission to a new plant in Tucson in 1980. Gerry Harries, the hard-driving manager who instigated the Dirty Dozen exodus, also ended up there. IBM Boulder became the center of printer development.

From that first task force on plug-compatible competition organized by Fred Brooks, IBM struggled with the best approach to compete. The most successful response was to move to the next plateau of technology, challenging its competitors to match the new machines. With disk drives, IBM also reduced their lease rates and contract durations. The most successful approach reduced the product cycle. IBM's traditional product cycle had been five to six years. This fit perfectly into leasing company standard lease terms. More frequent product introductions pressured the competition in two ways, reducing the time to make money on the old product and increasing the resources needed to develop the new product.

The obvious response to engineers leaving IBM to build an identical product for a competitor was the noncompete agreement. This would typically prevent an employee with confidential knowledge from using it at a competitor for a set period, usually two to three years. This introduced a sufficient delay in getting a competitive product on the market or forced the competitor to design and develop the product on their own, without IBM help.

IBM did get some measure of payback with plug-compatibles, a result of dueling lawsuits between the company and Telex. Telex was the direct beneficiary of the San Jose "Dirty Dozen" disk drive engineers, initially through its relationship with ISS and then later in starting its own PCM division. IBM made changes in its disk drive prices in part to address competition from Telex. Telex sued, accusing IBM of antitrust behavior, and was awarded $352 million in damages. IBM countersued. The resulting decision found Telex guilty of stealing IBM trade secrets and hiring IBM employees specifically for that confidential information. The original award was nullified, and the case was eventually settled out of court for nominal amounts. However, the decision shined a light on the business practices of plug-compatible companies.

# A MISS WITH TIMESHARING

Another major miss for System/360 would have dire consequences for Bob Evans. This was the failure to provide timesharing support. MIT had been a pioneer in timesharing, setting up facilities for multi-user access and multiprogramming on its IBM 704 in 1956, and subsequently moving operations to the 7090. In 1963, MIT had spearheaded a project called MAC (Multiple Access Computer) along with Bell Labs to explore enhanced use of timesharing. IBM had in fact set up a lab next to its existing MIT lab (the IBM Cambridge Scientific Center) in the presumption that it would win the MAC bid. In addition, IBM architects working on the NPL program had met with MIT prior to the System/360 announcement and discussed Dynamic Address Translation (DAT), a key requirement for virtual memory paging needed in a timesharing environment. The limitation of real memory necessitated this function. Back in Poughkeepsie, Gene Amdahl argued that DAT hardware and virtual memory were not needed as real memory had expanded to such a great extent. Fred Brooks disagreed and argued for the retention of virtual memory capability. The issue was hotly debated, but in the end, virtual memory was dropped due both to performance concerns and the impact on the announcement timeline.

Tom Watson was aware of MIT's work in this area as early as

1961 and told Bob Evans to "keep an eye on it.[1]" Evans delegated review of the NPL plans with MIT to his staff, principally to Brooks and Amdahl. Upon learning that DAT hardware and virtual memory management were not included in the new systems, both MIT and Bell decided to order GE 600 series machines. Watson was not happy. He was already disappointed with Evans regarding the general competitiveness of IBM's existing products as the NPL program developed. He had Charles Bashe do an analysis in late 1964 that concluded that IBM's existing product lines, particularly its peripheral devices, had fallen behind the competition. Evans would contend that he could not both launch NPL and enhance all of the current product lines, including input-output devices. Tom did not agree and shunted Evans aside to the Federal Systems Division, a role outside of the IBM mainstream. As a consolation, the Evans team at FSD would win the programming contract for the FAA air traffic control system secured by Learson and Watts Humphrey.

Learson turned to Watts Humphrey again. He asked him to put together a team to address the timesharing miss. Humphrey worked with Gerry Blaauw to design a hardware modification to the Model 65 that would support dynamic address translation. Called the "Blaauw box", it implemented virtual memory in 4K pages[2]. Humphrey's team created a prototype machine, the Model 65M, and installed at the University of Michigan. It was the first multiprocessor of the System/360 line. The university would create the Michigan Timesharing System (MTS) to run on the machine. Other universities and organizations heard about the machine and soon convinced IBM there was a market for such a system. An updated version, the Model 67, would see 18 installations.

For Watts Humphrey, his work on timesharing was yet another success. His stock had risen to new heights. Bob Evans counseled him that he was one of two technical executives that might be considered for IBM president or even CEO. The other executive was Jack Kuehler, who was busy burnishing his reputation in the San Jose lab. Evans suggested that Watts needed to run a lab to bolster his own resume. Watts took charge of the Endicott lab during the time when Vin Learson was the CEO. However, Evans' vision for him did not pan out. He left for Carnegie Mellon University to continue his work on the programming process. He would become known as the "father of software quality."

---

## BASIC

Timesharing and the coming wave of personal computers would need a suitable programming language. Machine code would certainly not do, nor would COBOL or FORTRAN. The task required a simple, accessible language. Two professors at Dartmouth College, John Kemeny and Thomas Kurtz, arrived at this conclusion in the mid-fifties. Surveying the Dartmouth student body, they saw only those in the mathematics and science disciplines who were experts or even conversant in computers. They believed that every student should be able to write a program and use a computer. They set about making that happen[3].

Kemeny emigrated with his family from Budapest to the United States in 1940. A number of his relatives remained and perished in the Holocaust. He started his studies at Princeton before taking a year off to work under Richard Feynman on the Manhattan Project[4]. He was one of Feynman's "boys" that developed applications on the IBM punched card machines[5].Returning to Princeton, he completed his degree in mathematics in 1946 and remained there for graduate studies, working under the supervision of the esteemed professor Alonzo Church. He was also a research assistant to Albert Einstein during this time. He attended a lecture on computers by John von Neumann, and came away a complete convert[6]. Von Neumann finally had his IAS computer up and running in 1952. The two would go for walks, presumably conversing in Hungarian. Von Neumann was a fellow "Martian." Kemeny moved to Dartmouth as an assistant professor in 1953 and was elevated to chairman of the department just two years later, at age 29.

Kurtz graduated from Knox College in 1950 and worked on Harry Huskey's SWAC (Standards Western Automatic Computer) at UCLA. Kurtz then left for Princeton, receiving his doctorate in statistics in 1956. He joined Kemeny's mathematics department. At the time, Dartmouth did not have a single computer. John McCarthy, a fellow professor at Dartmouth, arranged for Kurtz to use the IBM 704 mainframe at MIT. Kurtz took the 6:20 train out of White River Junction to Boston every two weeks to enter, debug, and run his statistical analysis programs[7]. He developed the code using the SHARE Assembly Program. He heard the buzz about IBM's new FORTRAN programming language, along with the rumor that it had to be much slower than machine code. This was before he toiled for several months on a complex statistical problem in assembler, getting progressively more frustrated. He decided to give FORTRAN a try. He completed the problem in a fraction of the time and ran it on the 704 without substantial debugging[8]. He learned the key lesson that it was now all about programmer productivity.

Dartmouth, for its part, aspired to catch up with the other Northeast colleges (Harvard, Princeton, Penn, and MIT) that had already made strides in computing. Kemeny and Kurtz wrote a mission statement the stated "every student on campus should have access to a computer, and any faculty member should be able to use a computer.[9]" Kemeny went on, "Whether a student will ever use a computing machine or not, his life is likely to be affected by such machines, and hence, he should know something about their capabilities and limitations. In this sense, contact with electronic brains is as essential as learning to use the library."

Dartmouth was a liberal arts college at the time, not a research university, but its administration agreed with his assessment. Kemeny and Kurtz set about creating a more accessible programming language. In 1956, they created DARSIMCO (Dartmouth Simplified Code), a simple mathematical compiler that produced SHARE Assembly Program statements[10]. In 1959, Dartmouth acquired its first computer, the Librascope General Purpose Model 30 (LGP-30). This was the system devised by the Los Alamos alum Stan Frankel while at Caltech. He sold the design to Librascope, which released the drum-based LGP-30 computer in 1956. The LGP-30 competed with the IBM 650. Kemeny financed the $40,000 purchase by declaring it as furniture. He installed it in the basement of College Hall[11].

Kurtz assembled a small undergraduate team and they developed an ALGOL compiler. ALGOL (its name comes from its focus on algorithmic syntax) was designed by American and European computer scientists (including IBM's John Backus) and released in 1958. It sought to improve the power and accessibility of FORTRAN. ALGOL was very influential, a forerunner of later programming languages such as C, C++, and Java. Kurtz called his version "SCALP", for Self-Contained Algol Processor. Kurtz's team marshaled their programming resources to track and predict the winner of the Presidential primary in New Hampshire in 1960. Their program on the LGP-30 correctly predicted the victory of Senator John F. Kennedy while the big mainframe computers at NBC wrongly forecast Nelson Rockefeller.

However, Kemeny was not a fan of ALGOL and set his sights on something far easier. He had used an English-based programming language on the JOHNNIAC computer at the Rand Corporation[12]. In 1962, he designed the Dartmouth Oversimplified Programming Experiment, with the unfortunate acronym of DOPE[13]. Dartmouth student Sidney Marshall wrote the majority of the code while Kemeny contributed key elements. The resulting language was much closer to the desired concept. In addition, the student participation demonstrated a better path forward for Kemeny and Kurtz.

Kemeny and Kurtz began the new programming language project in earnest in 1963. They would borrow heavily from FORTRAN II as well as their prior language experiments. The mantra was "simple, simple, simple." The guiding principles stated it should be easy to learn, general purpose, have the flexibility to add extensions, and, most importantly, not require an understanding of computer hardware. There were only 14 commands (e.g., LET, READ, DATA, INPUT, GOTO, IF, THEN)[14]. It was a compiler, not an interpreter. When you entered "RUN" at the terminal, the program statements compiled and the program executed. The completed compiler represented 3,000 lines of code. It just fit into the 4K drum memory of the LGP-30. They named the language BASIC, for "Beginner's All-Purpose Symbolic Instruction Code." The one-line program "PRINT 2 + 2" was the first BASIC program run.

The next challenge loomed—how to deliver BASIC to Dartmouth students and professors. John McCarthy and colleagues at MIT had invented timesharing on a DEC PDP-1 in 1959. He encouraged

Kurtz and Kemeny to deliver programming to the students and faculty with such an arrangement[15]. Kemeny secured a $300,000 grant from the National Science Foundation in 1964. The College sent requests for proposals to IBM, GE, NCR, Bendix, and Burroughs. GE won the business. The dean of Dartmouth's Thayer Engineering School, a former GE executive, helped secure at 60% discount on a GE system[16]. IBM representatives escalated the computer decision to the Dartmouth's President, John Dickey. He threw them out of his office. IBM was unwelcome on campus for years afterward.

Dartmouth installed a General Electric GE-235 system with a back-end processor to manage the system's teletype terminals. Work began on operating system software to manage a network that would span the campus. Kemeny's "Memo No. 1" outlined the plan. Undergraduates would develop the code, which went live as the Dartmouth Time-Sharing System (DTSS). The College and GE entered into an agreement to market the system. GE would upgrade Dartmouth's computer to the GE-635 at no charge, a tenfold increase in performance. GE marketed the DTSS system as the MARK-I timesharing offering. It formed the basis of their service bureau operations, earning millions for the company.

Dartmouth demonstrated its new timesharing system in the fall of 1964 at a large computer conference in San Francisco. Using a dial-up line, the audience of computing professionals saw BASIC for the first time. It was a scene reminiscent of the famous teletype demonstration of George Stibitz's relay calculator at Dartmouth in 1940. By mere coincidence, Stibitz had joined the Dartmouth Medical School faculty that year. He would use DTSS and BASIC to develop a cancer radiation analysis program.

About 80 percent of Dartmouth undergraduates took a BASIC course, fulfilling the vision of Kemeny and Kurtz. Your author happened to be one of them. In 1970, I sat at a Teletype terminal connected to the DTSS network and the new GE computer. There were terminals in the main computer center and throughout the campus. You keyed in your program one instruction at a time, with the teletype machine printing each line on a narrow roll of yellow paper. When you entered "RUN" your program was compiled and executed.

As a side note unrelated to BASIC, John Kemeny became president of the college in 1970 when I was a sophomore. The following year I joined the Sigma Nu Delta fraternity. Fraternities were very big at the time as Dartmouth was an all-male school located in the north woods of New Hampshire. Sigma Nu was next door to the president's house, and the fraternity was quite famous for its loud parties. Though I am sure there is no connection, President Kemeny took the 200-year-old college founded by Daniel Webster co-ed the following year. That did a lot to tame the fraternities, including Sigma Nu.

## OTHER MISSES

There were several other areas viewed as missed opportunities with System/360. Fred Brooks regretted going with 24-bit addressing. Though the decision was principally driven by cost, it would be a fundamental problem for years to come. Moving up to 32 bits or even 48 bits would have provided more than enough room for growth. Brooks also lamented the narrow definition of System/360 Job Control Language (JCL). The existing punched card job definitions provided the model. His team saw JCL simply as a series of control cards. Brooks felt it should have been a high-level programming language. This would have simplified the syntax and, more importantly, allowed for logic within the job flow, including subroutine and logical branching.

Chuck Branscomb would reflect on the decision to kill John Haanstra's 1401 Super. At the time, IBM owned 80 percent of the market for computers in the $2,500 to $10,000 segment, with the 1401 being by far the major contributor. That share would plummet to under 50 percent by 1969, but this was in part a redefinition of the midrange price points, which had come drastically down. Branscomb also cited the Model 30 pricing. System/360 compatibility and the need for a disk drive substantially increased its entry price. Branscomb concluded that doing both the 1401 Super and the Model 30 would have been a better strategy. This would have been a tough sale, given Haanstra's penchant for behind-the-back dealings.

# PART FOUR
# GOING SMALL

# EARLY EFFORTS IN EUROPE

The System/360 miss at the high end, and the subsequent failure of programs to address it, was certainly concerning. The main consequence, however, was the loss of prestige. Far more consequential was the miss at the low end. The fundamental premise of the NPL program would be as usual the sticking point. The challenge at the low end remained achieving compatibility with the System/360 line while reaching the required price point.

The WWAM program began work in this area. The 1401 was a welcome addition but left significant pricing room below its average $6,500 monthly rental and purchase price approaching $300,000. Groupe Bull and its Gamma line was no longer the main focus. The new challenge began right next door to IBM's Scientific Center at MIT. Digital Equipment Corporation (DEC) released its first computer, the PDP-1, in 1961 and essentially created the minicomputer market, with high-performance machines offered at a fraction of the cost of IBM's systems.

It did not help that IBM was complicit in the emergence of the minicomputer market segment, having a hand in birthing the two major players, Wang and DEC. Ken Olsen formed DEC in 1957. Olsen had graduated with an electrical engineering degree from MIT and joined the Whirlwind project at MIT's Lincoln Labs. He worked as an assistant to Jay Forrester on core memory and helped build the experimental TX-0 computer that tested the use of transistors in place of vacuum tubes. After IBM won the SAGE contract in 1954, Olsen spent two years in Poughkeepsie overseeing SAGE development and manufacturing. Seeing IBM's design, development, and manufacturing up close planted the seed for his own ambitions. When the SAGE assignment ended, he decided to transform the TX-0 computer concept into a small, single-user workstation computer. Such a system became more feasible as transistor prices continued to fall.

Olsen and fellow researcher Harlan Anderson left MIT to form DEC. The PDP-1 ("Programmed Data Processor") in 1961 was their first computer, priced at $120,000. With great success, DEC would release the PDP-8 in 1965 starting at $20,000 purchase. They would sell over 50,000 of them[1]. The PDP-8 was small. The entire machine weighed only 250 pounds and took up just eight cubic feet. DEC upgraded to 16-bit addressing in 1970 with the PDP-11, another smash success, selling over 250,000 machines[2]. By 1972, the company had 40 percent of the minicomputer market. DEC hit its stride in 1977 with the VAX line of compatible computers that featured large-system performance and virtual memory for a fraction of mainframe cost.

Most of the early DEC placements were in engineering shops, outside IBM's traditional markets. With no sign of competition from the computer market leader until IBM's General Systems Division entered in the early 1970s, DEC grew rapidly into the second largest computer company with 100,000 employees and over $1 billion in sales. In those days, DEC was the "anti-IBM." Dress was casual. The business organization was open and far less regimented, with employees encouraged to argue with one another until they reached a consensus. This was definitely not an atmosphere where Vin Learson would have flourished.

Down the road from DEC, An Wang had sold his patent for magnetic core memory to IBM and used the money to start Wang Laboratories. The company grew slowly at first. Its first computer arrived in 1971, but Wang's fortunes shot up with the VS line of compatible minicomputers in 1977. Wang made it clear that he was targeting IBM with blistering anti-IBM ads that aired years before

Apple produced its famous NFL Super Bowl commercial for the Macintosh. Wang hired salesmen that could talk directly to the computer department, where the feeds and speeds of his computers found a welcome ear. Wang and DEC spawned a minicomputer industry along the Route 128 corridor in Boston. IBM would take its time to fully grasp the threat and then respond.

Although rental price was not the complete barometer of low-end success, it does provide a good approximation. The target for the WWAM machine was $2,500, based on the typical cost of a small punched-card configuration. The 650 provided significantly more function with a stored program, but started at $3,500. The 305 RAMAC from the San Jose lab moved the bar slightly, down to $3,200. The 1401 provided much more capability, but again, a reasonable and effective system remained in the $3,500 range and above. The winner of the contest between the 660 and 750 contest completely fumbled the entry mission, providing the ill-fated 7070 at a whopping $18,000 per month.

At the System/360 announcement in April 1964, the low-end machine was the Model 30. At an average rental of $8,000 and purchase price of $350,000, this was not the planned entry model. That would be the Model 20. The German lab owned this model but work was not far enough along to announce it with the rest of the line. It was no accident that the Model 20 mission went to the Boblingen lab. When Endicott took over the WWAM program, the German team switched its focus to an even smaller machine than the 1401. This new concept apparently owed its start to a visit to the lab by Watson in 1953. One of the lab's junior engineers gave a presentation that proposed making machines less expensive by making them smaller. Afterward, Watson told the country manager to give the engineer whatever he needed to pursue this approach. The engineer was Karl Ganzhorn[3].

Ganzhorn was lucky to be present at the meeting. At the onset of World War II, he volunteered for the German Army at the age of 18 and became a tank officer in the Afrika Korps under Field Marshal Erwin Rommel. He arrived in North Africa in 1942, just after the battle at El Alamein. He fought in the battle at Kasserine Pass, where the inexperienced American forces were soundly defeated. With a year of desert fighting, the American forces under General Eisenhower and the British Army under Field Marshall Montgomery were resurgent. The Allies were also able to use key Enigma intercepts to their advantage. The two Allied armies caught Rommel's army in a pincer campaign in May 1943 and captured 275,000 German soldiers, including Karl Ganzhorn.

The British detained Ganzhorn but soon handed him over to the Free French forces. He spent the next four years in a prisoner-of-war camp in the desert of southern Algeria, hundreds of miles from any settlements. The French used the camp for forced labor, though Ganzhorn as an officer was exempted. His commander was also a prisoner in the camp and had been a POW in World War I. He suggested they use the time to prepare for life after the war. An American YMCA representative visited the camp and asked what assistance they could provide. The men asked for books. In response, New York's Metropolitan Library pulled some 4,000 German-language books from its shelves, copied them, and sent the material to the camp. The library of information included the essentials of a university education in several subject areas. Ganzhorn studied Italian, Spanish, mathematics, and physics.

The treatment of German and Italian POWs in Algeria and France soon raised alarms. The French held roughly one million POWs[4]. With rancor to the German and Italian prisoners, they put them to work in mines and treated them harsher than those at the American and British POW camps. The death rate in French camps (2.6 percent) was far higher than in American (0.1 percent) or British camps (0.03 percent), though all of these death rates paled next to the Russian camps (36 percent). Intense pressure from the United States failed to move the French. Ganzhorn was not released until 1947, after the Pope interceded and the Vatican arranged for a passenger liner—a French one, no less—to bring the POWs home. Years later, Ganzhorn would tell his story to *Reader's Digest*, which published it as "A University in the Desert Sands."

Ganzhorn weighed less than 100 pounds when he walked out of the French labor camp. However, he had been able to absorb the equivalent of a college degree during his time there. Back in Germany, he received acceptances from three universities. He opted to attend the one in Stuttgart. He related his story to the dean. He took an exam and the university awarded him a BA in mathematics before attending a single class. He would switch to theoretical physics and earn both his master's and doctorate.

Though he intended to go into university teaching, he realized that at 30 years old, he would be off to a very late start. One afternoon on campus, he met an IBM engineering manager, who hired him on the spot for the German lab. Walter Scharr, the lab director, started with IBM in Endicott and moved to Germany in 1937. He was especially interested in Ganzhorn's graduate work in solid-state physics, as transistors were just coming into play. He asked Ganzhorn to set up an electronics research lab in Sindelfingen, a short distance from the university. Ganzhorn had grown up in the town. His manager, an engineer steeped in mechanical punched-card equipment, gave him his marching orders: "Now, please, make physics for IBM.[5]"

After his presentation to Watson on the miniaturization concept, Scharr and Ganzhorn had the funds for additional engineers. This was not a simple task, as many European scientists and engineers left for the United States after the war. Nevertheless, Ganzhorn added 200 electronics hires to augment the 200 mechanical engineers already on staff. Ganzhorn would replace Scharr as lab director in 1958 and report to Ralph Mork, who was the overall World Trade engineering manager. His lab mission was small systems, printers, and semiconductor components.

Dick Watson endorsed the German concept of miniaturization and Walter Scharr got it going. The central feature of the program was a much smaller punched card, resulting in the project's code name - "Tiny." The German team developed a complete processing system around the smaller card, including a keypunch, sorter, and reader/punch. The French lab developed the application software. Meanwhile, San Jose was also working on a small system after its success with RAMAC. Ralph Palmer decided to merge the two efforts. He sent several San Jose engineers to Germany to work on the consolidated program. Ganzhorn's team was able to get assistance from Larry Wilson, the engineering wizard of punched-card machines[6].

World Trade announced Tiny in 1961 as the System/3000. The marketing rollout went well, and soon there were more than 2,000 orders. There were some technical problems prior to the announce date, including ink fouling within the printer and misreads of the new smaller card format. Product test felt those issues had been resolved and supported the announcement. Several small banks in the Stuttgart area signed up as beta accounts. They soon ran into the card read problem. The lab pulled the beta machines and Karl Ganzhorn started a frantic effort in late 1961 to solve the problem.

During this critical time, Ralph Mork moved to Kingston as lab director[7]. He commuted over forty times between New York and Paris but concluded in hindsight that he should have stayed on until the problems with System/3000 were resolved. Ganzhorn's engineering team tried a number of modifications, including changing the card stock, but the problem persisted. In hindsight, the multi-function card machine was more complex than anticipated. Even with Larry Wilson's help, it proved a difficult task to orchestrate the flow of cards between the two card hoppers, two read stations, and the card punch. In the spring of 1962, Tom Watson cancelled the System/3000.

The Boblingen lab conducted a post-mortem on the project. It highlighted a number of engineering mistakes during development. Designers finalized the card format before engineers tested the paths through the reader-punch. As an entry product, meager commissions incented salesmen to spend their time on bigger deals. Then there was the business case. As a standalone entry system, it was much tougher for the plan to come together and earn a profit. World Trade intended the system to be a loss leader, to get IBM computers into new companies and grow them. Quoting Mork, the business plan was "substantially sub-normal, maybe one-third the normal profit kind of a thing.[8]"

**Figure 41**. IBM 3000 System. Developed by IBM's World Trade labs in France and Germany, the system did not survive its beta program and was cancelled in 1960 by Vin Learson. Many of its designs were incorporated in the later IBM System/3. Reprint Courtesy of IBM Corporation ©.

The failure of the System/3000 program was a serious blemish on the World Trade resume, particularly coming on the heels of the WWAM cancellation. Ganzhorn was the lab director and was ultimately responsible. Tom Watson asked him to meet with him in New York. He fully expected to be fired, or be sent to some kind of IBM purgatory. Instead, Watson simply advised him to ensure that his products were fully tested and up to IBM quality before turning sales loose. Though the cancellation was painful for Ganzhorn and the German lab, it would prove to be a critical step in IBM's assault on the very low end of the computer market. The lessons learned with the Tiny program would contribute directly to the success of subsequent projects in San Jose and Rochester, and finally establish IBM as a player in the small systems market.

In the wake of the System/3000 cancellation, World Trade moved forward with two replacement programs aimed at the on-going Bull Gamma 3 threat. The French lab worked with Max Paley to integrate the 608 Multiplier into a punched-card system released as the IBM

628 system. Ralph Mork also had another program under development while work progressed on the System/3000. Called the Series
50[9], it took advantage of punched card machines returned to IBM
from larger customers. He had some of these models reconditioned
and with a change to a gear or pulley, reduced to half speed. He then
reduced their prices. The cobbled-together Series 50 featured a
keypunch, sorter, multiplier, and accounting machine. It was modestly
successful but bought some time.

Tom Watson spared Karl Ganzhorn the IBM penalty box and
offered him a chance at redemption. Prior to the adoption of the NPL
program, the German lab worked with Endicott on a very low-end
1401, the planned 1430[10]. The NPL program resulted in its cancellation. Tom Watson assigned the German lab to the low-end
System/360, the Model 20. By this time, the Boblingen lab had grown
to 1,000 employees, and Ganzhorn had picked up responsibility for
the smaller labs in Sweden and Austria. Ganzhorn would say that he
went into theoretical physics in part so he did not have to deal with
people. Now, this was his primary responsibility.

The SPREAD report envisioned a "very small" processor as the
entry point of the System/360 line. As was the case with the planned
high-end models, the Model 20 represented the far frontier of the
compatibility spectrum. Ganzhorn and his team traveled back and
forth to Poughkeepsie to consult on the design. Brooks and Haanstra
were at odds on the Model 20, repeating the same arguments over and
over. Brooks wanted a completely compatible System/360 model
while Haanstra wanted an inexpensive, low-end machine that was
1401 compatible. They reached a compromise, but it was far closer to
Haanstra's concept[11]. The cost of the required core memory,
input/output channels, a full instruction set, and complete set of registers was prohibitive, so Ganzhorn jettisoned full compatibility. All of
this transpired as the announcement date for the System/360 loomed.
By the time the German lab began the project in earnest, they had
only six weeks to meet their release checkpoint. They missed the April
1964 announcement and the Model 30 became the entry model.

Finally announced that November, the Model 20 was a mess of
compromises. It featured a much-reduced System/360 instruction set,
supporting just 37 of 143 instructions[12]. It also added several unique

instructions, which would complicate the upgrade of the Model 20 to any of the other models. It reduced number of hardware registers and replaced the I/O channel architecture with simple software function. It was a 16-bit computer, not the 24-bit design of the other models, and certainly not the 32-bit or 48-bit addressing that Fred Brooks wished he had embraced. The German lab had to develop three simple operating systems, for card, disk, and tape, as the DOS produced by Endicott was not ready in time.

Physically, the completed Model 20, designed by Ganzhorn's team, did not look anything like the other models in the line. Larry Wilson and his San Jose lab developed a new multi-function card device for the system, the 2560 MFCM (Multi-Function Card Machine), which was unique to the Model 20. It turned out to be unreliable, not unlike the multi-function card device on the System/3000. It was the source of a number of very uncomplimentary alternative interpretations of the acronym "MFCM." It seemed to violate Wilson's engineering maxim that said, "When you get the right design, it's gonna be simple, it's gonna cost less to build, it's gonna be more reliable, and that's what you want."

The Model 20 was also extraordinarily slow, able to calculate only 5,700 additions per second. This was only slightly faster than ENIAC. The high-end System/360—the Model 195—was rated at up to 17.3 million additions per second. The original System/360 performance range was 50 to 1. The Model 20 expanded the published range to 200 to 1 but only because it was so ploddingly slow. Much later, Ken Shirriff did an analysis of the two bookend systems of the System/360 line and ended up with a range of 3,000 to 1[13]. Of course, given the focus of the Model 20 on simple batch applications, performance was not a big issue. An entry Model 20 rented for $1,280 monthly, with a purchase price of $62,710. This was certainly WWAM territory, but a usable configuration rented for about $2,000, and was closer to $100,000 to purchase.

This litany of performance and compatibility issues would lead one to mark the Model 20 as a total bust. Quite the contrary, the Model 20 was by far the most successful System/360 model in terms of numbers, with over 7,400 installed in the United States and more than 20,000 installed worldwide[14]. Despite this success, work began almost immediately on addressing the Model 20 incompatibility. The Model 25 arrived in 1968, a fully System/360 compatible machine. However, it came with a far higher price tag, at $5,330 monthly or $253,000 purchase. The Model 22 followed three years later. It was an economy version of the Model 30, with a typical configuration priced at $5,600 rental and $246,000 purchase. Coming so late, it was able to use monolithic integrated circuits. Bob Evans would conclude that IBM missed the NPL entry target of $1,500 rent and $50,000 by double. The Models 22 and 25 made no pretense of hitting that original entry target.

The gaudy sales numbers for the Model 20 and Model 30 invited competition. Sperry Rand UNIVAC developed the 9200 and 9300 models in June 1966[15]. They were reasonable clones of the Model 20 and had some success. They had just 35 of the System/360 instructions, which included most of the Model 20 subset and some of their own. The complete system included the 2311 and 2314 clone disk drives from ISS, the "Dirty Dozen" company. The 9300 model was twice as fast as the Model 20, with a machine cycle speed of 600 nanoseconds, twice the speed of the Model 20. All of the UNIVAC clone models used small-scale monolithic integrated circuits rather than the SLT technology used by IBM. They also replaced magnetic core memory with thin film memory, a decision that would lead to reliability problems.

The fast speed of the 9300 had led UNIVAC to create the slower model, the 9200. As bizarre as it sounds, they slowed the 9300 cycle time to the equivalent of 1,200 nanoseconds by introducing a complete, nonfunctional "rest" cycle between every real functional cycle[16]. In addition, they slowed the card reader from 600 to 400 cards per minute and the printer 600 to 250 lines per minute. These were gimmicks that IBM used, but Sperry took the approach to a whole new level. RPG was the preferred programming language, though UNIVAC also included the MATHPAC library of arithmetic functions, a heritage of Grace Hopper.

When his temporary assignment as one of Learson's Four Horsemen settled down in early 1966, John Haanstra was in limbo. Learson asked Chuck Branscomb to stay on permanently as SDD president, taking Haanstra's old job. Bob Evans pushed to have

Haanstra added to his staff at Federal Systems. Though not necessarily IBM purgatory, Haanstra recognized that this was the end of his
run at IBM. He left the company in August 1967 and signed on as
Vice President of Strategy for General Electric.

Though he had fought the NPL plan from the beginning, he was
now in charge of GE's program for a System/360 compatible line.
The GE 200 Series was renamed the "Compatible 200s" to get some
of the press that IBM had created. A key developer on the GE 200
Series was Arnold Spielberg, the father of film director Steven Spielberg[17]. The GE 225 was the initial system at Dartmouth College that
spawned the Dartmouth timesharing system.

As Haanstra settled in at GE, he found work already underway on
Project Charlie[18], aimed at creating the microprogramming for full
System/360 compatibility. One element of the project was to build
and market a System/360 Model 20 compatible machine. Haanstra
worked with Richard Bloch on this project, who oversaw the business
side of GE's computer operations. This was the same Richard Bloch
from Harvard and the Mark I team. He had left Harvard in 1947 for
Raytheon. Raytheon sold its computer business to Honeywell in 1955,
and in 1968, GE lured Bloch to join its computer division in Phoenix.

GE held a planning meeting in Ft. Lauderdale in 1969 to work
out its strategy for the coming year. Haanstra took his family along for
the trip, flying in his private plane. On the return trip to Phoenix, his
plane crashed in New Mexico. Haanstra, his wife, and two sons were
tragically killed. Project Charlie and the Model 20 clone never saw
the light of day. The following year Honeywell bought GE's computer
business. Had Haanstra lived, he would have found himself working
for the company he had fought years before, with the Honeywell H-
200 and its Liberator software gunning for the IBM 1401. Ironically,
the new product plan for the combined Honeywell and GE computer
division was "NPL", also for New Product Line.

# THE ROAD TO SYSTEM/3

The System/360 Model 20 was supposed to anchor the low end for IBM, but substantial opportunity remained below its price points. Karl Ganzhorn and his team had tried to capture that space with the System/3000. His team continued to work on what they had defined as the WWAM market, proposing a System/360 Model 15. At the same time, some of the San Jose engineers that who had worked with his team returned to California with ideas of their own. However, both of these IBM labs would end up deferring to the growing aspirations of a new plant and lab constructed in the cornfields outside Rochester, Minnesota.

By the mid-1950s, IBM had the core company labs in Endicott and Poughkeepsie and a small but growing disk operation in San Jose. Tom Watson wanted Endicott and Poughkeepsie to focus on new system development and shed some of their legacy missions. That meant finding a new site that would work on punched-card equipment. With existing operations on both coasts, Watson focused on the Midwest. Costs were higher in the larger cities, so he narrowed the search to smaller towns. They considered more than eighty cities, and the search came down to Madison, Wisconsin and Rochester, Minnesota. Madison had the university and was the larger city[1]. At the same time, it would require higher wages and might result in unionization.

Rochester, a city of 30,000, was within an hour of Minneapolis and St. Paul, and home to one of the world's top medical facilities, the Mayo Clinic. Rochester had another factor in its favor. Tom Watson had been a co-pilot with Leland Fiegel on General Bradley's B-24 mission to Moscow. Fiegel was from Rochester. After the war, Fiegel visited Watson in New York. On his way back to Rochester, his plane crashed and he was killed[2]. That made the decision easy for Tom. The new facility would be in Rochester. Watson met with business and civic leaders that February to announce the plans. He stayed downtown at the Kahler Hotel, the grand dame of the city, and likely dined down the street at Michael's, the best restaurant in town. He outlined IBM's plan to have a facility built and operational with 1,500 employees by 1958. He also took the time to visit Leland Fiegel's father and tell him that IBM was building the new plant in his son's honor. Leland Fiegel's son would join IBM in Rochester, the start of a 30-year career.

Tom Watson had just recently taken over IBM from his father, so opening a new IBM site was still an exciting project for him. The company acquired two farms totaling 397 acres several miles northwest of the city, in the midst of rolling cornfields. Eliot Noyes, IBM's head of design, chose famed architect Eero Saarinen to design the plant. The modern structure made liberal use of the color blue, appropriate for the company now widely known as "Big Blue." The initial building measured some 400,000 square feet. It would grow over the years to over 20 interconnected buildings, earning the affectionate nickname the "Big Blue Zoo."

Watson traveled again to Rochester in January 1957 to welcome the first employees, who were working out of a temporary site. He returned a third time in September for the formal dedication of the plant, joined by Minnesota governor Orville Freeman. The site had already grown to 570,000 square feet of space. The company would add roughly 20,000 people to the city's population by 1970.

Charles Lawson was the original plant general manager. Clarence Frizzell would succeed him in 1959, the same Frizzell who would later become one of Learson's Four Horsemen. Endicott transferred

30 employees to the new plant to jumpstart the unit record mission. Engineers installed a 650 computer to manage plant operations. This would be later replaced with a 1401 and, still later, a System/360 Model 40. The first machine out the door was the 077 Collator. Originally announced in 1937, it played a key role in winning the Social Security Administration contract. The current version would read 80-column cards from two hoppers and after a field comparison would send them to one of five stackers. It used a plugboard to control the collation. Rochester's first in-house designs were the 085 numeric collator in 1958 and the 087 alphanumeric collator in 1959. These were completely modernized machines with more than double the processing speed. Rochester stepped up its game in 1962 with the 188 Collator[3]. Engineers replaced relays and vacuum tubes with transistors. Core memory eliminated the plugboard. Card reading was now done by photoelectric sensors, replacing wire brushes. This enabled the machine ramp up to 650 cards per minute. The advance in technology required a decided jump in skills. New hiring created an electronics-focused engineering team that would grow into a full-fledge lab. With Watson's reorganization of the manufacturing and development divisions in 1959, Rochester became part of the General Products Division, along with Endicott and San Jose. They now concentrated on computer systems under $10,000 rental per month. The plant and lab workforce grew to 2,300 in 1960.

**Figure 42.** Aerial view of the IBM Rochester plant with Highway 52 in the foreground, circa 1960. The site would grow to 34 buildings housing 8,100 employees by the mid-Nineties. Reprint Courtesy of IBM Corporation ©.

Harry Tashjian led the nascent lab team. He grew up in Johnson City, just east of Endicott. Tashjian had been a bombardier in World War II and graduated from Cornell University with a degree in mechanical engineering before joining IBM in Endicott. He moved to Rochester in 1959. He would be the lab director until 1977, shepherding a whole line of small systems that would include the System/3, System/32, and System/38. Hal Martin, who had hired Fritz Trapnell and moved him to Paris, was the plant general manager.

In 1964, Rochester picked up Endicott's software maintenance mission for the 1401 to relieve Earl Wheeler's System/360 programmers. By 1966, the plant had doubled in size to 1.2 million square feet and the workforce had grown to 4,000. Martin and Tashjian knew the System/360 work would be temporary and sought out a more expansive role. As part of the General Products Division, the small and midrange sector was a natural area. The immediate problem was that San Jose had the same idea.

## TINY BECOMES THE SYSTEM 3.7

The San Jose engineers who had worked on the "Tiny" program in Boblingen returned from Germany. Among them was mechanical wizard Larry Wilson. He had been very busy. He had managed the production of the 609 Calculator, a fast-transistorized successor to the 608. His team had produced the 1402 card reader for the 1401 in 1959, the 1442 card reader/punch in 1961, and the 2560 Multi-Function Card Machine for the System/360 Model 20 in 1963. In 1964, IBM named him an IBM Fellow, enabling him to work on any project he wanted.

He decided that he would take another crack at a small system along the lines of the ill-fated System/3000. He discussed the project with Karl Ganzhorn. Wilson enlisted Roy Harper, the engineer whose name should have been part of the "Wooden Wheel." Harper was named an IBM Fellow in 1968. According to Ralph Mork, Harper was a good example of what Tom Watson was looking for in a "wild duck," someone always looking for new angles and quite often coming up with a major invention for the company. Watson was big on "wild ducks," occasionally telling the origin story from Soren Kierkegaard, where a flock of wild ducks arrived on a farm and the farmer started feeding them. The ones that stayed lost their instincts and did not survive. The ones that left, remaining wild, flourished. Wilson added Greg Toben of Card Programmed Calculator fame to his team. Toben would work the mechanicals while Wilson would focus on the card processing and Harper would do the electronics. Wilson called the project the "3.7."

It would take advantage of advancements in technology in the years between 1956 and 1966, replacing electromechanical function with electronics. Wilson's team was able to use photocell sensing in card reading, eliminating one of the two major issues that led to the System/3000 cancellation. Wilson decided to go even smaller with the punched-card format, designing a format just three inches square. It featured 96 columns arranged in three tiers, reminiscent of Remington Rand's three-tier 90-column card from 1930 that tried to compete with IBM's 80-column card. Though the concept of a small card had ended badly with the German System/3000 project, Rochester had both the card sensing technology and the experience to make it work The new small IBM card had three times the density of the standard IBM card, but the photocell sensing took care of sensing the tightly spaced holes.

Back at Rochester, any plans that Hal Martin and Harry Tashjian had to establish the site as the center of small and midrange systems was complicated by the work Wilson and his team were doing in San Jose. Besides having five IBM Fellows on staff, San Jose had engineers with small system experience, working with the 305 and 310 RAMAC programs. Compared to San Jose, Rochester's resume was a little thin. The big plus for the site was its staff's significant experience, skills, and technologies relating to card processing. Disk drives turned out to be the deciding factor. They were some of IBM's most profitable products and continued to be the target of multiple clone manufacturers. San Jose had to maintain its focus on disk drives. Both development sites were part of the System Development Division, and the president of the division at this time was none other than Chuck Branscomb. Despite his long relationship with Wilson, Branscomb directed the small-system mission to Rochester.

Dick Trachy, a systems manager who transferred to Rochester in 1965 from Poughkeepsie, invited Wilson to Rochester in 1966 to review his program and determine if Rochester had the skills and resources to take it over. Wilson drove to Rochester in his 1961 Jaguar XLE. It was an unusual car to be seen in the Rochester parking lot. It was definitely a summer car, as it would be hard-pressed to function in the frigid Minnesota winters. The car, however, was involved in a later discussion on where the program name "3.7" came from. Frank Soltis, who would figure prominently in the design of future Rochester systems, had his own theory. Soltis was a car guy. His father worked for General Motors, and Soltis quite naturally became a GM enthusiast. That included racing Corvettes. Soltis knew the Jaguar had a 3.7-liter engine and speculated that Wilson simply used that as his project code name.

When the time came to name the new system, Soltis suggested "System/3.7.[4]" Corporate was not enthusiastic about that and shortened it to "System/3." Of course, this is simply speculation on the part of Soltis. An alternative theory was that the new system became System/3 because it had three main components—the system unit, the printer, and a new multi-card device. Whatever was the true explanation, Wilson notified Dick Trachy after a few weeks that he was satisfied Rochester could handle the project. He drove the Jag back to San Jose before winter arrived, and the car resumed its rightful place as year-round transportation.

Small and midrange competition flourished with IBM's continued absence from the space. IBM had no computers under $100,000. The System/360 Model 20 was closest at nearly $200,000 purchase. It was clear that Rochester could not adopt the System/360 architecture due to cost, starting with the hardware and software prices. However, those figures represented only the beginning ante for a small computer system. With the System/360, you were working in a complex programming and operational environment. There was also a hidden cost, the lack of ease of use. In addition, and even more important, was the dearth of available software applications. The small- to medium-size business that purchased the System/3 was unlikely to develop its own applications in COBOL or assembler. The experience of unit record applications, the 650, the 1401, and especially the accessible languages like RPG, showed the way. A new concept emerged to guide the philosophy of Rochester's new system, the Total Cost of Ownership, or "TCO" for short. It more broadly defined what it would take to purchase, install, and operate a new system. It included the hardware, software, programming, applications, operations, maintenance, staffing, and support.

In 1967, shortly after picking up the 3.7 program, Rochester held a task force with field participation to review its product plan. Though there was some pushback on the new 96-column card, the general reaction was positive. The field review would be the first in many interactions with the IBM field, business partners, and customers. The new system targeted small- to medium-size businesses without a computer, with a complete package that rented from $1,000 to $2,500 monthly. This was similar to the targets set for WWAM, the System/3000, and the System/360 Model 20, but Rochester was committed to getting below $1,000 with a usable system. That system would feature ease of use, rapid development of applications using the RPG programming language, and a physical design that would neatly fit in a typical business office. Finally and most important, the plan included the development of application software.

Dick Trachy led the project with Tashjian handling hardware engineering and Bob Webster managing software. Carl Gebhardt in product planning had the critical job of creating a salable business case for the program. He was prepared to get creative, as Sheldon Jacobs had done with the 1401. Rochester changed the formal name of the program to AURS, for Advanced Unit Record System[5]. The name obscured the fact that a new and incompatible computer was in the works in the Midwest.

In fact, the program had a high profile. Hal Martin was used to dealing at a distance with IBM Corporate, based on his experience in Paris with World Trade. He described the relationship: "IBM HQ was well aware of our program. Frank Cary was Group executive at the time, and all systems development reported to him. I met with him personally, and with his staff, weekly to review all aspects of the program[6]."

Rochester was not alone in this endeavor. The new system's disk drive would come from the Hursley lab, the printer from Boca Raton, and the tape drives from Boulder. The system's central logic, Monolithic System Technology (MST), was the responsibility of the lab in East Fishkill. Rochester, Boca Raton, and plants in World Trade would manufacture the system.

Erich Bloch worked on MST prior to and during the System/360 project. MST replaced the discrete transistors, diodes, and resistors of the SLT modules with integrated circuits etched in silicon. The "monolithic" in the name meant a single piece of silicon contained a complete set of circuits. Each MST chip initially contained six circuits and replaced one SLT card. That density increased to 40 circuits per chip. The transistor design used a new lower-cost variant called MOSFET (Metal Oxide Semiconductor Field Effect Transistor). Bell Labs invented MOSFET in 1959. It represented the first transistor that could easily be scaled down and mass-produced. MOSFET transistors soon became the standard in virtually all electronic devices. The system used core memory in sizes ranging from 8K to 32K.

The Rochester architects were under no obligation for compatibility, despite the ongoing work on System/360. The low-end System/360 models, including the wildly successful Model 20, were not compatible, so no great barrier was broken. The complete instruction set had only 29 instructions versus 143 or more instructions on a fully compatible System/360. The designers optimized those 29 instructions for limited core memory and tuned them to the RPG II programming language.

Multiplication and division were handled in software, not in hardware registers. The designers did use binary computation and partial summing to reduce the total number of additions required. This was a commercial machine, so there was no pretense of addressing scientific performance. The address space was two 8-bit bytes, so it was a 16-bit machine, providing a maximum addressable memory of 64K bytes. This would become a problem, complicating later models of the family, just as the System/360 decision to use 24-bit addresses would take years to resolve. However, the immediate task was to produce a small, affordable batch machine and this involved some trade-offs.

Rochester did not repeat the mistakes that Ganzhorn's German lab had made with the 2560 Multi-Function Card Machine (MFCM). The new Multi-Function Card Unit (MFCU) was a small, compact combination with MST logic circuits to control its operation. It had two input hoppers, four output stackers, a punch station, and print station. It read the cards with a photoelectric beam, something not available to Herman Hollerith or Karl Ganzhorn. Though it was an outstanding piece of engineering, it was both the apex of card-processing as well as the end of the line. The photoelectric technology provided another connection to Germany. Gustav Tauschek, the German inventor that IBM had bought out and hired in 1928, developed a photoelectric sensing device. He failed to divulge his invention and IBM threatened to sue him. He fled to Germany where he designed a machine gun for the Nazi war machine. He died in 1945[7].

Rochester offered a card-only model of the System/3. This was very much like the entry card-based 1401, a base offering added to ensure a low entry price point. A disk drive dramatically improved the system. It used the 14-inch ADF disk drive, renamed the 5440. It was configured in two office-style drawers, each capable of containing one fixed disk and one removable disk. Each disk stored 2.5 million bytes, for a maximum total capacity of ten million bytes. The drive, developed by the Hursley Lab featured Winchester-type flying heads. By coincidence, the Hursley lab was located in Winchester, England. The 5440 disk program was known as Dolphin, named after a pub in the area[8]. The system's printer featured a print chain mechanism much like the 1403. This was no surprise. The same team that produced the 1403 moved to Boca Raton and engineered the new printer.

Central to the program objectives was a compact physical presence. The system's units were combined in one compact L-shaped arrangement and sported more color than its System/360 big brother. Designers concealed the connecting cables behind panels, fixing a major mistake that Chuck Branscomb had complained about with the 1401. No raised floor or special air conditioning was required. The entire footprint of the system was just 150 square feet. The ergonomic layout of the system drew many praises, which flowed to the industrial design team in White Plains[9].

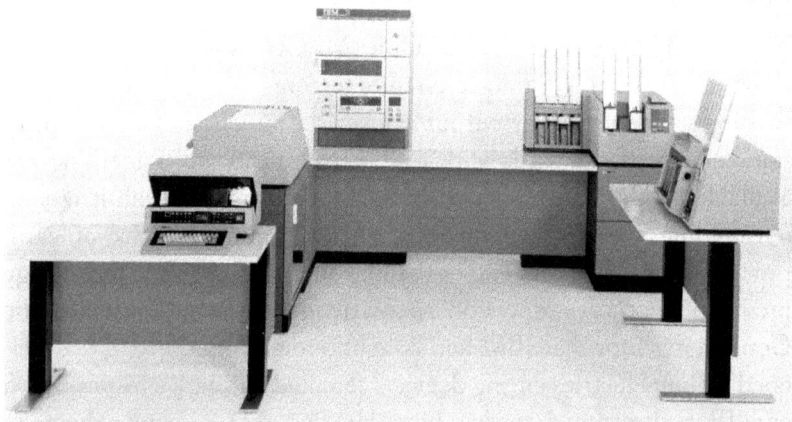

**Figure 43.** System/3 Model 10 (1969), the first IBM Rochester system, shown with the 96-column keypunch on the left, sorter on the right, and the L-shaped core system with printer, system unit, and MFCU (Multi-Function Card Unit) in the middle. Reprint Courtesy of IBM Corporation ©.

RPG (Report Program Generator) was the primary programming

language. It was the obvious choice for environments that had less staffing or perhaps employees new to programming. The language had progressed significantly since the days of FARGO. Many customers would be clueless that RPG had its roots in the plugboard processing of unit record equipment. However, they understood the basic flow of business accounting, and RPG would nicely mirror that. Rochester's RPG would not be a carbon copy of the RPG released with the 1401 or System/360. It would undergo a significant transformation from RPG to RPG II.

The central feature of RPG would not change—the built-in processing cycle. RPG took care of all the coding associated with processing transactions. You could skip directly to defining the application data, the calculations, and the output. Many customers would tap accountants in their back office to pick up the language and program their applications. These people knew the company's data and business flow and would just need to master filling out the RPG specification sheets.

There were five major coding sheets: the program header, file descriptions, input fields, calculations, and output. In addition to the built-in processing flow, a number of automatic constructs were built into the language. RPG used indicators to communicate certain conditions. For example, RPG will tell the programmer when it located a record, when it reached the bottom of the printed page, or when the control fields changed. That last function was termed group totals in 1921 when IBM was struggling against Powers competition. The totals function had not changed, but the coding certainly had.

The RPG language developers in Rochester focused on the new functions needed to make RPG easy to use but highly functional. They add some 37 extensions over the System/360 definition. They tested the new language on high school and college students in the Rochester area[10]. Rochester would shrug off the comments that RPG was not a "real" language, and its widespread success would prove the detractors wrong.

The following simple ten-statement RPG program reads an inventory file and prints out those items whose on-hand quantity is below 100:

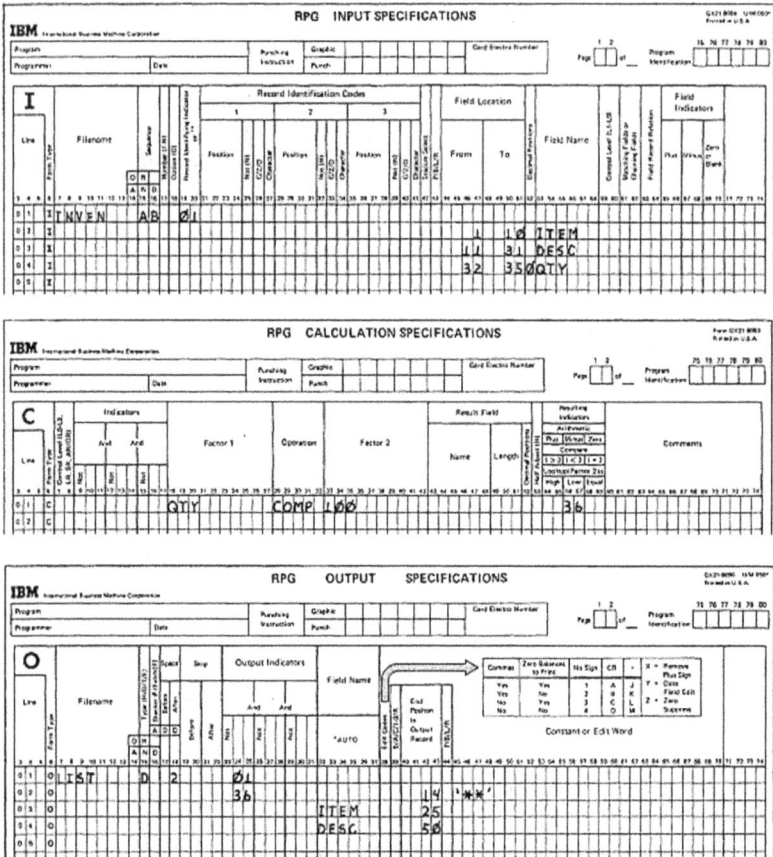

**Figure 44.** The input or "I" sheet defines the fields and locations in the inventory file. The calculation or "C" sheet compares the "QTY" field to the value "100" and sets on an indicator (36) if it is below that figure. The output or "O" sheet lays out the report called "LIST", printing the item and description for those items with a quantity under 100.

The operating system was similar to DOS. The System/3 was a single partition batch machine and system control could piggyback the work done in Endicott on the System/360. Operator Control Language (OCL) provided the interface between the application program and the operating system. It was largely based on the JCL (Job Control Language) of System/360. Rochester made the same mistake that Fred Brooks lamented -- the failure to make the job control a programming language. Rochester's focus at the time was a simple, inexpensive system. No one could foresee that Hal Martin's small team was building a $14 billion business for IBM.

## APPLICATION CUSTOMIZER SERVICE AND AUTOMATED PROGRAMMING

Rochester expected that applications for the new system would come from customers, IBM systems engineers, third-party programmers, and software companies. However, it also added a twist to the application plan tied to the new RPG: the Application Customizer Service, or ACS. One of the Holy Grails of computer programming that would emerge years later was the concept of "automatic programming," where an analyst created a description of a business process and the computer would write the program. The concept would blossom in the 1980s with relational databases and fourth-generation (4GL) programming languages, and would ignite anew with the advent of Artificial Intelligence (AI).

A new IBM software lab in Menlo Park, California, would be an early pioneer in automatic programming. Jack Mumford came up with the idea of producing RPG source code based on responses to a business questionnaire. He asked a small team, including lead developer Joyce Wrenn, to work on the concept[11]. Her team developed the processing code. A new applications group in Atlanta produced the questionnaires for common applications such as billing or accounts receivable.

The Application Customizer Service was a two-step process. A field systems engineer completed the application questionnaire with the customer. The engineer sent the completed responses to the processing center in Atlanta. The center returned RPG source programs which the customer compiled on their system. As delivered, the unmodified RPG programs might fit the customer's requirements, but more often than not, some tweaks would be required. The customer, systems engineer or a local third-party programmer could modify the programs to fit. ACS applications were developed for common business functions such as order entry, invoicing, inventory management, accounts receivable, accounts payable, sales analysis, payroll, labor distribution, and the general ledger. A typical service charge was $180 to $265 per application[12].

Wrenn's team in Menlo Park wrote the ACS processing code in System/360 Basic Assembly language. The center in Atlanta keypunched the questionnaire responses and ran them on a System/360. To some extent, the ACS approach echoed earlier approaches to implement applications. With punched-card equipment, IBM typically helped the customer build the first plugboard or even shared plugboard diagrams from other customers. Remington Rand's plugboard process resembled ACS even more. The customer sent the processing description to company headquarters, where a dedicated team would wire the panel and send it back to the customer.

The initial acceptance of ACS was encouraging, with up to a third of new customers opting for the service. However, the fatal flaw in the program soon became apparent. Running ACS production to generate RPG source statements required quite a bit of computing power, and at the time, Rochester's new entry system was not quite up to it. The ACS production machine, a System/360 Model 40, resided in Atlanta, and any changes required resubmission and reprocessing. On top of that, ACS required a large investment by IBM in development, marketing, service, and, most prominently, field support. ACS, so promising at the onset, was gone by 1977.

Within a few years, Rochester systems gained more than enough horsepower to run ACS, but the small-systems division was already down the road in a new direction. The Atlanta operation, formally the Atlanta Application Development Center (AADC), expanded to create a series of high-function industry applications. There were also other options for applications. One was field-developed programs, the approach that had produced the FARGO programming language. A more significant avenue for applications was programming partners. It was clear early on that the new system would be a success and here was a substantial opportunity to develop applications for the new platform. Many software companies ported their application packages, a development actively courted by Rochester. In addition, local IBM branch offices encouraged and supported a local stable of third-party programmers to provide custom programming for customers.

The new system debuted on July 30, 1969. The interim names for the system were retired and replaced with System/3. The 3.7 code name would live on at the bottom of the new IBM 96-column card where "IBM 3700" was printed.[13] The full name of the first system was the System/3 Model 10, which not so subtly indicated that there were more models in the line planned. During the development of the System/3, Frank Cary had risen from group executive to IBM president. He was on hand in Rochester for the formal announcement, calling it the "most important product announcement since System/360." Tom Watson Jr. arrived in Rochester two months later to congratulate the 2,000 employees assembled at the Mayo Civic Arena. The plant held a follow-up celebration at the ever-expanding IBM site, where employees danced to the tunes of Ralph Marterie and his orchestra. The formal announcement came to the field in the form of a peach letter[14], the color chosen as the new small-system division was headquartered in Atlanta. Before peach, announcement letters came in blue from the Data Processing Division in White Plains.

The entry System/3 started at $1,185 per month or $54,400 to purchase. This was a card-based configuration without a disk drive. It was roughly half the cost of the System/360 Model 20. A Model 10 disk system rented for $1,390 per month, and a full-featured configuration with 24K of core memory, ten megabytes of disk, and the new 200-lines-per-minute chain printer rented for $2,300 a month. The first System/3 went to Lasko Metal Products in West Chester, Pennsylvania, in January 1970[15]. Rochester shipped more than 1,500 Model 10s in the first year, a good start to what would eventually be 25,000 System/3 installations and over 600,000 installations of follow-on Rochester systems.

# IBM IN THE SOUTH

After the System/360 announcement, John Opel wrote to Vin Learson about the continued low-end gap. Though the Model 20 was in the works, he concluded that it would fall short of addressing the low-end market. Frank Cary, now senior vice president and former president of the Data Processing Division, knew more than most that the "big iron" division in White Plains dominated IBM marketing and strategy. In fact, IBMers called the route connecting Armonk with White Plains the Watson Freeway[1], reflecting the mainframe's central importance. Cary decided the only way for IBM to address the low-end market was a focused organization, one completely separate from the Data Processing Division in White Plains. His plan went even further than simply breaking out sales and creating new branch offices. He wanted a fully integrated company, with research, development, manufacturing, marketing, sales, and service under one roof. It was, in essence, a company within a company. Most critically, it was a concept that Learson and Tom Watson supported. Cary's new organization was the General Systems Division (GSD).

Cary realized that the new division needed to be physically separate from both Armonk and White Plains. A neutral site was needed, perhaps somewhat central to the new division's plants and labs located in Rochester, Boca Raton, and San Jose. That turned out to be Atlanta, in part because there was no shortage of executives who wanted to be there. Jack Rogers was first in that line. He was a dynamic executive who managed small systems in the General Products Division, the division that handled systems below $10,000 in rent. Rogers's roots were in the South and he lobbied hard for Atlanta. He tapped Jim Dezell to lead sales and marketing for the new division. Dezell had famously come to Learson's aid not once but twice. As a systems manager, he had convinced a Honeywell H-200 customer with 1401 Liberator software in his territory to give a demo to IBM. Later, Learson asked for his input on 1401 emulation, specifically whether 80 percent of Haanstra's Super performance was sufficient. Dezell handled that as well.

Chuck Branscomb was also from the South. He had moved from Georgia to Poughkeepsie in what was supposed to be a short-term assignment. It lasted 33 years. He went to Frank Cary and asked for an assignment with the new division. His request was typically low-key, telling Frank "GSD is doing some interesting things headquartered down in Atlanta. And I don't know if they're interested in me getting involved, but I would be receptive to going down to Atlanta.[2]" He got a call from Jack Rogers the next day, who tapped him as vice president of Manufacturing and Development.

The General Systems Division opened for business in 1969. It was a humble operation in the beginning. Rogers cobbled office space across multiple locations in North Atlanta. It would be a number of years before the division had a proper headquarters. In the field, Cary transferred 52 Data Processing Division branch offices to the new division. Jack Rogers would comment, "We were on thin ice for a while," meaning it was questionable whether such a key division could survive outside the core New York area[3].

The IBM branch office in Miami opened in 1940[4]. With Tom Watson's 1956 reorganization of IBM, the branch became part of the new Data Processing Division. Then, with the formation of the General Systems Division (GSD), a new "metro" branch opened at 2125 Biscayne Boulevard, just north of downtown. It was thankfully separate from the "big iron" office located across town. Your author joined IBM in 1973, working out of this Biscayne office. The early days and weeks were all about getting to know the branch, tagging along on sales calls, and working through study materials to prepare for the first marketing training class in Endicott. I already mentioned the quick wardrobe change to a power suit, buttoned-down shirt, striped tie, and wingtips.

Dave Ziska headed the Miami branch. He had made his mark by selling large systems to the utilities sector. There were roughly 100 IBMers in the branch, not counting the service engineers. Ziska organized his office into three marketing units, three systems engineering units, and an administrative team. Each marketing manager had eight to twelve sales representatives. The corresponding systems engineering units had roughly the same staffing. I was fortunate to work for an outstanding systems engineering (SE) manager in Shirley Kappeler.

The big iron branch office had perhaps two hundred large named accounts, with examples like Ryder Systems and Eastern Airlines among them. The GSD branch had everything else. This came to perhaps 100,000 establishments in a territory that ranged from Daytona Beach to Key West and from the Gold Coast to the Everglades. At the outset, the only existing customers were 1401 accounts and small unit record installations. Initially, the branch had just the one product to sell, the new System/3.

The sales representatives ran on an annual quota of 10,000 to 20,000 points depending on their experience. A point represented a dollar of monthly rent. The typical System/3 configuration totaled between 1,000 and 2,500 points. If a salesperson replaced machines in an existing account, IBM netted the point values for commissions. A sales rep achieving his or her quota would typically earn 150 percent of their stated salary. The systems engineers were on full salary. The main recognition for sales, aside from earnings, continued to be the 100% Club. The top achieving 10 percent of the sales force qualified for the Golden Circle, a far more lavish event that would include the representative's spouse. The Eagle award, given for unique and stellar performance, joined the Hundred Percent Club and Golden Circle for sales recognition. Its winners represented perhaps the top one percent of the sales force. The best-performing

Systems engineers, selected by the branch SE managers, went to their own event, the Achievement Conference.

To keep the focus on annual quotas, the branch had monthly "all-hands" meetings. They featured a review of progress toward the branch targets, recognition of individual achievements, and liberal amounts of "rah-rah." The annual kick-off meeting in January was far more elaborate and took place at an off-site hotel.

As a trainee in 1973, I was oblivious to the standards and culture that Watson Sr. had created. Now that I am educated on the relevant history, I can comment on the IBM I encountered that first year. The dress was very similar, though not as stiffly formal, and with the occasional blue shirt. The standard IBM "uniform" could vary given the Cuban influence in Miami. For many local businesses, the common attire was the guayabera shirt, open-necked, not tucked in, and very casual. The sales structure was nearly identical, with defined territories and the point system. The systems engineering role had expanded as the company transitioned from punched cards to computers and installations became more complex.

There remained a few "Think" signs scattered around the office, but no pictures of the Watsons, and no singing whatsoever. By the time I joined the company, both Watson Jr. and Learson had retired and Frank Cary was the CEO. There were still family dinners with CEO attendance, but given how big the company had grown, they were infrequent in any one location. The three Watson Basic Beliefs—respect for the individual, customer service, and excellence in all endeavors—remained. IBMers held them in high regard. One had to work hard to be fired, although the risks increased as you ascended the corporate ladder. The age of mass layoffs was still 20 years in the future.

The idyllic life of our sequestered GSD branch on Biscayne Boulevard ended in 1975 when all of the Miami sales branches consolidated into a new high-rise office building in Coral Gables. The GSD office was located on the 10th floor and the "big iron" branch was on the 11th floor, representing a fitting metaphor for the relationship.

After the first month in the branch office, it was off to marketing training at the IBM School in Endicott. The syllabus in Endicott was pure small systems. The scope of the investment by IBM in its new hires was impressive. The complete cycle took over a year. Sales and SE trainees attended the same classes. In fact, many branches did not decide which way a new hire would go until they had completed the full series.

The program consisted of three 4-6 week classes interspersed with time back in the branch office. A two-week sales school at the end finished the program. Those who became systems engineers then had an array of technical education to get up to speed in their focus areas. There was a priority in keeping the classes together through the entire program, which enhanced camaraderie and made for lifelong friends. The instructors were typically marketing and systems engineering managers fresh from the field who hoped to move up to the next level.

My class stayed at the North Street Motor Lodge for the first class. It was certainly not the Hilton but was just down the street from the Art Deco Schoolhouse that Watson Sr. had erected in 1933. The focus of the first class was IBM background and culture, computer fundamentals, and basic selling techniques. The instructors taught the standard sales call process by using a pencil as your product. You left each class with the "Endicott suitcase," a hefty cardboard box to take home all of the class materials.

After several months back in the branch office that included working on prerequisites, you returned for the second class. Now, the content was purely technical and included RPG programming, the Application Customizer Service, and even a lecture on wiring plug-boards. Thankfully, no actual wiring was required. After another stint back in the branch, the third class focused on typical customer scenarios and far more sales call role-play. A nicer hotel awaited those that made it to the third class. My third class took place in July 1973, at the same time that a Woodstock-like rock festival was held at Watkins Glen, a Grand Prix racetrack 70 miles northwest of Endicott. A number of us decided to go. There were enormous traffic jams that forced many, including us, to abandon our cars and walk to the venue. The Allman Brothers, the Grateful Dead, and many other bands provided a nice weekend break from the rigors of IBM training.

An essential technical class in IBM training was the "Ideal Milk Bucket" school, formally known as Manufacturing and Distribution

Applications. It was a two-week course hosted at the Poughkeepsie site. The case studies used in the class revolved around the fictitious Ideal Milk Bucket Company. The fact that Poughkeepsie was hosting this training was no surprise, at least to someone familiar with company history. Back in 1955, Poughkeepsie installed one of the early 702 systems to run its plant operations. It processed the payroll for 8,000 employees in 24 minutes flat. More critical were the manufacturing applications, which included management of a 40,000-parts inventory and advanced production control[5].

After a full year of training and branch work, it was time for the final class, Sales School. This was a two-week finishing program, with a heavy focus on practice sales calls. There had been a significant drop-off in the class composition from the first class. A number did not make it this far, and others switched to different schedules. The North Street Motor Lodge was ancient history by this time. For this penultimate class, we stayed at Watson's IBM Homestead. It featured a complete dining facility for breakfast and dinner. In 1973, the kitchen and wait staff were IBM employees, and many had been there for years. The rumor that circulated among the class was that quite a few of the cooks and servers were actually millionaires due to their long service and the IBM stock purchase plan. This was before IBM hit hard times in the 1990s and contractors assumed many such roles within the company.

Staying at the Homestead was a real treat after completing the lengthy training program. In earlier times, one of the Watsons might drop in on the last day of sales school to congratulate the graduates and extol the opportunities that lay ahead. Alas, both were gone. We took our diplomas, packed up our Endicott suitcases, and headed home. My class was the last group of GSD hires to go through Endicott training before a GSD education center got underway in Atlanta. It lacked the history of Endicott and certainly the luxury of the Homestead.

There was a standard IBM joke on the respective roles of the marketing representative and systems engineer. The salesman would go hunting for a computer prospect. He would burst back into the IBM offices with a large tiger nipping at his heels. He would breathlessly yell to the systems engineer, "You skin this one, I'll go find another." Well before Sales School, it was clear that the engineering role was a good fit for me, and I would be skinning more than a few tigers.

During the training program, I had already spent many hours in the Miami branch data center working on the System/3. Learning RPG was a key focus. At the time, I was not aware that my colleagues in the "big iron" branch did not consider RPG a real programming language. It was the Rodney Dangerfield of programming languages. However, it was the perfect fit for GSD's small and intermediate businesses. It would not be long before thousands of software companies would create applications in RPG, and a very high percentage of the Fortune 1000 corporations would use Rochester systems and RPG. Even Microsoft would run its company for many years on IBM small systems and RPG.

My own experience with RPG did not have a very auspicious beginning. An early project involved developing an RPG program on the branch office System/3. A slight complication was the lack of a disk drive. This early configuration was card only. The task seemed straightforward. I coded the program by filling out the RPG specification sheets and then punched the program source into the new 96-column cards. It was now time to load the RPG compiler. It was contained in two boxes of cards, with 2,000 cards per box. The first box processed your source deck, checking for errors. If it ran cleanly, then the second box did the actual compile and punched out the object deck. If these steps went well, you added the application's transaction cards after the object deck and pressed the blue start button on the Model 10.

Well, it did not go all that well. Two thousand small cards in a box are somewhat unwieldy. I had the not-so-rare experience of dropping one of the boxes, spilling the compiler cards on the data center floor. The next step had long-since acquired a name, the "floor sort." Each card had a sequence number, so you could put the deck back together in proper sequence. Even better, you could assemble the cards off the floor and run them through Rochester's high-tech sorter. If you acted quickly, no one would be the wiser about your gaffe. I was not that lucky. The entire episode unfolded before a busy data center. Of course, the rookie mistake would later become a badge of honor.

Once I recovered from the dropped compiler and floor sort, I proceeded to run the RPG compiler until I had a clean compilation. This was, of course, just the first milestone. You had to run the program with data cards until it produced the correct results. This involved the iterative process known as "debugging," thanks to Grace Hopper. Debugging could require running the entire compile and test processes any number of times. In many cases, the errors produced with testing were obvious. You modified the source program, recompiled, and ran the program again.

More challenging errors sometimes required a "core dump," a printed listing of the program contents in the system's main memory as it was running. The core dump printed the binary information in hexadecimal format (Base 16) for "easier" reading. Surprisingly, one got very good at deciphering hex after a short while. The alternative was reading long strings of binary ones and zeros. The name "core dump" originated with magnetic core memory, which was nearing the end of the line as a technology. The term would live on in IBM as it came to mean a complete brain dump, where someone would rattle off absolutely everything they knew about a specific subject, whether you wanted to hear all of it or not. The term was not to be confused with "opening the kimono," where you revealed what you knew about something that was confidential. I was able to get past the debugging and the core dumps and get my early RPG programs working. In any case, the most important takeaway from the early RPG experience was that System/3 customers absolutely had to have a disk drive.

I had spent many hours in the Miami data center. At the same time, I worked with System/3 customers many months before the training program was complete. I had two Application Customizer Service accounts. Working with ACS was great for the customer and even better for one's business and programming skills. I was able to plan the flow of the application with the customer and work with them on the installation. Even better, it was live training in RPG. Both ACS accounts required minor changes. I was able to look at professionally created RPG source code and figure out how to make the desired modifications.

The Fontainebleau Hotel, on Miami Beach at Collins Avenue and 44th Street, was one of my early accounts. This was definitely a case of a raw beginner under fire. The hotel's Model 10 system was in an office off the main lobby and used for the hotel's back office applications. I should add that just two years prior to walking into this luxurious hotel in an IBM suit and tie to support its new computer installation, I had been outside the hotel working a jackhammer in the median as part of a sewer replacement project. It was a very tough summer job. The hotel computer work was a breeze compared to that.

# THE FIRST PERSONAL COMPUTER

I joined IBM at the tail end of the Model 10 program. The next design in the System/3 line would go even smaller. For this new project, Rochester would take advantage of its sister plant and lab in Boca Raton, Florida. In 1967, IBM leased several buildings in Boca and began manufacturing the System/360 Model 20. Two years later, the site became part of the General Systems Division, with the mission of small systems development and manufacturing. The new facility worked with Rochester on the Model 10.

Tom Watson moved quickly to build another uniquely IBM site, handing the task once again to his head of industrial design, Eliot Noyes. Noyes had met architect Marcel Breuer at Harvard in 1937 and worked for him early in his architectural career. Breuer designed the IBM facility in La Gaude, France, in 1961[1]. Noyes hired him for the Boca Raton project. The new plant and lab would sit on 550 acres, just west of Interstate 95 at Yamato Road. The completed complex consisted of two large double-Y shaped buildings situated around a central circular lake. Breuer used Y-shaped supporting columns to evoke trees, similar to the abstract rendering of trees at the IBM pavilion at the 1964 New York World's Fair. This layout served to concentrate core functions at the nexus of the building wings. The buildings were massively concrete in structure, an example of the

Brutalist school of architecture. It was, in fact, an "over the top" use of concrete. Of course, this being the hurricane alley that is South Florida, the design was functional, more than able to survive a Category 5 hurricane. Though everyone knew the facility as "Boca," the buildings actually sat in the neighboring town of Delray Beach.

Watson Jr. was at the dedication in March 1970, with 3,500 in attendance. The Boca facility would grow to four million square feet of space and peak at 10,000 employees during its later IBM PC days of glory. There was an agreed division of mission between Rochester and Boca. Boca would focus on small scientific computers and Rochester would work on commercial ones.

The Model 10 introduced the very compact layout of the three interconnected system components. The layout of the new system, the Model 6, would go much further, combining all the elements into a single large desk. The Boca team eliminated the MFCU card device and added an integrated keyboard and display. A contingent of Jonie Dayger's 1403 printer team transferred from Endicott to Boca and developed a small dot matrix printer for the system. The Model 6 retained the System/3 architecture, disk drives, and core memory.

Since Boca had the small-systems scientific mandate, they adapted the machine for limited scientific applications. The Model 6 would acquire BASIC programming for this role. Boca tracked down a young engineer with the right credentials for this work. He would be in the center of IBM's small-systems success in the coming years. His name was Glenn Henry[2].

Henry grew up in Berkeley, California, and took classes at the university during high school. This included a field trip that featured a look at Cal's newly installed IBM 650. It was Henry's first encounter with a computer. He graduated from high school at age 15 and went on to Cal in 1958. He dropped out after three years and became a commercial helicopter pilot for two years. He developed eye problems that forced an end to his flying career. He went to work for Shell Oil in 1963. The company had just installed an IBM 7040 computer, the bargain-basement version of the high-end 7090. They drafted Henry to program it, writing in FORTRAN and assembler. His manager at Shell was a great mentor and encouraged him to return to school and get a degree[3].

He decided to attend Cal State East Bay, not Berkeley. Cal State had an IBM 1620 computer, a small scientific system, which he programmed in assembler[4]. After he completed his degree require- ments in mathematics, the department head suggested that Henry continue with a research program that would lead to a master's degree with just one additional year of study. His research work led him to a local company (Fiberboard Corporation) that needed computing help. Henry taught the company's engineers BASIC programming and then worked on their milk carton manufacturing process, which was dealing with the problem of moisture accumulation. He devised a BASIC program to produce the needed analysis, but the company really needed a real-time process computer to run the program.

This led to discussions with IBM, which recommended its current scientific computer, the San Jose-developed 1130. The 1130 shared 80-column card devices with the System/360 as well as SLT logic technology. It was very successful, with volumes approaching 10,000 systems. IBM was impressed with Henry's work at Fiberboard. They were working on a replacement for the 1130 and asked Henry if he was interested in joining IBM to work on the project. IBM actually offered him a choice between the System/360 programming team, working with thousands of other programmers, or working in San Jose on the operating system for the planned 1130 replacement, the 1800. Henry chose the latter and started with IBM in 1967. He quickly became the lead programmer on the project, in charge of designing the operating system.

Shortly thereafter, San Jose decided it would not develop another operating system after all. Instead, it would adopt the System/3 archi- tecture from Rochester. Don Castella traveled from Rochester to San Jose to brief Henry and his team on the System/3. Henry saw a simple architecture focused on the batch process of read a card, do some calculations, and print some output. This was quite far afield from the needs of a scientific computer. However, that was not the only change in the plan. Since Boca Raton had the small-systems scientific mission and was already familiar with the System/3, the work to add scientific function moved from San Jose to Boca. Newly hired engineer Glenn Henry would not be working on the 1800, or even working in San Jose. He moved to Boca Raton in early 1969, becoming a first-line manager after only two years with the company, in charge of a team to develop BASIC on the upcoming System/3 Model 6.

Henry and his team took a closer look at the System/3 architecture and did not like what they saw. Unlike the System/360, Rochester did not design the system with both commercial and scientific functionality in mind. The System/3 processor lacked floating-point, multiply and divide instructions, and shift instructions. In fact, there were just 28 instructions. It was limited to 16-bit addresses, with an eight-bit-wide data path, and no virtual memory support. Without physical registers, the machine was ploddingly slow. On top of all that, there was only 8K of memory available to work in the new BASIC interpreter.

Their options were moaning about the limitations or taking it on as a challenge. They decided on the latter approach. The team developed a complete virtual memory system, tapping into the disk drive to provide fully 64K of usable memory, more than enough to run BASIC programs on such a small, interactive system. Henry personally wrote the virtual memory subsystem. It turned out that the System/3 architecture with the Henry team enhancements worked well. They also included a "disk calculator mode" where one could run calculator functions from the keyboard interactively, without a BASIC program. That worked fine, though it was a very expensive calculator. For his part, Glenn Henry learned a lot about small systems, architecture, and virtual memory that would serve him well later.

This second System/3, the Model 6, arrived in 1970. It was one of the last IBM computers to use core memory. The Model 6 used the same 5440 Hursley-designed disk drives as the Model 10, providing ten megabytes across four platters and two drawers. As was the case with the Model 10, the real keys to success for the Model 6 were the same as the Model 10—RPG, applications, and ease of use. The Application Customizer Service was still available. Glenn Henry would call the Model 6 running BASIC the "real first IBM Personal Computer.[5]" A system that weighed 1,300 pounds, required 220 volts of power, and cost in excess of $50,000 would belie this claim.

**Figure 45**. System /3 Model 6, the single-user, interactive model of System/3 architecture. Highly successful, it would be Glenn Henry's inspiration for the System/32. Reprint Courtesy of IBM Corporation ©.

The new System/3 model was a great success. In stark contrast, the quasi-scientific BASIC version of the system was a flop. Glenn Henry surely felt he had made the wrong career choice in coming to Boca Raton, that his fast track at IBM was over. He might even find himself in an IBM "penalty box." However, his BASIC failure was nowhere near the worst of BASIC disasters for IBM. The highest it rated on the disaster scale would be three. The top two disasters, the theft of BASIC from its inventors and the use of BASIC in Microsoft's rise to dominance both involved Bill Gates. Rather than a total disaster, Henry's work on the Model 6 program was instrumental in laying the foundation for enormous successes that would follow.

The Model 6 was a welcome addition for the Miami branch office and kept me busy in my early years with IBM. One of my first accounts was Sydney Bag and Paper. Located in northwest Miami, the company manufactured custom paper products. Its highest-profile product was the small paper bag for McDonald's French fries. Sydney Bag used the Application Customizer Service to generate RPG programs for its back office accounting, replacing NCR bookkeeping machines. Besides general customer technical support, I would be called on to make minor changes in the company's RPG programs.

Sydney Bag was an important account for another reason. The back office staff responsible for the Model 6 were all women, and mostly older women. After several visits, one of them suggested that I might consider asking the young receptionist at the front desk out. That thought had not crossed my mind, but the fact that the receptionist was very attractive and personable certainly had. Long story short, I asked her out for a "low intensity" date, a bike ride down to Coconut Grove. She accepted, and it was the start of a beautiful friendship, well quite a bit more than that, as we shall soon see. In the meantime, Sydney Bag became one of my favorite customers.

# GENERAL SYSTEMS GROWS UP

I BM announced the System/360 as a comprehensive family of computers spanning a sweeping performance range. Even so, it was several years before all the models were up and running. Rochester did not have the resources to emulate the System/360 approach and roll out multiple System/3 models at once. Nevertheless, they had a strategy in place to build on the System/3 architecture and expand the breath of the offering. Going lower with the Model 6 was just the first step.

Bob Evans was leery of the small computers coming out of the new division. He predicted the new customers of the GSD would outgrow their original systems and pressure would mount for larger System/3 models, with more and faster memories, larger disks, and faster printers. This came as no surprise as the ease to develop and deploy applications on the platform soon led to calls for increased performance. A major additional driver was the increase in online applications where interactive support of many terminals required substantially more computing power.

System/3 customers facing a performance shortfall found an amplified voice in a user organization called COMMON. The organization began in Chicago in 1960 for IBM 1401 and 1620 users. Supposedly, the name "COMMON" came from a "common area" on the 1620 scientific computer that enabled FORTRAN programs to communicate with each other[1]. The COMMON organizers patterned the group after SHARE, the large computer user group. COMMON rapidly expanded with the arrival of the System/3. Like SHARE, it became a force for the customers to exchange ideas, attend training, and, most importantly, influence IBM.

The COMMON meetings featured a broad array of training classes, sometimes over 1,000 in number. It also had a main tent exposition hall that was wall-to-wall with IBM and business partner booths. COMMON meetings lasted a week and featured a feedback session on the last day. Here, IBM executives received pleas for future enhancements. Attendees jostled to grab the open mike and make their pitch. A common refrain was the need to grow the architecture to keep pace with expanding applications. There was always a feeling that big-iron corporate IBM would artificially cap the growth of Rochester systems. COMMON was also an opportunity for networking. After a full day of sessions, it was time for CUDS—Customer User Discussion Session—which was simply a happy hour. SHARE had a similar session irreverently called SCIDS, the SHARE Committee for Imbibers, Drinkers, and Sots.[2]

Speaking of growth, the next System/3 was the high-performance Model 15. It would not use core memory. Magnetic core had a good run. It was reliable, and IBM automation reduced the cost per core bit down to about a penny. IBM core production peaked at 20 billion bits in 1970. Several System/370 models and the System/3 Model 6 were the last to use core. The new memory technology used capacitors, though it represented a significant reimagining of the device. John Atanasoff had used capacitors for his rotating drum memory. In a similar vein, Bletchley Park produced a specialized code-breaking machine called Aquarius that used a large bank of capacitors to store a series of message characters against which they applied various tests and cribs. The central issue with capacitors was the rapid deterioration of the charge state. They needed to be continually refreshed, as Atanasoff and Berry did with every rotation of their drum.

In 1966, Robert Dennard, working at the Watson Research Center in Yorktown Heights, developed a memory cell using a capacitor to store the bit and a transistor to control the charge. His approach reduced the requirement for a storage bit to one capacitor and one transistor. His memory cell was more widely known as Dynamic Random Access Memory, or DRAM[3]. Intel perfected the first commercially successful DRAM, the Intel 1103, in 1970. The new memory provided increased performance, used less power, took less space, and rapidly eclipsed the pricing of magnetic cores. The most common DRAM format was a 64K chip. Japanese manufacturers soon came to dominate the industry, as U.S. manufacturers like Intel opted for higher-margin microprocessor products. IBM was a technology leader with DRAM memories and was the first to produce one-megabit DRAM chips in volume. Core memory lived on in NASA Space Shuttle computers until 1990. Core memory retained its bit status even after the loss of power, an advantage that NASA relied on.

The Model 15, as well as the low-end System/370 Models 115 and 125, used the new MOSFET DRAM modules for main memory. MOSFET was reputedly the most-copied manufactured device ever, with an estimated 13 sextillion MOSFET units produced. A sextillion is represented by a one followed by 21 zeros. The MOSFET success was due to its essential characteristics: It was denser, used less power, was easier to mass-produce, and less costly. It was also a key to the continued progress of Moore's law, which projected transistor density would double every two years, as well as Dennard's Law of Scaling, which stated that power consumption drops proportionately with increased density. The actual manufacturing process used for MOSFET transistors on the Model 15 was CMOS (Complementary Metal Oxide Semiconductor), a cost-effective process that would later have dire implications for IBM's mainframes.

Besides larger and faster memories, the Model 15 had to solve another problem, the architecture's 16-bit addresses that limited total addressable memory to 64K. The Model 15 would end up supporting up to 512K bytes of memory. The problem was resolved in a manner much like the System/360, with the addition of an address translation table. This allowed the swapping in and out of 64K pages between main memory and disk. This required more disk capacity. A revised version of San Jose's ADF drive became the 5445 drive. Ken Haughton's Winchester disk drive (IBM 3340) provided even more capacity. The venerable 1403 printer was still around, but it now had a much larger, faster model that would print at up to 2,000 lines per minute. It was an impressive machine, fully enclosed, with motorized access at the push of a button.

Main memory remained a challenge. Rochester endured the same challenges as Poughkeepsie had with OS/360. The operating system continued to take up more and more space, growing from 3K to 20K. File sharing added 4K, spooling added 14K, and the new Communications Control Program (CCP) took up 26K. Thus, a customer lost 76K of main memory before loading a single application. As a consequence, online programs needed to be as small and efficient as possible. It also helped to make each program reentrant, allowing one copy to be shared by multiple terminal users. All of this required some forethought and creativity.

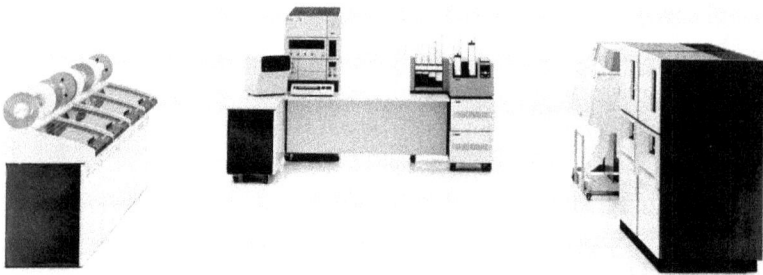

**Figure 46.** System/3 Model 15 (1975), the high-end model that moved the new architecture into interactive applications. Tape drives on left, the venerable IBM 1403 printer and IBM 5445 Winchester "pizza oven" disk drives on the right. Reprint Courtesy of IBM Corporation ©.

The Model 15 debuted in July 1973. An early installation at Caressa Shoe in Northwest Miami was one of my more complex projects. The customer was a major importer and distributor of women's fashion shoes from Italy. Leonard Taicher was the CEO. They purchased a maxed-out Model 15 configuration to handle online inventory, sales, and back office functions. I did not know at the beginning, but this customer would soon rise to the top of my list for sheer drama. For openers, it was located in a very tough area of town. A third-party programming team, led by Barry Greibel and Frank

Smotherman, handled the bulk of the applications. I had worked with both of them. Taicher wanted to have the online inventory application ready for the upcoming fashion season. This task fell to IBM and, in turn, to me. The timing on this systems engineering work requires a digression on IBM bundling of services and the Justice Department's ongoing antitrust actions.

## MORE ANTITRUST

Fred Brooks signaled early in System/360 development that the innovation and complexity of IBM systems would require far more SE technical resources. By 1969, the company had concluded that it could not continue to absorb those costs bundled as part of the hardware and software price. The U.S. Justice Department was of the same opinion. On the final day of the Johnson administration in January 1969, the Justice Department filed an antitrust suit against IBM. Its investigation started in 1967 and IBM had met voluntarily with the Justice Department over the ensuing two years. Tom Watson went to Washington to meet with Lyndon Johnson's attorney general, Ramsey Clark. There was a striking parallel to the meeting: Tom had traveled to Washington in 1951 with his father to meet with the then U.S. attorney general on the same subject. That attorney general happened to be Ramsey's father, Tom Clark[4].

Though IBM's market share was certainly over 70 percent, the issue was the same that IBM had faced in the 1930s and again in the 1950s—IBM's tie-in practice of bundling hardware, software, services, maintenance, and education into one bottom-line price. This had been the very successful sales approach since the days of Hollerith. Though Tom convinced his father to accept an antitrust resolution in 1956, he reacted just as his father did with the new lawsuit. He resented the action and geared up to fight it. He felt that IBM resolved the principal issues involved with the 1956 agreement. He was already incensed about plug-compatible manufacturers. He called them "parasites" and chafed that IBM had to coddle them[5]. The company had to tiptoe warily on product plans, announcements, and pricing so as not to appear to be driving them out of business.

The Justice Department's new goal was to break up IBM into smaller companies, possibly as many as seven billion dollar entities. Tom considered this untenable, though he went ahead and planned for a scenario where he split IBM two companies, one large and one small. In actuality, competitors actually feared IBM being broken up, contending that they would then have multiple IBMs to deal with.

Besides the Justice Department, an organization called the Computer Industry Association (CIA) also lobbied to restrict or even punish IBM, starting with breaking the company into pieces. With the ultimate chutzpah, the CIA wanted IBM to divulge the specifications for new products with sufficient lead-time so that the PCM manufacturers would be ready with their clones on IBM announcement day. To back up its claims of IBM malfeasance and predation, the CIA cited the 1971 exit of RCA from the computer business. In rebuttal, a *Datamation* survey of computer customers at that time showed poor marks for RCA while IBM received the top marks across all categories except price/performance[6]. This seemed to belie the CIA message that IBM was the cause of RCA's demise.

In the meantime, Tom geared up to defend the company. His focus hit an immediate roadblock when his lead general counsel, Burke Marshall, weighed in. Marshall, who had been the assistant attorney general under Bobby Kennedy, told Watson that IBM's practice of bundling was the very definition of a "tie-in sale," a practice expressly prohibited by the Sherman Antitrust Act.

Then, there was the lawsuit from the Control Data Corporation. CDC sued IBM in 1968 over the System/360 Model 92 announcement, arguing that the Model 92 was a "paper" machine used to freeze the market while IBM scrambled to build a competitive supercomputer. Bill Norris of CDC further complained that the IBM action forced him to reduce prices on his CDC 6600 and CDC 7600 machines in order to combat the IBM disinformation. Tom Watson considered "paper machines" a standard practice in the computer industry. However, the practice took on a different meaning when you were as big and dominant as IBM was. Watson as much as admitted the "paper machine" process when he ended his famous janitor memo by concluding, "We had embarrassed ourselves by announcing four different versions of our supercomputer and still not delivering a one[7]."

IBM believed that the Justice Department would struggle with an antitrust action as complex as the one IBM presented. The department could only afford to have a staff of perhaps 20 assigned to the case. The situation changed dramatically, however, when CDC sued. CDC had the resources to build a giant, incriminating database through the legal mechanism of discovery. And, much as Honeywell had done years before in fighting the Sperry UNIVAC patent lawsuit, CDC moved forward aggressively with discovery, compiling and indexing upward of 40 million IBM documents. The existence of this database was a problem for IBM, not only if the CDC suit went to trial but also if the Justice Department was able to subpoena this information.

With the antitrust exposure outlined by his legal team and the CDC cloud hanging over the company, Tom decided to look at a couple of preemptive actions. The first focus was on bundling. At the time, only two competitors charged separately for their software (Burroughs and Xerox). Watson created the New World task force to look at all aspects of bundling and make recommendations on how to unbundle software. He wanted to know which products to select, at what prices, and what the resulting impact would be on hardware sales. The task force started in December 1968, working at the Data Processing Division headquarters at 1133 Westchester Avenue in White Plains. Eventually, over 100 people would be working on New World[8].

Burton Grad was responsible for addressing application software. He had graduated in engineering and physics from Rensselaer Polytechnic Institute and joined General Electric in 1949, working on punched-card equipment. He moved to GE's new plant at Appliance Park in Louisville, where he coded manufacturing applications on their new UNIVAC I computer. He could not convince GE to continue his this work, so he left in 1960 and joined IBM. With experience at both companies, he would later comment on the quality of people he encountered. Though both companies had strong training programs, Grad concluded that while GE hired smart people, IBM was able to hire the "cream of the crop[9]."

Grad's New World team evaluated 17 key IBM program products, estimating their market potential and setting prices. CICS (Customer Information Control System) and the IMS (Information Management System) database software ended up the big winners. The sales estimate for CICS was 100 licenses. The final total was closer to 15,000. The story was the same for IMS. The New World team recommended that operating software continue to be bundled with the hardware. IBM had done this since 1956 even though it technically violated the terms of that decree.

This unbundling of software naturally led to a greater focus on software as a business. Grad's team recommended splitting IBM software into large and small customer sets. The large customer focus was financial, insurance, communications, utilities, and transportation. Though the company had been successful with SABRE, this did not always translate into success with other applications. At the time, large customers would simply not buy critical applications code from a software company, even IBM. They would develop it in-house. IBM did not grasp that this perspective would soon change and the market for application software would explode. IBM cut staffing in this emerging area from 400-500 down to 100. This left the door open for companies like SAP, Oracle, J.D. Edwards, and PeopleSoft to grow and prosper. The story for small customer software was more promising. Cary moved the mission to Atlanta and the General Systems Division where it thrived, at least for a time.

The New World assessment of Systems Engineering Services represented a different story. The systems engineers in the field had a dual role, sales support and installation support. Many system engineers did no programming at all but assisted in marketing and provided guidance after the sale. Most system engineers had never written a contract or billed for their services. Unbundling required the separation of SE activities into billable and nonbillable. Though it would take time to work through this transition, the billable content in time would grow into a significant segment for IBM. With the New World work complete, IBM announced the change on June 23, 1969.

At the same time, there was a 3 percent decrease in hardware prices to reflect the removal of the bundled software and services content.

Starting in 1970, IBM began providing pretrial documents to the Justice Department, which advised IBM that it also planned to use the massive trove of catalogued documents held by Control Data. The trial finally began in 1975. The government took three years to present its case. During this time, its definition of the market that IBM monopolized changed no less than five times. IBM asked for a mistrial but the trial judge denied the motion. IBM did not call its first witness until 1978. Watson had retired in 1971, and new CEO Frank Cary spent many days giving depositions and providing direct testimony on the stand. He spent far more additional time in discussions on contingencies if the case was lost.

Tom Watson, in his new role as chairman of the IBM board of directors, worked with Cary on the CDC lawsuit. The CDC database on IBM represented a clear and present danger to the company. Tom made clear that he intended to trade an agreement with CDC for the database. In addition, it was his intention to destroy it immediately. Tom queried and re-queried his legal team on any exposure this might create. They agreed there was none except for some public griping. IBM and CDC settled in 1973. IBM would sell its Service Bureau Corporation subsidiary to CDC for $16 million and stay out of the services business for six years. IBM also paid CDC $60 million, representing the damage done by the Model 92 "paper machine." IBM estimated cost of settling with CDC at $115 million. CDC provided the discovery database and Watson had it immediately destroyed.

With the loss of the Control Data files, the Justice Department requested a huge range of additional information. IBM estimated that this would require "five billion pages of material from 2,000 locations throughout the world, entailing almost sixty-two man-years of effort at a cost to IBM of more than $1 billion.[10]" Pleading the unreasonableness of the request, IBM asked for removal of the judge for bias. The appeals court denied the motion but quashed the subpoena asking for the additional documents.

The entry of the Reagan administration in 1981 represented a sea change in the government's view of monopolies and the viability of the Justice Department's IBM action. In January 1982, William Baxter, the head of the Antitrust Division, issued a joint notice signed by IBM's legal team that declared the Department "has concluded that the case is without merit and should be dismissed.[11]" Just short of 13 years "with 2,500 depositions and 66 million pages of documents[12]," the longest antitrust trial in history was over. The discovery process and trial cost hundreds of millions of dollars, all for nothing. However, it also brought significant changes at IBM, a more cautious approach to doing business and a toning down of the aggressiveness that had served the company so well. Robert Bork, who was the solicitor general during the Reagan administration, would call the IBM case "the Antitrust Division's Vietnam.[13]"

With the unbundling of IBM's Systems Engineering Services, we can return to Caressa Shoe's online inventory project. Caressa agreed to the application specifications at the end of January 1975 and signed a services contract. It was my first experience programming for a fee. Used to the old bundled days, the customer felt that the contract was more of a suggestion, and less a binding financial document. The inventory programming was a significant effort that soon stretched to two months, at a time when I was still holding down an extensive customer territory. The specifications constantly changed and the scope expanded.

In addition, production work was running on Caressa's Model 15, so most programming and testing occurred at night or on weekends. Barry Greibel of the third-party programming team was a very big guy, and given Caressa's location in Northwest Miami, he packed a shoulder-holstered gun under his suit jacket. The jacket came off for the night shift to ensure maximum intimidation. However, the real drama came from Greibel's team. One of his programmers was upset about her pay and decided to take the customer's main 3340 removable disk pack and leave the building. Caressa's entire business ran on that disk, so this was in effect a criminal action. Fortunately, cooler heads prevailed and the disgruntled employee returned the pack.

The moving target of the inventory application and the odd hours continued to be a challenge. After one frustrating test session, I left the plant in a hurry but stopped before getting into my car to check on some papers in my briefcase. I set the briefcase on the roof of my Ford Mustang, then absentmindedly got into the car and proceeded to drive off. I realized too late that the open briefcase was still on the roof. Then it was no longer on the roof but its contents were quite visible in the rear view mirror. The project dragged on with the finish line constantly redefined. Fortunately, I had a very good manager in Shirley Kappeler. She was tough as nails and negotiated an acceptable path to conclusion. The new application was complete and up and running in July 1975. The project was a mixed blessing. The Caressa environment was far beyond reasonable, but the experience in online program development was immensely valuable.

Within a month of the end of the Caressa project, I had another complex and demanding account. Like Caressa, they had a small IT staff and leaned heavily on IBM. Aero Systems was a broker of aviation parts located at Miami International Airport. It was another high-end Model 15 needing an online inventory application. Robert Holmes was Aero's president, his brother Ed Holmes was in charge of finance and data processing, and Earl Ashton was the data processing manager. Of this crew, Ed was the most demanding one. Their Model 15 had 256K of main memory but after the overhead components loaded, only 100K remained for online programs. From day one, we focused on memory usage and performance. As the online parts inquiry application took shape, Ed Holmes was particularly insistent about one feature: The list of selected parts had to display in part number sequence. As the file was not set up that way, the program required an on-the-fly sort. By this time, I was very comfortable with RPG, but this was one function where the language came up short. Fortunately, I worked with another systems engineer, Rema Summers, who had recently transferred from the IBM NASA team at Cape Kennedy, where she worked on the Apollo and Space Shuttle programs. She introduced me to the bubble sort, a variation of the sort John von Neumann suggested in 1945. It was also the process used by IBM's card-collator machines. I needed to retrieve all the relevant parts, throw them into two columns, and run the bubble sort process, all before displaying the list for the salesperson.

The programming contract at Aero Systems played out just like Caressa Shoe. There came a time for my no-nonsense manager to make a call on Ed Holmes, She patiently explained to Ed how the contract was going to end. Aero Systems ran into problems after I left[14]. In 1992, the company was convicted of selling fighter jet parts to Iran in the mid-1980s when there was an embargo in place. The resulting action forced Aero into bankruptcy. In 1994, Ed Holmes forced his older brother out and then sold the company.

A third Model 15 customer provides insight into the spread of applications within GSD. Variety Children's Hospital installed the System/3 Model 6. They hired Dave Pomerance, who had a CFO background, as a consultant to make recommendations on use of the new computer. Pomerance's next-door neighbor happened to be one of our sales representatives. He told Pomerance that IBM small computers were the "next big thing" and encouraged him to take an RPG course[15]. Pomerance was not one to just dip his toe in the water. He dove into programming. He developed back office applications for Variety in both RPG and assembler and soon had the hospital ordering a System/3 Model 15 to take its operations online.

Pomerance called the system he developed for Variety the Hospital and Patient Management System (HPMS) and it became an IBM Installed User Program (IUP). IBM sold the application to hundreds of hospitals. Pomerance soon formed his own company (Dynamic Control) and sold his RPG software package to over 200 additional hospitals. Over time, he would port the System/3 code to follow-on IBM small systems (System/34, System/36, System/38, and AS/400). He eventually sold his company to Baxter International in 1983, at a time when he had 450 employees and was doing $30 million in business annually.

## THE MEGA-BATCH MACHINE

With the Model 15, it appeared that GSD and Rochester had moved on to online systems, leaving batch systems behind. That was not quite the case. Two new technologies coming out of San Jose would pave the way for a new high-performance batch system. This would be the System/3 Model 12, which debuted in 1975. It would feature the 3340 disk drive as well as the new floppy diskette, leaving the 96-column punched card behind.

The breadth and success of the System/360 depended heavily on the ability to emulate prior systems. Microcode made this happen. With the System/370, the Mylar packets used for System/360 gave way to a simpler microcode storage medium. The System/370 designers wanted an inexpensive method distribute microcode updates. They also could use such a medium to ship program updates to customers. In San Jose, Alan Shugart assigned the project to a small team under David L. Noble. The code name for the program was "Minnow[16]." This was at a time when "Gilligan's Island" was a popular television series, with the SS *Minnow* the ill-fated ship. Much more likely, "minnow" was a small fish and the new medium was smaller version of the hard disk. The key was substituting an inexpensive ferromagnetic material for the rigid hard disk aluminum platters. Noble's team tried several approaches before designing a thin 8-inch flexible disk and added a sleeve for protection. They called it the floppy diskette. The first version, completed in 1971, held 80 kilobytes of data.

Shugart was gone by now, taking his product ideas and IBM engineers to Memorex. The floppy disk group had grown to 25 engineers, led by Jack Harker. He was in charge of some of IBM's most profitable products, including the 3330, 3340 Winchester, and 3850 Mass Storage System[17]. He was interested in expanding the floppy disk program but could not get funding. A chance meeting between Harker and Don Stephenson, the site general manager of Rochester, provided a path forward. Stephenson could provide the funds but he had far more ambitious plans. He wanted to develop a complete system around the floppy drive to replace punched cards. This was a concept that Shugart had been working on at Memorex but was never able to see through.

In Rochester, Lou Blenderman led the team expanding the floppy disk[18]. Engineers incorporated MOSFET transistor technology in the design of the new drive. Rotating at 360 RPM, the eight-inch diskette held 1,898 128-character records for a capacity of 240K. This was the equivalent of 3,000 80-column cards or 2,500 96-column cards. They inserted the new drive into a new data entry station, the 3741, that had a keyboard and CRT displaying six lines of 40 characters each. In quite a novel approach, the team also created a dual data entry station

(the 3742) by splitting the mirror and giving half to each of two operators facing each other. IBM released the diskette-based system as the
3740 family. The 3340 hard disks and 3740 floppy disk system, along
with new processor technology, enabled the original Model 10 to
morph into a batch production monster. Transaction data keyed into
the 3740 data stations transferred directly into the system. At a
purchase of roughly $125,000, the Model 12 represented quite a bit
of performance for the price.

**Figure 47.** System/3 Model 12 (1975), a cost-effective, high-
performance batch machine, with IBM 3340 Winchester disk drives
on the right and the new IBM 3740 key to diskette station in the
foreground. Reprint Courtesy of IBM Corporation ©.

Before going any further on the Model 12, permit a digression on
the receptionist who I met at Sydney Bag and Paper. I discovered that
she had graduated from the University of Massachusetts, just a short
road trip from my alma mater at Dartmouth. Her degree was in
teaching though she quickly recognized this was not her calling. She
hired on as a receptionist at Sydney Bag as a holding pattern until she

got her bearings. I suggested that she might consider a career in computers. She enrolled in COBOL classes at Miami-Dade Community College where she keypunched her programs onto 80-column IBM cards and submitted the decks to the school's System/360.

The next step was practical experience. I had a friend who was the data processing director of the Cuban Refugee Center in downtown Miami. It was another System/360 shop. The center, called the "Ellis Island of the South," processed incoming Cuban refugees. I connected Cynthia with Clifford Harpe, and he hired her as a keypunch operator. As a bonus, the center was located in the Miami Freedom Tower, just a mile south of the Biscayne Boulevard IBM office.

There remained another step. Get COBOL out of her system and move her to a real programming language. That would be RPG, of course. Shirley Kappeler had assumed the added responsibility for the IBM Datacenter and Education Center in the branch. She slipped Cynthia into several of the RPG classes. It was her first exposure to IBM and to IBM classes. The instructor, long-time IBMer Ralph D'Angelo, worked through stacks of foils before turning the students loose on the systems. Cynthia emerged as a card-carrying RPG II programmer.

Now, she needed a place to exercise her new skills. This turned out to be Southern Underwriters, an insurance company that sold reinsurance contracts to primary insurance companies in order to buffer them from large losses such as hurricane damage in South Florida. The company started in 1942 in Coral Gables. Ryder Systems acquired them in 1970. Southern's operations were a textbook case in insurance company computerization. As it grew, it acquired IBM punched-card machines, then upgraded to the 1401 in the early 1960s with a typical configuration comprised of the processor, card reader, and 1403 printer with the keypunches and a sorter/collator offline. The core application was reinsurance accounting. The data processing staff kept insurance company master cards and the associated transaction cards in tub files. The operators collated the cards and placed them behind the program deck for execution. All of the programs were in COBOL.

In 1975, Southern Underwriters installed one of the first

System/3 Model 12 systems in Miami. Our General Systems Division branch office had moved to a high-rise building in Coral Gables that was next door to Southern Underwriters. The account was not mine. Rema Summers, that sharp systems engineer who schooled me on the bubble sort, was the assigned systems engineer. Since I worked closely with her on several accounts, I might have let slip that I knew someone who programmed in RPG and could be a fit for Southern. When she learned that Cynthia was skilled and available, she discussed the possibility with Southern's CFO. The timing was such that Cynthia's skills and Southern's needs were a very good match.

Southern decided that porting its COBOL programs to the new system was not an attractive option. The new system would run far better in RPG. The company's unique reinsurance business model meant they were unlikely to find canned applications that fit their business. They also wanted to avoid bringing in the IT team from the mainframe shop at parent Ryder Systems. Southern decided to convert their tub files into disk files and write new programs. As Rema Summers had mentored me at Aero Systems, she jumped into working with Cynthia. It was not your normal tutoring approach. Summers would often work side by side on a knitting project then jump into a programming discussion. She was in her element whenever special programming requirements surfaced. Southern wanted to recap all of its division totals on the last page. This meant you had to step out of the RPG program cycle briefly, prime intermediate totals in arrays, and then execute a "do loop" at the end of the program to print the totals. Summers stepped Cynthia through the coding. The Rema method was the best possible apprenticeship for a new programmer.

# GLENN HENRY'S ENTRY SYSTEM

The Model 6 was a success, though the use of Henry's BASIC on the machine was not. Despite this, Glenn Henry's stock had still risen within IBM. He would publish a research paper on his team's design and implementation of virtual memory. Henry worked on a new system concept aimed at addressing the flaws he saw in the Model 6. His called the design "GSD Entry"[1]. He presented the plan to Hal Martin at GSD headquarters in Atlanta. At the conclusion of the pitch, Martin asked one key question: "Is this a commercial system?" Henry responded "yes." Martin said it would be developed in Rochester as Rochester owned commercial small systems. In addition, Martin wanted Henry to move to Rochester and manage the development of GSD Entry. Henry agreed, and after two years in Boca Raton, he was off to Rochester in late 1971. This was his fourth promotion in four years with IBM.

The General Systems Division focused on the small business computer segment, with purchase prices ranging from $10,000 to well over $100,000. By the late 1970s, there were about 80-90 companies selling such systems. With the low-end gap in the System/360 line and IBM's focus on high end, it was slow to respond to minicomputer makes like DEC, Wang, and Data General. These minicomputer companies had their own set of issues. They had limited or no experience in selling commercial solutions, had limited sales forces, and sold principally through VARs (Value-Added Resellers). Aside from basic operating systems, their systems came with limited software. The initial sweet spot for these companies had been scientific and engineering customers that did their own programming.

The System/3 continued to sell well. It would have over 30,000 installations worldwide by the time IBM announced GSD Entry. The division realized that the product needed continual expansion. The GSD sales force in the United States would grow to 4,500 sales reps in 67 branch offices by 1975, along with 3,000 systems engineers[2]. These sales people were hungry for a wider portfolio. If the price for the new system dipped below $1,000 per month, research said there were 500,000 businesses in the United States alone that could afford it.

Henry started his low-end System/3 mission in September 1971. He managed a team of more than fifty programmers as second line manager for the new system's software. He had a solid first-line manager in David Schleicher[3]. Schleicher grew up in Zumbrota, just 25 miles north of Rochester. After graduating from Mankato State, he joined IBM in 1964, working for Federal Systems tracking spy satellites. Schleicher put in a transfer request for Rochester and arrived in 1971. He worked on manufacturing control applications before joining Henry's team. He had fifteen programmers in his department. He would be Rochester's lab director 20 years later, managing a team of 2,700 programmers[4]. Quite a bit of computing history would transpire in between.

Upon arriving in Rochester, Henry formed a task force to define his GSD Entry concept in more detail. He tapped Frank Soltis, Dick Bains, and Roy Hoffman for this work. The high-level roadmap for Henry's system was a simple operator-oriented system that would extend the one-piece workplace that the Model 6 had pioneered. The system would retain the System/3 architecture but move it into the background with the use of microcode. The emulation in microcode would consume many processor cycles, but as a single-user system, performance was not a priority. The focus would be on the interactive user interface. Though a second-line manager, Henry decided to rewrite the disk operating system code microcode himself. He would comment that this was "a later theme of mine at IBM—moving stuff from high software levels into the hardware." He remembers writing this code at home while watching the 1972 Olympics on television[5].

Early on, the planned name for the new machine was System/3 Model 2. However, it ended up as the System/32. It represented a significant hardware upgrade from the Model 6. Core memory was history, replaced by the far denser Monolithic Systems Technology (MST) logic modules, each containing over 100 circuits. The entire processor resided on a single card. To meet cost objectives, the processor had to run the application program as well as manage input/output. There would be no outboard processor or channel. Henry addressed one of the major issues he encountered with the Model 6, the lack of hardware registers. He added eight 16-bit general-purpose registers that allowed for faster logic and arithmetic operations without using up main memory[6].

The Hursley lab produced a Winchester-style internal hard disk called Gulliver. It stored nearly 14 megabytes of data on a single 14-inch platter. It would be the first disk drive to ship more than 100,000 units. It used a simple low-cost arm actuator. The actuator design became a universal disk drive standard and was used on perhaps ten billion drives over the following 40 years. It was cited as the most-licensed electromechanical patent invention of all time[7]. Ken Haughton had a group from his Winchester team in San Jose travel to the Hursley lab to assist in the Gulliver program. The main input/output medium for the system was the 8-inch floppy disk. The System/32 also borrowed the display from the 3740 floppy disk machine, providing six 40-character lines through a mirrored inter-face. Boca Raton contributed built-in matrix and line printers.

Though the hardware changes were dramatic, the software drove the system's success. The 1969 unbundling of software and services had spawned a new focus on programming for the low end. With the demise of the Application Customizer Service, the Atlanta program-ming center moved ahead with Industry Application Programs (IAP). The IAP mission was established in 1971 as the Atlanta Application Development Center (AADC), under the direction of Raynor Moore.

The IAP concept depended on creating industry-specific applications with such comprehensive function that they could be adapted by most companies in that industry. The programs introduced more jargon. That included "BICARSA", for Billing, Inventory Control, Accounts Receivable, and Sales Analysis; and "GLAPPR", for General Ledger, Accounts Payable, and Payroll. The initial series of IAPs included the Construction Management Accounting System (CMAS), the Distribution Financial Accounting System (DFAS), the Distribution Management Accounting System (DMAS), and the Manufacturing Management Accounting System (MMAS). At the beginning, there were 14 IAPs, including a series of variant for individual segments such as wholesale food, electrical, plumbing, appliance dealers, tire companies, lumber, and sporting goods. Pricing was attractive. The Wholesale Food Distribution IAP was typical, selling for $3,120 plus $147 per month in support[8]. The programs were of course developed in RPG. Moore's staff in Atlanta grew dramatically and soon totaled more than 300.

GSD announced the System/32 on January 7, 1975. Like the Model 6 it replaced, it was a one-piece desk system. However, the similarity ended there as the new system was a far superior piece of technology. It quickly acquired an alternative name, the "bionic desk," a reference to the television series "The Six Million Dollar Man" that ran from 1973 to 1978. Within just over three years, the System/32 surpassed the System/3 record of 25,000 systems installed, and kept going. It was the most-installed IBM computer of the 1970s.

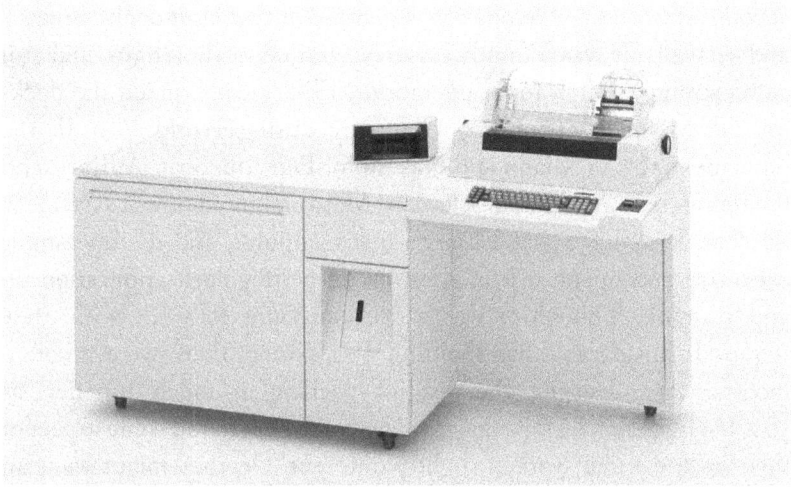

**Figure 48.** IBM System/32 (1975), the "bionic desk", a worthy successor to Glenn Henry's Model 6. Reprint Courtesy of IBM Corporation ©.

The system provided leading-edge technologies and a wide catalog of affordable application programs in a compact and non-intimidating package at a monthly rental price starting at $770 and purchase at $33,100. It was a good fit for the small- to medium-sized company looking for its first business computer. As with IBM's later entry into the PC market, the System/32 served to legitimize the small business market for computers.

It was also a shot across the bow to the minicomputer companies that IBM sought to address with the creation of the General Systems Division. DEC was first to respond to the new system—just ten days later—with the Datasystem 310. Wang followed with a similar offering. Both companies lacked IBM's sales-and-marketing firepower, and, more importantly, they lacked the applications that small businesses would need. The System/32 helped lift IBM's share of the minicomputer market to 37 percent by 1978[9].

Your author would have quite a few construction accounts using the System/32. Most were upgrading from manual accounting or bookkeeping machines. They would start with basic billing and financial applications, and then advance to revenue and labor costing as they became more sophisticated in running the system. My first account was Ebsary Foundation, a contractor that built docks, bridges, and seawalls. It was a family business, started in the 1930s, and still going strong through four generations. Ebsary constructed the docks for the Miami Shipbuilding Company during World War II, the seaplane ramps in Miami and Nassau for Pan American Airlines, and the foundations for Miami's elevated Metrorail rapid transit system.

The System/32 was Ebsary's first computer, and the operations executive and operator leaned on me in getting each application up and running. Though it was a great account to work with, they required considerable handholding. In addition, there were program modifications. Despite the goal of packing as much function as possible into each IAP package, many customers found some aspect of their business that worked slightly different. Or, they might want an additional total on one of the reports.

The IAP programs shipped in both RPG source and object formats. I did not have any IAP accounts that made the changes themselves. This was somewhat puzzling, as it seemed to be a golden career opportunity for a person selected as the System/32 operator. The desired modifications sometimes fell to me. In most cases, they were minor and I was happy to make the change. According to the guidelines of the 1969 software and services unbundling, any change required an assessment, the development of a System Engineering Scope of Effort contract, negotiation, agreement, and periodic progress reports before completion and billing. At Ebsary, it was far simpler to make the change and join them in a cup of coffee.

I had a large number of System/32 customers. Most were like Ebsary, a completely different breed from my high-end Model 15 accounts. They were making a huge change in their business and were happy to have an expert to ensure that it went smoothly. Several of these customers stand out for one reason or another. Bared and Cobo was an HVAC (heating, ventilation, and air conditioning) construction company. Like Ebsary, this was the company's first computer, and like Ebsary, they had a sharp operator with no interest in learning programming. They needed a substantial amount of training and support to get operational. At the time, the company was growing rapidly and as it turned out, perhaps a little too rapidly. Bared and

Cobo ran into tax issues and was out of business by 1978, only a year after they were up and running.

Durbin Paper was a paper recycling company in Northwest Miami, not far from Caressa Shoe. Again, they required significant support. Waste Management acquired the company. The System/32 and staff did not make the transition. A far more satisfying customer was Hi-Fi Associates, a South Florida chain selling stereo equipment. Fred Parce, one of the partners and vice president of operations, oversaw the installation. At one point, he asked if I was in the market for a stereo system. He configured a Pioneer receiver, Garrard turntable, and KLH speakers, a very nice combination and one that I still have.

## AN EVEN BETTER ENTRY SYSTEM

While Glenn Henry and his team were working on the System/32, work was already in progress for a multi-user version of the concept, again based on the System/3 architecture and incorporating Glenn's microcode enhancements. While small businesses like Ebsary managed just fine with the single-station System/32, larger companies would need multiple terminals. The task for Rochester was to deliver the online functionality of the System/3 Model 15 with the pricing and ease of use established by the System/32. This would not be an easy task. Rochester remained bound by the System/3 architecture and would need to tackle the 16-bit addressing limitation while providing a multitasking and multiprogramming system that utilized multiple processors for performance.

The new system would reach the market as the System/34. A key mover in bringing it to market was Bill Lowe, the director of Manufacturing and Development, who reported to Chuck Branscomb. Lowe joined IBM in 1962 as a test engineer in Endicott. After several promotions, he ended up an engineering manager working in Chuck Branscomb's GSD organization out of Boca Raton. As a GSD executive, he was known for his stony comportment, earning a comparison to then-President Gerald Ford. However, behind that exterior was a creative executive who would be intimately involved with some of GSD's most successful products.

With the System/32 and virtually all subsequent systems, Rochester was committed to smaller, integrated disk drives. It sent 40-50 engineers to San Jose to learn the design-and-manufacturing process. With that indoctrination, Rochester became a major manufacturer of small-capacity drives. The standalone 3370 would be the last disk drive developed in San Jose for GSD. The division would need smaller, more compact drives for its line-up of small to medium systems. The workhorse was an eight-inch 64-megabyte hard drive known as the Piccolo. It was developed in Hursley. Over the next decade, roughly 360,000 Piccolo drives would be sold[10].

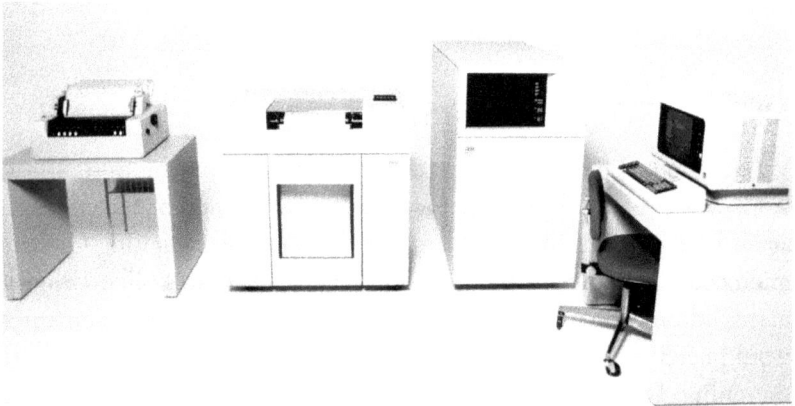

**Figure 49.** IBM System/34 (1977), the multi-user version of the System/32. Shown from left – the desktop printer, the "famous" 5211 line printer, system unit, and the famous "green screen" standard display on the right. Reprint Courtesy of IBM Corporation ©.

Rochester continued to have trouble getting a supply of 3270 displays, as they were also used on IBM's mainframes. The 3270 display format was comprised of 24 lines of 80 monospaced characters each. Each character appeared in one of two colors, green or bright green. Green happened to be chosen because it was least costly. Rochester decided to create a completely new ecosystem of online devices, both displays and printers, to ensure adequate supply for its rapidly increasing customer base. The new displays were formally designated the 5250 series but were universally known as the "green screens."

Rochester also changed how the new devices connected to the system. Using twinax cabling provided for higher transmission speeds and automatic configuration, both big pluses for System/34 environments that had little or no staffs. A new series of twinax-attached printers joined the line-up. This included the famous 5211 "washing machine." It was a high-speed dot matrix printer that "shaked and baked" its way to producing 160 print lines per minute.

**Figure 50.** IBM 5250 "Green Screen" Display. Reprint Courtesy of IBM Corporation ©.

Atlanta scaled back the range of IAP applications provided for the System/34, with just construction (CMAS II), distribution (DMAS II), and manufacturing ("MAPICS" for Manufacturing, Accounting, and Production Inventory Control System) making the cut. This would turn out to be the first step in getting out of the application software altogether. By the 1980s, there were plenty of good options among business partners and less of a business need to have IAP programs support hardware sales. GSD sold MAPICS to Marcam in 1992. Support for CMAS was contracted out in 1987, and the product was withdrawn in 1991. In hindsight, with software eventually taking over the lead role in computer decisions, IBM's exit of the business seems ill-advised at best.

The System/34 arrived in April 1977. The performance range from smallest to largest model was 16-fold. Prices ranged from $20,000 to more than $200,000. The largest System/34 model would support up to 256 "green screens." GSD would sell more than 60,000 System/34 systems before pressure for even more performance and technology advances would bring on a follow-on machine[11]. As had been the case with the System/32, the new system was a big hit within GSD and with customers in South Florida. I had another great construction account in Miscott Construction. Located in Hollywood, Florida, the company built large condominium towers in Dade and Broward counties. Its principal, Barney Lombardi, was a well-built former ironworker who moved from the job site to his own offices. He started with the System/32 and the construction software (CMAS) and with the business expanding, upgraded to the System/34 and CMAS II in 1978.

Barney Lombardi liked the work that I did for his company. In 1979, he invited my wife and I to dinner at Horatio's, a fancy rooftop restaurant at the Coconut Grove Hotel. It turns out that the reason for the dinner was not to celebrate the success of the new computer system but to propose that I come to work for him. It was certainly tempting and very difficult to turn him down. However, I enjoyed what I was doing with IBM and could not see making a change. It turned out to be the right decision. The condominium business was tough. Barney had a major project where he was not paid and ended up filing for bankruptcy in 1984.

Speaking of Coconut Grove, it was the home to another of my System/34 accounts. Howard Hughes, the wealthy businessman and celebrity, set up the Howard Hughes Medical Institute as a way to reduce his tax burden. He transferred stock ownership of his core company (Hughes Aircraft) to the nonprofit institute. Upon his death, the Institute sold Hughes Aircraft to General Motors, netting $5.5 billion. In the 1970s, the institute operated in a secluded and highly secure 12-acre estate in Coconut Grove. It decided on a System/34 in late 1979 for financial accounting. It was one of my most unusual accounts. It was always a pleasure to leave the concrete warrens of Miami and enter the lush institute grounds.

The GSD office in Miami covered quite a bit more than just Miami. The branch's territory ran from the cattle ranches south of Orlando down the Gold Coast to Key West, including everything east of the Everglades. This huge expanse provided another account that would give the Howard Hughes Institute competition for most unusual customer. That would be Monroe County. It sprawled across the Florida Keys with its county seat 170 miles south of Miami, across many miles of open water, in Key West. The county ordered a System/34 in 1979. That meant quite a number of trips to Key West.

This is a five-hour drive, each way, assuming all the bridges were open. An alternative presented itself -- Air Sunshine. The airline provided twice-daily service from Miami International to Key West. In 1980, it was still flying DC-3 aircraft, a plane that first flew in 1935. Following World War II, surplus DC-3 aircraft were not competitive on mainline routes but found a home on short hops like this. Flying a nearly 50-year-old plane over the vast expanses of open water to Key West was perhaps above and beyond the limits of account support. Monroe County, which tried to secede from the United States in the tongue-in-cheek Conch Rebellion in 1982, was very appreciative of its IBM support.

The General Systems Division was riding high, with one success after another. Operations in Atlanta spread across many locations on the north side of the city. When it was became clear that the new division was going to be around for a while, GSD president Jack Rogers initiated the search for a new, central headquarters. IBM purchased a 30-acre wooded site north of downtown Atlanta, on a bluff above the Chattahoochee River. The company announced plans to build a headquarters building in 1975. The building would span a small lake within the dense woods. CEO Frank Cary joined Rogers and Atlanta mayor Maynard Jackson for the dedication of the 11-story structure in

July 1977. The Lakeside building as well as several satellite locations at Northcreek and Perimeter Center would house over 5,000 IBMers. The new headquarters was just eight miles south on Highway 41 from the Big Chicken[12].

Figure 51. General Systems Headquarters in Atlanta, the Lakeside building. Dedicated by CEO Frank Cary, GSD President Jack Rogers, and Atlanta mayor Maynard Jackson in July, 1977. The Hillside companion building was added in 1987. Reprint Courtesy of IBM Corporation ©.

Ensconced in deep woods, Lakeside was hard to find. That is where the Big Chicken comes into play. We are talking about a very big chicken, a 56-foot-tall chicken. The chicken sat in Marietta on Highway 41. Most anyone from the Atlanta area knows the Big Chicken. It is a directional landmark as in, "You go to the Big Chicken and turn right." Pilots use it as a landing beacon for Dobbins Air Force Base, which was next door. Tubby Davis built the huge chicken in 1963 for his Johnny Reb's Chick, Chuck, and Shake restaurant. He hired Hubert Puckett, a Georgia Tech architecture student, to design the novelty structure to stand over his restaurant. Puckett's creation had eyes and a beak that moved, driven by a huge electric motor whose vibration would sometimes shatter windows in the restaurant.

The restaurant quite naturally became a Kentucky Fried Chicken

(KFC) outlet in 1974. KFC executives, including Colonel Sanders himself, wanted to take the chicken down and make it conform to the style of the thousands of other KFC restaurants. The Davis family advised them to do otherwise, and the Big Chicken survived. In 1993, it was severely damaged in a storm and once again slated for demolition. The ensuing public outcry convinced KFC to spend $200,000 to make repairs. Now a local treasure, it had another renovation in 2017. This time around, KFC spent $2 million to give it a modern restaurant look, though it remained a restaurant with a giant chicken looming overhead. The Big Chicken continued to be the reference point to guide IBMers and customers to GSD offices.

Figure 52. The Kentucky Fried Chicken (KFC) restaurant at the corner of Route 41 and Roswell Road, a local landmark known as the "Big Chicken."

Working in the new headquarters with Jack Rogers and Chuck Branscomb was the vice president of Sales, Lew Gray[13]. Gray had grown up in Ottumwa, Iowa, and attended Iowa State in Ames. He joined IBM as a salesman in 1961 in the Des Moines office before a series of promotions followed, starting with branch manager in

Jefferson City, Missouri. He came to Atlanta in 1977. Gray's tenure at GSD overlapped with mine in Miami. He visited the branch on several occasions, the most memorable time a trip to give an Eagle award to one of our top salesmen. The Eagle went only to the top one percent in sales, in the rarefied zone above both the 100% Club and the Golden Circle. It usually meant a salesperson had attained 250 percent or more of their assigned quota. I was one of two systems engineers that had helped the salesman reach that lofty level.

Lew Gray was electric, a bundle of energy, and very personable. It was immediately clear how he quickly ascended through the ranks of a sales-oriented company like IBM. His memory impressed me the most. He had thousands of people working for him, in Atlanta and all across the country. He had met me once before, quite briefly. Yet, he remembered my name and our previous conversation. He was committed to GSD and Atlanta, and would underscore that with the remark, "As long as I am around, there will be an IBM division headquartered south of the Big Chicken." Now, a 56-foot chicken was a lodestar for the small systems division.

# FUTURE SYSTEM (FS) CRASH
# AND BURN

As early as October 1964, there were discussions about what would follow the System/360. IBM tended to work on five-year product cycles, so despite the size and breadth of the NPL program, something should happen by 1970. Dick Watson pressed for a blueprint for the System/360 follow-on, a plan that became the Next Generation System (NGS). The urgency was driven by rapid changes in the mainframe market. A recession in 1969 had depressed mainframe sales. Advances in logic technology would force a shift from safe, reliable SLT modules to integrated circuits. Fast and inexpensive DRAM memory replaced core memory. Armonk's was concerned that this rapid acceleration of computing power would lower prices and drive down IBM revenues. Dick Watson also wanted to get the company out in front of the technology curve. Despite the System/360 success, competitors were nipping at IBM markets. A plan like NGS sought to reassert IBM dominance and take another giant leap forward.

Market analysis of the computing landscape done in 1970 underscored the concern[1]. The average CPU utilization was just 30 percent. Where once machine time was the most precious, it was now user's time. Programming had shifted from coding in binary or assembler to procedural languages like COBOL, FORTRAN, and RPG. Programmer productivity was a constantly recurring theme. Computer operations were moving from card-oriented batch, with one job running at a time, to multi-user, interactive, multiprogramming environments. Within IBM, the earlier batch-processing era was called the Chad Age, reflecting the chads punched out of 80-column cards.

Bob Evans convened a group in early 1969 to develop a plan, called the Higher Level System. John McPherson, the "only IBM vice president who could write a computer program"[2], chaired the effort. His group completed their work in February 1970 and presented the recommendations to Evans. Their report summarized the general conclusion that "we have come close to the limit of the von Neumann instruction-based machine. . . the more direct, powerful Higher Level System approach may be needed to sustain IBM growth in the years ahead." It went on to describe "higher level" in more detail. The machine-man interface would move from binary coding, past the current programming languages, and on to new structures that provided easier coding and debugging. Elements of this approach included self-defining files and such innovations as single-level store, where the programmer no longer was concerned with where the data resided. The focus of these new human interfaces would be on interactive computing. The report concluded that a "higher level" system was indeed feasible and should be pursued. It recommended further study.

Soon after leading the Higher Level System task force, McPherson retired from IBM. His legacy was secure. Born in 1911, the year that Charles Flint acquired Hollerith's company and combined it into C-T-R, McPherson had presided over many of IBM's key greatest successes. He had Steve Dunwell work with Wallace Eckert at the Astronomical Bureau. He assigned Clair Lake to design the Aberdeen relay computers and then had Lake and Hamilton build the Harvard Mark I. He listened to Watson's tirades about the Mark I and directed Wallace Eckert, Rex Seeber, and Frank Hamilton to design the SSEC. He supported the 603 and 604 Multipliers and sketched out the design for the IBM Card Programmed Calculator. Herb Grosch's view: "Eckert had the scientific vision, Hurd had the marketing vision, but John put it all together.[3]" McPherson was no slouch technically, either. He had authored the paper "On Mechanical Tabulation of Polynomials" for the Annals of Mathematical Statistics in 1941[4].

It took some time for any action on the Higher Level System recommendations to coalesce. The launch of System/370 took precedence. Then in early 1971, Carl Conti, who had participated in McPherson's task force, launched a follow-up study group[5]. Conti earned a degree in physics from Case Institute of Technology and joined IBM in 1959 as an engineer at the Poughkeepsie lab. He worked on Stretch and developed the cache ("muffer") concept for the System/360 Model 85. Conti, who had worked with John Cocke while on the Stretch project, enlisted him for his initiative and, in particular, his assessment of the use of RISC architecture. John Sowa and Steve Zilles drafted the group's final report, titled "Advanced Future System." Conti distributed the highly technical and confidential 128-page report along with the caveat that his staff would be available to assist in digesting the contents. As with the prior High Level System analysis, the report recommended that IBM pursue such a program.

About the same time, one of the designers of NPL approached Bob Evans with some radical ideas about where to take computing after System/360. His name was George Radin[6]. He had a somewhat unusual start for someone now espousing cutting-edge ideas on computer design. He obtained his BA in English literature from Brooklyn College, followed by a master's in English from Columbia in 1952. He belatedly decided that this was not to be his passion. He switched to mathematics and received his master's from the City University of New York in 1961. He then managed the computing center at Columbia, the center that had so much IBM history. He joined IBM in 1963 in the Research Division, when it was still located in the Time-Life Building in New York. The FORTRAN, COMTRAN and COBOL programming teams were also housed there. He worked on OS/360, helped design the PL/1 programming language, and then designed the timesharing software (TSS/360) for the Model 67.

Radin reviewed his new computer architecture with Evans in

1970. Evans was division president at the time and had just launched System/370. He asked Radin and Watts Humphrey to form a small group to assess the ideas. Though receptive, Evans was concerned whether it could be done in six to eight years. Radin and Humphrey concluded that it could. One major issue would be the low end. The planned architecture appeared too big for the small-systems groups to embrace. DEC had just released the wildly successful PDP-11, and Rochester was focused on this low-end minicomputer threat.

The High Level System and Advanced Future System investigations had both been small groups with the mission to analyze the general feasibility of an expansive new program. John Opel was now President of IBM and decided to convene a third and much larger group in May 1971 to determine whether to move forward on any of these ideas. His mandate was to identify the set of technologies that would raise IBM's "game."[7] He asked Evans to be the executive sponsor. Evans had returned to his banishment to the Federal Systems Division (FSD) after the System/360 miss on time-sharing and had just shepherded the release of System/370. Evans tapped System/360 veteran Richard Case to lead this new group. Case had just finished leading the design of the System/370 architecture. Case would have a team that included Conti, Phillip Estridge from Boca, John Fairclough from the Hursley lab, and George Radin.

Opel wanted the initial focus to be on market trends, as a guide to where the problems and opportunities lay. Market analysis noted that the average customer was spending $2 on computation for every dollar spent on IBM products[8]. IBM's competitors would continue to ride the technology curve, focusing on price/performance. IBM could rise above this pack and focus on advanced function. When the System/360 and System/370 reached the end of their product cycles, used machines would likely depress prices. The System/370, announced the previous year, was tracking below plan. Jerry Haddad completed a study that showed the costs for circuitry and memory plummeting. An architecture that enabled rapid development of new applications could soak up all that expanding compute power. Absent this, the company would struggle to sell an increasing volume of processing power and revenues would fall.

The task force worked for two months during the summer of

1971. This was in the midst of a worldwide recession that led to GE selling its computer business to Honeywell and RCA selling its computer division to Sperry Rand. Though the System/370 started to rebound, there was still the concern that falling chip prices would depress margins. Then there was Gene Amdahl. He had left IBM the previous year and taken what was in his head—the complete architecture for the System/360, his plan for System/370, and his work on the ACS-360 supercomputer—to Fujitsu in exchange for funding to create a mainframe clone. He also licensed his mainframe technology to Hitachi[9].

The Case team reached their conclusion in June 1971, recommending that IBM proceed with a major program to replace the System/360 and System/370 architectures. The program would be called Future System or "FS" for short. The project formally began in September 1971. It sought to take a bold step forward from von Neumann's stored program model. The planned budget for FS would be roughly $1 billion[10], with a projected announce in 1975.

FS would become a vast enterprise, tapping the best engineers from all parts of the company. Those engineers would be assigned to one of three architectural layers: the New Machine Interface (NMI) which focused on logic circuits and microprogramming, the Execution Discipline Interface (EDI) that managed program execution and virtual addressing, and the Application Development Interface (ADI) which addressed programming productivity. The Endicott and Boblingen labs pressed to focus solely on the programming interface but George Radin, for one, was not in favor. Fred Brooks, retained as a consultant to Case, was decidedly negative of the three-layer approach, noting, "complexity is the fatal foe.[11]" However, Radin's position won out.

New and faster technologies would drive the hardware definitions. This was perhaps the easiest part of the task. At announce, there would be four FS models slotted by increasing performance. FS1 would be rated at .08 MIPS, to be developed in Hursley. FS2 would be .40 MIPS and owned by the Boblingen lab. FS3 would be 2 MIPS and be produced by Endicott. At the high end, FS4 would be 10 MIPS and developed by Poughkeepsie.

George Radin was the chief architect for FS, considered the "intellectual giant[12]" of the undertaking. The project labeled individual component areas after colleges (Vanderbilt, Hofstra, Tulane, and Ripon). Radin's architecture design featured a high-level machine interface with many functions handled automatically in microcode below the user's interface. The design included single-level store, 48-bit addressing, and an integrated database. Many of these elements were at the cutting edge of computer design but Radin had skillfully combined them. They were designed to produce a five-fold improvement in programming productivity.

Single-level store was the "most revolutionary new feature" of FS[13]. The concept had originated at the University of Manchester, with the Ferranti computer's "one-level store" feature. The concept simply required that every system object have a unique and permanent address, an address that was of no concern to the programmer. The operating system and machine microcode would take care of physically locating any object and bringing it to the program when requested. As John Sowa described it, "With infinite storage or infinite speed, no one would care where the data happened to be located or how the system happened to be moving it around.[14]" Of course, the system would need to have adequate address space to assign each object a unique address, something that 48-bit or 64-bit addressing would be able to handle.

FS would have an advanced relational database, and the team would look to Ted Codd for its design. Codd worked for Fred Brooks on Stretch, developing a multiprogramming OS. Brooks suggested that Codd put his IBM career on temporary hold and get a PhD in computer science[15]. Codd took him up on the offer and studied at the University of Michigan from 1961 to 1965. Brooks says this was one of his best moves as a manager. Codd returned to IBM, moved to the San Jose lab, and developed his ideas for a relational database, producing a design far superior to IBM's existing database (IMS).

John Sowa, who was a member of the AFS task force chaired by Conti, was not afraid to press certain issues as FS got underway. Sowa had degrees in mathematics from MIT and Harvard and received a doctorate in computer science from the Vrije Univeriteit ("Free University") in Brussels. He joined the Applied Math team at IBM in 1962. Sowa argued early that the primary architecture for FS needed to be fully defined well before any work on secondary elements could proceed. He cited Fred Brooks and the OS/360 experience. In hindsight, Brooks concluded that there should not have been 150 members on the original OS/360 design team, but rather a small, talented 12-person team. Sowa described the lack of a complete architecture up front as "a movie with a cast of thousands, but no script.[16]"

There were many other issues to resolve, including the complexity of the instruction set and the significant migration effort required for System/360 and System/370 customers. In addition, a program this radical was bound to have issues surface later. In the end, however, the fate of FS boiled down to performance and, in particular, the performance of the high-end Vanderbilt model. When FS prototype hardware was available for testing, the System/370 emulator on Vanderbilt ran many times faster than in native FS mode. This was especially true of COBOL programs.

If this sounds familiar, the news was an eerie repeat of the benchmarking of the 1401 Super by John Haanstra's team. However, this time the gap was far too large to be acceptable to Marketing, which lined up in favor of killing FS. In addition, the Vanderbilt technology only increased performance three-fold over the current System/370 high-end, the Model 168. Given the time to develop FS, Sowa argued that IBM's largest customers would top out and could not wait. Sowa stated it more plainly: "Would any customer that bought an 8-mip machine in 1977 move to an incompatible 4-mip machine in 1980?" The answer was clearly "no[17]." Going forward with FS would thus also require an interim high-performance System/370 machine plus a reasonable performance upgrade path to FS.

Sowa also noted that the missing "horsepower" for the higher-end models could have been supplied by John Cocke's RISC architecture or by adopting the revised System/370 architecture developed by Gene Amdahl. With those changes, the resources that had to be used for the eventual System/370 replacement (the 3033 "Big One" system) could just as easily be applied to a set of machines with FS architecture and technology.

In addition, the FS architecture was complex. The sheer breadth of FS documentation comprised many separate highly confidential documents, each of which a person had to demonstrate a need to know for access. This internal "red tape" inhibited any one person from understanding the overall structure and flow of the entire system. Many of IBM's engineers on the program, thus compartmentalized, did not fully understand the complete FS concept.

Sowa put many of the issues he raised about FS in a missive called Memo 125 in November 1974. Carl Conti had raised many of the same issues during the 1971 Opel task force. However, he was too low in the IBM hierarchy to carry the day[18]. In addition, there was CEO John Opel's nontechnical background. It would have taken a Herculean effort to communicate the complexities of the proposed new architecture to him and his senior executives.

Bob Evans had the FS plan reviewed by the IBM Cambridge Scientific Center at MIT. Its recommendation was to start small and add function incrementally. This, of course, was the approach with OS/360, with customers taking the brunt of the pain as OS/360 versions slowly trickled out. By now, it was becoming clear that FS was doomed, but the project had a momentum of its own, in part due to how much money had been already spent. An even greater factor was the reputations of the executive sponsors. As issues surfaced, those who did not support the project found themselves deemed "not team players.[19]" It was politically dangerous to contest FS. Particularly with Opel, dissent was not advisable. Tom Watson suffered a heart attack in 1971 and retired as CEO. Vin Learson had replaced him, but he retired at age 60 in 1973, paving the way for Opel. John Akers, who was now president of the Data Processing Division and would be the prime beneficiary of FS, would brook no dissent on the program. This was certainly not the Watson way. Tom had valued internal competition, debate, and discussion. However, he was gone.

There was also an external issue with FS. The Justice Department had initiated an antitrust suit in 1969 against IBM. It was lurking in the background. FS would be a huge, dominating market move, one that would underscore the monopolistic contentions of the Justice Department case. Bob Evans was one executive who had the standing to voice concerns about the trajectory of FS early on. The projected target date for FS had slipped from 1975 to 1978, and now it was looking closer to 1980. He was looking at a completely missed product cycle with the System/370. In 1974, he began researching ways to enhance the System/370 without FS. By 1974, nearly three years into the effort, the S/370 was selling well. The slump that IBM had predicted did not materialize. According to Richard Case, this drove the motivation for FS to near zero[20]. Case told management he did not believe that FS was going to make it. He was told to stay the course and asked to "do my best to sell FS" to a review committee[21]. Evans made the decision to extend the System/370 architecture and formally recommended cancellation of FS in late 1974.

The end came on Valentine's Day, 1975. It was the most expensive failure in IBM history, estimated to have cost upwards of $1 billion, even before the bulk of development had even begun. That does not include the lost opportunity represented by the drafted army of brilliant engineers who could have been working for current and more viable projects in the IBM labs.

Though FS might have reached the light of day as the System/380 in 1980, Sowa argued for an evolutionary alternative, which he called System/375. He referred to the 28-month turnaround to produce the 3033 "Big One" machine after FS cancellation, contending, "IBM could have delivered a machine with similar or better performance in 1975 instead of 1977, if they hadn't killed all the System/370 design projects to avoid competition with the FS fantasy." His System/375 could have been delivered simply by adding 16 of the latest FS circuit boards to the existing Model 168[22]. Richard Case came to the same conclusion, that IBM could have cut its losses earlier and proceeded with the hardware enhancements on the existing product lines[23].

Sowa also argued that the impact of FS continued into the next mainframe cycle, with the 3081. He viewed it to be technologically inferior to Amdahl's machine. The 3081 required price reductions in order to be competitive. Sowa wrote, "The Amdahl machine was indeed superior to the 3081 in price/performance and spectacularly superior in terms of performance compared to the amount of circuitry.[24]" Gene Amdahl had simply implemented the design he had proposed for System/370.

It was no surprise that the failure of FS made IBM gun-shy about shiny new architectures. The risk takers, Watson and Learson, were both gone, and perhaps the appetite for big moves had exited with them. Though Opel had commissioned the FS program, he was certainly not happy with the way it turned out. The mainframe group was apoplectic about FS. It had delayed normal System/370 product turns by up to five years. The Japanese mainframe competitors had a five-year respite to catch up. Richard Case ended up out of the company mainstream at IBM Research[25].

# FROM THE ASHES COMES PACIFIC

I f FS ever got the go-ahead, the small FS1 model would have been at a significant disadvantage, not unlike the 1401 program's exposure with the NPL program. However, in the ultimate of ironies, the promise of FS would actually be fulfilled, but not at the high end. It would emerge from the small systems division. All of the IBM labs had participated in FS, including Rochester. The Rochester engineers who participated in FS understood the performance issue. At the same time, they were under the gun to replace the outmoded and restrictive System/3 architecture. The ability to expand the System/3 line, including the innovative System/32 and System/34, demonstrated what could be accomplished with technology advances and in particular, with microcode.

A key feature of FS was intriguing in this regard was high-level machine interface. Onrushing enhancements in technology could be potentially contained "under the covers" and separated from the operating system and user interface. Back in Rochester, Glenn Henry had completed the work on the System/32 and turned his attention to the System/3. He already had a small team on the case.

Work had been underway on a replacement architecture for System/3 since its introduction in 1969. Its early driver was a young engineer named Frank Soltis. Soltis studied electrical engineering at

North Dakota State, graduating with both BS and MS degrees. He had a lead on a summer job in California with an aerospace company, but then interviewed Harry Tashjian, who at the time was a Rochester engineering manager working on a banking terminal. Tashjian convinced Soltis to work the summer in Minnesota rather than Southern California. At the end of the summer, Tashjian further advised Soltis of the growing importance of computers at Rochester and the need for good people knowledgeable in computer architecture. He encouraged him to take a personal leave to secure an advanced degree in electrical engineering, with a focus on computer science.

Iowa State had become a leading institution in computer design and use, and Soltis enrolled in Ames that fall. One of his professors was Dr. Robert M. Stewart, Jr., the same Stewart who as a graduate student at the school in 1948 had been asked to dismantle a desk-size electronic device, saving any useful parts and disposing of the rest. This contraption was, of course, the Atanasoff-Berry computer[1].

Soltis' doctorate dissertation at Ames was titled "Automatic Allocation of Digital Computer Resources for Time Sharing" and delved into virtualization of computer memory and the performance of paging segments in and out of real memory in a multiprogramming environment[2]. He came to view many current computer architectures as "East Coast" designs with operating systems such as MVS and DEC VMS being outgrowths of the research at MIT with Multics and UNIX[3]. He was not a big fan.

There was another connection with Iowa State at about the same time. Rex Rice had been one of the driving forces for the Card Programmed Calculator at Northrop before joining IBM in 1955. He worked on the transistorized 608 Calculator and the Wooden Wheel. After shepherding the four prototype "Wheel" machines to Northrop, he worked on the high-end 750 proposal. With the demise of the 750 design, Rice moved to IBM Research in 1958 and studied the implementation of high-level programming languages in hardware.

When these efforts flagged in 1966, Rice left IBM and joined Fairchild Semiconductor, working for Gordon Moore. He continued the work he had done at IBM Research, developing a high-level language computer called SYMBOL based on the new and fast VLSI

(Very Large Scale Integration) chips[4]. Rice wanted to maximize the use of hardware by placing both the operating system and programming language into those hardware circuits. Gordon Moore left Fairchild, along with the "Traitorous Eight" engineers in 1968 to start Intel. Rice remained at Fairchild and continued working on Symbol. Fairchild decided in 1971 that the project was not commercially viable. Professor Stewart at Iowa State was aware of the project. He secured a National Science Foundation grant to do further research on Symbol at Iowa State[5]. Fairchild transferred the prototype Symbol machine, now roughly 90 percent complete, to the university. Besides being less than fully operational, the design presented other problems. When Iowa State discovered a bug in the operating system or programming language, the fix had to be in the 20,000 circuits that comprised the hardware. This required soldering the planar board or replacing individual logic chips. In addition, in spite of the fast system architecture, performance was actually slow. After four years, Dr. Stewart put the machine in storage.

Frank Soltis was aware of the Symbol machine but was prevented from working on it due to his employment by IBM. However, he would soon apply some of Symbol's concepts. He graduated in 1968 and returned to Rochester with his newly minted doctorate. His first assignment was the complete revamp of the System/3 architecture. Within six months, he had a team of nine that included Dick Bains and Roy Hoffman. They worked for Roy Klotz, who had been the CPU designer for the 1440 system in Endicott and then an engineering manager on System/360. Klotz reported to Glenn Henry, though in the beginning, Henry was far more preoccupied with the System/32 program.

Soltis and his team developed an initial design and reviewed with Klotz in December 1969. Klotz relayed progress to Harry Tashjian, now the Rochester lab director, who asked to see a presentation. This was done in early 1970 on "a bitterly cold winter day,[6]" no surprise for those familiar with Rochester winters. Soltis would observe, "There are only two seasons in Minnesota; winter and getting ready for winter.[7]" The presentation went well and Tashjian approved funding to explore the concept further.

With space at a premium in Rochester and the desire to keep the effort on a low profile, Soltis and his small team moved into the former Wells department store in downtown Rochester. It was a sprawling one-story building that IBM gutted to add cubicles and offices. The original 18-foot ceiling remained. The various IBM teams took the names of the former Wells building departments. The performance group was in the toy department, the input/output team was in automotive, and the Soltis architecture team was in ladies lingerie[8]. The building was officially "648" but affectionately referred to as the "single-level store."

With the announcement of the System/32 in January 1975, Glenn Henry could turn his full attention to the System/3 replacement project[9]. He reported to Brian Utley[10], the overall system manager. Henry thought about what a clean-sheet architecture would look like, evolving the concepts he used on the Model 6 and System/32[11]. Virtual memory, single-level storage, and most importantly, moving the machine interface higher in the structure to adapt to technology changes were high on his list. He came to many of these ideas independently. He had also read up on trends in computer architecture, including an Association for Computing Machinery (ACM) article that forecast the next ten innovations to look for in computer design. Many of his ideas were on that list. He was also familiar with comments made by Fred Brooks, who concluded that the second system you designed was the one where you tried to correct all the mistakes you made on the first one. Once he saw the FS blueprint in 1974, it simply reinforced his own ideas.

Frank Soltis joined the FS program in 1971. He saw Radin's concepts of a high-level machine interface, single-level store, and object orientation. He worked on the architecture team with John Cocke and Bob Tomasulo (famous for devising out-of-order instruction execution for Stretch)[12].The Endicott and Rochester contingents pushed for the more radical Radin design but Poughkeepsie pushed back. Soltis and his fellow Rochester engineers were excused from the program as Poughkeepsie felt there was too much small system influence. Soltis had a different take on the decision. At the time, there was a real concern that the Justice Department would split up IBM. The small systems part of IBM would be the logical part to jettison. He viewed the rejection not as a problem but an opportunity. Soltis proceeded to incorporate many of the key FS elements into the design of the next-generation Rochester architecture, which blended in with Henry's thoughts. The new project would shed the name "FS" in favor of "Pacific."

The task of programming such a radical design was daunting. It was a contributing reason for the FS cancellation. Rochester's new system would have five million lines of new code. Henry would soon learn what Fred Brooks realized a decade earlier -- the futility of adding resources to a late project. He also recognized that there were regular programmers and star programmers. The stars could do far more work, but the company did not compensate them for their productivity. Programming was done in PLS, the variant of PL/1. Early on, the hardware group produced a prototype called Linus for testing.

Soltis had conflicting impressions about his new hands-on boss. His first reaction was, "Who is this wild man? Why does he introduce himself as one of the world's greatest authorities on virtual memory?" According to Soltis, there was nothing remotely shy about Glenn Henry, "the most brilliant and totally outrageous manager I have ever known in IBM." Henry was certainly not in the typical IBM mold as he featured "the scruffy beard, the often mismatched clothing, and the six-packs of Tab.[13]" However, Soltis also recognized the intensity and drive that inspired those who worked for him. Henry was never bothered by seemingly intractable problems, and Soltis felt "without his talent and vision" the Pacific program would not have lasted long.

Rochester would be on its own with this project. According to Henry, there would be no help forthcoming from other parts of the company[14]. IBM Research did not contribute at all. The mainframe folks were borderline antagonistic. Henry describes them as "the enemy[15]" as they were constantly talking about driving the S/370 line down and encroaching on Rochester territory. Though supported by CEO Frank Cary, it was clear that Rochester was under the microscope in running such a large and risky program. In the early stages, Henry focused on the design. Later, his role would transition to one of "marketing," which translated into managing Corporate and keeping the program moving forward.

The new system needed the latest logic technology to compensate for its complex architecture and high-function software. It would share LSI (Large Scale Integration) circuits with the 4300 midrange mainframe series. The main circuit board for the 4300 contained 28,000 circuits, composed of 40 LSI chips[16]. The complete Pacific processor resided on one 10"x15" planar board with 29 LSI chips, each containing 704 circuits, providing 20,000 total circuits[17]. Reduced wiring distances illustrated the increased density. In the System/370 Model 148, the combined length of all the processor wiring came to 330 meters. With the new LSI circuit package on the 4300, the total wiring length shortened to 8.4 meters, a 40-fold improvement.

Pacific appropriated many instructions from the System/370. It was a CISC (Complex Instruction Set Computer) system. This term came in usage with the rise of RISC (Reduced Instruction Set Computer). There were two layers of microcode, one layer translated instructions and the other dealt directly with the hardware. Together, they provided the high-level machine interface that was at the core of the FS design. The two layers comprised the operating system kernel, which was considered part of the hardware. This was done in part to discourage clone compatibles. The microcode layers handled many of the standard functions of an operating system such as memory management and process control.

Pacific would end up very close to George Radin's FS design. The key elements were all there: single-level store, object-based structure, high-level machine interface, and integrated database. Single-level store was the core and most powerful idea. Virtual addressing had started with the Atlas computer at University of Manchester, considered the most powerful computer in the world in 1962. Observers said when Atlas went down, "half of the United Kingdom's computer capacity was lost.[18]" Atlas had one view of memory across the magnetic core and magnetic drum storage. Tom Kilburn used the term "single-level store" in his 1962 paper. A plan for hardware somewhat short of infinite storage would require a very big address space. The original plan for Pacific was 32-bit addressing, but the design team lobbied for 64-bit. Roy Klotz compromised at 48-bit.

Glenn Henry was especially sensitive to having a large address space, given his experience with the 8K limit on the System/3 Model

6. The 48-bit space could handle addresses up to four gigabytes. Microcode managed this large address space automatically and transparently. There was no separate Blaauw or DAT Box as had been required with System/360 virtual addressing. The programmer did not have direct access to the data. There was no need. In addition, the system spread the data across multiple disk drives in order to get the best possible performance. The down side of single-level store was the loss of all access to data if the system had a severe crash. A complete backup of the "store" data was essential. It is interesting to note that Sowa, though supportive of single-level store, backtracked a bit when talking about the large, complex installations. He still felt the need for programmers to have some control over file placement and data access. Clearly, Rochester did not agree.

It was customary when a new product design was complete to distribute the specifications to IBM Research and to other labs within IBM. Rochester scheduled a two-week technical audit on the Pacific plan. Soltis recounts one of the objections raised during the review. When using virtual memory on the mainframe, the transition from one program task to the next required an address reset. Roughly 1,000 instruction cycles were needed to clear the virtual data and associated pointers from the current task to prepare for the new task. In a large interactive system with many online tasks, substantial processing power was expended on this switching. A reviewer asked why this processing requirement was missing. Soltis replied that with single-level store, every task and object has its own unique address so there is never any need for this type of conversion. One could imagine how chastened the questioner was with the Soltis response.

Corporate gave the formal go-ahead for the project in August 1973, with a target for completion in 1978. It was no longer a skunkworks. Henry expanded the development team beyond the core group led by Soltis. He added Roger Taylor, who had also been part of FS. Taylor was an architect and nearly always the smartest guy in the room. Henry was one of the few who could challenge Roger technically. Soltis was the architect on the hardware side. Henry and Taylor focused on the software and microcode. Henry's team would eventually grow to 600[19].

Giving single-level store a contest for the most important architectural innovation was the high-level machine interface. This meant that the operating realm of the programmer and user was set very high. The application program has no knowledge of the underlying hardware. In traditional computer architectures, the programmer had direct access to the hardware through assembler or machine code. The hierarchical structure began with the application program at the top. From there, the layers drilled down to the database, compiled languages, and operating system before reaching the two levels of microcode that sat directly on top of the hardware. Even Rochester programmers developing the operating system did not code to the hardware. They coded to the interface defined by the microcode layers. The "bit twiddlers" like Roy Nutt could not get anywhere near the machines physical registers or memory. Glenn Henry would note that the concept of the high-level machine interface led to heated debates among his core technical team of eight. The main issue from the hardware perspective was the potential poor performance with this architecture sporting two layers of microcode. Roy Hoffman took particular exception to this design. Henry was adamant about this element and offered to sign transfer papers for anyone that wanted to leave the group[20]. Henry would receive a $50,000 IBM cash award for the machine interface design.

The advantage of this structure was amply demonstrated in 1990 when Rochester systems moved from the 48-bit CISC architecture to a 64-bit RISC architecture[21]. The change was virtually automatic, with little impact on customers. Glenn Henry relied on this when performance issues surfaced during development. He knew the hardware technology would catch up, thanks to the onward march of Moore's Law, and the revised microcode layers would take care of the change.

Pacific also implemented an object-based architecture. Every discrete element within the system, whether it was a data file, application program, or printer definition, was an object. The essential descriptive information about each object was stored in the object itself. For example, the customer master data file would contain the actual customer data fields but would also include a map of the field structure of the data. An application programmer would not need to define the fields in the program. The system would import the field definitions automatically when the program opened the file. Once a database file was defined and created in this manner, any number of different sequences or "views" could be created automatically. For example, consider a customer master file indexed by customer number. A new sequencing by salesman or annual sales could be created with a simple operating system command. Another advantage of the object structure was security. The system could secure access to each object down to the level of the individual user, meaning the machine was virtually unhackable.

Pacific would fully embrace Ted Codd's relational database concept. The prevailing database structure up to Codd's work was indexed files where keys were first retrieved and then used to navigate through the data to find the desired record. Codd's relational concept was to structure the data as simple tables and let the computer figure out the best access path. Skeptics at the time felt his approach would be too slow and that programmers could devise the fastest retrieval method. The rapid advances in processing power and disk speed played perfectly into Codd's concept.

Perry Taylor led database development. He studied Codd's work, including his 1970 paper, "A Relational Model of Data for Large Shared Data Banks.[22]" He called Codd to review his plans for Pacific. He was taken aback when Codd brushed him off, stating that relational databases were only for large systems[23]. Despite the rebuff, Perry and team pressed on. John Sears, a technical expert on the Rochester marketing support staff, was a strong acolyte for Ted Codd and his database concepts. He was equally fussy with regard to creating compliant, non-redundant table structures. Codd had developed a set of twelve rules to ensure that a database was structured correctly, and thus was fully relational. With Sears, you knew when he rolled up his shirtsleeves, he was about to launch into the gory details of Codd's twelve rules.

IBM's mainframe community did not share Rochester's embrace of Codd's concepts. They were concerned that any new database product would endanger the revenue stream of IBM's IMS database product. Working out of the San Jose lab, Codd worked his relational concepts into a design called System R[24]. He also created a relational database access language called Alpha. IBM did decide to develop

Codd's design into a product called SEQUEL, for Structured English Query Language. Don Chamberlin and Raymond Boyce of the San Jose Research Lab programmed SEQUEL. Chamberlin was aware of the work by Codd and was one of the few that understood Codd's complex constructs. He teamed with Boyce to create the new database manager in 1981. Codd was not entirely happy with how the product turned out as it fell short of his twelve rules. IBM named Codd an IBM Fellow in 1976. He was also selected for the Turing Award, the Nobel Prize of computing.

IBM's dalliance in embracing Codd's relational invention would have severe long-term repercussions. Bob Miner, a software engineer, left Ampex in 1977 to form a software company in California with Ed Oates. Miner had read Ted Codd's seminal 1974 paper in the *IBM Research Journal* and immediately saw the import of his relational database. Here was yet another case where an IBM trade secret was published in a journal accessible to non-IBMers. Miner recruited Larry Ellison, who had worked for him at Ampex, and the three founded Oracle. Miner led a small programming team in developing the product, produced in C language to be portable across system platforms. Though the Oracle product fell short of Codd's full definition of relational, it was on the market in 1979, a full two years before any IBM product.

One of the big misses Fred Brooks regretted with OS/360 was the lack of logic and function in Job Control Language (JCL), the principal communication layer between the application program and the operating system. In contrast, Henry's team designed the Pacific OS as a dynamic language. It included process logic, parameter passing, and even database access. And like most programming languages, OS programs were compiled. Roger Taylor would manage OS development. This programming task alone comprised well over a million lines of code.

The least controversial aspect of Pacific was the application programming language. From FARGO to RPG and then RPG II, Report Program Generator was established as the core programming language for all of Rochester's systems. With Pacific, Frank Soltis entertained the concept developed by Rex Rice in the SYMBOL computer, a programming language embedded in the hardware. The team tested this idea on the Linus simulator.[25]. The performance was unsatisfactory so Glenn Henry's standard approach was adopted, putting key elements of RPG in microcode.

Equally important, the RPG language underwent significant changes aimed at matching the new system's advanced capabilities. The new version that emerged was, unsurprisingly, called RPG III. It took advantage Pacific's object-oriented data files, with embedded field definitions. It offered some off-ramps from the standard RPG processing cycle, implementing procedural elements found in languages like PL/1 and COBOL. Advanced programming facilities such as do loops and logic subroutines were added. You could also dispense with RPG coding sheets altogether, as programmers now used an interactive tool. .

There was one further change made with RPG III. Traditional compilers converted source code into executable machine code, or binaries. In many cases, these binaries would not survive a significant new release of the system. Given Pacific's high-level machine interface, new releases would be far easier for IBM to develop and potentially come with greater frequency. But, what about the customer? An installation with hundreds or thousands of programs would have a significant task in recompiling and testing those programs, assuming that additional changes were not required to adapt to new hardware. The scope of this task might even lead a customer to skipping an upgrade or even moving to a competitive system.

The new RPG system introduced a two-stage compile process to address this issue. In the first phase, the compiler generated a low-level program template from the source. The second phase converted this template into the actual binary for execution. The program template was stored along with the binary in the program object. With new hardware, the system simply retrieved the program template and produced that new binary automatically. For a customer with thousands of programs, a new release was now something to welcome, not dread.

# TROUBLE IN ROCHESTER

A program of this size and scope would not be without issues. The operating system was designed to fit into 64K of memory. This ratcheted up to 120K and eventually to 192K in order to achieve reasonable performance. Additional memory caused the projected price points to go up and endangered the business plan. The healthy profit margins originally predicted for the new system helped in holding the line. At the same time, the programming scope was clearly undershot. When Corporate approved the program in 1975, the target for announcement was 1977. Within a year, the committed date moved out to 1978.

**Figure 53.** System/38 (1978), IBM's Future System design "in the flesh", a revolutionary system. Shown from left with line printer, diskette drive, system unit and console, and the very rare and little-used card device. Reprint Courtesy of IBM Corporation ©.

Fortunately, Rochester had a buffer – the System/34. Development on the multi-user extension of the System/32 continued. The two development groups, Pacific and System/34, worked in competition with one another. The System/34 was ready and announced in 1977. Its great success took some of the pressure off Pacific. The big issue with the System/34 was its slim profit margins.

Pacific debuted on October 24, 1978 as the System/38. Jack Rogers, president of the General Systems Division, said of the new system, "the System/38 is the largest program we've ever introduced in GSD and it is one of the top three or four programs ever introduced in IBM[1]." The first delivery dates were still a number of months off, but the latest performance results were disappointing. There was a view shared in the lab that function was primary and performance would come around. This led to the alternative interpretation of the "PACIFIC" code name, to "Performance Ain't Critical If Function Is Complete." However, early test customers quickly corrected IBM of that notion. Even a top-end System/38 with two megabytes of memory was painfully slow. This should have come as no surprise with so many complex FS concepts coming together in one machine. Performance had sunk FS, and now it threatened the same with System/38. Another contributing factor was the rapid scale-up of the programming team. It had quickly ballooned from less than 100 to 450, resulting in the same issues that Fred Brooks had identified with System/360 development in The Mythical Man Month[2].

Brian Utley, the overall system manager, was very aggressive in setting dates. He was a bit naïve about the complexity of the system. He had told Henry to just bull over the System Assurance testing team. For his part, Henry was no angel and was equally aggressive in trying to skip steps and get machines out the door. Hal Martin, the general manager of Rochester, stepped in and counseled the team to avoid taking small slips in the schedule just to please Corporate. Henry told him the slip would be at least a year. Henry and Utley traveled to Armonk to explain the projected slip to the Corporate Management Committee. Chuck Branscomb was the GSD development executive in the meeting. Utley led the presentation, saying the delay would be at least three months and a maximum of six months. Branscomb spoke up, telling "Brian, keep quiet. Let Glenn answer.[3]" Henry said the slip would be a year. As the overall systems manager, Utley would do the IBM "penalty box" for two years but would later return in good shape, becoming the lab director at Boca Raton in 1988. Henry survived, but he had to produce weekly status reports for Martin the rest of the way.

The delay was a big problem. The System/38 was an exciting system and addressed significant System/3 pent-up demand. Tom Watson had railed about a 90-day slip with System/360. . It would be a year and a half before the System/38 began shipping in volume. Jim Dezell, who had saved the System/360 program from John Haanstra's Super plan, was now head of marketing for GSD. He worked with Rochester on a contingency plan. The division would create a new service in Atlanta for customers waiting for their on-order system. This would provide remote access to System/38s for conversion, testing, or new development.

The transition program, called Remote Test Service, consisted of a room full of System/38s that connected via leased phone lines to each customer that signed up for the offering. There was one other critical element, a staff of System/38 "top gun" experts to educate and guide those customers. Your author was one of those top guns. By 1980, I logged seven great years in Miami, done most everything, and was ready for a change. Shirley Kappeler was not happy with the thought of me leaving but orchestrated a nice send-off. I spent four weeks in Rochester coming up to speed in the new system and then joined the Remote Test Service staff in Atlanta. Robbie Weir had moved from Rochester to Atlanta to manage the team, which included Joann Nickels, Phil Baker, John Bowman, and John Hall.

Customers would begin the service by mailing program and file tapes to the center, which we would load and set up. Each customer would have one or more remote displays and printers in their offices. E.D. Smith, a maker of jams in Winona, Ontario, was an early Remote Test Service user. Their data processing manager, Beverly Russell, was a joy to work with. She would later become president of the COMMON user group.

The System/38 finally started shipping in July 1980. GSD sold over 20,000 systems in the first five years. The volume was significantly less than System/34 sales but consisted of far larger systems with dramatically higher margins. It addressed those high-end System/3 customers that had run out of gas. Most importantly, it provided the architectural platform for all future Rochester systems. The new system represented a significant achievement, essentially the FS system in the flesh. It was a supremely productive machine. The integrated database made advanced applications easy to develop. However, at the same time, it planted the seeds of future turmoil for Rochester. It was not compatible with any of the System/3 architected machines, including the System/32, System/34, and the later System/36. Moreover, even with its productive development environment, it was more complex to use. And at the start, there were very few applications written for the platform, though many applications were available in RPG. Yet overall, the system's strengths far outweighed the initial issues. Rochester would have to contend with a bigger issue that rose internally. IBM Corporate, anxious to protect low-end System/370 systems against encroachment, was concerned about the ease of growth in the System/38. It was not long before rumors began to circulate among System/38 customers that the platform was not strategic, and that its performance ceiling was being capped.

Glenn Henry reflected on the Pacific program, "I am most proud of the System/38 and I've done lots of things since then, but this is a very innovative system done by a bunch of young kids in the cornfields of Rochester, which was a backwater of IBM.[4]" Looking back on the ACM list of the ten highest potential advances in computer science, his team had delivered on seven of them. Richard Case downplays the role that FS played in the System/38. As leader of the FS project, he speaks with some authority on the subject. He felt that it would not have been good politics to make any claim of FS heritage, given its ignominious ending. In addition, there were many other ideas that contributed to the Pacific design. This certainly included Henry's experience and the work of the Soltis team. Case concluded, "I think they got to the place that they got to by their own process.[5]"

Roger Taylor developed an interesting summary of the progress made between the System/3 in 1969 and the System/38 in 1980[6]. It highlights the technology progress but probably shortchanges the dramatic advances in architecture and software.

|  | System/3 (1969) | System/38 (1980) |
|---|---|---|
| OS | Single batch partition, CCP interactive communications an add-on | Integrated, interactive multi-user system |
| RPG | RPG II tuned to the system, compiler to produce efficient code. | RPG III takes advantage of advanced architecture. Key programming constructs (Files, labels, fields) handle by the system. |
| System control interface | Modified version of JCL | System control a complete programming language, much like what Fred Brooks wanted to include in the System/360 |
| Processor Technology | Monolithic System Technology (MST) | Large Scale Integration (LSI) |
| Logic density (circuits/chip) | 3 | 704 |
| Memory Density | 100 bits per chip | 64K bytes per chip (524,000 bits) |
| System Hardware | 3,000 circuits, 24K core memory, 10 megabyte disk | 20,000 circuits, 512K DRAM memory, 130 megabyte disk |
| Disk | 20 kilobits/square inch | 6 megabits/square inch (over 300x increase) |

Frank Soltis was the early architect and led the small team at the Well's department store, incorporating FS concepts. His focus was primarily on the hardware architecture. After the announcement, Soltis transitioned into a spokesman for the new technology, working with fellow IBMers and customers. He had a unique ability to explain the complexities of the system in simple layman's terms. He soon became known as the "father of the System/38." Others would stress that Pacific was a massive group effort. David Schleicher opined that if there was to be any one "father," it had to be Glenn Henry. Of course, Schleicher was slightly biased. He worked for Henry and was his next-door neighbor. He considered Henry "absolutely brilliant," going toe to toe with Roger Taylor and keeping the huge program on track. He took all the arrows and never let the pressure fall on his team. He felt that Henry's technical excellence and management tenacity stood behind the program's success. Schleicher summed up by saying "he truly was the father of the System/38.[7]" Henry had spent seven years in Rochester, bringing the System/32 and System/38 to market. He moved to the IBM lab in Austin, Texas, in 1980 after the System/38 was released, in part to avoid the long and brutal Rochester winters.

Remote Test Service was a temporary operation from the start. Once systems started shipping in volume, and certainly when the branch data centers received their own System/38, the need for the service faded. Robbie Weir looked at all the high-powered System/38 talent working for him and advanced an interesting idea. The General System division's branches were all about new business. A branch territory could contain tens of thousands of companies, quite different from the big iron branches. The branches relied on company databases such as Contacts Influential, Dun and Bradstreet, Standard and Poor's, and local sources to develop prospect territories. Once defined, the sales representatives blanketed their "patches" with marketing campaigns that included direct mail, telemarketing, customer center events, and plenty of cold calls. This generated a substantial volume of useful sales information to keep track of including key contacts, installed systems, competition, and lead status. The branches also had to manage events, classes, and scheduling. Weir proposed to automate these branch office operations. Every major GSD branch had a raised-floor data center with a System/38, the ultimate programmer produc-tivity machine. Why not develop a complete marketing management system on it? Who better to do this than the experts just signing off Remote Test Service?

Robbie's concept became the Marketing and Coverage System, or "MACS" for short. He tapped several of his experts, including your author, to develop the system. The team of just five was fun, competi-tive, and productive. The System/38 architecture made it work. Each GSD branch office installed the application.

I was gone but not forgotten in Miami. Royal Caribbean Cruise Lines was an early System/38 account. Its new reservation applica-tion was ponderously slow. Shirley Kappeler called and asked for my help. Once onsite at the cruise line's Miami offices, I made all of the usual tweaks to tune the system performance. The terminal response times continued to be well over two minutes. Their reservation program was a monster RPG application with over 6,000 lines of code. I turned to the suite of debugging tools supplied with the System/38. They were more advanced than the light bulbs on ENIAC or the hexadecimal core dumps on the System/3. A trace run of the program showed the exact flow of RPG statements within the program.

Each time an agent accessed cabin availability, the program ran over 500 disk accesses against the single disk drive in their system. In computer lingo, the system was "thrashing." I recommended changes to the program that would store available cabins in one place. I also recommended adding a second disk drive to split the load. Since the System/38 automatically controlled the placement of data, it would naturally spread the accesses across the two drives. Those two changes resolved the performance issue. As of 2017, Royal Caribbean was still running its reservations on an FS machine, having upgraded to two AS/400s, the follow-on system platform.

With my move to Atlanta in 1980, my better half had to leave her dream job at Southern Underwriters. She joined Johnstown American Properties, a large Atlanta-area property manager. It was converting from a System/3 Model 15 to the System/38. Cynthia was an expert in conversion and here was an opportunity to learn the new system. After five great years in Atlanta, I had the opportunity to move into management as a Systems Engineering manager in Denver. I would no longer be a "top gun" but I would manage some. It would be another great opportunity for Cynthia. She became a programming consultant for Total Petroleum, a company installing a System/38 with oil and gas management software from a company called J.D. Edwards.

With the System/38 and System/36, GSD and Rochester became even more focused on applications and, in particular, supporting companies that brought their programs to these platforms. J.D. Edwards was Exhibit A for this outreach. Ed McVaney was the driving force. He had worked at Western Electric in 1964, programming the 1410 in machine code. He moved to Peat, Marwick, and Mitchell in New York City and later to their office in Denver. In 1970, he joined the Alexander Grant accounting firm, where he met Dan Gregory and Jack Thompson. The trio decided there was a future in computer consulting and formed J.D. Edwards in 1977. The "J" was Jack, the technical whiz. The "D" was Dan, the sales expert. The "Edwards" was for Edward McVaney, who managed the company[8].

Though initially focused on the System/3, they made the critical decision in 1981 to standardize their application offerings on the System/38. They used the system's unique programming environment to create a suite of business applications known Enterprise Resource Planning, or "ERP." This was a fancier name for the BICARSA and GLAPPR applications that Raynor Moore and his team in Atlanta had produced. J.D. Edwards software was installed on 15 percent of the System/38s that Rochester shipped. This being Denver in the 1980s with oil and gas production booming, it was natural for J.D. Edwards to develop an oil and gas variant of its ERP set. Total Petroleum installed that J. D. Edwards variant. This presented a slight conflict of interest. The systems engineers on my team supported the J.D. Edwards account, located right across the street from the IBM offices in the Denver Tech Center.

With the delays and early start-up issues with the System/38, Rochester decided to move forward on a replacement for the System/34. The larger customers were running out of "headroom." Jeff Robertson managed the program. He was a key member of the original System/3 team. The System/36 provided about twice the performance of the System/34. It retained the architecture of the prior System/3-based models, including the 16-bit processor and extensive microprogramming. The main processor ran at eight mega-hertz, significantly faster than the personal computers of the time. The RPG II programming language added the SORTA function, which was, in essence, the long sought-after bubble sort. The platform supported nearly 3,000 applications from partner software companies. The System/36 was a workhorse system. It could run for months without much attention or even an IPL (Initial Program Load), with no errors or performance degradation. The System/36 arrived in May 1983. It continued the scorching pace of sales established by the System/34. Where the IBM 650 was the first computer to sell over 1,000 and the 1401 the first to top 10,000, the System/34 sold 60,000 and the System/36 would rocket up to 200,000.

# FORT KNOX AND TOO MANY
# COMPUTERS

B y 1982, the General Systems Division with an assist from the "big iron" division managed to re-create the situation that IBM faced in 1961 -- a proliferation of incompatible computer product lines. Rochester had been very successful addressing the low-end market space and proceeded to grow into the mainframe space. Now, five different systems from four different labs overlapped the midrange -- System/36, System/38, 8100, 4300, and the Series/1. Looming on the horizon was IBM's entry into the RISC (Reduced Instruction Set Computer) market, first with the RT/PC in 1986 and then with the RiscSystem/6000 in 1990. The overlap was most acute in the area of $2,000 per month rental. IBM competed not only with minicomputer companies but also with itself. The significant market share that Rochester had claimed early on started to fade in the early 1980s, with market share falling into the single digits. For GSD, the stakes were high. The System/36 was running out of gas, and the System/38 was increasingly associated with the word "nonstrategic."

Bill Lowe, who had been leading the development of the IBM personal computer in Boca Raton, became general manager of Rochester in 1981. Hal Martin had been the lab director from the early System/3 days and general manager of the site since 1977. Frank Soltis called Martin "one of the best liked executives Rochester ever had.[1]" Martin was an avid marathon runner. While on vacation in Mexico City in 1980, he went for a run and suffered a fatal heart attack. Rochester lost a giant in Martin but gained a dynamic replacement in Lowe.

Lowe took stock of IBM's midrange systems and began ranking them. At first, he concluded that only the System/34 had ongoing prospects. Its replacement product, the System/36, would arrive the following year. Two years later, his rankings were System/34 and 36, Series/1, 4300, System/38, and 8100, in that order. The bottom two platforms would receive far less investment and could be dropped in IBM's future plans. A year later, his ranking changed again. He listed the System/36, System/38, and 4300. The 8100 and Series/1 fell out due to poor sales. Another year passed, and the rankings were a single Rochester follow-on, a 4300 follow-on, and new system to replace the Series/1. The most advanced and often maligned System/38 rose from nearly last to the top in four years.

Lowe's rankings were a part of an expansive program aimed at consolidating IBM's midrange lines. The effort came under the executive purview of John Akers, who was the IBM vice president in charge of all midrange products. The program was not quite as grandiose as FS but it was ambitious. Fort Knox was the program name. Four labs were involved—Rochester, Boca Raton, Poughkeepsie, and Endicott. As with FS, each lab spent considerable time trying to meld the design of Fort Knox in its favor.

All Fort Knox systems would use the new 32-bit Iliad processor designed by IBM Research. This was based on the RISC technology that John Cocke prototyped with his 801 machine[2]. This common hardware plan suffered early on with defections from Iliad. Rochester developed a System/36 and System/38 co-processor while Endicott produced a System/370 co-processor. This left only the planned native Fort Knox operating system running solely on the Iliad processor. For Rochester, the Fort Knox architecture would not be a step up from the System/36 and would be miles behind the advanced architecture of the System/38. The only plus would be Cocke's RISC reduced instruction set, which could provide a significant performance boost.

IBM Corporate drove Fort Knox. Alan Sheer was the headquarters liaison for Rochester[3]. David Schleicher suspected that Sheer's mission was less about helping Rochester and more about ensuring the lab stayed on track and moved expeditiously to Fort Knox. In other words, it was about reining in Rochester. Frank Soltis went even further. He thought Fort Knox was all about solving IBM's perceived problem of too many product lines, when the principal manifestation of the issue was internal competition between the two IBM sales divisions. There was no broad cry from customers for consolidation.

There was also a budget issue. Monies allocated to System/36 and System38 follow-ons now went to Fort Knox. Midway through the program, with the 8100 and Series/1 eliminated, Fort Knox limped along with just three systems—the System/36, System/38, and the 4300. Rochester was of the mind, "fool me once, shame on you . . . fool me twice, shame on me." With the FS debacle firmly in mind, the lab continued development of new models of the System/36 and S/38 in parallel. Tony Mondello, the lab director, had assumed system management of the System/38 and now oversaw all three competing development teams—System/36, System/38, and Fort Knox. As was the case with FS, the Fort Knox effort commanded Rochester's best engineers. This would not last long. The Fort Knox designers were unable to combine all three architectures into a single line with huge expense and an extended timeline. The program ended in 1985.

The cancellation of the 4300 and 8100 components left Endicott and Poughkeepsie without a converged plan for a low-end mainframe. The focus was on the 4300, a very successful line that had been around since 1979 and was overdue for a refresh. In the space of only a year, the labs were out the door with a replacement, the 9370 line of "baby" mainframes. Poughkeepsie and Endicott actually worked with one another and got something done. The new product line used rack-mounted hardware like the Series/1, with prices ranging from $68,000 to over $900,000. The media referred to the new series as the "VAX killer[4]," a designation that IBM did not dispute. The target was indeed the DEC VAX, but the 9370 failed to make much of a dent in VAX sales. The 9370 had some success in large companies that were already committed to the System/370 architecture. However, any attempt to market outside that segment faced the two inherent issues with mainframes—the total cost of managing a System/370 installation, even a small one, and the lack of application software. The *New York Times* would opine that the 9370 was far more important than a rumored Rochester system called Silverlake[5]. This take on internal competition between the large and small divisions seemed to miss the thousands of applications available to Rochester customers. The 9370 was a modest success, with roughly 6,000 installations[6]. Hidden from view and rarely mentioned was the RISC-based Iliad processor under the 9370 covers[7], a legacy of Fort Knox.

Rochester was not opposed to consolidating product lines, but including the 4300 introduced all sorts of compromises that would lessen customer value. An alternative idea started to take hold, the possibility of combining the System/36 and System/38. However, this was not a message to communicate to headquarters. This left Rochester in a weak position, with incomplete and somewhat underwhelming plans for new System/36 and System/38 models based on work done in parallel with Fort Knox. Rochester realized that maintaining two different architectures would be difficult. Funding would be a major issue, as the two large programs would have to split any available monies. A crisis was developing in Rochester.

A group of engineers involved in Fort Knox broached the idea of a System/36 and System/38 combination, a "mini Fort Knox." Such a pairing could then run on the far more powerful Fort Knox hardware. This group included Pete Hanson, Brian Clark, Dick Bains, Frank Soltis, and Dale Dahl[8]. Hanson had been the Rochester representative on the Fort Knox project. David Schleicher, who returned to Rochester as the programming lab manager, reviewed the concept. He asked Hanson to lead an effort to develop the concept. Though all of them had ongoing responsibilities, he gave them a mandate to present a proposal by Christmas 1985. He wanted a product that could be ready in two years, one that salvaged as much from Fort Knox as possible. Ideally, he would see an actual working prototype to prove the design. The group met in the basement of the new white buildings north of the main site.

There were issues with both platforms that had to be resolved, though most were on the System/36 side. It continued to use Larry Wilson's System/3 architecture designed as a punched card system, an architecture continually patched and modified in the ensuing years. It did not scale up well. Application programs were limited to 64K. The obvious solution was to eliminate the System/3 architecture altogether and use the advanced architecture of the System/38 as the foundation. They could easily integrate the Fort Knox hardware, as it would fall below the high-level machine interface and into the microcode layers.

Getting System/36 applications to run in the new environment represented a key technical test. Working with System/3 limitations, many System/36 applications resorted to a variety of tricks and techniques, including assembler routines. These "nonstandard" applications had to run along with all the straightforward ones. The prototype group adopted the IBM manufacturing application system MAPICS as the System/36 test bed. If MAPICS programs ran on the new box, then anything would run[9].

An even greater challenge was political. There were over 200,000 System/34 and System/36 installations in place while the System/38 had just 20,000 machines installed[10]. It was a replay of John Haanstra's argument about the exceptional success of the 1401 when confronted with the NPL plan. Schleicher was, of course, on board with the effort, but many other executives needed to be convinced.

Hanson secured a makeshift machine room in the basement of Building 30. The relatively easy first step was to set up the Fort Knox rack hardware and implement a System/38 on it. Hanson had Roy Hoffman and Frank Soltis on the System/38 side and Dick Mustain from the System/36 side. The team worked nights and weekends in order to meet Schleicher's deadline. They added a final step, dressing up the machine room for the executive demonstrations.

The president of the Systems Products Division, Mitchell Watson, would not agree to the go-it-alone plan endorsed by Schleicher. However, Steve Schwartz replaced him in late 1985 and immediately authorized the plan. He brought in Tom Furey the following year to replace Tony Mondello. Furey had started with IBM in 1963 as a systems engineer and moved to Kingston to manage mainframe operating system development. He helped launch SNA (Systems Network Architecture) in the 1970s. He was a technical aide to John Akers in 1982 before Schwartz tapped him for the Rochester job. Once on board, Furey reviewed the plan, agreed to it, and presented it

to the Corporate Management Committee for sign-off. He received a
$1 billion budget with the caveat that this was a Rochester program
and a Rochester commitment. The other labs would not contribute.
This was, of course, just fine with Rochester.

Tom Furey received a call from Earl Wheeler, who was now head
of all IBM programming. Wheeler wanted Rochester to have a separate programming center. He wanted autonomy for the programmers,
separate from hardware engineering and the system managers. He did
not want his developers "being commingling out there with undesir-
ables like engineers[11]." He wanted consistency across all of the company's programming centers so that he could direct overall efforts and
inject uniformity. He also insisted that the director of a programming
center be an executive. Furey had selected Schleicher to head the
programming lab, thinking that Schleicher would be a 62 level, just
below the first IBM executive level (E1). Wheeler disagreed. Schleicher rose to an E1, managing the 1,200 programmers in the lab. With
the programmers no longer reporting to the System/36 and
System/38 system managers, he would have a unified organization.

Furey had an exciting new mission and significant funding, but he
faced daunting obstacles. He had two teams that did not seem ready to
work together. The System/36 team was skeptical of the new project
and viewed it as just as bad as Fort Knox. The System/38 contingent
looked down on the System/36 architecture. Furey addressed the
issue head-on. He pressed the vision that Rochester was more than
just some functional unit of the bigger IBM. They had to think like an
independent business and focus on the customer. He started by reorganizing the lab. Jeff Robertson, the system manager for the
System/36, would lead the team supporting existing System/36 and
System/38 products. Schleicher would lead new software development, with the challenge to develop seven million lines of code[12]. Jim
Flynn was the hardware manager. He had opposed Fort Knox, but
Furey gave him another chance to embrace the revised Fort Knox
plan. Furey also created his own Rochester version of the Corporate
Management Committee, staffed with the 18 top stakeholders. He
tapped Frank Soltis as his technical assistant. Furey made one addi-
tional organizational move, one aimed at managing the greater IBM.
He installed Jack Bell in White Plains to be Rochester's onsite repre-

sentative with Corporate in Armonk and the Data Processing Division in White Plains.

Furey was serious when he pressed the heavy focus on the customer. He leaned on market research to understand the different markets of the System/36 and System/38. He instituted extensive customer and business partner councils to ensure that development plans met their needs. The planned system was one of IBM's worst kept secrets due to this openness. There was an upside to making this information widely available. If existing System/36 and System/38 customers knew something good was cooking in Rochester, they would be less likely to switch to competition. Furey also relied on the COMMON user group. As Rochester systems proliferated, the membership and programs of the group kept pace, with an average of 5,000 attendees to each of two major conferences a year. Furey often presided over the executive feedback session at the end, wanting to get customer input firsthand.

The program name was Silverlake, named after a lake formed by a bulge in the Zumbro River near downtown Rochester. Before the Fort Knox cancellation, the System/36 and System/38 teams were separate and focused. Now, they had to come together. It was a rerun of the lab battles back east, with the System/36 team playing the Endicott role and the System/38 team, the Poughkeepsie part. Getting on the same page was critical, as Fort Knox resulted in the loss of a full product cycle. As the two groups interacted, they gradually saw the features of each platform combining to create a better system. The System/36 had a superior user interface, better input/output management, and far more applications. The System/38 brought a highly productive development environment and the benefits of its leading-edge architecture.

Silverlake progressed in record time, just 26 months from inception, representing about half the time of previous large projects[13]. The most significant changes were required in the kernel, the layers of microcode that served as a buffer between the physical hardware and the operating system. This effort alone was three million lines of code, with 200 programmers assigned to the task. The crunch came in the years 1987 and 1988. Furey assumed the role that Glenn Henry had played with the System/38. He kept the naysayers at bay.

In the midst of this massive endeavor, IBM Europe asked for a very small entry model of the System/36. Furey asked Victor Tang to take the lead and evaluate the business case. With a positive assessment, Furey gave his approval. The "baby" System/36 would sell 50,000 units, starting at an entry price of just $5,995[14].

The new hardware and software provided a complete solution for the markets that Furey and Rochester targeted. Applications were central to the offering. There would be 1,000 applications ready at announce, a number that would grow to 2,500 by the time the first shipments began. There had been quite a bit of discussion on what to name the new system. The next logical jump in system sequence would have been to "System/40." The "40" became "400." Supposedly, "400" was greater than "360" or "370" and an order of magnitude greater than "36" and "38". Whatever the thought process, "400" it was. This led to the formal name of the new system, Application System/400, or AS/400 for short.

RPG remained central to the new system. There were minor changes to it, the most visible being the name change to RPG/400. Furey and his team opened up plans for the new system to select customers, application developers, systems engineers, and VARs (value-added resellers) far ahead of the announcement date. This included customer councils to validate content and an event called the Migration Invitational where customers and business partners came to Rochester to port their applications, essentially providing field betas. Rochester shipped 4,755 systems to software companies and customers before the June 1988 announce date, an unheard-of move for IBM[15]. J.D. Edwards received several systems, vindicating their early embrace of the System/38.

Missing from the thousands of applications available for the new system were IBM's own Industry Application Programs (IAPs). IBM would exit the applications market entirely in 1992. The logic seems to be that business partners provided more than enough choice. There was no need for IBM to compete with its own partners. There was more going on than that simple explanation. IBM was not fully committed to the applications market. Customers, though generally very happy with IBM software, felt the company was slow in delivering new function. IBM's application problem was a reflection of the company's thinking at the time. The traditional focus had been on hardware, and application software existed to sell hardware. IBM subsidized the IAPs to fulfill that goal. The company did not view software as a separate and potentially lucrative business. However, the shift was already underway where the higher margins would move from hardware to software. Rochester's IAP customers learned this firsthand when they looked at alternatives to their IBM applications. They were typically far more expensive.

The flagship of IBM's applications programs offers a good example. The Manufacturing, Accounting, Production, and Information Control System, or MAPICS, debuted in 1977. IBM sold MAPICS to Marcam[16]. Richard Cook, who had started as an IBM programmer in Boca Raton and rose to take over the Atlanta Center from Raynor Moore, left IBM after 25 years to join Marcam and head its newly acquired MAPICS division. Taking a page from Alan Shugart, Cook took over 100 programmers from the IBM Atlanta Center. Four years later, Marcam made the MAPICS an independent company with Cook still at the helm. In the first year, revenues were $130 million with nearly $20 million in net profit. It is puzzling why IBM would not want a business like that. The hardware to run the software would be icing on the cake. It represented another cautionary tale, that IBM's focus on hardware, and later services, would be missing a central market opportunity. More than that, the software companies with applications designed to run on multiple platforms would often control the hardware decision.

The AS/400 family, announced in June 1988, included six models (B10 to B60), spanning a 24-fold range in memory, 48-fold range in disk storage, and 10-fold range in performance. The top-end model was twice as fast as the System/38 and five times the performance of the System/36. The new processing technology and the pre-announce programs ensured that there were no performance surprises as experienced with the System/38. The new family moved to a standard rack-mounted physical structure, the same arrangement used by the 9370 small mainframe two years earlier. Prices ranged from $19,000 to $230,000.

The operating system (OS/400) included standard OS functions plus integrated database, security, network management, and application development programs. A comparable System/370 mainframe environment required DOS or MVS, CICS for terminal management, TSO for interactive programming, DB2 for database, VSAM for file management, Netview for network management, RACF for security, and various other tools for programming. In addition, this environment also required a technical staff[17]. OS/400 included all of that with one simple, standard interface.

On the eve of announcement, Rochester had grown to well over 7,000 employees working at a site composed of 32 buildings spanning 586 acres under one roof, all interconnected with eight miles of indoor, winter-resistant corridors. There were a total of 3.5 million square feet of working space. By the launch date, GSD had sold over 250,000 System/3-heritage systems[18]. Tom Furey and his team had successfully completed Silverlake development and the AS/400 announcement. The pressure did not end there. The year 1988 would be challenging for IBM, with a heavy focus on revenue and profit. John Akers had taken over as CEO in 1985 and struggled to maintain the year-on-year growth of his predecessors. With the openness of the Silverlake plan, many customers decided to hold off on purchases or upgrades of System/36 or System/38 and wait for the new system.

Furey decided to ship about 400 systems early to help make 1988 numbers. This predictably produced a high level of software failure calls from customers. Steve Schwartz, the IBM senior vice president overseeing Rochester, saw only the problems and felt Rochester had let him down. He turned the whole lab into a support team to fix the problems. Schleicher commented that customers became hesitant to report software bugs for fear that "within twelve hours, I've got three people in my lab area," as one customer noted. Schleicher had set up a separate team of programmers called the "Bug Stompers", ready to descend on a customer in a moment's notice.

The successful early ship program in 1988 was a prelude to far greater things. The AS/400 moved the bar for success far higher. GSD sold 100,000 AS/400 systems within the first 18 months, bringing Rochester's cumulative install base to 450,000 systems.[19] In the first four years, over 200,000 AS/400 systems shipped, while roughly 300,000 System/36 and System/38 computers remained installed. While the System/34 and System/36 surmounted the

100,000 plateau, the AS/400 would top that and keep going. The 200,000th system shipped to Heineken in the Netherlands. By 1994, Rochester reached the 250,000[th] plateau with an AS/400 installed at Coca-Cola in Belgium[20]. By 1995, the application portfolio for the system reached 25,000 applications.

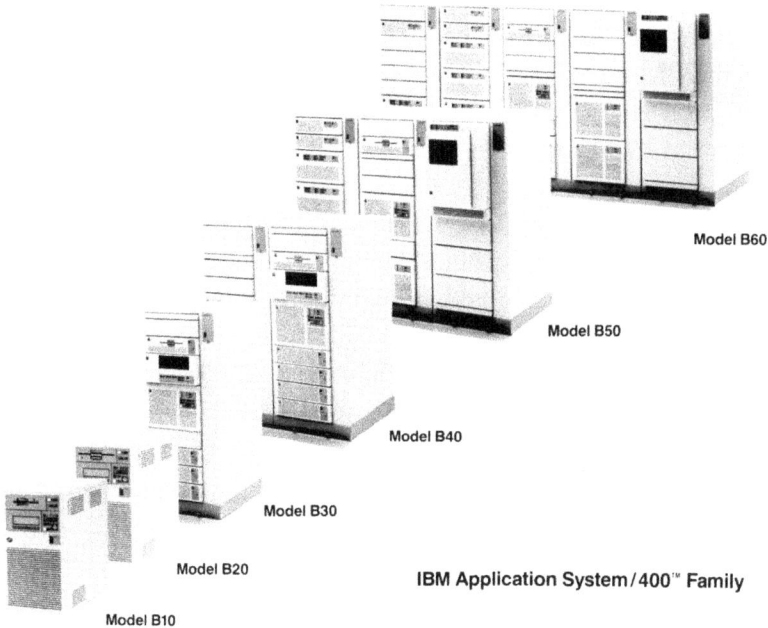

Model B60

Model B50

Model B40

Model B30

Model B20

Model B10

IBM Application System/400™ Family

**Figure 54**. AS/400 family of computers (1988), combining the System/36 and System/38 architectures, an outgrowth of the Fort Knox project. Reprint Courtesy of IBM Corporation ©.

The ease of use, security, and price-performance of the new line resulted in a significant number of customers installing more than one system. State Farm had well over 10,000 AS/400 machines, one in each field office. Allstate made the same move, with 13,000 systems running in each of its branches. American General Finance had nearly 1,400 small loan branch offices, each with an AS/400. J. D. Edwards had thousands of customers on the platform. For many years, Microsoft ran all of its corporate applications on twelve AS/400 systems[21]. This was certainly an embarrassment to Bill Gates and

Steve Ballmer, who had targeted IBM systems with Windows servers. The list of companies that used AS/400 systems included a very high percentage of the Fortune 1000.

The AS/400 was a significant factor in IBM's continued growth. Bob Unterberger, the site general manager, confirmed that annual AS/400 revenue was $14 billion (as reported by *Datamation* in July 1991)[22]. This would make the AS/400 Division, if a separate company, the second-largest computer company in the world, after parent IBM and ahead of Hewlett-Packard (HP) and DEC. The AS/400 averaged 10-12 percent annual revenue growth with gross margins of 56 percent[23].

Rochester recognized a number of individuals for their contribution to the new system. Al Cutaia, a senior engineering manager who had a hand in 121 IBM patents, became an IBM Fellow. Brian Clark and Dale Dahl, who had both been on the Fort Knox team with Pete Hansen, received large (for IBM) cash awards. Frank Soltis had a much more visible role with the AS/400, particularly at COMMON and key customer meetings. He became known as the "father of the AS/400." David Schleicher reiterated the substantial contributions Soltis made but noted, "people that refer to him as the father of the AS/400 will raise up a thousand eyebrows[24]."

Like the System/38, the new system was a massive team effort, led by Furey and Schleicher, and the management team. Pete Hanson was responsible for the concept and with driving a small working team to produce the prototype and secure approval. The scope of Silverlake was far less in terms of radical architecture than the System/38. The architecture eased the task of melding the System/36 into the same box. Furey earned a victory lap. He drove the customer and partner focus, getting them involved early, and ensuring a tidal wave of applications was ready at announce. David Schleicher returned to Rochester and now had a part in two of its biggest successes.

The COMMON user group cheered the new system. The angst about "headroom" disappeared. Members switched to comments about IBM's attempts to convert larger AS/400 customers to small mainframes. There was generally zero appetite among COMMON customers to migrate to System/370. One leading AS/400 consultant, Al Barsa, achieved minor celebrity by producing lapel pins that read simply "400 > 370." Barsa was also famous for being first in line at the open-mike sessions with IBM executives. A common refrain voiced there was the lack of marketing by IBM for the AS/400. It continued as "IBM's best kept secret." These customers applauded when an IBM executive reviewed the plan to advertise the AS/400 on the television shows "Boston Legal" and "Lost" during the upcoming holiday season.

Commenting on the AS/400 announcement, the *Computer Business Review* noted that DEC remained strong and "IBM will have a very tough time expanding its base with the AS/400." In 1989, DEC had 125,000 employees and revenues of $11.5 billion. Yet, two years later, IBM's Rochester division replaced DEC as the second largest computer manufacturer in the world. As for DEC, so great was its rise, and greater still was its fall. Ken Olsen, known as the "Grand Dragon," decided to go all out in an attempt to overtake IBM. He made huge investments in plant, equipment, and workforce to support dramatically higher revenue targets. Unfortunately, this massive infusion came at the same time that the company's spectacular growth leveled off, the PC revolution gained steam, and IBM was fighting back with competitive machines. The over-investment bubble burst and the company spiraled downward. It hit rock bottom in 1992 when Olsen was pushed out. Compaq acquired DEC at a fire-sale price[25].

The other minicomputer leaders endured similar fates. Wang's forte had been word processing. The company elected to follow IBM's DisplayWriter lead with the proprietary Wangwriter before making a belated attempt to enter the PC market. However, the Wang PC was not compatible with the IBM PC, and sales went nowhere. More poor decisions followed in the wake of An Wang's succession, an echo of the experience of NCR when John Patterson died in 1922. An Wang was not your typical tech company CEO. He noted, "I only own two suits, which I replace when they wear out." He combined a personal approach with a laser focus on results. His nickname was the "Velvet Hammer.[26]" He passed the company to his son, Fred Wang, who had none of the capabilities of his father. The company went into free-fall and declared bankruptcy in 1991. Data General managed to last a bit longer by diversifying into disk drives.

# THE RISC REVOLUTION

Rochester was flying high with the success of the AS/400. As a $14 billion business, it represented 15 percent of IBM's total revenues. The Rochester workforce peaked at 8,000 in 1991. With IBM under John Akers struggling, Bill Gates was asked if he was interested in buying the company. He responded that he was only interested in the AS/400 division[1]. Rochester was not about to rest on its laurels. There was the normal pressure for more performance from larger accounts. More and more customers replaced batch applications with online functions, soaking up CPU cycles. Technology continued to march forward. As Glenn Henry predicted, that technology was easy to implement with Rochester's advanced architecture. In addition, Rochester would receive a corporate "nudge." Despite prior efforts by Armonk to limit the System/38 "ceiling," a voice from FS days would push Rochester to new performance heights.

John Cocke's RISC (Reduced Instruction Set Computer) would take center stage. Cocke had been thinking about RISC since his days on Stretch, and on through FS, Project Y, and the ACS supercomputer program in Menlo Park. Fred Brooks, who had shared an office with him during the Stretch days, recounted Cocke's great skill at wandering around, diving into technical discussions, and enlisting collaborators. Cocke got the idea of pipelining instructions from

Harwood Kolsky, compiler optimization from Fran Allen and Jack Schwartz, and elements of RISC from George Radin[2]. These experiences led him to believe that an architecture that could execute an instruction with each machine cycle, or even execute multiple instructions, would be a powerful step forward.

In 1974, he worked on telephone switching project[3]. The system needed to handle one million calls per hour, or 300 per second. A rough estimate of the compute power needed with current CISC (Complex Instruction Set Computer) machines was 12 MIPS (million instructions per second). At the time, the fastest IBM machine was the System/370 Model 168, a 3 MIP system. However, the application had little need for the expansive System/370 instruction set. It did not require extensive arithmetic or logic. Cocke started looking into a radically reduced instruction set. Though the switching project ended the following year, his design approach seemed to hold promise.

Cocke, named an IBM Fellow in 1972, decided to keep working on the idea. He looked to combine his reduced instruction set work with some tricks employed on Stretch. He added branch prediction, look-ahead buffering, and pipelining to his emerging concept. Pipelining facilitated multiple instructions executing in parallel. He revised his goal of one instruction per cycle to multiple instructions per cycle, a concept termed superscalar.

He needed to decide which instructions to keep and which to jettison. He wanted only simple instructions capable of running in one machine cycle. This required frequency analysis to identify the most used instructions and create the optimum reduced set based on them. Program traces run by engineers at Poughkeepsie and Endicott vividly demonstrated the skew of high-frequency instructions. While the System/360 architecture had well over 200 instructions, the analysis showed that about 40 instructions comprised 90 percent of the executions. Another study showed a typical program used a group of just five instructions 50 percent of the time. Cocke combined this analysis with winnowing the complex instructions that required multiple CPU cycles. This made pipelining and parallel execution far more effective.

The programming language compiler determined the machine language instructions. Therefore, the next step required changes to the compilers to favor simple, high-frequency instructions. John Backus and his team used this process in ensuring the FORTRAN compiler produced machine code that competed favorably with hand-developed machine code. Cocke enlisted another member of the Stretch and ACS teams, Frances Allen, for these optimizing compilers. She first placed any required complex instructions in high-speed microcode, then used the instruction frequency to guide which instructions she selected to optimize the resulting program.

In 1975, Cocke built a test RISC machine called ServiceFree that ran at 80 MIPS, fifty times faster than the fastest System/370 model. This prototype led to the 801 project. "801" was the name of the building at IBM Research in Yorktown Heights where Cocke worked. He reported to George Radin, the key architect of FS. Radin kept the 801 project moving forward and ensured the funding was in place. Cocke continued his practice of wandering around, talking with colleagues, bouncing off ideas, and getting feedback. Along with his project team, he took this approach even farther afield. He talked with Seymour Cray on his use of a simple instruction set[4]. His team would also use Cray's compact wiring designs to keep electrical distances to a minimum, and thereby boost performance. Cocke also met with two researchers who were studying RISC on the West Coast—Dave Patterson at Berkeley and John Hennessey at Stanford.

Cocke completed the 801 prototype in 1978. His team reduced the instruction set to 120 instructions. They increased the address space to 32-bit, perhaps a nod to Fred Brooks and his "miss" on System/360. The 801 used advanced chip circuits supplied by Motorola that provided a fast machine cycle time of 66 nanoseconds. Performance was outstanding. The team ran a comprehensive analysis that put 801 performance at 15 MIPS, roughly the same performance as the high-end System/370 Model 168[5]. At the outset of the program, Radin concluded that if RISC techniques on the 801 yielded a 10-30 percent performance improvement, they were not worth pursuing. IBM would be better off just waiting for faster chip technology to catch up and increase the performance of existing CISC processors. However, Cocke's RISC techniques resulted in performance gains of one to two orders of magnitude, an enormous jump.

These performance numbers were also a problem. Vicky Markstein, who worked for John Cocke for 15 years, put it, "You're going to go out and put out a computer for $15,000 purchase that's faster than our $25,000 a month computer," referring to the Model 168[6]. Despite the outstanding performance, the 801 fell short of one goal, that of executing at least one simplified instruction per machine cycle. Subsequent advances in circuit density would solve that problem, enabling expanded pipelining to occur during machine waits. The team arranged a demo of the 801 for IBM president John Opel and Data Processing Group executive John Akers. There was a slight problem with cooling on the 801. It would not run until it warmed up and then, there was a 20-minute window when it ran fine. Then, it grew too hot and shut down. They were able to get Opel and Akers in and out in that 20-minute window.

Cocke's RISC architecture relied heavily on the compiler's ability to produce efficient RISC instructions. In 1982, IBM Research and the Austin lab developed the Austin Compiler System[7]. It would compile source written in PL/1, FORTRAN, or C and produce compiled objects that would run fast and efficiently on the 801. It would also compile object code for the System/370 and the Motorola 68000 microprocessor. This allowed the same program to run on a RISC machine, the System/370, or a personal computer using the Motorola chip. Programs built with the Austin compiler ran faster on the System/370 than native CISC compiles. The Austin system could serve as a bridge to a RISC-based architecture for the IBM mainframe. The Austin system had one flaw: It required ten megabytes of memory to run. The System/370 remained limited to just 16 megabytes of memory total. The mainframe team chose to pass on Austin's compiler and the innovative system was shelved.

IBM Research continued to experiment with RISC. The Cheetah project in 1982 achieved a superscalar extension of the 801, with multiple instructions executing in parallel per machine cycle. In 1985, a RISC prototype called "America" deployed new, higher density circuits. John Cocke would joke that he selected the program name as no one, not even an IBM executive, dared kill a project named America.

IBM's work on RISC was secret until 1982. IBM's initial response to RISC was lackluster. The graduate students at Stanford and Berkeley, John Hennessy and David Patterson, knew of the 801 and pushed their own RISC concept forward. Sun Microsystems and Hewlett-Packard recognized the potential and invested in it. Based on Patterson's work, Sun introduced a workstation in 1987 with a RISC chip, which evolved into its proprietary SPARC (Scalable Processor Architecture) RISC series. It looked like IBM had once again invented a breakthrough technology only to let competitors beat it to the market.

## GLENN HENRY AGAIN

Commercial work on RISC began in an unlikely place for IBM. Austin was the center of development for the company's Office Products Division (OPD). The division was a legacy of Watson's purchase of the Electromatic Typewriter Company in 1933. Its product mainstays were the Selectric typewriter and a series of dictation equipment. In June 1980, the division announced the DisplayWriter, an 8086 microprocessor-based, dedicated word processor priced between $10,000 and $14,000. The announcement of the IBM Personal Computer the following year significantly dimmed its success.

Office Products Division worked on a more ambitious word processing project in collaboration with IBM Research. This was a shared word processor based on RISC designed to compete with Wang. It had the decidedly unexciting name of Research OPD Miniprocessor[8], or "ROMP." The ROMP processor chip contained 45,000 transistors. In January 1981, John Opel took over as CEO, and one of his first actions was a massive reorganization that effectively eliminated the Office Products Division, leaving ROMP in limbo.

Glenn Henry left Rochester and its brutal winters in 1980 and moved to Austin. He reviewed the ROMP program and concluded that the focus was on the wrong set of users. He believed that engineering users were the more appropriate targets. He developed a proposal for such a machine. He received rejection after rejection, working his way up the IBM management chain with the company's non-concur process. He finally reached John Akers, who as IBM president was second in command. He presented to Akers in Armonk. His pitch ended with a foil that summarized what he was going to build, what it was going to cost, how many he expected to sell, and when he would deliver it. Akers approved the funding but grabbed that last foil and stuck it in his pocket. He wanted to remember Henry's commitment.

Henry built a 50-60 person team and started working on the engineering workstation. With a clean-sheet design and the benefits of RISC, he hoped to address his earlier failure with the System/3 Model 6 and BASIC. He was familiar with John Cocke's 801 prototype and readily adopted RISC as the appropriate internal architecture for his hardware. The new machine was the RT/PC, for "RISC Technology Personal Computer." IBM's Microelectronics Division in Burlington produced the machine's processor chips. Henry decided to forego developing another operating system and converted UNIX to an IBM variant called AIX[9].

The RT/PC arrived in 1986. The naming of the RT/PC was a miscalculation, as it was clearly not a PC. The entry price of $20,000 did not attract the attention of many IBM salesmen. The sales of 23,000 machines, including 4,000 installed internally, did not impress Armonk and fell short of the numbers that Henry had put on Akers' foil. The failure of the RT/PC did not help the cause of RISC within IBM. IBM's chief scientist, Ralph Gomory, stepped in to champion RISC, and enable the company to take a second crack at a RISC workstation.

The RT/PC paved the way for IBM's real performance workstation, the RISC System/6000 (RS/6000), announced in 1990. The ROMP architecture evolved into the POWER (Performance Optimization with Enhanced RISC) architecture, an instruction set architecture that would eventually become the basis for all IBM systems. The RS/6000 used POWER1, the first generation of the architecture. IBM would produce an integrated circuit chip containing 6.9 million transistors that featured pipelining, out-of-order instruction execution, branch prediction, and multiple instructions per machine cycle. The RS/6000 featured a one million-pixel display, one million instructions per second (one MIP) of processing power, and one million bytes of memory[10]. The RS/6000 was one of the fastest workstations of its time. It ran at two to three times the performance of competitive

machines (Sun, DEC, MIPS, and Apollo) and obliterated the performance of CISC machines[11].

The hot performance of the RS/6000 did not go unnoticed. In
1991, Apple was looking to replace the Motorola microprocessor that
powered its Macintoshes, as they had fallen behind Intel microprocessors in performance. Motorola wanted a way out of the constant
struggle to match each new Intel chip generation in order to stay relevant. IBM now saw the potential of RISC architecture to transcend
Intel's CISC-architected PC designs. All three companies were interested in any avenue to blunt the duopoly that was Intel and Microsoft.

Apple and IBM represented two dramatically different corporate
cultures. Apple had sarcastically "welcomed" IBM when the IBM PC
was announced, and then followed that with the 1984 Super Bowl ad
casting IBM and its PC as Big Brother[12]. However, IBM's lead technical executive, Jack Kuehler, concluded that the promise of RISC
was so great that the potential pitfalls of working together represented
an acceptable risk. He had an assist from Phil Hester, an engineer on
the RT/PC and the leader of the RS/6000 project. Hester pressed for
a broader coalition to promote RISC and worked with his counterparts at Apple and Motorola to establish an alliance. The goal was to
create a complete new range of RISC-based microprocessors. In
October 1991, the three companies agreed to a joint venture called
AIM, for Apple, IBM, and Motorola. The announcement of AIM sent
shock waves through the industry[13].

IBM and Motorola created a new lab in Austin staffed with 300
engineers. The team first focused on changes to the POWER architecture to accommodate the use and potential growth for each of the partners. The changes included moving from 32-bit to 64-bit addressing
and adding Stretch techniques to produce superscalar results. The
revised RISC architecture debuted as the PowerPC, with the first
microprocessor releases as the PowerPC 601. The chip contained 2.8
million transistors with a density spacing of 600 nanometers. Apple
would initially use the chip in its high-end Macintosh computers.

The other part of the duopoly, Microsoft, also needed to be
addressed. With AIM, there was now increased dialog between IBM
and Apple. The leader of an Apple programming team developing an
Intel version of the Mac OS contacted Phil Hester about IBM's

interest in such a product. Hester was not interested, but the conversation segued to another Apple operating system project, called Pink. Pink had its start in the late 1980s when a small group of programmers became bored with the current Mac OS and threatened to leave the company[14]. At the time, the Mac operating system, called System 7, was inferior to Microsoft's DOS, especially in multitasking. The threat of losing key engineers and the need to improve Apple's operating systems led to a planning session where the team discussed future options. As they identified features, they captured them on blue index cards for enhancements to the current Mac OS, on red cards for some future OS, and on pink cards for a major new release of Mac OS. The features listed on the pink cards became the "Pink" OS project, and the disenchanted programmers, now the "Gang of Five", started work on it in 1987, with an initial release targeted for 1989.

The Pink OS was to be an object-oriented system able to run on any hardware platform. Pink would be a controlling software layer that could sit atop any other operating system. Despite ramping Pink up to over 100 programmers, progress was slow. The OS was nowhere near completion when Phil Hester caught wind of it. He was impressed and pressed Jack Kuehler to explore a software alliance with Apple to compete with Microsoft. Discussions between IBM and Apple led to a joint venture called Taligent. The goal was a new desktop operating system that would complement what IBM, Motorola, and Apple were already doing with RISC and the PowerPC. The first IBM-Apple planning meeting took place in Austin in February 1991. The Apple contingent showed up in three-piece suits while their IBM hosts dressed down in denim. The dual mismatch reflected both parties' motivation to bridge the culture gap and make the new venture work. In 1994, HP joined the Taligent venture. Eventually, there were over 400 people from the three companies working on the project.

## THE MEASUREMENT OF COMPUTER SPEED

Grace Hopper was still around and still talking about computers. Morley Safer of "60 Minutes" caught up with her in 1983 and asked about the state of the computer revolution[15]. Hopper replied that it was still in the "Model T" stage. The torrid pace of Moore's Law and ever-increasing transistor density reflected that assessment. Grace Hopper still had that piece of wire 11.8" long to illustrate how far light, and thus electrical impulses, would travel in one nanosecond (one billionth of a second). If you halved that distance, then performance would double. Early on, the microsecond (one millionth of a second) was the better distance to illustrate. Hopper had a visual for that as well, a coil of wire 984 feet long. She did not have an aid for the millisecond (one thousandth of a second), as electrical impulses traveled 186 miles in that time span. As miniaturization continued, she scrambled to find an aid for the picosecond (one trillionth of a second). This turned out to be a finely ground grain of pepper.

The early history of computer performance largely depended on the pace of miniaturization of computer circuits. Going back to ENIAC, the vacuum tube represented a single data bit. The machine's electrical cycle read or wrote that bit in 14,200 nanoseconds. The electrical impulses still travelled 4,000 meters during that time slice. Ralph Palmer recognized this dynamic and sought to produce a faster, more reliable vacuum tube in-house. His pluggable units for the IBM 604 Multiplier reduced the size of the base vacuum tube and the associated resistors and capacitors down to one compact tube. At about the same time, the trio of Bardeen, Brattain, and Shockley perfected the transistor. The new device represented a 50-fold decrease in the space required to represent a bit of data. IBM deployed discrete transistors in Ralph Palmer's new packaging, the SMS card. In 1966, integrated circuits, the etching of components on silicon wafers, put computer technology on a completely different performance track. The density of transistors on the silicon layer defined the technology. The first implementations, termed Small Scale Integrated Circuits (SSIC), contained up to ten transistors per chip. Large Scale Integrated Circuits (LSIC), with 1,000 or more transistors, was the next stage. Rex Reed tried these circuits on his Symbol program. Gordon Moore, who led the Traitorous Eight and was one of the founders of Intel, laid out "Moore's Law" in 1965, predicting a doubling of circuit density every two years. Intel's first chip, the 4004, contained 2,300 transistors. Less than two years later, the Intel 8008 had 3,500 transistors. Transistor density reached 29,000 per microprocessor chip in 1981 with the IBM PC. Intel pushed the transistor density to 134,000 with the 80286 and 275,000 with the 80386. They crossed the million-transistor threshold in 1989 with the 80486 microprocessor chip. It contained 1.2 million transistors with 800-nanometer spacing between the elements. The following generation, the Intel Pentium, had 3.1 million transistors at 600-nanometer density.

| Machine | Year | Tech | Components | Adds/sec | Mult/sec | Add/Mult Ratio | Cycle Time (ns) |
|---|---|---|---|---|---|---|---|
| Burroughs Adding | 1911 | Mech | - | 0.40 | - | - | - |
| IBM 601 | 1931 | Mech | - | 1 | 0.16 | 6.3 | - |
| Zuse 1 | 1936 | Mech | - | 4 | 0.25 | 16.0 | - |
| ABC | 1940 | Tube | 300 | 30 | - | - | - |
| ASCC (Mark I) | 1944 | Hybrid | 2,225 | 3 | 0.17 | 17.6 | - |
| Colossus | 1944 | Tube | 2,400 | 5,000 | 50 | 100.0 | 200,000 |
| ENIAC | 1946 | Tube | 17,468 | 5,000 | 357 | 14.0 | 200,000 |
| EDVAC | 1947 | Tube | 6,000 | 1,160 | 340 | 3.4 | 862,069 |
| Manchester Baby | 1948 | Tube | 550 | 3,330 | 50 | 66.6 | 300,300 |
| SSEC | 1948 | Tube | 12,500 | 2,000 | 70 | 28.6 | 500,000 |
| EDSAC | 1949 | Tube | 3,500 | 3,500 | 1000 | 3.5 | 285,714 |
| IBM 604 | 1949 | Tube | 1,250 | 3,330 | 333 | 10.0 | 300,300 |
| BINAC | 1950 | Tube | 700 | 8,333 | 555 | 15.0 | 120,005 |
| SWAC (Huskey) | 1950 | Tube | 2,300 | 15,000 | 2604 | 5.8 | 66,667 |
| UNIVAC I | 1950 | Tube | 5,200 | 1,905 | 465 | 4.1 | 500,000 |
| LEO I | 1951 | Tube | 6,000 | - | - | - | 500,000 |
| Whirlwind I | 1951 | Tube | 5,000 | 16,000 | 1400 | 11.4 | 62,500 |
| EDVAC | 1951 | Tube | 6,000 | 1,160 | 340 | 3.4 | 862,069 |
| IAS computer | 1953 | Tube | 1,700 | 1,430 | 100 | 14.3 | 699,301 |
| IBM 650 | 1953 | Tube | 2,000 | 15,000 | 2200 | 6.8 | 66,667 |
| IBM 701 | 1954 | Tube | 4,000 | 16,000 | 2000 | 8.0 | 62,500 |
| IBM 702 | 1954 | Tube | 1,700 | 3,950 | 833 | 4.7 | 23,000 |
| IBM 704 | 1954 | Tube | 4,000 | 66,000 | 32000 | 2.1 | 12,000 |
| IBM 705 | 1954 | Tube | 1,700 | 8,400 | 1250 | 6.7 | 23,000 |
| SAGE | 1955 | Trans | 60,000 | 50,000 | 33000 | 1.5 | 20,000 |
| NORC | 1958 | Tube | 9,800 | 250,000 | 50000 | 5.0 | 4,000 |
| UNIVAC II | 1958 | Tube | 5,200 | 5,000 | 526 | 9.5 | 200,000 |
| IBM 7090 | 1959 | Trans | 50,000 | 229,000 | 39500 | 5.8 | 4,367 |
| IBM 1401 | 1960 | Tube | 10,600 | 250,000 | 125000 | 2.0 | 4,000 |
| LARC | 1960 | Trans | 62,000 | 250,000 | 125000 | 2.0 | 4,000 |
| DEC PDP-1 | 1962 | Trans | 2,700 | 715,000 | 400000 | 1.8 | 1,399 |
| IBM 7030 Stretch | 1964 | Trans | 168,000 | 3,300,000 | 2000000 | 1.8 | 303 |
| CDC 6600 | 1966 | Trans | 400,000 | 1,000,000 | 5500000 | - | 100 |
| 360-91 | 1968 | Trans | - | 1,600,000 | 5500000 | - | 60 |
| 370-168 | 1972 | IC | - | 1,250,000 | - | - | 80 |
| 3033 "Big One" | 1977 | IC | - | 900,000 | - | - | 58 |
| 801 RISC | 1978 | IC | - | 1,600,000 | - | - | 66 |
| 3081 "Adirondack" | 1980 | IC | - | 5,000,000 | - | - | 26 |
| Pentium PC | 1993 | MP | 3,100,000 | - | - | - | 15 |

**Figure 55.** Machine technologies and computing speeds. Note the ratio addition times to multiplication, a reflection of progressing algorithms starting with James Bryce's 601 Multiplier.

IBM's POWER technology was able to ride this wave of increasing density. The Microelectronics Division packed 45,000 transistors into the ROMP processor of the RT/PC. The POWER1 processor used in the RS/6000 arrived in 1990 with 6.9 million transistors, each spaced 1,000-nanometers apart. Chip design began with single-thread, single-CPU processors like the pioneering Intel 4004. This changed when designers etched multiple logical processors, called cores, into a single piece of silicon. Each core was a separate CPU, able to execute program instructions. The technology jumped dramatically with the process of multi-threading, the ability to run more than one program task on a single core. The Power4 generation of the PowerPC had two cores, each capable of running two threads. A POWER4 machine was called a "4-way" processor. When the POWER10 generation of the PowerPC arrived in 2021, the processor could have up to 30 cores with eight threads each, providing a 240-way system. Some specialized microprocessor chips contained up to 10,000 cores while a supercomputer might sport a core count in the millions[16].

There was another lever to accelerate processor performance. That was the clock speed, the internal machine cycle time of the system measured in hertz, or pulses per second. The ABC computer had a cycle time of 60 hertz, determined by the rotation of its small electric motor. The ENIAC ran as high as 100 kilohertz, or 100,000 pulses per second. Moving to transistors produced an order of magnitude improvement in clock speeds. Both Stretch and the CDC 6600 ran at ten megahertz, or ten million cycles per second. Integrated circuits ramped clock speeds up to gigahertz levels, or billions of cycles per second. The Intel Pentium III was the first processor to cross one gigahertz. The POWER microprocessors went through a similar acceleration. POWER1 clocked at 30 megahertz, POWER2 at 160 megahertz, and POWER3 at 450 megahertz. The POWER4 chip crossed one gigahertz. POWER speeds continued to increase until POWER8 topped out at 5.6 gigahertz.

Multiple cores and higher clock speeds generated dramatically more heat. Later POWER architectures dialed back to four gigahertz and used other means, such as massive banks of parallel microprocessors, to increase performance. As of 2023, the terahertz level, or a trillion cycles per second, remained unbroken.

## ROCHESTER GETS RISC

The Rochester campus acquired the moniker "Fortress Rochester", a reflection of its insularity. It had served the General Systems Division quite well. However, Carl Conti was not happy with Rochester's progress in one important area. As you recall, Conti had chaired the AFS (Advanced Future System) group in 1970, looking at an FS-type leap for IBM. Conti pressed an approach that included John Cocke's RISC innovation but was too low in the IBM hierarchy to prevail. In the ensuing years, Conti moved up in the company. He became the lab director of Endicott in 1975, president of the Data Systems Division in 1983, and senior vice president and president of Enterprise Systems in 1990. In those 20 years, Rochester had embraced the FS architecture, first with the System/38 and then with the AS/400. However, Rochester failed to implement RISC. As a senior vice president, Conti was finally in a high enough position to make that happen. It was not a corporate "nudge." He decreed that Rochester would get the AS/400 off its proprietary hardware and on to the PowerPC RISC platform. His decree also included Austin's RS/6000. It was one of his last official actions before retiring[17].

Rochester viewed a conversion to RISC as expensive and counterproductive. However, two big pluses would change its initial opinion. First, the high-level machine architecture of the AS/400 made for a far less disruptive conversion. Second and more important, it would position the platform on PowerPC microelectronics and enable the AS/400 to ride the technology curve to higher and higher performance. It would ensure the long-term survival of the system. Gone would be the days when System/38 customers worried about the "headroom" of their system.

David Schleicher had been back East for only two years, working for Earl Wheeler on Systems Application Architecture (SAA). He would note, "I was kind of like, the darling boy of Earl Wheeler's kingdom[18]." In his time in Rochester, Wheeler felt he did everything right. Rochester enticed Schleicher to return to Rochester to lead the RISC transformation. He would return as an IBM vice president in charge of the entire lab, with Bob Unterberger his hardware peer as vice president of manufacturing.

The RISC programming task involved a rewrite of the microcode layers just below the high-level machine interface. Mike Tomashek led the programming project.[19]. A small team that included Bill Berg, Bill Armstrong, and Dick Mustain recommended a switch to C++, an object-oriented programming language. [20]. They stressed several key advantages to the change. First, programmers could define C++ objects into different categories, each with its own set of properties. Related objects could automatically inherit those properties, reducing total development time. The resulting code structure also dovetailed nicely into the AS/400 core definition as an object-oriented system. Second, the C++ language was fast, scalable, with a rich standard library of routines. It was portable across system platforms and adapted well for both low-level and high-level functions. Its low-level capability was especially important in working with the microcode layers. Third, with the wide availability and desirability of C++, Mike Tomashek would be able to attract the additional 150 programmers he needed.

The task to convert the AS/400 CISC to RISC ran from 1992 to 1994. There would be two million lines of C++ code integrated within the 20 million lines of the existing microcode. As with so many IBM projects, the hardware was not available at the outset, and testing would proceed with simulators. Actual RISC AS/400 hardware would not be ready for a year and a half into the effort. Tomashek and most of his managers had experience in prior programming projects and were able to avoid many of the pitfalls that plagued OS/360 development. The science of software engineering had also progressed, and many of the project programmers had computer science or electrical engineering degrees.

The task on the hardware side of the RISC conversion was a different story. There was more discussion at the outset, trying to assess the best strategy to pursue. The initial focus of PowerPC was high-performance RISC workstations like the RS/6000. The resulting instruction set not a great match for a commercial system, let alone one with such a unique architecture. The AS/400 depended on complex instructions that enabled functions such as database access, decimal arithmetic, virtual paging, and fast online program swapping.

The first Rochester hardware proposal outlined a custom RISC design called C-RISC where the where the "C" stood for "Commercial". Rochester would design the C-RISC processor. IBM's lead technical executive Jack Kuehler vetoed this approach. He wanted all IBM systems to be strictly PowerPC based. Frank Soltis led a ten-person team to come up with an alternative. It would use PowerPC processors, upgraded to 64-bit addressing, with extensions to provide the needed commercial functionality. This hardware design, named "Amazon", received Kuehler's approval. The Soltis team had to bridge the gap between the 228 instructions of the PowerPC and the 392 instructions in the CISC AS/400. They ended up in the middle, with an Amazon processor called "Cobra" that had 253 instructions.

Rochester completed work on the PowerPC-based AS/400 in June 1995. The conversion from CISC to RISC was difficult, but the change to 64-bit addressing was a breeze by contrast. For the customer, reloading their microcode produced a smooth and near painless migration to both RISC and 64-bit addressing. The 64-bit address space activated up to 18 quintillion bytes of data. That would be 18 followed by 18 zeroes. The architects designed the AS/400 to advance to 128-bit addressing, if needed. That would require a relatively simple change in the microcode.

The AS/400 was not the first computer platform to implement 64-bit. The Stretch system used a combination of 32-bit and 64-bit addressing. Seymour Cray used 64-bit in his Cray-1 supercomputer in 1975. However, the AS/400 was the first system with a complete 64-bit implementation, spanning the hardware, operating system, and applications. Many others struggled to make the transition to 64-bit processing and, in many cases, ended up with less than full 64-bit implementations. DEC was under the gun to upgrade its 32-bit VAX minicomputer line. "VAX" stood for "Virtual Address eXtension," so clearly addressing was central. The DEC Alpha instruction set architecture provided the semblance of a 64-bit implementation. Arriving in 1992, the operating system remained as 32-bit code. In addition, roughly 15-20 percent of VAX applications had to be rewritten.

At the time, Intel's 486 microprocessor supported 32-bit addressing, but many applications remained in 16-bit from the prior generation. AMD, a maker of clone Intel microprocessors, released a 64-bit version in 1999. Intel had announced plans for a 64-bit architecture called IA-64 that it would jointly develop with HP. Whereas RISC relied on the hardware to make decisions on branch prediction and out-of-order execution. Intel's architecture would rely on software. This would prove to be a mistake. IA-64 was also not full 64-bit. It supported only 48 bits for virtual addressing, limiting the address space to 256 terabytes. Furthermore, the new architecture would not make it into Intel microprocessors until 2001, when it debuted as the Itanium series[21]. The choice of "Itanium" was unfortunate as it was phonetically too close to "Titanic." The new line soon acquired the nickname "Itanic." A litany of performance problems with Itanium grounded in its composite architecture sunk its prospects.

There was one on-going problem with the AS/400 program. Many System/36 accounts continued to resist making the change. The reasons were typically price and the complexity of the new operating environment. Bob Schmidt began a skunk works in Rochester to develop a new machine that would run System/36 natively on the 64-bit PowerPC RISC hardware. Long-time System/36 designers Dick Mustain and Steve Dahl joined Schmidt in this effort[22]. System/36 applications ran natively, without recompilation. At the same time, the customer could still migrate to the AS/400 environment at a time of their choosing. The new machine, the Advanced/36, debuted in October 1994. The entry price was just $12,000.

With RISC bringing new levels of RISC performance to the AS/400, it became increasingly important to translate all that processing power into real-world impact. The standard measurement for mainframe and scientific performance was MIPS (millions of instructions per second). However, this number bore a limited relationship to commercial applications, particularly mission-critical online applications. Rochester needed a new way to gauge a system's capacity for work. The first tool used was RAMP-C (Registered Approach for Measuring Performance—COBOL). It was also known as "Rochester's Attempt at Measuring Performance.[23]" The CPW (Commercial Processing Workload) performance estimator replaced RAMP-C in 1994. CPW provided more accurate modeling of workloads across the AS/400 line. The baseline for all CPW values was the original AS/400 entry model (B10). Its CPW rating was arbitrarily set at one.

Using the CPW metric, the high-end RISC AS/400 with the POWER1 processor in 1995 (Model 650) rated at 650 CPW. Customers grew to expect annual product announcements, usually in the summer, with significantly increased performance. In 2001, POWER4 chips with 44 million transistors spaced at 180 nanometer density broke the CPU clock speed of one gigahertz, or one billion cycles per second. It was also the first POWER microprocessor to use multiple cores and threads. With this release, IBM combined the two system platforms, AS/400 and RS/6000, completing the FS ambition long sought by Carl Conti. The high-end AS/400 (Model 890) was a 32-way processor with a CPW rating of 37,400. In 2019, the POWER9-based E980 model contained 192 cores with up to eight threads per core. It had eight billion transistors with circuit density down to 14 nanometers. Its CPW rating was 2,743,000.

Timothy Pickett Morgan, a frequent observer of all things Rochester, compared the pricing of an entry AS/400 model in 1988 with a comparable Power Systems model in 2010[24]. The B10 system processor, rated at one CPW, sold for $6,552. That is $6,552 per CPW. The 2010 Power Systems 720 model sporting a three- gigahertz clock rate was fairly close to the same price but rated at 6,000 CPW. That translates to $1 per CPW. Memory was $6,000 for four megabytes on the B10, or $1,500 per megabyte. On the Power Systems 720, eight gigabytes went for $1,065, or 13 cents per megabyte. Disk was $5,500 for a 315-megabyte drive on the B10, or $17 per megabyte. The Power Systems 720 disk was priced at $5,200 for a 69-gigabyte drive, or seven cents per megabyte. Those figures represent an 88-fold reduction in memory, 243-fold reduction in disk, and a 6,552-fold reduction in the price per CPW. Of course, the flip-side of these gaudy numbers was the challenge for IBM to maintain its revenues in the face of such dramatic reductions in the cost of computing.

The AS/400—at least the name—lasted until 2000. Rochester explored rebranding in 1997, considering "PowerStorm" and "Cyber-System/400" before settling on the far more modest "AS/400e.[25]" By 2000, Corporate rolled out far more ambitious naming across all product lines. The company's systems became the eServer zSeries, eServer pSeries, eServer iSeries, and the eServer xSeries. The "z" designation stood for "zero downtime" and referred to the mainframe models, the "p" for performance with the RS/6000 workstations, the "i" for integrated with the AS/400, and the "x" for Xenon, the processor class for the PC server line. In 2004 with the arrival of the POWER5 microprocessors, Rochester fine-tuned its brand to "eServer i5." The platform became "System i" in 2006 and "IBM i" in 2008. In 2020, IBM Corporate replaced all of the platform names with "Power Systems." Through all of these changes, the platform remained "AS/400" for most customers.

Increased performance was important, but the changing nature of competition required much more. The biggest change came in RPG, the venerable Report Program Generator. The language that started by filling in specification sheets to mimic the unit record cycle of read a card, do some calculations, and write some output, had progressed over the years. However, a language geared to batch applications was ill suited to the world of online, interactive applications. In this environment, the fixed flow of RPG was more of a hindrance than help. The most far-reaching changes occurred in 1994 with the release of RPG IV. RPG became a procedural language, joining the ranks of modern languages like C. RPG designers added freeform calculations in 2001 and took the language fully free format in 2013 when the last connection to its unit record past was broken. Through all these revisions, existing programs dating back to FARGO continued to run. Though RPG was by far the dominant AS/400 programming language, it was not the only one. COBOL had been long been supported. The transition to interactive and Internet applications brought new languages to the platform, including SQL, Java, Java-Script, PHP, JSON, Python, Perl, Pascal, C, C++, and Ruby. In addition, the merger of the AS/400 and RS/6000 within the same platform enabled customers to run both AS/400 and AIX applications interchangeably.

The 1990s were years of heady success for Rochester. The AS/400 program reached its high point in revenue in 1990 at $14 billion. The following year the site's workforce peaked at 8,100. The number of systems shipped was staggering. By 1994, 250,000 AS/400 systems servers had been shipped[26]. The 400,000th AS/400 went to Greg LeMond (Tour de France champion), who used the system to manage his bike business[27]. AS/400 volumes were approaching 500,000 by 1997. In 1998, Rochester produced an AS/400 every 12 minutes of each workday. By 2004, the count of AS/400 servers shipped worldwide reached 750,000[28]. Rochester ensured the system capability kept pace with technology, as price/performance increased at a rate of 30 percent per year, driven by the RISC-based POWER architecture and ever-increasing transistor density on the microprocessors. The frequent injections of increased POWER performance served to dispel any notion that the platform had an imposed ceiling orchestrated from Corporate.

There was one person less than impressed by the success of the General Systems Division and Rochester's systems. That would be Bob Evans. Under the title of "A Strategic Blunder" he would contend that "GSD bolted off to develop one product after product, each one basically competing with the low-end and mid-range on the mainstream products.[29]" By "mainstream", he was talking about System/360 and System/370. Though this author highly respects Evans—he is one of the IBM greats—he is dead wrong here. The mainframe line never came close to an entry price and ease of use that was at all attractive to small and medium companies. GSD was successful in cracking all of the price barriers where other IBM systems fell short. IBM watched as midrange competitors like DEC and Wang fell by the wayside. Evans was accurate that early systems suffered with an architecture that did not scale well. Ironically, it was based on the System/360 blueprint. There were a series of less than compatible models at the start, but it was RPG and the laser focus on usability that led to Rochester's success. The fact that these new customers had no interest at all in moving to System/370 architecture speaks for itself. Evans also neglects to mention how the problem of incompatible architectures was resolved. After he cancelled Future System that Valentine's Day in 1975, Rochester picked up the FS advanced architecture and made it work. The FS blueprint with the System/38 and AS/400, along with the easy transformation to 64-bit RISC, continues to power the platform decades later.

As for your author, after working on the System/38 Remote Test Service and the MACS (Marketing and Coverage System) program, I was promoted to Systems Engineering Manager in Denver in 1985.

Once selected for management, you had to attend New Manager's School within 30 days. The school was better known as "Charm School" and taught at IBM's Palisades Conference Center, a 106-acre wooded retreat 25 miles north of New York City fronting the Hudson River. You learned how to delegate. The instructor demonstrated this by trying to get a stuffed monkey off his back. A more serious section was on leadership and used the 1949 movie "12 O'clock High", with Gregory Peck leading an Army Air Force bomber group in England during World War II. Though that sounds unusual, the movie is one of the most widely used aids in teaching leadership in business. I had seen the movie before, but watching it from a leader's perspective was powerful and influential. Tom Watson had described the management training he received at Fort Leavenworth after the war ended. IBM's management program, including this movie, was likely patterned after that experience.

Once "charmed," I took up my new role in Denver, responsible for midrange systems engineers. I subsequently became unit manager in Boulder, responsible for both marketing representatives and systems engineers. With the arrival of the first ThinkPad in 1992, I led the rollout of the new laptop and IBM's initial foray into working remote as the chief information officer for the Rocky Mountain Region.

## CHANGING OF THE GUARD IN ROCHESTER

David Schleicher left Rochester for a second time in 1994. He followed Glenn Henry to Austin. He would tackle a project similar to SAA (Systems Application Architecture) called Workplace OS, which would attempt to build a multi-OS operating system for the PowerPC[30]. There were probably other factors at play in the move. He had returned to Rochester in 1991, the first year of three years of disastrous losses for IBM under John Akers. Despite its great record, Rochester had to share in the painful actions aimed at reviving the company. The first layoffs ever at the site came in 1993. Further cuts eliminated the disk drive production called ADSTAR at the site. By the end of 1994, employment at Rochester had plummeted from 8,100 to 5,500[31].

Schleicher's replacement at the lab was Steve Ladwig, who was returning to Minnesota after having started his IBM career there as a programmer. He was counseled that the days of Rochester running its own show were over. Rochester persevered and even thrived after the elimination of GSD in 1981 and the loss of a dedicated sales force in 1986. It remained the focus for small systems, with everyone at the site working on the same mission. That would no longer be the case. Corporate split the site into a number of unrelated IBM missions, with Austin, San Jose, and Poughkeepsie assuming responsibility for different areas. There was no backfill when Ladwig left in 1998. Armonk was the new manager.

# WATSON, LEARSON, AND SYSTEM/370

After Watson's heart attack in 1971, he decided that living was more important than running IBM. He stepped down as CEO but remained on the IBM board until 1984. He had plenty to keep him busy. He and Olive had six kids and numerous grandkids. He was able to enjoy his passions for sailing and flying. He took off on a major cruise in 1974, sailing 500 miles inside the Arctic Circle off Greenland. He would go through seven sailboats. He had a veritable fleet of flying machines. He owned a Lear jet, a King Air Twin, a Taylor Cub (the forerunner of the Piper Cub), and a Bell Jet helicopter. He also had "Breezy," a contraption that was a step up from a hang glider and looked like the Wright Brothers' first airplane[1]. In 1987, he decided to retrace his wartime B-24 flight to Moscow, spending 16 days traveling around the world in his Lear jet. He took a course in stunt flying, which led to the purchase of a stunt airplane that he could perform the "Hammerhead Stall." He racked up 16,000 flight hours in 50 years of flying. He teamed up with James Rockefeller to organize the Owls Head Museum in Maine, stocked with classic automobiles and airplanes. As an added bonus, the museum had an attached runway.

Tom Watson's record at IBM was a surprise. He was a poor student, a screw-off who fought with his father and struggled with his anger. The early signs pointed to a repeat of the disastrous experience of Frederick Patterson. Though he had many faults as the leader of IBM, his wartime maturation and the pressure he experienced in taking over the company in a time of great transition were essential to his growth and success. He had the vision to see where IBM needed to go and led the charge.

That charge produced a consistent and utterly amazing record during his time from 1956 to 1971. He grew IBM's revenue from $696 million to $8.3 billion, a 12-fold increase that represented a 17 percent compound growth rate year in and year out. He never had a down year[2]. He raised income from $73 million to $1.1 billion, an annual rate of growth of over 18 percent. During the high-flying 1960s, IBM grew at nearly 30 percent per year. His father had taken the company's market value from next to nothing to $2.8 billion. He took it from there to $39 billion, an annualized growth rate of 19 percent. IBM stock multiplied 16-fold. That required seven stock splits during his tenure just to keep the share price within reason. One hundred shares in 1960 became 375 shares by the end of the decade. Watson increased the size of IBM's workforce 373 percent, from 56,000 to 265,000, an average boost of 10 percent per year. IBM climbed to the top of the Fortune 500, leaving traditional rivals like Remington Rand, GE, RCA, and Honeywell far behind. In an analysis done by *Fortune* magazine, IBM created more wealth for its shareholders during Tom's tenure than any other company[3]. The poorer record of many of his successors shines a glorious light on his time at the helm.

His record is more than just financial accomplishments. He propelled IBM to the perennial #1 "Most Admired" company. He compared IBM's education programs with General Electric and saw a company that was behind in a number of areas, particularly in management training. Strong managers would be essential as he transformed the company from a one-man show into a modern corporation, adopting an organizational structure that would enable it to grow dramatically. He also hewed closely to IBM's core values, going so far as to encode them in a book, *A Business and Its Beliefs*[4]. Though he had made the comment very early on that the emerging computer market could only support a handful of ENIAC-size computing machines, he soon recognized his error. He also saw the need for many thousands of programmers to support the many thousands of new computers. He viewed his push to get more programmers trained as one of his foremost accomplishments.

His record in transforming IBM into a worldwide powerhouse was richly rewarded. He received 19 honorary degrees from colleges, including Columbia, Harvard, Yale, Brown, and Oxford. He received the Presidential Medal of Freedom, French Legion of Honor, Order of Leopold II (Belgium), Order of Merit (Italy), and the Royal Order of Vasa (Sweden)[5]. He sat on a number of corporate boards, including Time, Mutual Life, Bankers Trust, and Pan American Airways. In a nod to his wartime experience, President Jimmy Carter selected him as his ambassador to the Soviet Union.

In 1987, *Fortune* called him the "greatest capitalist who ever lived[6]"—certainly a bit of hyperbole but still high praise. In 1999, *Fortune* revisited its analysis[7], pitting Tom Watson Jr. against Henry Ford, Alfred P. Sloan, and Bill Gates. Though the winner was Henry Ford, Watson finished second in this rarefied company.

The company he led repaid him well. He retired with 366,000 shares of IBM stock[8], with a value well over $300 million[9]. Most of his inner circle, including Albert Williams, Vin Learson, and Red LaMotte, also did well. The largest IBM stockholder in Tom's last years was Sherman Fairchild, George Fairchild's son[10]. For Tom, it was a well-earned retirement. In addition to his planes, automobiles, and sailboats he had homes in Greenwich, Stowe, Vail, New York City, Maine, and Antigua. He would muse about everyone wanting to have one "great adventure" in life. People would assume that his 15 years spent leading IBM was that great adventure. However, he would say that though his work at IBM was extraordinary, it was "somewhat dwarfed[11]" by his experiences during his war years.

Watson and Learson worked magic together, but they were not close. Al Williams retired in 1966 and Tom promoted Learson to president of IBM. Now, the tension between them increased. Learson preferred communicating with Tom by memo[12]. Learson's office was a floor below Tom's in Armonk. There were Management Committee meetings 4-5 times per month, and Learson did not like the grillings that were common in those reviews. He was far more comfortable giving the grillings than taking them. As president, he made his decisions and did not like them to be discussed or second-guessed. Tom would say that Vin was "never exactly insubordinate" but he certainly came close. He described Learson as someone who worked best by himself, had no interest in meetings, and did not like to have large staffs or have staff work to review.

Tom Watson had planned to step down as IBM's CEO in 1974 when he turned 60, establishing that as a guideline for future IBM chief executives. He would have Frank Cary take over the company at that time. However, he suffered the heart attack in 1971. Suspecting the continued rigors or running IBM would probably kill him, he decided to retire immediately. Despite his often contentious relationship with Vin Learson, he concluded that Learson had more than earned the right to the top job. He had been his "most able executive" for decades, proving his worth repeatedly. His fingerprints were on every major IBM success from the 701 to System/360.

At the same time, he was apprehensive about what Vin might do as CEO. He decided to limit Learson's time at the helm. He had Vin sign a letter stating that he would resign at age 60[13]. The new guideline that IBM CEOs retire at age 60 became the "Vin rule." As a further measure, Tom remained on the board for several years to watch over him.

Learson retired as promised at age 60 in January 1973. The two giants of the company were gone. Bob Evans mused about the loss of these two individuals so quickly and the impact it had on the success of the company in the ensuing years. There was a reason for Watson to retire but not Learson, who was still at the top of his game. We can ponder what Learson thought of his career—and only ponder, as he left next to nothing of his own; not a book, oral history, or collected papers.

## SYSTEM/370

As initial work proceeded on FS, Bob Evans was busy managing the rollout of System/370. He had been out of the mainstream at Federal Systems for four years when Watson asked him to return to the

Systems Development Division (SDD) and lead the System/370 program. According to Evans, he received significant incentives to make the move. He replaced Chuck Branscomb in October 1969 as Branscomb was bound for Atlanta and the General Systems Division.

As the head of SDD, he oversaw all the IBM labs worldwide. He brought in John Fairclough from the Hursley lab to manage the new lab in Raleigh. He had Byron Havens, of the Naval Ordnance Research Calculator (NORC) and the microsecond delay circuit, report directly to him. Havens had managed all the labs in Europe. He asked Jerrier Haddad, the head of the Poughkeepsie lab, to lead work on an advanced architecture for the System/370. The planning for the new line had started in 1965 as the NS (New Systems) program, with 1970 as the target date. The System/360 standardization of architecture and model compatibility freed up IBM resources to work in new areas, such as supercomputers, peripherals, and applications.

The new line would be natively compatible with System/360. Though originally intended to be a fourth-generation family of computers, it ended up being incremental. This was a stressful time in IBM, not unlike the System/360 push. This was evident in Evans's request to San Jose for a new disk controller. San Jose pushed back, but to no avail. Jack Harker, the disk product manager, advised his team succinctly, "Bob Evans does not run a democracy.[14]" San Jose delivered the 3330 disk drive along with a new controller as per Evans's "request."

Virtual storage was a key centerpiece of the new machines. Evans spent time in IBM "jail" because of his miss on timesharing, which required address translation and virtual storage. In the hand-off lunch with Branscomb, he learned that Dynamic Address Translation (DAT) was no longer in the plan. Evans was not about to make the same mistake twice. He immediately set up a task force to plan for DAT. He also reviewed a proposal from Scott Locken, who led the design of OS/360, to develop an entirely new OS that was tuned to DAT but decided against it. Even so, adding DAT to the existing plan represented a billion-dollar decision, one that could potentially delay the System/370 announcement. Evans asked Martin Belsky to design the implementation of virtual storage. With the existing 24-bit addressing, virtual storage raised addressable memory per to 16 megabytes per user.

Coming into the program late, Evans wanted to delay the high-end models (155, 165) until the DAT hardware was complete and the software was ready. He was overruled, and both models announced in 1970 as planned. The DAT versions arrived two years later with the models 158 and 168. Evans took the same approach with the midrange 135 and 145 models. Evans attributed his success in getting DAT done to support from Vin Learson.

The System/370 machines used integrated circuit chips (MST) that were 70 percent cheaper and five times faster than SLT modules. All of the electronic elements—resistors, transistors, and diodes—were etched on a single wafer of silicon. Erich Bloch and his team designed the new logic chips. The packaging was outstanding, as usual. The circuit density increased dramatically, resulting in heat that had to be dissipated. The architects switched to water cooling, leading mainframes to be called simply "water-cooled." The Model 145 was the first mainframe to substitute magnetic cores with DRAM chips for main memory. As was by now standard practice by the press, they accused IBM of being too timid with the new line. Had FS actually come to fruition, the press would have lambasted the company for taking foolhardy risks.

Endicott was once more up to its old tricks. It was developing a new low-end operating system called LEOS (Low End Operating System) to replace DOS, with a planned release in 1970[15]. The lab pursued LEOS because the DOS code was viewed as such a mess. Evans got an unlikely assist from Jim Frame, the 1401 programming manager who had resisted the NPL software plan and had worked in secret on Haanstra's 1401 Super program. Frame sided with Evans, recommending enhancing DOS instead of developing LEOS.

Tom Watson announced the System/370 on Tuesday, June 30, 1970. The initial models were the 155 and 165. A typical Model 165 sold for $4.6 million and rented for $98,715. The models 135, 145, and 195 followed later in the year. The Boblingen lab did most of the work on the Model 125, an upgrade for System/360 Model 20 users. They followed a year later with the entry Model 115. The market planners, in their dreams, considered it an upgrade for Rochester's systems.

Just as the standardization of input/output devices on System/360 led to the plug-compatible market, so did the standardization of the instruction set and system architecture lead to cloned System/360 processors. Moreover, the IBM antitrust settlement required the company to sell its software separately and at market rates. This enabled a company to buy its computer from someone other than IBM and order system software such as MVS directly from IBM.

Gene Amdahl got the ball rolling. His proposal for the System/360 follow-on featured an enhanced architecture and fourth-generation LSI (Large Scale Integration) semiconductor technology, though he continued to spurn the need of virtual memory hardware or software. After leaving IBM for the second time in 1969 with his vast knowledge and next-generation mainframe plans, he secured funding from Fujitsu and formed Amdahl Corporation in 1971. He raided IBM, hiring a number of engineers that had worked of the ACS super-computer program. Fritz Trapnell would later join him as vice president of software. His first Amdahl mainframe, the 470/V6, launched in 1975. As was the case with plug-compatible peripheral manufacturers, a company like Amdahl could simply piggyback on IBM's invention, development, and marketing. Yet in this particular case, the Amdahl Corporation was getting a design by Gene Amdahl, the lead designer of System/360.

The success of Amdahl's venture led many competitors to adopt some version of the System/360 architecture. The MITI industrial consortium funded the establishment of the Japanese computer industry. Fujitsu, Hitachi, and Mitsubishi entered the mainframe market with System/370 clone machines. Existing IBM competitors also joined the fray. This included the UNIVAC 9000 series, the RCA Spectra 70 line, and the English Electric System 4. Though these systems mimicked the System/360 instruction set, they were not 100 percent compatible.

The failure of Bob Evans to delay the System/370 announcement in order to include virtual addressing would work in IBM's favor. Amdahl designed his first system to compete with the System/370 Model 165. However, as his team was completing that first machine,

IBM announced the Model 168 in 1972 with virtual addressing and hardware. This forced Amdahl to scramble to match IBM's innovation.

Frank Cary was now at the helm of IBM, and he added another tactic to blunt the impact of plug-compatibles like Amdahl. He shortened product cycles in order to devalue the leasing companies' inventory and leave them with outmoded systems that fell short of their profitable payback. The Models 155 and 165 replaced by the 158 and 168 in just two years in part to stiff the leasing companies with old iron. Existing customers could upgrade the older models with the addition of Gerrit Blaauw's DAT address translation box. However, this upgrade was very pricey. The Model 155 and 165 machines became "boat anchors" and the required upgrade box was known as "dat joke.[16]"

Cary did not stop with the Models 158 and 168. Soon after Amdahl released his first model in 1975, IBM replaced the Model 168 in just ten months with the more powerful Model 168-3 to match his performance. Cary added IBM's first multiprocessor models, the 158MP and 168MP, soon thereafter. Nevertheless, Amdahl still got traction. His first install was at NASA, replacing an IBM System/370 Model 165. When MassMutual installed the Amdahl 470/V6, it signaled to other large companies that they could go forward without IBM.

Cary continued his onslaught on mainframe clones in 1977 with the surprise introduction of the "Big One" just two years after the debut of the 158 and 168 models, Formally the IBM 3033, the "Big One" essentially doubled the performance of the 168. The 168 had taken 40 months to develop. The 3033, developed under the code name of "Meridian,[17]" announced just 28 months after development began. *Datamation* magazine called the 3033 the "Big Bombshell." Though the new system was more of enhanced Model 168 than a completely new machine, orders for it were off the chart, well into the hundreds.

Jack Bertram, who abandoned the ACS program when Amdahl took over the project in Menlo Park, was the lead designer of the 3033[18]. He no doubt thoroughly enjoyed trumping Amdahl. Besides the increase in performance, IBM had moved parts of the operating system microcode into firmware, which made it part of the hardware. This meant Amdahl systems could no longer run MVS software until they reverse-engineered IBM's firmware. Amdahl went back to the drawing board and designed his next general machine, the 470/V7. Released in 1977, it surpassed 3033 performance and sold over one hundred systems.

By 1979, Amdahl Corporation had 6,000 employees and crested at 24 percent market share in mainframes[19]. However, such success would not linger. The 1979 announcement of the midrange 4300 system followed by the high-end 3081 in 1980 severely cut into Amdahl's business, with revenues plummeting from $321 million to just $21 million. The company tried to merge with Memorex or Storage Technology, but neither was interested. Gene Amdahl would leave, this time from his own company. Though he was a genius at computer design, Amdahl was quick to exit when the going got tough. He teamed with his Amdahl CFO (Clifford Madden) and his son Carl to form Trilogy, a company aiming to produce VLSI (Very Large Scale Integration) chips for yet another mainframe clone. He had no trouble getting venture capital funds as Bull, DEC, Unisys, and Sperry all invested. Trilogy went public in 1983. Amdahl left the next year amid rising difficulties in chip manufacturing. Trilogy became one of the largest financial failures in the history of Silicon Valley[20].

Amdahl had single-handedly created the plug-compatible mainframe industry. He had done so by using highly confidential information, but information that he carried in his head when he left IBM. The nascent Japanese mainframe industry did not have the same advantage. Fujitsu and Hitachi did have funding from the MITI industrial consortium. In addition, the 1956 Justice Department antitrust ruling against IBM required the company to sell its patents at a "reasonable" price. The two companies introduced their M-Series of System/370 compatible machines in 1974. They soon took share away from IBM in the Japanese market. Like Amdahl, they were caught flat-footed when IBM announced the 3033 "Big One." This surprise announcement underscored the exposure plug-compatible mainframes had whenever IBM came out the next generation. Hitachi decided to do something about it. IBM's planned follow-on to the 3033 was code-named "Adirondack", scheduled for announcement in 1981 as the 3081. A disgruntled Poughkeepsie engineer named Raymond Cadet left IBM in 1980 with the highly confidential Adirondack plan books[21]. He sold the material for $600,000 to Kenji Hayashi, a senior engineer for Hitachi.

Amid the furor in Menlo Park when Gene Amdahl's took over the supercomputer project with his ASC-360 design, the then lab director Max Paley left IBM and started anew as a consultant. Hitachi became one on his clients. He was actually working with Hayashi at the time. The Hitachi engineer let it slip that he had a number of the secret Adirondack books, but lacked a complete set. Paley immediately contacted Bob Evans, who enlisted IBM's industrial security group and the FBI. They asked Paley to serve as an undercover agent in a sting designed to determine how far up the conspiracy went[22]. Paley flew to Tokyo to meet with Hayashi. He said he could deliver the missing books, and alluded to even more detailed information about IBM's Adirondack plans. The sting operation identified eleven Hitachi executives involved in the trade secret espionage. IBM filed suit against Hitachi, settling for $300 million. That was a small price for Hitachi to pay, given that it was able to use the purloined information to develop their 3081-compatible model. Raymond Cadet evaded responsibility for his involvement on a technicality. The big takeaway of the whole affair was the loyalty shown by Max Paley despite the circumstances of his leaving IBM.

IBM announced the Adirondack machine, the 3081, in 1980. It was a dual-processor machine running at 5 MIPS. Erich Bloch and his logic team used LSI (Large-Scale Integration) semiconductor chips with 704 etched circuits packaged in a Thermal Conduction Module (TCM) that was six inches square and weighed five pounds. Each module comprised 800,000 circuits. A single TCM had the computing power of the System/370 Model 148. The 3081 had 26 of these modules. The machine cycle time was just 26 nanoseconds. By way of comparison, the cycle time on the ABC machine was one million nanoseconds. ENIAC and the Colossus had cycle times of about 200,000 nanoseconds. The Havens-designed NORC system reduced cycle time to 4,000 nanoseconds. Stretch brought the machine cycle to 303 nanoseconds. IBM still commanded a premium for its mainframes. The top end of the 3081 line was priced as $285,000 per MIPS (Millions of Instructions Per Second) while the midrange HP3000 was $170,000 per MIP, and the DEC VAX780 came in at $150,000 per MIP[23].

IBM finally tackled the small 24-bit address space designed into System/360. In 1983, IBM released System/370 XA (Extended Architecture). XA expanded addresses to 31 bits, with one bit retained to distinguish between the new architecture and the prior design. The range of addresses expanded from 16 megabytes to four gigabytes.

Endicott and Boblingen decided to take another shot at the low end. The System/370 Models 115 and 125, though modestly successful, missed the most fruitful target. The labs concluded that a new line was needed, one specifically designed for a smaller computing environment. The result was the 4300 family, consisting of four models, with the high-end 4381 sporting a machine cycle of just 68 nanoseconds. All models used VLSI circuits, ran the DOS operating system, and included virtual storage. The new line had 40 percent better price/performance than the 303x product line, and up to eight times the price/performance of the Models 115 and 125.

The business plan was aggressive at 20,000 systems[24]. This led to equally aggressive pricing. With the rest of the System/370 line, the ratio of purchase price to rent was roughly 56 to 1. However, the dramatic increase in performance afforded by the VSLI chip technology pushed the purchase-rent ratio for the 4300 down into the forties[25]. Such a huge jump in price/performance for a new product line made it very attractive to leasing companies. They figured it would be at least five years before this level of price/performance would be obsolete, providing plenty of time to maximize their leases.

The new machines announced in January 1979. There were 42,000 orders in the first few weeks. A very high percentage of those were duplicate orders from customers and leasing companies attempting to get the best delivery date. The rollout was an utter mess. It took time to sort through the orders and schedule those systems that were real. This incensed both customers and leasing companies. In addition, the sheer popularity of such fast, new machines depressed sales of the other System/370 models. In addition, IBM had to finance more systems than it planned to, creating a cash-flow problem. It was several years before things stabilized. Though marketing priced the 4300 to sell 20,000 machines, the final tally was just 6,300 systems.

# PART FIVE
# IBM AFTER THE WATSONS

# TALK TO FRANK

The post-Watson and Learson era began in 1973. Tom Watson decided quite early on that Cary would be the first non-Watson to run the company, though his heart attack resulted in Vin Learson taking that honor. He would describe Cary as "a brilliant business analyst, cool, impartial, and totally self-confident.[1]" Watson and Learson had driven the company to great heights. They had set a very high bar for success. As IBM got bigger and bigger, the job of CEO got harder. Tom Watson would say, "No textbook in the world can tell you how to be the chief executive of IBM.[2]" He had the advantage of being schooled by his father for years before he assumed the reins.

Cary and the seven CEOs that would follow all came from sales and marketing. Could they steer a technology-driven company to continued success? Cary would make all the right moves and come very close to that high bar. He graduated from UCLA and served in the U.S. Army during World War II, followed by an MBA at Stanford in 1948. He joined IBM as a salesman in Los Angeles. On his way up, he handled the introduction of the Selectric typewriter. Continuing his rise through the ranks, he was named president of the Data Processing Division in 1964, where he played a central role in the System/360 program. He was promoted to IBM president in 1971 before being selected for the top job.

**Figure 56**. Frank Cary, IBM CEO 1973-1981. Reprint Courtesy of
IBM Corporation ©.

Cary was decidedly low-key but thoroughly up to the task. He was well built and athletic. He fostered an open environment, prompting the wide use of "If you don't like it, talk to Frank.[3]" This indicated an openness to walk in and talk with Cary or any of his senior executives. This was tempered by the fact that he would lead a company with 341,000 employees by the time he retired.

Frank Cary's strategies differed from the common goal and structure that Learson had set up. He recognized that IBM needed to decentralize to provide better focus on markets beyond the mainframe. He embraced the relatively new General Systems Division despite the product overlap and the potential for multiple IBM representatives calling on the same customer and proposing different solutions. He supported the mainframe ratcheting down with the 4300 and Rochester growing up with the System/38. He was aggressive in addressing Amdahl and the Japanese plug-compatible mainframes. IBM's gross margins slipped a bit, down to 67 percent in 1974 mainly due to price erosion at the high end[4]. He shortened the average product cycle from four years to three and supported accelerated technology to put more pressure on mainframe competitors.

Unfortunately, Cary would spend an inordinate amount of his time on one ongoing issue, the Justice Department antitrust suit. Tom Watson had taken preemptive action in 1969 by unbundling hardware from software and services, then orchestrating the resolution of the CDC lawsuit. The Justice lawsuit still took a good deal of Cary's valuable time, either giving depositions or discussing strategies should IBM lose the case. The Justice Department suit dragged on for six years, with 2,500 depositions and 66 million pages of documentation[5]. Cary would spend 45 full days doing depositions.

IBM had a roughly 70 percent market share in 1969. It would trend down to 37 percent share by 1982[6]. The cancellation of FS in February 1975, though costly and disappointing, was a plus from an antitrust perspective. There remained a strategy to cleave IBM into separate companies should the Justice Department ax ever fall. Cary reviewed plans to spin-off operations as independent subsidiaries and concluded that it would be too expensive, even for IBM, to do that[7]. It was also not necessary. The Justice Department concluded in 1982 that its suit against IBM was without merit and dismissed it.

Cary was a professional manager with a focus on continuing IBM's product leadership. He had risen up through IBM during heady times in the 1960s. He led the rollout of System/360. Given the company's performance, there was little need to manage IBM's stock performance, other than to decide on dividends and stock splits. However, given his MBA training, Cary did experiment with one financial tool that would play a far larger role in future IBM administrations: the practice of buying back IBM shares. Stock buybacks were aimed at reducing the total number of outstanding shares and thus increasing earnings per share (EPS), which typically resulted in an increased stock price. In addition, IBM touted the stock buyback as an efficient way to compensate shareholders as the buyback was an expense and thus avoided corporate taxation, and therefore double taxation for shareholders. Cary experimented with the technique in 1977, purchasing $905 million in buybacks, and again in 1978, with $440 million, for a total of $1.4 billion in the program. That was the extent of it[8].

Cary received criticism from Tom Watson for one of his key initiatives. Tom Watson noted that, "he and his eventual successor John Opel rapidly phased out the rental system, shifting billions of dollars of business to outright sales[9]." Watson considered the rental relationship with customers a cornerstone of IBM's success.

Cary was CEO for eight years, from 1973 to 1980. During his tenure, he increased revenue from $9.5 billion to $26 billion, an increase of 175 percent or at a rate of increase of 13.5 percent per year. Income grew from $1.3 billion to $3.4 billion, an increase of 162 percent, at a pace of 10.1 percent per year. The market value of IBM stock actually decreased, from $47 billion to $40 billion, perhaps a reflection that the go-go days of the System/360 were waning. IBM was selling at 60 times earnings in the 1960s. Cary was disappointed that he could only get the stock price to 15-20 times earnings during his time. He increased the IBM workforce from 262,000 to 341,000. This represented a gain of 30 percent, at a clip of 3 percent per year. Though Cary's performance did not match the gaudy numbers put up by Tom Watson, his financial record is still outstanding.

Cary was a cool, calm professional manager, perhaps what was needed at the time. He faced an IBM that was dominated by executives who had come out of the Data Processing Division and lived through the System/360 and System/370 time. He pushed the company in many new directions. He championed the personal computer and saw that the IBM PC was developed and launched. He supported the Aquarius program for distributed computing that became Systems Network Architecture (SNA). He endorsed the growing impact of small systems and established the independent

General Systems Division, separate from Armonk and White Plains, and I might add located south of the Big Chicken. He encouraged competition between GSD and DPD, with overlapping product lines.

The biggest bet under his watch was FS, though it was driven by John Opel, who would be his successor. With its failure, Tom Watson asked Cary to review the company's development processes in order to prevent such a massive failure in the future. The shadow of FS and a new conservatism introduced in product strategies probably meant that the bright success of System/360 would not be repeated. He left office with his product overlap strategy producing a muddle in the midrange, with the S/34, S/38, Series/1, 4300, and 8100 occupying overlapping spaces.

IBM dominance continued to erode. IBM had as much as 90 percent market share of the nascent computer market in 1956. This trended down to 60 percent in the 1960s and fell to 40 percent in the 1970s. This was probably the natural order of things. Cary was not flashy, but competent and steady. As his years receded in time, his ranking as an IBM CEO continued to rise. Tom Watson felt, "he ran IBM as well as it had ever been run.[10]" He left the company in good shape. Cary would step down at age 60 in 1981, making way for John Opel.

# FRANK AND THE IBM PC

Though Cary spent much of his time focusing on IBM's mainframe business, he was well aware of the emergence of new machines at the other end of the computing spectrum. Intel had developed the first microprocessor chip with the Intel 4004 in 1971. Early hobbyists combined the 4004 with other electronics, perhaps from Radio Shack, to create an inexpensive personal computer. Steve Wozniak and Steve Jobs were part of that cohort, joining the Homebrew computer club in Menlo Park in 1975. The Altair computer with Intel's 8080 microprocessor arrived that same year. Jobs and Wozniak's Apple I was on the market the following year. In 1977, the Tandy TRS-80 and Commodore PET expanded the personal computer choices.

IBM's work with personal computers goes back much further, starting with John Lentz and his Personal Automatic Computer[1]. Lentz was one of the early hires by Wallace Eckert for the Thomas Watson Lab at Columbia University. Lentz worked on the design in the 1948-1954 time frame. He started at the former Delta Phi fraternity house, working in the attic, then moved to the new lab location at 612 W. 115 Street[2].

Under the code name of "CADET", Lentz's concept was vacuum tube based, with relays for the computing circuits. It featured a typewriter printer/keyboard, a small plugboard, and magnetic drum storage. Programming was via paper tape in a simple BASIC-like language. The lab contracted with ElectroData in Pasadena for several parts of the system, including the magnetic drum. It was Clifford Berry who had pushed Consolidate Electrodynamics into computers that resulted in the Electrodata spin-off.

Lentz's system was not announced until 1957, a delay that made it virtually obsolete on arrival. The formal name was the IBM 610 Auto-Point Computer, a designation that referred to its ability to deal directly with floating point calculations. It weighed in at a hefty 750 pounds and sold for $55,000, or leased at $1,150 a month. This was a tough sell with the far more capable IBM 650 already on the market. A total of about 180 machines were produced[3].

Engineers at Burroughs, which had acquired Electrodata, interpreted the "CADET" program name as "Can't Add, Doesn't Even Try" because it used tables in memory as opposed to CPU addition circuitry. Perhaps, this was merely a nod to the Napier's bones. The far more successful IBM 1620 replaced the 610, yet inexplicably retained the "CADET" functionality.


Cary pressed for a renewed foray into personal computers. He found an early proponent of single-user computing in Bill Lowe. Lowe wanted to create a small portable computer to highlight the APL programming language. In 1972, he linked his engineers in Boca with the IBM Science Center in Palo Alto. A team led by Paul Friedl cobbled together the machine, using a display from Bell Brothers, a tape unit from Norelco, memory cards from IBM Boblingen, a keyboard from IBM Raleigh, and a microprocessor from IBM's lab in Boca Raton. The completed device was called SCAMP. It was not the SCAMP of John Fairclough's Hursley lab, but a SCAMP as in "Special Computer APL Machine Portable"[4]. Some would call it the first personal computer.

With the success of SCAMP, Lowe asked a team in Rochester to produce a commercial offering with all IBM content. The effort, named Project Mercury, resulted in the IBM 5100 portable computer, arriving in 1975. It featured the IBM-developed PALM ("Put All

Logic in Microcode") microprocessor and added BASIC programming to the APL function of SCAMP. That BASIC used Glenn Henry's BASIC from the System/3 Model 6 as a base. IBM advertised the 5100 as portable, but at 52 pounds, it would be more aptly termed "luggable." Boca released an enhanced model with compact diskette drives, the 5110, in 1978.

Lowe then asked Bill Sydnes to develop a machine with full-size eight-inch floppy disk drives, one that would be compatible with other IBM systems. The Sydnes team produced the IBM 5120 in only 90 days. Announced in early 1980, it was the first IBM computer with an entry price under $10,000. The machine won industrial design awards for its compact, all-in-one desktop layout. At 99 pounds, it made no pretense of moving from whatever desktop it sat on. The success of the 5120 led to a more substantial system that targeted to small businesses. This was the System/23 Datamaster.

Cary was familiar with Apple computers and killer applications such as VisiCalc[5]. It was clear that the personal computer market was soon going to explode. Though IBM's 5100 series and the planned System/23 Datamaster were promising, they were too big, too expensive, too corporate, and decidedly not "personal." Bill Lowe suggested to his boss Chuck Branscomb at GSD headquarters in Atlanta that IBM develop a personal computer. Branscomb concurred, and Lowe presented his ideas to Cary in July 1980. He included a business plan that recommended selling the IBM machine through ComputerLand and Sears retail stores[6]. Lowe outlined a development approach where IBM partnered with a company like Atari. However, Cary wanted to work the project in-house. Lowe stressed that such a project had to be a skunkworks, that trying to produce such a machine within the traditional IBM bureaucratic framework would simply not work. It would inevitably produce machines like the 5100 and the Datamaster that were too big and too expensive for the consumer market. In addition, the IBM way would "take four years and 300 people.[7]" He described an in-house project with a small team using an open architecture with mostly non-IBM components and software. Cary agreed with the concept and gave Lowe 30 days to come back with a prototype. He urged Lowe to "teach the Big Blue elephant to dance.[8]"

Lowe recruited a small team of engineers, working under the

overall direction of Phil Estridge, to produce a prototype, under the banner of "Project Chess." This core team included Lew Eggebrecht, Jack Sams, and David Bradley[9]. They were all personal computer enthusiasts and most had Apple II machines. Bill Sydnes would lead the hardware engineering. Sydnes was a local, having gone to college at Florida Atlantic University in Boca Raton. Lowe decided to do the work at GSD headquarters in Atlanta. He had a few walls knocked out in the Lakeside headquarters building. Lowe and his team anticipated Cary's approval by tearing apart existing competitor machines and designing the basics of what would become the IBM PC architecture. The prototype was ready by Thanksgiving 1980. Lowe was back at Armonk, along with Bill Sydnes, to present the machine to Cary and the Corporate Management Committee. Cary asked Jack Rogers, the GSD president, if he could fund the project. When Rogers hesitated, Cary made his most critical decision. He would fund the effort himself. Furthermore, he wanted Lowe to bypass the IBM management chain altogether and report to him directly.

Lowe assembled an expanded team to begin work on the program, with the code name "Acorn." His team would later be called the "Dirty Dozen[10]" but without the connotation of the infamous San Jose "Dirty Dozen." Phil Estridge, who had grown up in Florida and attended the university in Gainesville, would lead Acorn. He joined IBM in 1959 as a junior engineer at the Kingston lab. He worked for Federal Systems with NASA and the Apollo program, and participated in FS until its cancellation in 1975. For the next four years, Estridge was programming manager for the Series/1. This system was targeted directly at the DEC PDP-11. As such, the program was all about hardware. A customer could even order a Series/1 without an operating system. This was a new frontier for IBM sales, one dominated by the minicomputer makers. IBM was simply too late to this market. The Series/1 was a modest success but failed miserably in its core mission, to be a dent in DEC's business. The Acorn project would be a chance for redemption for Estridge.

Lew Eggebrecht was the lead hardware architect for Acorn. He had joined IBM in Rochester as an engineer. He was responsible for evaluating third party microprocessors such as the Intel 8080 and the Motorola 6800. Jack Sams, who had worked on the Datamaster, would be responsible for the software. David Bradley worked on both the Series/1 and the Datamaster. He would design the BIOS (Basic Input/Output System), the core microcode responsible for booting and running the computer. He would become famous for the required CTL-ALT-DEL sequence to restart the PC, usually after the much-feared blue screen. He chose the keys for a reason, in that they were spaced far apart to avoid an accidental reboot[11].

Patty McHugh was the only female member of the team. Her mission was to design the main planar board. This core part became the "motherboard" and McHugh was dubbed the "mother of the motherboard.[12]" As Acorn progressed from design into development and release, hundreds of additional engineers joined the project. After the amazing success of the program, David Bradley observed perhaps 50 or 60 engineers now claimed to have been part of that original Dirty Dozen.

The design of the Datamaster provided an initial roadmap for the Acorn team. They used the internal bus, keyboard, and monochrome display from the Datamaster. The team moved from the eight-inch diskette drive to the more compact 5.25-inch drive. They went to Intel for the microprocessor, selecting the Intel 8088. They chose it over the Motorola 6800 mainly due to price. It was also similar to the Intel 8085 that went into the Datamaster. It was a 16-bit processor, but the data paths were only eight bits wide. It was a conservative but probably correct choice. The microprocessor provided for one megabyte of memory, but the engineers reserved space for microcode, video control, and adapter card interfaces. The result was the familiar 640K space of usable memory.

IBM would be haunted by the decision to base Acorn on the Intel microprocessor. The agreement was very one-sided in Intel's favor and represented the first of several missed opportunities. IBM's Research and Components Divisions had made great strides in advanced integrated circuits, including RISC implementations, but the die was cast with Intel, which would be in charge of the PC microprocessors for many technology generations to come.

Jack Sams was tasked with the operating system and BASIC programming software[13]. The latter should have been a no-brainer for IBM. It had substantial experience in BASIC, including Glenn

Henry's work on the Model 6. However, in a situation that would prove disastrous for IBM, the needed programming resources to port IBM's BASIC to Acorn were already committed to consolidating the BASIC on the System/34, a platform that really didn't need it. This had already caused a delay of a year in getting the Datamaster out the door. Sams wanted to look outside of IBM. Watts Humphrey was now head of programming for IBM. He pressed Cary and Lowe to use IBM programming teams for Acorn's software. However, Cary wanted it too fast for that. Instead, Sams received the green light to look outside IBM. Sams was familiar with Microsoft and aware that the company had recently ported its MITS BASIC to the Intel 8086. The company was barely five years old. Sams called Bill Gates.

Gates was born in Seattle in 1955[14]. His father was a prominent attorney in town, and his mother was a director at Pacific Northwest Bell and a regent at the University of Washington. He learned BASIC, the language Dartmouth had invented for academic environments, on a time-sharing setup at his Lakeside prep school. He was also able to connect to a GE computer via a teletype terminal that the school had purchased at a rummage sale. When his allotted computer access time ran out, he joined with his childhood friend Paul Allen to seek a computer elsewhere. They were able to secure time on a PDP-10 at Computer Center Corporation in exchange for identifying bugs in the company's software. Gates and Allen developed a software product called Traf-O-Data based on the newly available Intel 8008 microprocessor that recorded road volume data and produced reports[15]. With near perfect SAT test scores, Gates was admitted to Harvard in 1973, where he took a pre-law curriculum.

Once at Harvard, Gates and Allen kept in touch. They were both excited by the MITS Altair microcomputer announcement in 1975[16]. They decided it was now or never to jump into this fast-moving new world. The Altair would need a BASIC interpreter to be truly useful, and they decided they were the ones to supply it. They had no running code or even access to an Altair machine. They developed an Altair emulator on Harvard's PDP-10 system and started work on BASIC. In an early preview of Gates's future tactics, he called the owner at MITS, Ed Roberts, and said they had BASIC ready for his machine, though they were still frantically working to finish the code and ensure it fit in the 4K memory of the Altair. Allen traveled to Albuquerque to demo their program, coding the paper-tape bootstrap initializer program on the way. Since this was his first encounter with a real Altair machine, Allen was unsure whether it would work. To his relief, the program loaded. He keyed in "PRINT 2 + 2." The machine came back with "4" and went to "READY." Roberts was impressed and hired them to complete the work on BASIC. Gates took a leave of absence from Harvard. He and Allen soon had wider aspirations than MITS could provide. They changed the name of Traf-O-Data to Microsoft with the goal of providing their BASIC interpreter on other platforms. With sixteen employees, the company moved to Seattle in 1979.

John Kemeny and Thomas Kurtz had made the BASIC compiler code widely available in order to spur widespread use. They also made the Dartmouth time-sharing system available to many high schools and colleges in the Northeast. BASIC grew in acceptance far beyond their expectations. Yet unfortunately, it also grew unchecked by any control or standards. DEC made substantial changes to Dartmouth BASIC, providing much more low-level access to its machine functions. Working on DEC PDP-10 machines at Harvard, these changes allowed Gates and Allen to adapt the language for the Altair. MITS charged $500 for BASIC, marked down to just $75 if you bought the Altair computer. With many early PCs, BASIC was the interface. If you turned on the Radio Shack TRS-80, you were not interacting with an operating system. You went straight to the BASIC interpreter.

Microsoft BASIC was one of the first indications of the ruthlessness of Bill Gates and the path that he would use to build Microsoft into a billion-dollar company. He had the nerve to rant at microcomputer hobbyists who copied his Altair BASIC without paying for it. None of the financial rewards of Microsoft BASIC came to Kemeny, Kurtz, or Dartmouth College. The lack of financial rewards for BASIC did not greatly upset Kurtz and Kemeny. It was the bastardization of the language that irked them. It cut against the central philosophy of the language. They regarded renegade BASIC versions such as Microsoft's as "street BASIC." They attempted to wrest control of the BASIC standard with their definition of "True BASIC," but the die was cast with the Altair and there was no turning back[17]. To add insult to injury, Microsoft released Visual BASIC in 1991 with Windows. Gates himself would observe, "Over the years, BASIC in all its forms has been the key to much of our success[18]."

Jack Sams had three companies on his search list for BASIC—Digital Research, SofTech Microsystems, and Microsoft. With an assist from Bill Gates's mother, he would end up with Microsoft. She was on the United Way board in Seattle and through this organization, she was able to meet John Opel, the new CEO of IBM. She lobbied Opel on behalf of her son's fledging 31-employee company. Opel, having recently taken over the PC project from Cary, mentioned the conversation to some of his senior executives. Though Sams was not impressed with Microsoft, he continued to discuss acquisition of their BASIC interpreter. As much as Glenn Henry was disappointed by the failure of his Model 6 BASIC project, this would be a far more costly BASIC "disaster" and one that would have far-reaching repercussions for IBM.

Jack Sams still needed to find an operating system. Gates had no operating system and suggested talking with a friend, Gary Kildall of Digital Research, who had the highly regarded CP/M operating system (Control Program for Microcomputers). Kildall attended the University of Washington in Seattle. The Navy drafted him and posted him to Southern California, where interest in microcomputers was heating up. He wrote programs for the Intel's initial microprocessor, the 4004. After completing his doctorate in computer science in 1972, he developed the CP/M operating system. In 1974, he started Digital Research and soon licensed CP/M to a number of early personal computers, including the Apple II and the Osborne 1.

Jack Sams set up a meeting to discuss use of CP/M in the Acorn machine. He met with Kildall's wife, Dorothy, who handled the business end of the company. When he arrived, Gary Kildall was reportedly out flying his airplane. Dorothy refused to sign IBM's standard disclosure agreement, which was critical considering the secrecy of the project[19]. In addition, IBM wanted a 16-bit version of CP/M to run on the selected 8086 processor. They could not reach an agreement for Digital Research to produce it in the needed time frame. The negotiations reached an impasse, and Sams and the IBM team returned to Boca Raton.

Still without a plan for an operating system, Jack Sams went back to Gates, who suggested that he could supply an operating system. He did not actually have one, but Paul Allen knew where one could be found. Tim Patterson, a programmer at Seattle Computer Products just down the road from Microsoft, had developed a simple microcomputer OS he called QDOS, for "Quick and Dirty OS". Allen was able to reach a deal with the company to license QDOS for roughly $75,000 with limited restrictions[20].

Tim Patterson was familiar with CP/M and in fact, had written code for it to run on the Apple II. QDOS was patterned after Kildall's CP/M, though Patterson would argue that he just took the concepts and then wrote QDOS. Patterson would join Microsoft and further develop QDOS into MS-DOS, working alongside Gates and a team of forty programmers. They would produce DOS 1.0, a very primitive operating system that took only 30,000 bytes of code.

Gates and his new business manager, Steve Ballmer, were then ready to reengage with Jack Sams and IBM. Ballmer had lived just down the hall from Gates at Harvard. He went on to graduate and after a brief stint in the MBA program at Stanford, Gates hired him as Microsoft's 30th employee in 1980 to focus on the financial end of the fledgling company. Gates gave him $50,000 salary plus an 8 percent share of the company. Allen strenuously objected to the arrangement.

Gates and Ballmer negotiated with Bill Lowe to add their new 16-bit QDOS software to the other programs they were to supply. The final agreement with IBM totaled $430,000, with $310,000 for BASIC and several additional programming languages, $45,000 for DOS, and $75,000 for modifications[21]. However, it was not the price that was notable but the terms. First, Gates and Ballmer wanted a nonexclusive license that allowed them to market the software to other companies making an IBM personal computer. Even though IBM had suffered for years with clones, no thought was apparently given to what might develop. Microsoft built their business by selling the BASIC interpreter to personal computer makers. Nonetheless, IBM agreed. Second, they asked to retain the source code. Jack Sams was already of the opinion that if IBM started writing software, it would just screw it up. In addition, Microsoft was a small company and the relationship at the time was very good. Gates and Ballmer also sought a royalty for each copy of DOS sold. IBM was without a backup plan at this point and agreed to $10 to $50 for each DOS license sold.

In hindsight, it is hard to fathom why IBM signed this agreement. Each of the elements came back to severely burn the company. The entire PC-compatible market would rest on MS-DOS. The lack of access to the source code would keep Microsoft in control for many years to come, limiting IBM's flexibility in the new market it alone created. Finally, IBM would be paying Microsoft for every DOS license. IBM would spend years and many billions of dollars trying to get around the agreement and reestablish control of the PC market. The total cost in development expense and lost opportunities is certainly in the hundreds of billions of dollars.

The year Bill Gates signed the agreement, Microsoft's revenues were $16 million. The $75,000 that Microsoft spent with Tim Patterson would turn Microsoft into a software monopoly. In just five years, Microsoft revenues were $197 million. It would not be long before Gates would be the richest man in the world, with Allen and Ballmer cracking the top five.

Meanwhile, Kildall was contemplating suing IBM, Microsoft, or Tim Patterson for the code in QDOS and MS-DOS he considered stolen from CP/M. IBM reached an agreement with Kildall to market CP/M with the IBM PC. However, with the price set at $240 versus $40 for DOS, Kildall's software went nowhere. Kildall would comment on his relationship with Bill Gates, "I have always felt uneasy around Bill. I always kept one hand on my wallet and the other on my program listings.[22]" The missed opportunity of Gary Kildall and CP/M would become the stuff of legend in the computer industry.

The damage the Microsoft agreement did to IBM was not apparent for several years. In defense of Bill Lowe, some raised the specter that a big and dominating IBM running roughshod over a small company might trigger an antitrust action by the Justice Department. The more logical explanation was that the IBM team was desperate for the software components and did not have time or prospects for a Plan B. This does not excuse Lowe, who agreed to the completely one-sided deal with Gates and Ballmer. It is easy to consider how this might have turned out quite differently if the team had listened to Watts Humphrey and developed the software in-house. A key sticking point was the unavailability of the BASIC developers who were committed to porting BASIC to the System/34. The idea of BASIC on the System/34 was even more of a stretch than Glenn Henry's BASIC on the Model 6. It was an idea that went nowhere and yet set in motion the sequence of events that would eventually shatter IBM's place in the market it created.

Once the core Acorn work was complete, development started in earnest under Estridge. More than 150 IBMers were working on the project by the end of 1980. The team demoed a prototype in January 1981. Despite all the hurdles, the IBM PC announced in August 1981. H.L. "Sparky" Sparks led the product launch, including the highly effective Charlie Chaplin "Little Tramp" marketing campaign. He also orchestrated the retail channels for the PC, signing up Sears and Computer Land. A typical system was not cheap. An entry machine with 16K of memory, a cassette drive, without a monitor or floppy disk drive sold for $1,565. A more functional system fitted with 64K of memory, a 160-kilobyte diskette drive, and monochrome monitor cost nearly $3,000.

All during development, there was limited appreciation of the market potential for the Acorn machine. The original plan was to sell 250,000 PCs in the program's first five years. The Datamaster arrived just two weeks prior, but it paled next to the frenzy that engulfed the IBM PC. Actual sales were 200,000 the first year and ramped up to a rate of 200,000 a month in the second year. IBM captured over 20 percent market share in the personal computer market, generating upwards of $2 billion for the company in 1983[23]. It soared past Apple. Time Magazine named the IBM PC its "Man of the Year."

In the four years after the announcement, Boca grew from 4,200 employees to 9,200, and would eventually top 10,000. With revenues of $4.5 billion in 1985, it would have been the third-largest computer company, after IBM and DEC[24]. The press lauded Frank Cary for his cultivation of the IBM PC despite the bureaucracy of the company. He would step down as CEO several months before the PC announcement. Though he would remain on the board of directors, the day-to-day management of this new business would fall to his successor, John Opel. Cary watched as his creation grew beyond all expectations.

However, in the rush to get the IBM PC to market, IBM fumbled two critical decisions that would plague the company for years -- ceding DOS to Gates without access to the source code, and signing the one-sided deal with Intel. The error with DOS was especially serious. In earlier IBM systems, software had been only 8 percent of the total configuration. It was now in the neighborhood of 40 percent, if not higher. The secret of IBM's hegemony in markets like the mainframe was its control over the hardware and software architecture. That was glaringly missing with the PC. These two mistakes would wreak havoc and bedevil the next four IBM CEOs.

# JOHN OPEL AND IBM'S GREAT EXPANSION

Frank Cary relinquished the CEO office when he turned 60, as had Vin Learson before him. The board selected John Opel as the next chief executive. Opel took over in January 1981. During his rapid rise, he would take on 19 different positions in 32 years in order to reach the top job. Opel was already 56 when he became CEO. He would have only four years to make his mark.

John Opel was the quintessential button-down IBMer. He was at times overly cautious and bureaucratic. On the other hand, he knew how to navigate the company's far-flung divisions and significant "red tape." He had ably demonstrated that in finding Al Williams' missing cash in the manufacturing plants during System/360 development. He could reach across organization lines to solve problems, a trait he shared with Learson. Learson supported his elevation to CEO.

Opel was clear about his strategy from the very beginning. He wanted IBM (1) to grow with the industry, (2) exhibit product leadership, (3) be the most efficient manufacturer, and (4) sustain profitability. These were certainly worthwhile acceptable objectives, but there was no mention of the customer or the IBM employee. Moreover, these objectives were in place to serve his central metric, to put IBM on a 15 percent annual growth track.

Opel entered the CEO's office with the company in very good

shape. The Justice Department suit was resolved. The PC business was about to explode. Rochester and the General Systems Division were killing the midrange. The mainframe business remained strong. By 1984, IBM ranked #1 by huge margins in all major computer hardware categories[1]. In mainframes, IBM's sales of over $13 billion were nearly nine times that of second-place Fujitsu. In minicomputers, IBM's growing $3 billion business was twice that of DEC. IBM's startling $5.5 billion in personal computers was roughly three times the revenues of Apple. IBM's peripheral business, racking up sales of $11.7 billion, was almost five times that of runner-up DEC.

Opel had risen to the highest levels in IBM during its years of unbridled success. He may have assumed that this was normal for IBM, and that he should strive to meet Tom Watson's unparalleled record. He translated his 15 percent growth goal into revenue targets of $100 billion by 1990 and $180 billion by 1994. He then set about making the massive investments required to attain them.

Opel would oversee a huge expansion of the company. He built new plants and labs, and upgraded existing ones. He funded an automated facility in Lexington for Selectric typewriters, aimed at making IBM the low-cost producer of typewriters. With Rochester booming, the site expanded three times during his watch. His capital spending of $16.5 billion represented most of the entire computer industry's capital outlays. The company spent a total $42 billion on plants and labs during the 1980s. Opel added 50,000 employees to staff the expansion.

It was customary for new CEOs to reshuffle the organization upon taking over. Opel was no different. Frank Cary led a yearlong analysis that culminated in the plan Opel announced in January 1981. He restructured all of U.S. marketing into two new divisions. The National Accounts Division (NAD) would handle IBM's largest 2,200 accounts, and the National Marketing Division (NMD) would handle all other customers and prospects. Opel located NAD in White Plains and NMD in Atlanta. Jack Rogers, the former GSD president, would lead both divisions. He would have an organization composed of 15 regions, 149 branches, and 16,000 employees. Rogers announced Lew Gray as his vice president of Operations. Gray vowed to stand by his promise that an IBM division would continue to operate south of the Big Chicken as long as he was around. In 1982, IBM announced plans for a showcase 50-story IBM tower in midtown Atlanta. The following year, IBM announced plans for a second head-quarters building, called Hillside, next to the Lakeside structure.

At the same time, Opel moved to rein in the autonomy of the remote divisions, centralizing more control in New York. Many marketing and planning functions formerly done in Atlanta and White Plains moved to Corporate offices in New York. It spelled the end of the integrated and "one team" General Systems Division that had seen such great success. Chuck Branscomb worked on the reorga-nization plan and saw in the GSD breakup the loss of the "secret sauce" that had contributed so much to the division's outstanding performance. He decided to move on and took a group executive job in White Plains working for John Akers, a prelude to retirement.

Opel decided to accelerate the Cary strategy of converting IBM's rent/lease customers to purchase. He moved IBM from 85 percent rent to 88 percent purchase. By 1987, when IBM's revenue was $54 billion, only $3 billion came from rentals. This created a bubble of additional revenue each year during the program, overstating the real business that the company was doing. In addition, IBM sales represen-tatives needed to sell the equivalent of the rental annuity every year. They now started the year at zero. The change had the negative side effect of breaking the traditional IBM relationship with its customers, one in which you had to earn the business year in and year out. In addition, the purchase windfall would overstate IBM's progress towards Opel's revenue targets of $100 billion by 1990 and $180 billion by 1994. Tom Watson would write how concerned he was about dropping the rental system. He said, "I probably would have been deeply depressed if IBM had been the only outlet for my energy. Fortunately, I knew not only how to sail, but how to fly.²" Watson understood the need to focus on real growth, not financial gimmicks.

Opel sponsored the task force that led to the FS program. Condi-tioned by the great success of System/360, he felt there was no reason why IBM could not make the great technology leap forward that FS envisioned. He was bitterly disappointed in the FS failure, particu-larly considering its great cost and the loss of critical product cycles. The experience certainly colored his thinking when the next big opportunity surfaced, one that could potentially aid in his aggressive revenue goals.

That opportunity was RISC computing. Bob Evans was shunted aside in 1975, when he was effectively demoted from president of the front-line Systems Product Division to president of the much-lower-profile System Communication Division. He characterized the move as "being put out to pasture.[3]" However, he rebounded two years later when named vice president for all of IBM's worldwide engineering, programming, and technology. He was familiar with John Cocke's work on RISC. He felt that RISC computing, with its dramatic increases in performance, offered a unique breakthrough moment for IBM. He pushed hard for the development of RISC chips to be used across all of IBM's product lines. It would impose one consistent and leading hardware technology much in the way the System/360 had done 20 years earlier.

Evans continued to press this vision with John Opel between 1980 and 1983. He apparently pressed too hard and now, he no longer had Vin Learson at his back. Opel was unwilling to bet on a single technology across the company. He was particularly concerned about the impact on the critical mainframe business. On top of that, Opel invested $650 million in Intel in 1982, an investment that helped Intel in its fight against Japanese DRAM competition. The RISC issue came to a head in a Corporate Management Committee meeting. Opel brutally shut Evans down. Evans retired soon after. His vision of a complete RISC-powered product line would eventually come to pass, though long after he was gone. Opel missed the chance for a decade head start in high-performance RISC technology.

IBM's mainframe business continued to be a central focus of the company. Though IBM's mainframe share dwarfed those of the Japanese clones, its encroachment was concerning. Fujitsu and Hitachi could reproduce mainframe hardware but not the software. There was a growing suspicion that the Japan competitors were simply copying IBM software without licensing it. Opel finally decided to stop dealing with Fujitsu until IBM completed an audit on its software use. The audit revealed massive appropriation. Fujitsu resolved the issue with an $833 million payment to IBM, a huge figure but one that likely understated the scope of the theft. In 1983, IBM stopped distributing source code with the software to discourage its use by competitors.

Opel's strategies to grow with the industry, exhibit product leadership, be the most efficient manufacturer, and sustain profitability represented a challenge to IBM's sprawling bureaucracy. Opel decided to sharpen the focus of selected product divisions. He started the Independent Business Unit (IBU) plan in 1982. He granted much greater autonomy to selected business units in exchange for the promise of better results. The IBU approach also established separate operating statements for IBUs, should IBM decide to sell a particular unit. Opel's strategy would be embraced and substantially expanded by his successor.

## RISE OF THE CLONES

Opel received an unexpected gift from Frank Cary—the new and rapidly expanding PC business. Cary stepped down shortly before the announcement. Opel had little understanding of personal computers. He was not PC-literate, did not have a PC or terminal, and did not use email. His endorsement of Bill Gates's tiny company would sow the seeds of great calamity for IBM. His investment in Intel would further challenge IBM's new business. Now as CEO, he saw a rapidly growing organization in Boca that was untethered from Corporate. Cary understood that this was the secret of the PC's success. Opel did not. He immediately set about reining in the PC division. IBM's mainframe community endorsed the move. It was not interested in maximizing IBM's opportunity in the new, growing field but protecting the mainframe from it.

Opel's constraint of the PC division could not have come at a worse time. The central attraction of the IBM PC was also its Achilles heel—the open architecture. With the experience of plug-compatible competition of System/360 machines, IBM was fully aware of the clone menace. Opel, in the midst of his manufacturing buildup, felt IBM would be the low-cost producer and would benefit further by low pricing on its high-volume supplier contracts. Neither advantage materialized. There remained at least three additional lines of defense against clone intrusion -- the operating system, the BIOS internal code, and the unique array of chips on the motherboard. Unlike Apple and its Mac operating system, Bill Lowe had already ceded the operating system to Microsoft, including the source code and the license to market it to any computer manufacturer that wanted it. That first line of defense was gone before the PC announcement.

IBM fell back on the BIOS (Basic Input/Output System), the only proprietary aspect of the machine. BIOS was the "traffic cop" that resided between DOS, the applications, and the machine hardware. The BIOS was embedded on one 8K chip mounted on Patty McHugh's motherboard. IBM Legal had constructed a web of patents to protect any direct use of the BIOS. IBM even published the BIOS source code, but the company intended to sue any competitor that attempted to use it. There was, however, a legal loophole. The function of the BIOS could be duplicated, as long as there was absolutely no prior knowledge of the PC hardware and IBM's code. A competitor's pathway to an alternative BIOS had to be completely above reproach by IBM Legal. In addition, the resulting code had to be 100 percent functionally equivalent or the resulting machine was not fully compatible. At the outset, the industry considered recreating the BIOS either technically impossible or cost prohibitive, or both.

In 1984, one company accepted the challenge. Three engineers left Texas Instruments to form Compaq[4]. Leaning on a deep-pocket investor, they spent over $1 million painstakingly re-creating the BIOS function from scratch. Their process was transparent, and IBM chose not to legally challenge it. A challenge might have been worthwhile as Compaq proceeded to hire Sparky Sparks, who had developed IBM's distribution channels for the IBM PC and would do the same for Compaq. Compaq went from sales of $111 million in 1985 to $3.6 billion in 1990.

A second run at unraveling the BIOS would prove far more costly. Phoenix Technology set up a firewall to develop the code. It hired a Taiwanese programmer who had never seen or worked on the IBM PC. A separate team of engineers defined each BIOS function and passed it to the programmer. He would pass back completed code to be tested, If it did not exactly mimic the BIOS function, they sent it back. Phoenix documented the process step by step to ensure that any legal challenge would fail. With success, Phoenix offered its BIOS code to any company that wanted it for a flat $290,000[5]. This was a key milestone in opening the clone floodgates. HP, Tandy, and AT&T were early Phoenix customers.

There remained one hurdle to constructing a competitive IBM PC clone, the set of chips designed into the motherboard. Though all of them were available on the open market, IBM had negotiated volume purchase arrangements that resulted in a high cost hurdle for any potential clone machine. That hurdle came down in 1985. Gordon Campbell, who had left Intel, started Chips and Technology with the aim of reducing IBM's maze of 63 separate chips on the PC motherboard. His company ended up with a suite of five chips that replaced most of the IBM chips. It sold for just $72[6]. This completed the pathway to a viable 100 percent compatible clone, with software from Microsoft, BIOS from Phoenix, and the chipset from Chips and Technology. The number of clone manufacturers was about to explode.

A student at the University of Texas demonstrated the ease of developing an IBM clone. He was an Apple II user and had started selling hardware upgrades to the IBM PC out of his dorm room as a freshman. He soon dropped out of the university and incorporated his fledgling operation as PCs Limited, selling the upgrades by direct mail. It was only a very short set of steps from there to piecing together a complete machine and selling it as an alternative to the IBM PC. As the company grew, Michael Dell changed the company name to Dell Computer.

IBM still had an avenue open to address a clone onslaught. It was the same approach the company had taken with clone peripherals on System/360—push the technology forward and stay ahead. This would prove to be much more difficult in this market. The clone companies could move far faster than IBM, incorporating new technologies and advanced features. They could transition beyond competing solely on price and deliver a more advanced system. At the same time, Opel's transformation of Boca into a corporate unit replaced the fast-moving skunkworks environment that had created the new computer standard in the first place.

The IBM clone experience contrasted sharply with that of Apple[7]. After Steve Jobs was forced out, Apple initiated a clone program to expand its market presence. Apple owned the Mac operating system and licensed it. When Jobs returned to Apple in 1997, he reevaluated the program and decided to end it. As with the IBM experience, Apple footed the engineering, development, software, and marketing expenses while the clone makers had a free ride. Jobs eliminated most of Apple's clone companies by refusing to license new versions of the Mac operating system. He had to buy out its largest clone maker for $100 million, but otherwise, he moved on.

Opel did not realize where the growing clone market and the pull-back of the PC division would eventually lead. At announcement, IBM's share of the IBM PC market was 100 percent. Clones soon began eroding that share, though the overall size of IBM's business continued to grow. IBM reached its PC high point in 1985 with a 22 percent market share. The share and revenue headed downhill from there. One can only wonder if the same scenario would have played out with Cary still at the helm.

In his four years as IBM CEO, John Opel increased revenues from $26 billion to $46 billion, a boost of 75 percent at a rate of 15 percent per year. The company grew income from $1.3 billion to $3.4 billion, a 166 percent increase at a rate of 13 percent per year. The number of IBMers rose from 341,000 to 395,000, an increase of 30 percent. The market value of IBM stock gained $35 billion, to reach $75 billion. It was a good track record though it substantially benefited by significant tailwinds, including the one-time purchase program, the rise of the IBM PC, and the continued growth of Rochester's midrange business. Opel made the cover of Time Magazine in 1983, with the accompanying feature article called "The Colossus That Works.[8]" John Imlay Jr., a software executive was quoted, saying, "IBM is simply the best-run corporation in American history." James Marston, the Chief Information Officer for American Airlines, stated, 'you can take any specific piece of hardware or software and perhaps do better than IBM, but across the board IBM offers an unbeatable system."

However, Opel set the table for what was to follow. He set aggressive revenue targets and dramatically expanded IBM in order to make those targets. His goals forced the company to grow by the equivalent of a DEC each year. Ironically, DEC was on a similar strategy, with the goal to be bigger than IBM. It also made vast increases in plant and people to support that ambition. Both would end badly. Opel did very little to cut IBM bureaucracy. Opel traversed through 19 jobs to rise from salesman to CEO. This was a vivid demonstration of the layers built into the company he inherited. In fact, his aggressive expansion increased IBM's bureaucracy. Evaluating the three pillars of Opel's strategy, Robert Heller in *The Fate of IBM* concluded that IBM under Opel lost market share, lagged in technology, and remained the high-cost producer, a depressing hat trick. Instead of growth relative to the industry, IBM's market share dropped to 62% by 1982[9].

Even so, IBM remained the most admired company in the world. The influential book *In Search of Excellence* by Tom Peters and Robert Waterman came out in 1982 and profiled IBM as one of its great examples[10]. Their analysis was from the outside and was unable to ferret out failings on the inside that did not bode well for the future. Opel retired at age 60 in 1985 before the "overhang" of his ambition met reality and came crashing down.

# JOHN AKERS TACKLES AN
# OVERHEATED COMPANY

John Akers joined IBM in 1960 as a sales trainee in San Francisco. He had graduated from Yale University, where he was an All-Ivy hockey player. After graduation, he enlisted in the Navy and became an aircraft carrier pilot. Describing the IBM he joined, Akers recounted, "We were very square. We wore the blue suits, white shirts with button-down collars, striped ties, hats and wing-tipped shoes. There was no drinking—unless the customer wanted to. And then we were supposed to go home.[1]" Akers and his cohorts had up to 18 months of sales training. By the time they graduated and became quota-carrying salespeople, they were expert in IBM's products and schooled in the solution selling that Watson Sr. instilled in the company. Akers would excel and personify the gifted and over-achieving Watson salesperson.

Like his predecessor, it was a long climb to CEO. Akers had 15 promotions in 24 years. He sold punched-card machines and main-frames, earning 100% Club and Golden Circle recognition. Early on, his manager would have tagged his personnel file with fast track nota-tion. He was a marketing manager in Boston and a branch manager in New York City before the critical assignment as Frank Cary's admin-istrative assistant (AA) in 1971. Nine months later, he was vice presi-dent in the Data Processing Division, and President of the division a year later. By 1974, he was one of 37 men sitting just below the top job[2]. John Opel promoted him to IBM vice president and shortly thereafter, to group executive in charge of both the NAD and NMD sales divisions. He was now just one of seven men below John Opel. He became IBM President in 1983, a sure sign that he was next in line.

Akers became IBM's seventh chief executive on February 1, 1985. His salary was $750,000. He took over IBM at perhaps its high point. In the late 1980s, IBM could do no wrong, regularly rated the #1 corporation in the United States. There was a cottage industry in book publishing analyzing the "excellence" of IBM's management. At the time he took the reins, he had a one-time-only opportunity to set his own strategy. Instead, he signed up for Opel's revenue targets of $100 billion in 1990 and $180 billion in 1994, even though current revenues were just $46 billion. Anticipating that he could achieve those goals, he told *Fortune* magazine in 1986, "There have only been six chief executives of IBM. I hope that when my tour is over, people will look back and say, 'he deserved to be among them.[3]'"

Opel had left the company in relatively good shape, aside from the enormous shortfall against his growth targets. But, there were dark clouds on the horizon. The conversion of IBM's computer inventory from rental to purchase had run its course. The PC business was at risk from both outside and inside the company. The rise of open systems would put pressure on IBM's carefully constructed and proprietary computing ecosystem. Amdahl, Hitachi, and Fujitsu continued to pressure the mainframe business. By 1989, the estimated worldwide inventory of large computers (those over $100,000 in purchase) would increase to $260 billion, with the IBM share at roughly 50 percent. The company's bloated bureaucracy continued unabated, with upwards of 17 levels between an entry employee and the CEO. Opel had made a huge bet on IBM's future and bulked up the company to reach it. Was Akers, who had mastered IBM's bureaucracy and spent 20 years in the mainframe division, the right man for the challenge?

Akers started with a reshuffle of the deck chairs. He combined the National Accounts and National Marketing divisions and then split them geographically into the North Central Marketing Division and the Southwest Marketing Division. Each IBM branch would sell the full IBM product line to customers and prospects in its geographical territory. The North Central Marketing Division was based in White Plains. The Southwest Marketing Division was based in Atlanta, with Lew Gray in charge. His promise of an IBM division south of the Big Chicken was safe. The new divisions were soon rechristened as "Nasty and Cold" for North Central and "Sunny and Warm" for Southwest. Akers combined the separate headquarters staffs of the prior divisions, resulting in 2,000 individuals moving back to the field. This reorganization further diluted the small-systems focus that had been the General Systems Division's strength, as the branch offices in the new sales divisions would market the entire product line. Providing each customer with a single sales representative was a central motivation of the reorganization.

At this time, work continued on the 50-story building in midtown to consolidate Atlanta employees[4]. Tom Watson's desire to showcase the company in the designs of its plants, labs, and offices remained in force. The post-Modern structure erected in midtown Atlanta was a contemporary take on the church spires of medieval times, with neo-Gothic touches such as crockets, finials, and ogive arches[5]. The building at West Peachtree and 14th Street was known as the IBM Tower. It was also a landmark, the midtown equivalent of the Big Chicken to guide one's orientation near central Atlanta.

Akers was not through with reorganizing the company. In 1988, he split development into five new units in order to decentralize decision-making. He had Personal Systems under George Conrades, Application Business Systems under Steve Schwartz, Enterprise Systems under Carl Conti, networking under Ellen Hancock, and technology products including microelectronics under Patrick Toole. Akers called these changes the most "drastic" reorganization since Tom Watson's Williamsburg restructuring in 1956. Less than a year later, in July 1988, he took his pen to the organization chart once more, this time combining the North Central Marketing and Southwest Marketing divisions into one combined sales organization. He selected George Conrades, who had been leading the Boca Personal Systems unit for less than six months, to head this new team of roughly 100,000 employees.

Mainframes continued to be the most important element of IBM's performance. The Sierra project became the 3090 mainframe in 1985. With just twice the performance of the predecessor 3081 "Adirondack" machines, the performance was underwhelming. The year 1990 could not come and go without a major mainframe release. System/390 was the overall platform name, with the first machines labeled the ES/9000. Some of the ES/9000 models were air-cooled, taking a page from Gene Amdahl's playbook. The high-end models incorporated a number of the architectural elements of Menlo Park's Project Y. IBM Research recommended a IBM switch from expensive bipolar logic to less-expensive CMOS (Complementary Metal Oxide Semiconductor). This would make IBM's mainframes far more price-competitive with Amdahl, Hitachi, and Fujitsu. However, it required a billion-dollar investment in CMOS manufacturing. Notably, CMOS was a technology embraced by the FS project. Akers was president of the Data Processing Division during FS, and he scorned anything associated with FS. His decision to reject CMOS was a major mistake that would lead to continued price pressures and declining mainframe sales.

# THE PC CHALLENGE

John Akers had spent most of his career around big iron. In that light, Opel's strategy to rein in the renegade operation in Boca and integrate it with the rest of IBM seemed perfectly logical. A key piece of this strategy was the upgrade of DOS to a real operating system. IBM and Microsoft recognized early on that DOS was simply a stopgap program. The PC needed a new, far more advanced operating system. Though Microsoft was growing rapidly, Gates and Ballmer knew the future of DOS was limited. They wanted to be involved in developing the next-generation operating system. They felt Microsoft could gain valuable experience working with IBM. For its part, IBM wanted a more advanced operating system that would interface with its other computer lines and, at the same time, it desperately wanted the ability to regain full ownership of the PC operating system.

Despite the one-sided contractual arrangement on DOS, the relationship continued to be good. In fact, Gates and Ballmer secured an apartment in Delray Beach as they spent so much time in Boca[1]. The two companies agreed in 1985 to work jointly on the far more robust follow-on operating system. Originally called "Big DOS", it would be eventually released as OS/2. Gates would develop a close working relationship with Phil Estridge[2].

At the same time, IBM had a very good option to replace DOS, with a program called CP/88[3]. When the mainframe and PC-connected 3270 display was in development, IBM needed software to enable the machine to work in both PC and 3270 modes. Researchers at Yorktown Heights developed CP/88 that allowed the desktop to run either as a PC or as a 3270. When the more advanced PC AT came along, Yorktown enhanced CP/88 and renamed it CP/286. They had written CP/286 in IBM's PL/S language, making it portable. In contrast, Microsoft developed DOS specifically for the Intel 8088 and programmed it in Intel assembler. It was not portable.

CP/286 was the perfect antidote to both Gates and the DOS conundrum. There were many discussions in 1984 on what direction to take. It was Jack Kuehler, along with Bill Lowe and Mike Armstrong, who decided to ignore CP/286 and develop OS/2 with Microsoft. Many technical IBMers were aghast. Gates was aware of CP/286 and threatened IBM that he would never release DOS source code if IBM pursued CP/286. It was clear that Gates would not release DOS source code under any circumstances. It was a moment of ignorance, cowardice, or expediency on IBM's part. Gates did favor a simpler approach that enhanced DOS, adding multitasking and a graphical interface. In hindsight, it proved to be the right path but still left IBM without access to the source code. In any case, IBM declined the offer, and the two companies forged ahead with OS/2.

The development plan had IBM in the lead with Microsoft as the junior partner. IBM would design the architecture and assign selected programming sections to Microsoft. Though this seemed like a reasonable arrangement, it planted the seeds of future discord. The two companies represented vastly different cultures. Where Microsoft was young, aggressive, and attuned to the needs of the exploding personal computer market, IBM was slow, increasingly bureaucratic, and subject to its corporate masters in New York. As the project ramped up, the two companies combined to have 1,700 programmers working on OS/2. Their respective methods of "working" were very different. Microsoft developed simple, tight code to get the job done quickly and, in many cases, left the testing to customers. IBM measured its programmers on the lines of code developed, which could produce bloated and slow programs. IBM had extensive testing processes that served it well on much larger operating system projects but represented a drag with this program. Finally and most critical, large, mainframe-centric IBM was designing OS/2, and structuring it to handle the requirements of its corporate customers as opposed to the retail customer that Microsoft focused on.

The OS/2 program started after the shift to Intel 80286 microprocessors in 1984. IBM released the PC/AT that year based on that chip. However, OS/2 development was going to be a two-year project, at a minimum, and Intel was about to release the next-generation microprocessor, the 80386. It would hit the market in late 1985. The two microprocessors were dramatically different. The 80286 was a 16-bit microprocessor without virtual memory. The 80386 moved to 32-bit processing and significantly enhanced function, including virtual memory. It was a very good match for an advanced operating system like OS/2. However, IBM Corporate forced the OS/2 development teams to continue with the 286 microprocessor. The mainframe division feared the new advanced operating system running on far more powerful 386 machines would impinge on big iron computing cycles.

As a result, Boca was forced to design OS/2 around the inferior and soon to be obsolete 286 microprocessor. Boca could have overcome this poor decision later if Microsoft had not decided to code OS/2 in Intel-286 assembler as opposed to a portable language like PL/S or C. In the contemporary PC history *Computer Wars*, the authors speculate that Bill Gates had to understand the blind alley that IBM and Microsoft were marching down. They even surmise that Gates, with his focus on ramping up his MS-DOS monopoly with Windows, knew that basing OS/2 on Intel-286 assembler would ensure its demise[4].

Even with these issues, Gates offered to sell part of Microsoft to IBM in 1986 to finance additional development[5]. Lowe said "no". It was his last chance at correcting the mistake he had made with the initial agreement with Microsoft. According to James Cortada, the price would have been in the neighborhood of $100 million[6]. It would likely have yielded billions for IBM. More importantly, it would have avoided the many years of pain aimed at regaining the PC software standard.

Many of IBM's early missteps were a direct result of the rapid change of Boca Raton from a small skunkworks to a large IBM division. The Boca site workforce ballooned from 4,000 to nearly 10,000. Bill Lowe left Boca Raton in March 1981 to replace Hal Martin as general manager of the Rochester site. Lowe's mentor, Chuck Branscomb, moved back East when the Opel reorganization effectively eliminated the fully integrated General Systems Division. Phil Estridge replaced Lowe, and instead of directing the Acorn team of twelve engineers, he would now be in charge of the entire Entry Systems Division. He no longer had the autonomy he once enjoyed in GSD. With Lowe and especially Cary both gone, it would be very tough for Estridge to re-create the magic that had produced the IBM PC. The enormous new division and the overarching influence of the mainframe bureaucracy in New York would essentially swamp the kind of ideas that might build on the PC's runaway success.

The mainframe bureaucracy had grand plans for the PC. It would not be content to limit the speed of Boca's personal computers but wanted to tame the PC and make it subservient to IBM's main computer lines. This evolved into a scheme called SAA, for Systems Application Architecture, a blueprint that would fold neatly into the work on OS/2. Unfortunately, Earl Wheeler was back to lead the charge.

In 1971, IBM was struggling to get its various systems to talk to one another. Bob Evans was aware that the System/360 plan had not adequately addressed communications architecture. He commissioned work in this area in 1970. John Fairclough was now director of the IBM lab in Raleigh, North Carolina. He saw the immediate value of this kind of architecture as Raleigh's mission included communications hardware and software. He worked with Wheeler to define and develop the concept. The result was Systems Network Architecture, or SNA, announced in 1972[7]. Though it was mostly proprietary with a lack of openness to other computers and communication protocols, it was generally successful.

The threat of powerful personal computers replacing dumb terminals like the IBM 3270 and the experience of SNA led to the concept of SAA. The goal was a complete application environment that would harness personal computers to IBM's System/370 and midrange

computers in a structured client/server arrangement. A parallel focus
was on software developers, encouraging them to develop applications
using the SAA framework. Wheeler took the lead with SAA[8]. He
parlayed his success in Endicott during System/360 development,
along with ambition and political savvy, to become the director of the
Kingston lab in 1970. He was a very demanding manager, earning the
nickname "Earl Wants." Though he led the low-end operating systems
for the System/360 program, he was a mainframe guy and saw the
world from that perspective. He was the driving force behind the IBM
8100, an early attempt to respond to DEC with a System/370-
compatible but "subservient" computer to compliment a central mainframe. The 8100 earned the name "boat anchor[9]" by customers. It
would be replaced by the far more capable 9370 system.

Despite the utter failure of the 8100, Wheeler escaped the IBM
penalty box and ended up as head of the new Programming Systems
Division, in charge of all IBM software development. It was a position
he ran imperiously. He developed the concept of SAA in 1983, but it
took time to convince the corporation to embrace and fund it. SAA
had a common user interface, database access with SQL (Structured
Query Language), and a programming toolkit to develop SAA
applications. SAA aimed to address IBM's proliferation of system
platforms with a common client front-end. It was as ambitious as FS or
Fort Knox, and represented an undertaking as massive and complex as
either one of those programs. To address SAA, IBM "would add 6,000
employees to its software development staff, raising the total to 26,000
or about 7 percent of its total work force.[10]" The program got going as
Rochester's Silverlake program was in process. Wheeler liked what
David Schleicher was doing and enlisted him for SAA in 1988 after
the AS/400 had been launched. Much like FS, SAA sucked up first-
rate R&D and programming resources at a critical time for IBM. As
the SAA "Czar," Wheeler would eventually oversee up to 30,000
programmers working across many different software labs.

Presentation Manager was the SAA graphical common user interface. Boca had already developed a very serviceable graphical front-
end to DOS called Topview. Wheeler succeeded in deeming it
nonstrategic and had it killed. Presentation Manager was also central
to an application suite IBM would develop on its own -- office

automation, with email, calendaring, and meetings functions. IBM already had such application suites on the major system lines. The new SAA application intended to provide a common interface across all of these platforms, with the PC and OS/2 client as the entry point. The result would be a significant demonstration of the power of the Systems Application Architecture. This ambitious program was called OfficeVision.

The OfficeVision project started by surveying 500 corporate customers to gather requirements[11]. Roughly 1,500 programmers worked under Tony Mondello, the former Rochester lab director. Programming on the server sides progressed in parallel with OfficeVision/2, the OS/2-based client. Mondello planned to announce OfficeVision in May 1989. However, key elements of the program were delayed, not once but several times. Progress was slow. Releases were late. With slumping sales of OS/2, there was pressure to support Windows and UNIX clients. On top of this, OfficeVision required OS/2 Extended Edition, a far more complex and expensive version. The result was a price tag that translated to nearly $8,000 per user seat. Despite the problems, Wheeler insisted that OfficeVision remain completely aligned with OS/2 and SAA, sealing its fate. In 1991, IBM announced that OfficeVision was being "stabilized," a euphemism for impending cancellation. Meanwhile, a small 15-person team working at IBM Research developed a far simpler email program, but it was not going to see the light of day.

SAA and OS/2 represented only half of IBM's plans to regain control of the PC standard. The strategy also included a completely new line of personal computers. These machines would be the PC's next generation, called the Personal System/2, or "PS/2." Jack Kuehler would describe the PS/2 as the "porthole to IBM systems of the future.[12]" That was a not so subtle indication the new line would focus on corporate IBM customers and IBM systems. Bill Lowe returned to Boca in 1985, taking over the reins of the Entry Systems Division. Shortly thereafter, IBM transferred Lowe and two hundred of his senior staff to offices closer to Armonk in New Jersey. Even worse, IBM Corporate moved all of headquarters to White Plains.

Lowe concentrated on introducing proprietary elements into the PS/2 line. The centerpiece was a new motherboard bus called the microchannel. For IBM, the input/output channel had been a key area of computer design since the Stretch project. In the rush to market, there were many technical problems with the original PC bus. The microchannel intended to address most of them. However, the plan also called for the microchannel to be used on IBM's other computer lines, including the 9370, the AS/400, and the RS/6000. With such a broad mandate, it was compromised in its main mission, to replace the PC bus and regain platform ownership. There were more problems. The microchannel was more expensive than third-party add-on cards. It obviously required the PS/2, which itself was far more expensive. The add-on cards that plugged into the microchannel were also expensive. Moreover, there were very few available on the market. The microchannel was not backwards compatible with the original PC bus cards, requiring customers to purchase all new cards. Boca engineers were aware of this and had revamped the microchannel design so that it would support both existing and microchannel expansion cards. Jack Kuehler, Lowe, and Lowe's boss Mike Armstrong rejected the revised design.

On top of all this, IBM managed to step down the performance of the PC microchannel, which was one of its only selling points. Adding up all of these problems, it was hard to explain the value of the microchannel. Kuehler would tell the press that "there are many things inherent in the microchannel concept that you haven't seen yet.[13]" It was instructive that no one was interested in cloning the microchannel or the PS/2. Lowe made matters worse by killing the PC AT in the United States, which was still selling well, in an effort to force customers to buy the PS/2. IBM Europe did not follow suit and continued to sell the older PC model, with on-going success. The final setback was a six-month delay in announcing the PS/2 line. IBM spent three years and hundreds of millions of dollars in an attempt to regain control of the PC hardware standard. It only succeeded in losing those three years while competitors kept enhancing the existing PC standard.

IBM and Microsoft continued work on OS/2. At the same time, Gates and Microsoft worked behind the scenes on a graphical user interface (GUI) for MS-DOS. As was his standard procedure, he did not need to invent anything here. Why do that when you could simply appropriate someone else's invention? He tapped into concepts that Xerox had pioneered in its lab in the Silicon Valley. Xerox had set up the lab, the Palo Alto Research Center (PARC), in 1970. Robert Taylor, who first led the lab, focused on computer innovations while the main Xerox research lab in Rochester, New York continued work on the company's copier technology. PARC engineers developed the standard personal computer user interface elements of windows, icons, menus, and pointers, informally known as "WIMP". Taylor's team of engineers would also invent the mouse, Ethernet, and the laser printer. PARC incorporated many of these elements in a prototype personal computer called the Alto in 1973. This was a full two years before the debut of the Altair microcomputer from MITS. Xerox would go on to produce a commercial version of the Alto in 1981 called the Xerox Star. However, priced at $100,000, it missed the new market that the IBM PC would create.

PARC researchers were very accommodating[14]. They did not guard the secrets of WIMP and the Alto. On the contrary, they wrote research papers and conducted demonstrations for visitors at the lab. One group of visitors in 1979 included Steve Jobs and several Apple engineers, a result of Apple's investment in Xerox. Though they were already working on GUI elements for the Lisa and Macintosh computers, the demonstration at PARC was a revelation. At roughly the same time, Microsoft agreed to port some of its applications to the Apple Macintosh. Jobs provided Gates with Mac machines now sporting the graphical WIMP interface. Gates proceeded to copy the interface for Windows. Jobs was not happy about it, though Gates' view was that both Apple and Microsoft had borrowed liberally from Xerox. Jobs had no problem hiring a number of engineers from PARC. And Gates, not to be outdone, would hire Charles Simonyi, who had written Alto's graphical word processor program (Bravo). Simonyi started work on Microsoft Word[15]. Gates would also offer Ed Iocabucci, IBM's architect for both DOS and OS/2, the position of chief technical officer at Microsoft. Ed turned him down.

Meanwhile in Boca, IBM engineers had created Topview, a near graphical front-end to PC-DOS that featured multiple concurrent windows and multitasking. Topview announced in 1984, a year after Microsoft released Windows 1.0. With IBM and Microsoft still on relatively good terms, Phil Estridge invited Bill Gates to Boca to review the new interface. Though Gates was ahead in his GUI implementation, he was surprised to see how far IBM had progressed in multitasking. Gates returned to Seattle, bought a company with a similar front-end application, and moved immediately to shore up the multitasking capabilities of Windows.

IBM was unaware that Gates and Microsoft had been working since 1981 on an enhanced user interface for MS-DOS. Microsoft unveiled a new user shell for DOS called Interface Manager in 1983. An enhanced version called Windows arrived in 1985, again as a shell built on top of DOS. The new interfaces did not do well at first, but by Windows 3.0, released in 1992, the graphical DOS front-end had gained wide acceptance. The first release of OS/2 would not be available until 1987 and would feature the old DOS monochrome, line-mode interface. After Earl Wheeler killed the Topview program, Boca planned to use the SAA Presentation Manager to provide a graphical user interface.

The year 1985 was critical for IBM's PC business. Despite a few missteps, most notably the PC Jr., PC revenues reached a new high. The euphoria of success was short-lived. Phil Estridge and his family were tragically killed in the crash of a Delta L1011 at the Dallas-Fort Worth airport. His death robbed the division of a dynamic and stalwart presence. He was held in such high regard that Steve Jobs offered him a huge pay raise to come work for Apple, but Estridge turned him down.

Lowe faced far greater challenges than his first stint in Boca, with an SAA-centric strategy in place that would tie his hands in competing with Intel, the clones, and Microsoft. Despite this, Lowe would spurn an offer that represented a new path forward. Gates contacted CEO John Akers in July 1986 to discuss a plan where IBM could outmaneuver the PC clones with IBM's latest chip technology. Lowe heard of the planned meeting and arranged to meet with Gates and Ballmer first. Lowe made it clear that IBM would be going forward with Presentation Manager as its graphical interface, not Windows. Since Microsoft had already won the desktop battle, Lowe's decision essentially unleashed Microsoft. A cooperative and realistic approach might have yielded great dividends and placed IBM at the center of the PC market. Gates, Ballmer, and Microsoft had not yet reached the point where they could confidently charge ahead without IBM. However, Lowe's insistence on OS/2 and Presentation Manager forced their hand.

When OS/2 finally reached the market in 1987, it was expensive at $325. The effective price was even higher as it required upwards of $1,000 in additional memory to run properly[16]. Moreover, OS/2 was a separate purchase while the PC clone companies bundled MS-DOS and Windows with their machines. Thanks to the influence of IBM Corporate, OS/2 continued to run in outdated 16-bit 80286 mode. It would not support 32-bit programs until five years later. Though OS/2 had sophisticated multitasking support, it could only run one DOS application at a time. This was a big problem, as there were very few OS/2 applications on the market.

Boca rolled out an ill-advised scheme to jump-start OS/2 by shipping both Windows and OS/2 preloaded on IBM PS/2 systems. However, the long start-up time of OS/2 did not create a good first impression. Other issues with OS/2, including its voracious demand for memory and limited number of applications, further reduced the chances of a customer adopting the operating system. Once committed to Windows, customers faced the difficulty in removing OS/2 code from their hard drive to free up space. To top it off, IBM had to pay Microsoft royalties for both DOS and OS/2.

A revised agreement for OS/2 development had IBM working on the release 2.0 and Microsoft developing the next release. For a time, Gates remained behind OS/2, commenting at the Comdex show that "OS/2 is the platform for the 90s." Gates nurtured the IBM relationship because IBM was the PC standard-bearer. By 1987, he soured on OS/2 and was heavy in development of Windows. He had come to the realization that Microsoft, with full control of DOS and Windows, was the new standard-bearer.

The relationship between IBM and Microsoft spiraled downhill. Boca programmers tried to bridge the culture gap, writing programs in "blue jeans, T-shirts, and sneakers.[17]" Still, there remained the natural friction between the two sets of developers, where one was in charge and the other writing the code. The power balance shifted. The wild success of MS-DOS on IBM PC clones led to even greater success with Windows. Gates and Ballmer sensed that IBM's ambition for OS/2 was skewed to SAA, corporate users, and IBM's own PC hardware. Gates started taking his best engineers off the OS/2 project and putting them onto Windows[18]. Gates made yet another offer, to take over OS/2 development and split the revenues, but Lowe declined.

In 1989, Microsoft was readying Windows 3.0 and wanted to get rid of its contractual obligation to work on OS/2. Jack Kuehler, now a senior executive advising Akers, felt IBM could develop a better OS without Microsoft. He finalized the divorce at the 1989 Comdex exposition. The parting was not amicable. Microsoft immediately changed its Windows APIs so that IBM's version of Windows was not 100 percent compatible.

Gates would comment after the split with IBM, "Nothing makes me madder than the notion of somebody taking my code and competing with me.[19]" It would be hard to come up with a more ironic statement than that. The source of all Microsoft's billions was the imprimatur that IBM gave to the fledging company. Gates was a master of using people. The word "ruthless" commonly comes up in descriptions of him. He took computer time at Harvard for his commercial project. He took the academic-oriented BASIC language and adapted for his own uses. He used Tim Patterson and Seattle Computer Products to obtain the QDOS base code for the operating system he did not have but professed that he did. He enhanced QDOS liberally by copying elements of Gary Kildall's CPM operating system. He took the WIMP graphical-use interface of Xerox PARC and Apple and incorporated it into Windows. Gates would famously rail against PC hobbyists who copied his software without paying for it. Gates would later work to destroy Netscape, the first successful Internet browser.

Gates even tried to shortchange his co-founder, Paul Allen, not once but twice over ownership shares in Microsoft. Gates secured an initial split of 60-40. He pressed a change to 64-36 when Allen was still doing work for MITS and the Altair. He then granted Ballmer an 8 percent stake in the company over Allen's objections. Ahead of the Microsoft IPO in 1986, Microsoft issued additional stock that diluted Gates's share to 45 percent and Allen's to 25 percent[20]. At the time of the IPO, Gates's share was worth $350 million.

Bill Gates was not satisfied with domination of the DOS market with Windows. Even before the divorce from IBM, he had already started on a project to outdo OS/2, to be called NT, for "New Technology." Work began in 1987. His investment in NT would be huge for Microsoft, involving a team of 100 programmers and a cost of $400 million over its five-year development cycle[21]. NT would end up with over 4 million lines of code[22].

The key to his plan was Dave Cutler[23]. Cutler, who started programming on an IBM 7044 at DuPont, became a star architect at DEC. He led the design and development of the VMS operating system. He became bored with managing ongoing releases of VMS and threatened to leave the company. At the time, DEC was trying to figure out what to do about RISC. IBM and Sun RISC-based workstations were decimating its VAX workstation market share. DEC had taken several runs at RISC at the time that Cutler voiced his dissatisfaction. He reached an agreement where he and his key engineers would establish a satellite lab in Bellevue, Washington, and focus on RISC. Beginning in 1986, Cutler's team developed the PRISM (Parallel Reduced Instruction Machine) architecture and a new 32-bit operating system called MICA that would run on it. MICA would also run VMS applications. DEC canceled both PRISM and MICA in 1988, electing to simply port VMS to hardware it acquired from MIPS, with the resultant product called DEC Alpha.

Cutler was not the least bit happy and left. Much in the manner of Al Shugart, he aimed to make DEC pay. With Microsoft conveniently located nearby, he signed on with Gates and Ballmer. His mission was to develop the NT operating system. He brought many of his DEC engineers along with him. Cutler decided to discard the existing OS/2 base and start from scratch. However, he did not need start quite from scratch. What emerged from Cutler's lab was a blend of the VMS and MICA operating systems. At the time, Microsoft and IBM still had a joint licensing agreement in force. This allowed IBM to see Cutler's work as it progressed. Lee Reiswig, the executive in charge of OS/2, commented, "We have evaluated NT, and what we have seen so far has not made the cut.[24]" This would prove to be a less than astute assessment[25].

NT arrived in 1993, and Steve Ballmer publicly announced that

"NT is targeted at the core of IBM's business.[26]" Microsoft assembled a team of 100 NT "evangelists" and sent them out to developer companies to help them to port their applications to NT. Though NT required a huge system (eight megabytes of memory and 100 megabytes of disk) like OS/2, Microsoft positioned it as a central server rather than a consumer desktop machine. Like Windows, it was a huge success.

For IBM, if the challenge of tackling OS/2 was not enough, work started in 1991 in Boca Raton and Austin on an even more advanced operating system layer, called Workplace OS. It would harness the power of the RISC-based PowerPC hardware and was aimed squarely at the Microsoft and Intel duopolies. Workplace OS would be the ultimate "uber" OS and was called GUTS, for "Grand Unifying Theory of Systems" within IBM[27]. The Workplace teams looked at the Taligent program, with the object-oriented Apple's Pink OS at its core. They liked what they saw. So did John Sculley, the CEO of Apple who had forced out Steve Jobs in 1985. He said the unification of OS/2 and the Mac OS on common PowerPC hardware would "bring a renaissance to the industry.[28]"

Workplace was based on the concept of a microkernel, a thin layer of code that could sit on top of one or more full-blown operating systems. IBM adopted an existing microkernel called Mach 3.0 from Carnegie Mellon University. IBM would enhance and harden this code to transform it from an academic project to a commercial product. Renamed the IBM microkernel, it would be open-source software that companies and universities could license. The microkernel provided the core services that most operating systems needed to handle, including memory management, process control, input/output, and program interrupts. Additional OS "personalities" could sit on top of the microkernel, including DOS, OS/2, Mac OS, UNIX, and even Apple's Pink OS. The vast library of desktop applications represented a key driver for Workplace. With the "personality" layer, Workplace could run any desktop application directly and without modification.

IBM's ambition for Workplace extended to creating personalities for Austin's AIX and Rochester's OS/400 operating systems. This put the scope of Workplace in the rarified company of FS, Fort Knox, and

SAA. In fact, the IBM microkernel, sitting between the low-level PowerPC hardware and the high-level operating system, sounds a lot like FS. Furthermore, IBM already had this high-level machine interface with the System/38 and AS/400. It was called "licensed internal code" but was in effect an OS microkernel.

## GLENN HENRY ONCE MORE

IBM's PC business moved steadily downhill from 1985 on. In hindsight, the huge investment might have been better spent on IBM's mainframe business, which had upwards of 60 percent gross margins and accounted for nearly 90 percent of the company's profit. However, there was another potential lifeline to address the Microsoft and Intel duopoly. It involved Glenn Henry. He did not stay around for either the RT/PC or RS/6000 debuts. IBM named him an IBM Fellow in 1985. He chose to work on microprocessor hardware for his Fellow project[29]. He aimed to develop a high-performance version of Intel's X86 microprocessors[30]. His L86 program would combine IBM's RISC-based POWER architecture with full Intel compatibility. By 1987, Henry was able to demo a chip that could run in either RISC or X86 mode. In the latter mode, it was twice as fast as comparable Intel chips. Jack Kuehler was fully supportive of Henry's work. Henry was aware that IBM was already dabbling in this area. Terry Parks in the Burlington lab had developed a 386 chip called Blue Lightning. Henry's program passed a couple of IBM audits, but he sensed that IBM did not have the stomach to carry through on such a program. Henry's IBM Fellow manager was in New York. Henry met with him and asked him if he used a PC. The answer was "no." The manager thought the PC might make a good mainframe terminal. Henry replied that the PC would take over the world[31]. Jack Kuehler would have to make a choice at some point, between Intel and Henry's competitive microprocessor, and Henry likely knew what that choice would be.

It was during this time that a chance event interceded in Henry's life[32]. His wife told him she thought a drug dealer had moved into their expensive Austin neighborhood. There was a young single person coming and going at all times day or night. The "drug dealer" turned out to be 23-year-old Michael Dell. Henry would leave IBM in 1988 to become Dell's vice president of development. It was a very big deal to give up an IBM fellowship. It also was quite a change in corporate environments. At Dell, he would have one person doing what at IBM would be a whole department.

Kuehler had gone to school at Santa Clara University with Andy Grove, the third employee of Intel after Gordon Moore and Robert Noyce. Grove was now CEO of Intel. Though Kuehler endorsed Henry's project, he also recommended that IBM invest in Intel to ensure microprocessor and DRAM memory supply. Competition from Japan in the DRAM market was a dire financial threat to Intel. In 1982, IBM had paid $250 million for a 12 percent stake in the company[33]. That year, IBM's PC revenues were $500 million. In just three years, those revenues would balloon to $5.5 billion.

Kuehler's focus on protecting Intel missed a great opportunity to break up that half of the Wintel duopoly. Early on, IBM was far ahead of Intel in electronics development and manufacturing. Starting with Ralph Palmer's pluggable vacuum tube through SMS and SLT designs and magnetic core production, IBM had been at the leading edge in electronics production. Glenn Henry's L86 design, coupled with manufacturing expertise at IBM Burlington, could have broken Intel's monopoly with a more powerful, less costly, compatible x86 microprocessor. This was not hypothetical, as Henry would soon prove.

Henry worked at Dell for six years before leaving in 1994[34] to return to his pursuit of an Intel-compatible microprocessor[35]. He first went to work for MIPS Technologies, a company founded by Stanford University researchers into RISC technology, including RISC pioneer John Hennessy. Henry was hired by Tom Whiteside, who had previously worked for him at IBM.

Glenn Henry's mission at MIPS was a return to his IBM Fellow project, to design a plug-compatible replacement for the Intel 286 PC chip. After a short time at MIPS, he decided that starting a small company from scratch would be a better blueprint for this endeavor. He formed Centaur Technology in 1997. He chose the name as the company's product would be half X86 and half RISC. Based on his many years in engineering and development, Henry had come to the view that a small number of talented engineers could out-produce companies with large engineering staffs. He would note, "This is the company designed by me to be the kind of place I wanted to work in as an engineer.[36]"

The goal was a fully compatible Intel microprocessor chip that would undercut Intel prices and allow for low-cost PCs and similar devices[37]. It would not challenge Intel on high-end performance but would be lower cost, with lower power consumption, and good-enough performance. The first Centaur chip arrived in 1997. It was a 1 GHz Pentium chip that sold for $30 to $40 and plugged into the Intel socket on the motherboard[38]. Henry would direct the development of 14 additional Intel-compatible chips over Centaur's successful 20-year run. As usual, he was heavily involved in the design of each chip and, in many cases, writing some of the code[39]. As Henry and Centaur proved, IBM missed the opportunity to address the Intel monopoly.

# AKERS AND THE ROAD TO RUIN

John Opel retired as IBM board chairman in 1986, and Akers assumed the position in addition to the CEO role. Both of his predecessors, Cary and Opel, had carried both titles. Frank Cary, a key Akers supporter, retired from the board at age 70, as was customary. Akers had IBMers Jack Kuehler and Frank Metz with him on the board. The other 13 members were outside directors. Given the challenges the company faced, this new configuration of the board, with Akers in both operational control and oversight control of IBM, was the worst possible scenario.

Cary had appointed most of the board members and stayed in touch, but his impact was limited as Akers and the company struggled. He was concerned in 1991, but the rest of the board was not. The IBM board was typically "window dressing." It was not there to rock the boat. Vin Learson would comment on the board during his time saying, "IBM directors in those days were very quiet and agreeable, and knew how to say aye.[1]" Additionally, the architect of IBM's aggressive expansion strategy, John Opel, was still on the board.

Despite the bloat created by Opel's ambitious growth plans, the early years of Akers' tenure went well. The PC business was a big contributor, though it had just hit its peak as Akers took over. The AS/400 arrived in 1988 and would hit annual revenue of $14 billion, representing one-fifth of total IBM revenues. The RS/6000 followed in 1990, firmly establishing IBM in the fast-growing workstation market.

At the same time, there were indicators that all was not well. Exhibit A was the mainframe business, which drove upwards of 75 percent of the company's profits. Gross margins had approached 70 percent, but competition from Amdahl, Hitachi, and Fujitsu was depressing sales and squeezing margins. With prices up to 30-40 percent lower than IBM, there was no alternative but to reduce prices. IBM Research recommended a technology change to CMOS logic in 1986 that would have substantially reduced mainframe costs, but Akers rejected the switch because of its association with the FS project. Mainframe margins plummeted to 52 percent in 1990 and appeared headed even lower. The next mainframe generation, the 3090, was announced in 1985. At just twice the performance of the prior 308x family, it received a tepid response. Mainframe revenues had declined 50 percent by 1990[2]. In addition, mainframe power in general was under siege. The price per processing MIP was decreasing rapidly with the onslaught of UNIX and RISC machines.

The expense side of the business represented an even greater problem. Opel's expansion of plants and labs led to IBM inventories reaching an all-time high in June 1985, somewhat ironic considering that Opel had made his name by finding work-in-process inventories. The rapid growth of the workforce resulted in substantially higher Sales and General Administrative (SG&A) expenses. Between 1985 and 1990 those expenses increased by 40 percent while revenue only increased by 17 percent[3]. IBM's SG&A expense in 1990 was over 30 percent of revenue, roughly double other firms in the technology sector. IBM's revenue was growing but at a modest pace, far short of what was needed to reach Opel's ambitious goals.

Fiscal year 1989 should have laid bare the growing danger, that the company was hurtling towards an abyss. Revenue only grew 7 percent, but net income was down 36 percent. The following year saw revenue increase 10 percent to $69 billion and income nearly double to $6 billion. This performance reflected a momentary rebound from the prior down year. The true state of the company emerged in 1991, when IBM suffered the first loss in its storied history. And, it was not just a slight loss. Revenues dropped nearly $5 billion, and the annual loss approached $3 billion. For John Akers and for IBM, the game had irrevocably changed from Opel's full-throttle growth track to mere survival.

IBM announced first-quarter results for 1991 in May. They were not good. Akers called a meeting of his senior managers and proceeded to chew each of them out in front of the others. He zeroed in on sales performance. He had bulked up the number of sales people in the United States from 20,000 to 25,000. He asked his managers, "Where's my return for the extra 5,000 people"[4]? He expressed his frustration, saying too many IBMers were "hanging around the water cooler, waiting to be told what to do." It was not the typical manner that senior executives were spoken to, but the message was received. The next day, Akers was in front of IBM's management training program, with mid-level managers from the field in attendance. He decided to repeat the rant, hoping to ensure that his message got past his senior executives.

However, the tenor of his remarks shocked them. One of those managers in attendance took extensive notes and put them in a memo to everyone in his department. From there, the memo traveled on IBM's extensive internal communication system, making the complete rounds. Predictably, it was leaked to the newspapers. The *New York Times* was charitable in calling the rant a "hard-hitting talk." As a former salesman, it made perfect sense that Akers could view the company's malaise as simply a lack of focus on sales. For IBM employees, his comments about IBMers just "hanging around the water cooler" struck a chord. The *Times* story continued, writing about the raise Akers received earlier in 1991 and the fact that IBM had not fired any senior executive for lack of performance. It smacked of a typical story about IBM management: "IBM UK challenges Oxford and Cambridge to rowing contest. . . ." They finish last. "IBM had one man rowing and eight men steering. The report was sent to Armonk, which studied the findings and gave its orders: change the rower.[5]"

Akers decided to take action on executive accountability. He targeted George Conrades, who was running IBM's combined sales organization, with 100,000 employees and $27 billion in revenue, to take the fall[6]. Conrades had impeccable credentials, with degrees in mathematics and physics and an MBA from the University of Chicago. He was tall, personable, able, and highly admired. Through many promotions, he left unbridled success in his wake. He started the AS/400 division and ran it in grand style. An avid motorcycle enthusiast, he met with Harley-Davidson, a large AS/400 account, and suggested that he and the customer executives ride the 400 miles to Rochester on Harleys for a briefing on upcoming announcements. Harley-Davidson switched from an IBM mainframe to the AS/400 platform.

George Conrades was a leading candidate to replace Akers when he stepped down in three years. Conrades was just 52 years old and would have ample time to exercise his considerable talents at IBM's helm. Instead, Akers decided to make Conrades the scapegoat for IBM's woes. He demoted Conrades, the executive equivalent of firing him. Conrades got the message and left IBM. It is interesting that in May 1989, Conrades had told his team "Akers has told us that we'll be in our jobs for a long time." Little did he know that he would be the first to go. Frank Cary was irate, saying, "I'm a big fan of Conrades.[7]" It was a sentiment shared by many. Dick Gerstner replaced Conrades as head of the U.S. sales organization.

In the end, Akers had no choice but to cut the workforce. With the Opel buildup fully in place, the company had upwards of 100,000 employees too many for the revenues coming in. IBM's tradition of full employment was viable as long as the company earned healthy profits. Akers oversaw a revision in the employee rating system. The current system evaluated employees with five levels of performance, with most ranked as ones, twos, or threes. The new system imposed a system of four levels on each department with hard targets for each of the levels. Each manager was required to rate 10 percent of his or her department as 1, 40 percent as 2, 40 percent as 3, and 10 percent as 4. Managers put 4-rated employees on a detailed performance plan, to improve or be released. The new system also featured a "four check" rating that bypassed any improvement plan and immediately showed the employee the door.

At this point, IBM had never laid off anyone. During the cash crunch in 1921-1922, Watson Sr. had fired some, but there were no layoffs. In the current crisis, the existing employee rating system would not achieve the necessary cuts. Greater numbers would be primed by incentives. IBM Atlanta rolled out a very generous package that employees termed the "seven, two, one, and two" plan. IBM added seven years to an employee's service for pension purposes, paid up to two years' salary upon exit, gave you one minute to think about the offer, and asked you to take two others with you as you left. For many of the employees I knew in the Atlanta area, the deal was a no-brainer. If you had 23 or more years of service, you could take the program and return to IBM for one day, when you would have reached 30 years' service. IBM then tacked on seven additional service years, you were suddenly fully vested, and could retire with full pension and medical benefits. On top of this, there was a shift to filling IBM positions with contractors as opposed to full-time employees. A significant number who left under the incentive program would come back as contractors, often in the same job. This was called double-dipping.

The program was particularly attractive to top performers, those who could walk out the door and secure another high-paying job while receiving a huge severance check and full pension from IBM. It was a huge brain drain for the company. According to Wall Street Journal reporter Paul Carroll, "Of the 10,000 people who took the severance package, 8,000 were rated 1.8" Ernest von Simson would observe that 22,000 employees out of 220,000 in the United States were "1" performers. Losing nearly 40 percent of the top performers was not a great move. The program was also expensive. IBM spent at least $125,000 for each employee cut. The cost to cut 75,000 jobs in this manner came to $9 billion[9]. This huge cost led to future packages getting progressively skimpier. The various workforce programs reduced IBM's employment from a peak of 406,000 in 1985 to 302,000 by the end of 1992.

Work began early in closing the plants and labs that were to support Opel's revenue target. In June 1988, IBM had announced the closure of four manufacturing plants—Boca Raton, Tucson, Austin, and Santa Clara. John Opel commissioned the Tucson facility in 1980 as part of his expansion program. IBM Research also felt the pain. IBM invested $110 billion on R&D during the 1980s. Akers cut the annual budget to $6 billion, and that would be far from enough.

In concert with deep employee cuts, Akers adopted and expanded Opel's Independent Business Unit (IBU) strategy. He rechristened it the "Baby Blues". Aimed at reducing bureaucracy and increasing accountability, Akers split the company into 13 operational units. This included Application Business Systems (AS/400) in Rochester, the disk drive unit ADSTAR in San Jose, the large printer unit Pennant in Boulder, Personal Systems in Boca Raton, Enterprise Systems in White Plains, and Networking in Raleigh. In addition, he defined Application Solutions, Programming Systems, Technology Products, and the four major worldwide geographies as separate operating units. He delegated more autonomy to the general managers of each business line. For example, Steve Schwartz with the AS/400 unit was able to make most major decisions without the involvement of Armonk. One unit that did not make the Baby Blues plan was the printer subsidiary Lexmark. It was sold to a private equity firm in 1991 for $1.5 billion. This reduced the IBM workforce by 4,000 employees in one shot. Still, more employees needed to be cut

There was an additional motivation in pushing more independence to these units. The Baby Blues would produce separate financial statements, which would ease the process should IBM decide sell any one of them. Frank Cary contemplated a similar approach but rejected it because of the exorbitant expense. However, Cary did not face the dire situation of Akers' IBM. The printer unit (named Pennant) provided a good example of the spin-off approach. Lexmark produced low-end consumer printers while Pennant developed IBM's large and midrange printers, along with a significant set of software. It was a very profitable $2.3 billion business. Sensing that Pennant could be spun off like Lexmark, there was no shortage of IBM executives who wanted to be part of the unit.

Pennant had operations in Boulder, Tucson, and Endicott. Your author joined the division in 1993, moving from the sales organization in Denver to Pennant in Boulder. I managed a new team that provided printing services to customers. In 1995, I became the product manager for AS/400 printers and printing software. The AS/400 market represented a significant piece of Pennant's business. My new responsibility meant frequent trips to Rochester to interlock on both development and marketing plans. As it was not possible to make all of these trips in summer, I had some experience with the typically frigid winter at the site. The dead giveaway was the engine block plugs protruding out of the grilles of many of the cars in the site parking lots.

Pennant shared in the long and illustrious history of IBM printers. The first printer, developed by Eugene Ford, Claire Lake, and Fred Carroll in response to the Powers printer, arrived in 1921. Vertical slug printers dominated until Fred Demer's high-speed, horizontal chain 1403 printer broke the mold in 1959. IBM released the landmark 3800 printing system in 1976, providing very hi-speed laser printing. The 3800 and its follow-on models were continuous-form printers that ran at high resolutions and high speed. They reached speeds of 1,220 pages per minute, with the paper moving through the printer at 280 feet per minute. Companies used these massive systems for applications such as statements. In addition, the printers jump-started the print-on-demand book publishing industry. They printed a complete copy of Leo Tolstoy's *War and Peace*, with covers, in 60 seconds.

## THE FINAL SLIDE

The layoffs and plant closings failed to put much of a dent in IBM's downward spiral. The most critical business line, mainframes, cratered from $19 billion in revenues in 1990 to $9 billion by 1993[10]. The PC business lost $1 billion in 1992, while the duopolists Microsoft and Intel both earned $2 billion. Reduced sales amid a bloated cost structure resisted any short-term fix. There was no rental base to cushion the slide. Any continuation of the present trajectory would result in a company a mere shadow of what it once was.

The fourth quarter of 1992 was disastrous. Though this was traditionally the strongest quarter of the year, the company reported a loss. The loss from ongoing operations was just $45 million, but charges to the quarter for layoffs and restructuring resulted in a total loss of nearly $5 billion. It was the largest loss ever reported by any company. At the start of the year, Akers predicted profits of $4 billion for the year. The $5 billion loss represented a total swing of $9 billion. The

IBM board of directors had no choice but to act. Led by James Burke, they asked for Akers' resignation. They granted him a $5 million golden parachute[11]. Akers called a board meeting for January 26, 1993 where he announced a cut to IBM's dividend. The bombshell, though, was the news of his resignation.

The financial results for Akers' time as CEO paint a grim picture. In 1991, IBM stock was down 23 percent while the Dow Jones was up 194 percent. The company's market capitalization, sitting at $106 billion in 1987, collapsed to $23 billion as he left office. Over that period, the share price tumbled from $43 to $12[12]. In his eight years as IBM CEO, Akers did increase revenues from $46 billion to $64.5 billion, an rise of 40 percent, but only at a rate of 4 percent per year, far short of the 15 percent that Opel had baked into his expansion strategy. During this period, DEC was growing at 7.6 percent per year, Amdahl at 12.3 percent, and Sun Microsystems at 46 percent[13]. IBM fell out of *Fortune*'s list of most profitable companies. Akers instituted the first layoffs in IBM history. He left with that workforce at 302,000, a decrease of 23 percent. Though there were doomsday scenarios for the company, IBM was never close to bankruptcy. The company remained in a good cash position to the end, a reflection of the lesson that Watson Sr. had learned in 1921 and was absorbed by IBM finance in the succeeding years.

On the nonfinancial side of the ledger, many cherished IBM traditions fell by the wayside, with full employment being the first and possibly respect for the individual the second. Morale was at low ebb and once you lose morale, it is tough to regain it. IBM dropped to 45th place on the "Most Admired" companies list. IBM had been #1 in this ranking for the prior four years. IBM was not even the "most admired" technology company. HP took away that title.

Steve Jobs, the ultimate product guy, described Akers as "smart, eloquent, [a] fantastic salesperson, but he didn't know anything about product.[14]" His signature accomplishments were the Baby Blues decentralization and the actions he had to take when Opel's growth strategy proved illusory. He had, by far, the worst record of any IBM CEO. In the space of three years, IBM would lose close to $16 billion. This was from a company that was the second most profitable company in the world in 1990.

To be fair, there is plenty of blame to assign to Akers' predecessor. Akers had to deal with the effects of Opel's grand growth plan, though Akers was IBM president during the Opel years, so he was not entirely blameless. He had the opportunity to disavow Opel's strategy at the beginning but fully embraced it. In reality, the two should share the blame for IBM's fall. The greater share probably lies with Opel. His ambitious growth targets, the overexpansion to meet those targets, and the switch from rent to purchase were the main drivers of IBM's precipitous decline. It was also a bit presumptuous for Opel to establish goals so far into the future, knowing he had only four years at the helm.

His goal of a $100 billion IBM by 1990 required the company to grow by the 1980 size of IBM every two years. The 1990 goal translated into 14 percent compounded growth per year while the 1994 target required 15 percent growth per year. Opel felt these targets to be solidly in IBM's wheelhouse. Tom Watson had averaged 17 percent growth over his entire CEO period, from 1956 to 1971. He had averaged over 30 percent growth during the "go-go" System/360 years.

Akers turned out to be the wrong person to address the challenges presented by Opel's strategy. He spent too much time on IBM's PC business when the real damage was occurring with mainframes. He had his managers focusing on an extensive and bureaucratic quality program while the ship was sinking. He needed to be addressing IBM's bloat, not contributing to it. He certainly did not help himself by taking both the CEO and chairman's roles, which further reduced oversight as the going got tough. Their combined record was disheartening. IBM lagged in technology where it had once been the undisputed leader. The bold leadership and risk-taking of Tom Watson and Vin Learson was missing. The company's most critical asset, the mainframe plunged to new lows. After Cary had set the table with the IBM PC, savvy mainframe moves, and success in the midrange, the next two CEOs squandered it.

# ENTER THE "COOKIE MAN"

After John Akers announced his decision to step down in January 1993, the pressure mounted to find and name a new CEO. Despite the company's struggles, running IBM remained one of the most prized chief executive positions. Though the original list of candidates the board considered contained over a hundred names, there were only a few that board chairman James Burke and his fellow directors deemed capable of taking on the task. The board approached GE's Jack Welch, Apple's John Scully, Motorola's George Fisher, and even Microsoft's Bill Gates. Sculley expressed interest but wanted to bring his entire management team, a definite non-starter. Lawrence Bossidy, the CEO for Allied Signal, turned down the offer. He said IBM needed someone "who is 35 years old, knows computers, and can clone himself thirty-five times.[1]" James Cannavino was close to being that guy. He had replaced Dick Gerstner, Lou Gerstner's older brother, as head of the PC division. Dick Gerstner had been promoted over Bill Lowe to lead the PC group. He was a strong candidate to replace Akers as CEO but suffered from a chronic health problem, later diagnosed as Lyme disease. He retired from IBM in 1989[2]. Cannavino, still in his forties, was perhaps the leading internal candidate, but the board did not consider him sufficiently seasoned to take over. The board assigned the key candidates code names for secrecy. Candidate "Able" was the first interviewed. He was Lou Gerstner, the CEO of RJR Nabisco.

**Figure 57.** Louis V. Gerstner, IBM CEO 1993-2002. Reprint Courtesy of IBM Corporation ©.

Lou Gerstner planned to attend Notre Dame but received a scholarship offer from Dartmouth College[3]. He graduated in 1963 with a degree in electrical engineering and continued to Harvard for an MBA. In 1965, he joined the management-consulting firm McKinsey and Company. He rose fast and, in just nine years, became their youngest principal at age 28.Though a star at McKinsey, he jumped at the chance to run his own show. It would be at American Express. During his ten years there, he turned the Travel Related Services group into a huge success, adding the Gold and Platinum cards and boosting enrollment from 8.6 million to 30.7 million. He increased earnings at a 17 percent compound rate over those years. He became president of the company in 1985. Always supremely confident, he fully expected to replace CEO Jim Robinson while he was still in his early forties.

As that vision faded, Gerstner left American Express in 1989 for RJR Nabisco. This was a conglomerate formed in a $25 billion buyout by the hedge fund Kohlberg, Kravis, and Roberts. RJR Nabisco's sprawling businesses were the "4Cs," for the cookies, crackers, candies, and cigarettes. Though he did not know it at the time, his work at RJR Nabisco would be a dry run for the challenges at IBM. Gerstner's focus was on cutting expenses, laying off employees, and divesting unprofitable brands (including Planter's, Lifesavers, Butterfingers, Del Monte, and Chun King[4]). He had been there for four years when John Akers resigned and the search for a replacement at IBM was in full swing.

James Burke had an apartment in the same Fifth Avenue building as Gerstner. Gerstner held Burke, the former CEO of Johnson and Johnson, in high esteem. He admired his handling of the Tylenol cyanide-poisoning scare of 1982. Burke stepped up and faced the crisis immediately and expertly, even at a cost to his company of $100 million. According to Gerstner, Burke called him at home and asked to meet. Burke had heard the rumors that Gerstner was going to return to American Express as CEO. Gerstner assured him that this was not the case. He also told Burke he was not interested in the IBM job. Despite his code name of "Able", Gerstner was not the board's first choice.

In his time at McKinsey, Gerstner had learned the value of a deep dive into a company to understand its market position and problems.

At RJR Nabisco, he spent his early time there traveling throughout the enterprise, trying to get a deep sense of its strengths and weaknesses. He understood IBM's general challenges, that the mainframe and minicomputer businesses were under siege while the company was preoccupied with dealing with the Opel expansion. He also was aware of the other IT companies that reached the brink of financial collapse and could not be saved. That list included DEC, Data General, Wang, and Sperry-Burroughs.

At American Express, he embraced information technology and he relied heavily on IBM systems. However, IBM would sometimes give him pause. He related the story of a division manager who replaced an IBM mainframe with Amdahl[5]. The IBM representative on the account said IBM support would no longer be available. Gerstner was incensed at the lack of customer loyalty with such a large account as American Express. He sought a deeper view of where IBM was and how bad the situation was now.

He was familiar with the reports by Paul Carroll, the beat reporter for IBM at the *Wall Street Journal*. Carroll documented the troubles at IBM week by week. Carroll reported the general sense that IBM was in serious and potentially disastrous trouble. But, Gerstner wanted an insider view. Burke arranged a meeting with Paul Rizzo. Rizzo had been passed over for the IBM CEO job by Akers in 1985 and retired two years later. The IBM board brought him in as a senior financial executive as the business started to disintegrate. Rizzo told Gerstner that sales were declining and that mainframe revenue had dropped from $13 billion in 1990 to a projected $7 billion for 1993. If this trend continued, nothing could stem further descent. Gerstner left the discussion with the conclusion that going to IBM was a bad move.

At the same time, he was increasingly concerned about RJR Nabisco. He was well aware that Kohlberg, Kravis, and Roberts paid too much for the company, burdening it with mountains of debt. He could not see a path forward to profitability. Phillip Morris exacerbated the situation by starting a cigarette price war. That move reminded Gerstner of the quote from Ray Kroc of McDonald's, "When you see your competitor drowning, grab a fire hose and put it in his mouth.[6]"

Gerstner started to waver about his decision to turn IBM down. His older brother Dick encouraged him to take the job. Two weeks after the initial discussion with Burke, Gerstner was at his Florida home. Burke and fellow board member Thomas Murphy (CEO of Capital Cities/ABC) flew down to try to convince him to take on IBM. They surprised him with a new argument, that IBM was a "national treasure" and that "you owe it to America to take the job[7]." To his own surprise, he decided to take up the challenge.

With both parties interested, Gerstner and the IBM board began negotiations. Gerstner was prepared going in. His first account at McKinsey was an executive compensation study for the Socony Mobil Oil Company. McKinsey had a long and lucrative history consulting on executive pay, a veritable franchise started by Arch Patton[8]. Therefore, Gerstner came well-armed to the discussion. The negotiations would go on for ten days. He drove a hard bargain. For openers, he wanted both the CEO and chairman positions. He also wanted a new board of directors. The existing board had 18 members, including John Akers, Jack Kuehler, John Opel, and Paul Rizzo. He tactfully encouraged some departures and reconstituted a smaller board with just 12 members. He retained an executive pay firm to negotiate his compensation package. As described by Doug Garr in his book *IBM Redux* the final tally included "a signing bonus of $4 million, first-year base salary of $2 million, performance bonuses of up to $1.5 million, and 500,000 stock options.[9]" In addition, Gerstner had 2.4 million shares of RJR Nabisco stock plus options for an additional 2.6 million shares. He asked IBM to guarantee the $9.9 million value of his unexercised RJR Nabisco stock options. This turned out to be an inspired move. He left RJR the day before "Marlboro Friday," the start of the price war with Philip Morris. RJR Nabisco stock cratered.

Burke made the announcement to the press at the Hilton Hotel in New York, with Akers and Gerstner on hand. Gerstner and Akers had a good relationship. The two had met when Gerstner was at RJR Nabisco. Burke spoke first and defended the search process. He said that Lou Gerstner was the first candidate he spoke to and the only one offered the job. Coming from RJR Nabisco, financial pundits viewed Gerstner as a mere "biscuit salesman" without the necessary background in technology. After the announcement, Gerstner was flown by helicopter to Armonk headquarters. He noted the closed, sterile atmosphere of the building. He commented on the bright orange "go-go" carpeting, which Tom Watson had confused with a Howard Johnson motel. Gerstner took over Akers' corner office in Armonk, overlooking an apple orchard. He noted there was no personal computer. IBM provided Akers with a transition office in a drab building in Stamford.

Gerstner started on April Fools' Day, 1993. The financial picture was bleak. After three years of ever-increasing losses, the stock price had sunk as low as $12 from a high of $176 in 1987. He met that morning with the Corporate Management Committee (CMC), the top 50 executives in the company. They were all wearing white shirts. Gerstner wore a blue shirt. He was startled that each executive had his own Administrative Assistant (AA) and would lean on the AA whenever questioned. Unlike his normal informal style, Gerstner had spent considerable time preparing his remarks for the meeting. He outlined three major points: First, he wanted to get the "right-sizing," including the additional layoffs, done by the third quarter. Second, he stressed that everyone owned employee morale, though he would take the lead. Third, he signaled that he would bring in some outsiders but each executive would have a chance to prove his or her worth. He also previewed his current thinking on a core IBM strategy. He was leaning toward reversing the Akers Baby Blues direction and embracing the value of one united IBM Company.

He ended the meeting with an assignment. He wanted a five-page report on each of their respective businesses, detailing the market, customers, and competitors while summarizing their division's strengths and weaknesses. Jim Cannavino would later note, "five pages is not what we do.[10]" Rather, it was usually a stack of several hundred foils. Gerstner also wanted travel recommendations, as he planned to get out and meet with customers and IBM teams, as he had done at RJR Nabisco. At a follow-up meeting, Gerstner wore a white shirt and most of his executives arrived in blue.

Gerstner would take some time evaluating the management team that Akers had left him. He had to determine who to keep and who to cut loose. Gerstner commented there was "no mention of culture, teamwork, customers, or leadership" in Akers' game plan. He was also puzzled with Akers' negative comments about his senior managers. "What struck me was how he could be so critical but still keep some of these executives.[11]"

Gerstner looked outside IBM for critical positions. At the top of his list was his new CFO, Jerry York. York learned a demanding work ethic at the U.S. Military Academy, an engineering degree at MIT, and an MBA at the University of Michigan. He started his career in engineering but switched to finance. York was hired away from Chrysler, where he had established a reputation as a tough guy when it came to cost cutting. IBM would put that skill to the test. York saw the challenge at IBM and actually lobbied Gerstner for the job. He would only be at IBM for two years, but they were the two critical years in the company's turnaround. Gerstner also hired Abby Kohnstamm as his chief marketing officer. She had been his AA at American Express. Her father had been an IBM marketing executive and was responsible for the Charlie Chaplin ad campaign for the IBM PC.

Gerstner did have a valuable and trusted resource from day one— his brother. Early on, he scheduled a session with Dick, who came well prepared. Dick warned of the almost religious fervor around the PC, and taking back what was rightfully IBM's from Intel and Microsoft. Dick also recommended immediate reduction of mainframe prices. Lou had already come to that conclusion. Dick suggested that Lou get a personal computer in his office and start using PROFS, the IBM email system, to communicate. He warned of the need to fight IBM's bureaucracy at every turn. He should cultivate a small cadre of unbiased and reliable advisors. His final piece of "brotherly advice" was to "call your mom.[12]"

Though IBM was not in mortal danger, emergency surgery was required. The immediate focus was to stop the bleeding and get the financial house in order. The timing of his hire would work to Gerstner's advantage. The current year was a free year for him where turmoil and financial shocks were expected. In addition, given the company's momentum, the year would be considered more a part of Aker's legacy than his own. In this context, he made every effort to concentrate the pain and losses in 1993, and start the following year with a clean slate.

Gerstner took his role to shore up employee morale seriously. He logged upwards of a million miles on the corporate jet visiting IBM locations. His stump speech netted out to the following: "We lost $16 billion in three years. Since 1985, more than 175,000 employees have lost their jobs. The media and our competitors are calling us a dinosaur. Our customers are unhappy and angry. What we were doing was not working." He would then focus on the "bad guys," IBM's competitors. He quoted Larry Ellison of Oracle, who said, "IBM? We don't even think about those guys anymore. They're not dead, but they're irrelevant.[13]" Gerstner would say that people like Ellison had taken IBM's market share, resulting in more IBMers laid off. He used the PROFS email system and started regular "Dear IBM Colleague" communications to the workforce.

One morning in early April 1993, Gerstner opened the door to his limousine and was surprised to see Tom Watson sitting inside. Watson was distressed by what had happened to "his company." Gerstner sensed that Watson really wanted to jump back in and save IBM, despite his 79 years. Tom would not live to see IBM fully recovered. He would die on the last day of 1993. About the same time, Gerstner ran into Joshua Lederberg, a Nobel laureate in genetics whom he had known from the board of directors at Memorial Sloan-Kettering Cancer Center. Lederberg told Gerstner about IBM, "It's a national treasure. Don't screw it up.[14]"

Gerstner was not a nice guy. The board did not hire him for nice. He could be rude, blunt, impatient, intolerant of dissent, and manage by intimidation. He focused on getting to the root of IBM's problems and did not suffer fools. He also had a vindictive streak, as evidenced when he cut off *Fortune* magazine access to IBM after they published an unflattering article. He had a big ego. Yet, he was also a very private person. He would enter IBM headquarters at Armonk by the back door to avoid running into anyone. He was laser focused on turning IBM around, and he outworked just about everyone while on this mission.

Though a professed nontechnical executive, he had an engineering degree and was intimately familiar with the workings of computers from his time at American Express. He was a voracious reader. He liked all relevant information distilled and written down. When traveling, he would always carry two briefcases loaded with mountains of material to read. He was very smart and quickly zeroed into the heart of each issue.

Gerstner's first meeting with the CMC had been an anomaly. More typically, he waited until everyone had arrived, started abruptly, and aimed to finish within 15 minutes. If there seemed to be a consensus on the issue under discussion, Gerstner would frequently take the opposing side of the argument. The subsequent back and forth usually surfaced the best decision. Once they reached a decision, Gerstner did not entertain additional debate and adjourned the meeting. Gerstner would relate a story concerning an early presentation by Nick Donofrio, his senior technical executive: "Nick was on his second foil when I stepped to the table and as politely as I could in front of this team, switched off the projector. . . It was as if the President of the United States had banned the use of English at White House meetings.[15]" The company had come to rely obsessively on foils, and lots of them. There would be a main set of foils and perhaps a number of backup sets that were at the ready in case of objections or the need for a deeper dive. In contrast, Gerstner wanted recaps prepared ahead of key meetings that all participants could review, allowing the team to go directly into discussion. He did not want to sit through a hundred foils. He did not want to see a presentation of more than ten pages. He virtually banned foils from his senior executives, forcing them to be conversant with issues and able to argue them directly without the crutch of such aids.

Gerstner soon disbanded the Corporate Management Committee. He concluded that too many issues reached this group that should have been decided several levels below. He also railed about the IBM practice of non-concurrence, where anyone could block a decision and force it to the next level. Within months, he realized he still needed an executive body of last resort. He created the Corporate Executive Committee, a smaller panel that handled only the most urgent escalations.

Though he favored a centralized corporate structure, he also liked to reach down into the organization to find the individuals who could solve a given problem as opposed to listening to high-level pitches from managers. He would comment, "I look for people who work to solve problems and help colleagues. I sack politicians." He was laser focused on business outcomes. His would caution, "Never confuse activity with results.[16]" Customer needs drove results. His view was that "the marketplace dictates everything." He tried to talk with a customer every day. And above all, his mantra was "move fast."

Gerstner did use email and would communicate with IBMers outside the management chain. As an example, he heard that Hitachi was going to make a disk drive announcement in June, 1993. He wanted IBM to announce a disk product days before that, to steal Hitachi's "thunder.[17]" They told him IBM only made announcements on the first and third Tuesdays of the month. This was Tom Watson's heritage. Gerstner immediately changed that. IBM announced a new model of the high-end 3390 disk drive on May 20, 1993, ahead of Hitachi announcement and on a Thursday.

Gerstner used compensation to drive business results. At IBM, he encountered a system centered on salary, with little variance between the top and marginal performers. There was a heavy emphasis on employee benefits. Tom had started a stock options plan in 1956 but limited it to the top 1,300 managers. Gerstner would grant options to far more IBMers, reaching 72,500 in 2001[18]. He had one and only scoreboard, the IBM share price. Besides generous stock options, he also asked his most senior executives to have "skin in the game." Gerstner spent more than $1 million of his own money on IBM stock and expected his managers to do the same. He also announced in late 1993 that he would evaluate all business-unit executives on overall company results, to avoid parochial strategies.

For the wider IBM workforce, he introduced variable pay tied to performance and cut back on benefits. The key reduction was the IBM pension system. In 1999, York split the IBM pension plan in two, between a traditional program for older workers and a new cash-balance plan with IBM contribution for newer workers. This change invited no small degree of controversy. Bernie Sanders, then a congressman from Vermont, bought five shares of IBM stock so that he could voice his displeasure at an IBM stockholders meeting. Sam

Palmisano, as IBM president, would later say that changing the IBM pension was the hardest thing he ever did at IBM.

Early in his tenure, analysts asked Gerstner what his vision was for IBM. He replied that a vision was the last thing the company needed at this time. The vision comment was the main takeaway by the press, even though he went on to describe IBM's strategy as regaining profitability, winning customer mindshare, moving aggressively into client/server, and becoming the full-service provider. However, he was already converging on a vision, or central strategy. The consensus among analysts was that he should accelerate Akers' Baby Blues strategy and split IBM into separate companies. This would provide complete accountability for each descendant company. It would also increase market capitalization, as the total worth of the individual companies would be greater than the current IBM.

Gerstner adopted the opposite tack, his "One IBM" strategy. He argued that a fully integrated IBM was more than the sum of its parts and better positioned to provide its customers with the best expertise and unequaled solutions. The strategy would also best leverage IBM's size and breadth. He was surprised at the huge sums IBM spent to set up the existing Baby Blues. He spent much of his first year visiting IBM customers and getting their feedback. He wanted their endorsement for his "One IBM" strategy. This was the business model that Watson Sr. had constructed, with one-stop shopping and full customer support. Customers were generally supportive of this approach. He also discovered that most customers truly wanted IBM to be successful.

Gerstner decided that IBM would return to its traditional role as integrator of the total solution and take responsibility for "translating the pieces into value.[19]" He called this vision "IBM singleness[20]," and he considered this his greatest accomplishment as IBM CEO. This vision would directly translate into a substantial focus on services. He wanted the company to be a leader in the emerging client-server and e-business markets. Like many customers, he favored open standards, something unheard of within IBM.

He was quite aware of the work done by the 2,800 scientists at IBM Research in Yorktown Heights, working with a budget of $5.6 billion[21]. He would press for more practical applications of their inventions. He was quoted as saying, "Good ideas don't come out of IBM Research, they escape." The Research director, James McGroddy, would use PowerPC microchips as an example. He focused on getting them into more IBM products. IBM would open five new labs during Gerstner's tenure, including those in Austin, Beijing, and Delhi. Gerstner was not one for big bets such as the System/360 or FS. He was far more comfortable doing acquisitions. This would be an huge ongoing problem, as IBM had always led by its own invention and technology.

When Gerstner took over the company, IBM was a mainframe-centric company. Gross margins were still over 50 percent. The market share was at least 30 percent. However, the technology landscape had changed. UNIX systems were an attractive alternative to high mainframe prices. The PC revolution was siphoning off applications. Though he was openly dismissive about the dot-com mania in the 1990s, he believed that real growth would come from traditional companies embracing the Internet. When he saw the first mock-up of web pages for ibm.com, he asked, "Where's the buy button?[22]" His focus was to establish IBM servers as the focus in client/server and lead the transition to network and Internet applications. This vision led to the term "e-business." He would compare the growth of the Internet to other technological innovations by the metric of the time it took to reach 50 million people: "Radio took nearly 40 years to hit that threshold; television thirteen years; cable TV ten years; Internet in five.[23]" This rapid embrace would play to IBM strengths, with big mainframe computers, supposedly relegated to the "old economy," at the forefront.

Gerstner found a trusted advisor to help him make e-business a commercial reality. Irving Wladawsky-Berger, born in Cuba, immigrated to the United States with his family when he was 15. He earned masters and doctorate degrees in physics at the University of Chicago before joining IBM Research in 1970. He would be a central figure in IBM's adoption of the Internet, supercomputing, and Linux. Gerstner asked him to set up the Internet division in 1997 and lead the way in what he called IBM's "moon shot" to catch the Internet wave.

Gerstner was initially dismissive of company culture. IBM's was unique and insular. Though he would change a number of elements of IBM's customs, he would come around to believe "that culture isn't just one aspect of the game, it is the game.[24]" His greatest challenge would be to refocus IBM's culture to the tasks that lay ahead. The company was under siege, between the layoffs that sapped morale and Akers' Baby Blues strategy that had split the business in half, with some wanting the old IBM and others excited about their prospects as a spin-off company. He knew he had to win the respect of his fellow IBMers, some of whom called him the "Cookie Monster," a reference to Nabisco. He would start with simple things, like abolishing the IBM dress code.

Vision and culture were long-term projects. The most pressing need was to stop the bleeding, and the first priority was reducing the workforce. Akers had laid off 30,000 in 1991 and 43,000 in 1992. York recommended cutting 35,000 employees immediately. Gerstner agreed. IBM released 45,000 employees in 1993 and an additional 36,000 the following year. Akers had started the cuts when the company had roughly 400,000 employees, with half in the United States. Gerstner reduced IBM's workforce to 220,000. The days of lavish layoff packages were over. The new, leaner rounds of layoffs were termed "right-sizing." York took a $9 billion charge in the second quarter of 1993 to reflect these actions. He also reduced costs in other areas. He eliminated duplicate IBM data centers and sold IBM real estate, including the new 50-story IBM Tower in Atlanta and the 41-story IBM World Headquarters at 590 Madison Avenue. The New York headquarters had preserved intact Watson Sr.'s office from the previous 590 Madison Avenue building[25]. That bit of history went by the wayside.

Gerstner committed to divesting unprofitable businesses within IBM. He started by selling the Federal Systems Division to Loral for $1.5 billion. This division had an illustrious history within IBM but was unaffordable with the company in crisis. IBM reported an $8.1 billion loss for 1993. IBM had paid dividends of $2.8 billion ($1.21 per share) in 1991 and 1992. York reduced the dividend to 40 cents per share in 1993 and cut it to 25 cents a share in 1994 and 1995.

Gerstner kept Paul Rizzo's comments about mainframe revenues at the top of his mind. Nick Donofrio led the mainframe (System/390) division. After Gerstner switched off Donofrio's foil presentation, Nick proceeded to paint the picture of the market. He said it was time to reassert the mainframe's central position. It was not "dead," as some in the media contended. It was, however, under siege from Amdahl, Hitachi, and Fujitsu, all of whom offered mainframes at 30 to 40 percent below IBM prices. Gerstner heard the same story when he visited IBM operations in EMEA, the World Trade acronym for Europe, Middle East, and Africa. Greatly expanded by Dick Watson, it now consisted of 44 countries with 90,000 employees. He met the European executive team in Paris at the headquarters building known as La Cage Aux Foils for the many overhead presentations given there. He saw duplication in functions across each country, what he called "fiefdoms." He also heard the same consistent message, that mainframe prices were too high. Gerstner hosted a meeting of the CIOs of the largest 175 U.S. companies in May, 1993 They were angry that IBM let the "PC bigots" run wild with the myth that they could do what mainframes could uniquely do. Gerstner took advantage of the meeting to announce price decreases on IBM mainframes. The price per MIP would decrease over the next seven years from $63,000 to just $2,500[26].

The secret of this dramatic reduction was CMOS, for Complementary Metal Oxide Semiconductor. Irwin Wladawsky-Berger of IBM Research recommended to Akers in 1986 that IBM switch from the far more expensive bipolar chip technology to CMOS. It would cost $1 billion to make the change, and Akers decided against it. ES/9000 mainframes rolled out in 1990 with the expensive bipolar circuitry. Akers went even further and said that research in microprocessors was not affordable. He even started negotiations with Motorola to sell the business. Gerstner authorized the investment in CMOS the minute Donofrio briefed him on it. IBM Microelectronics produced the new chips, and Donofrio had new System/390 mainframe models sporting CMOS running at 800 MIPS by 1997[27]. Mainframe shipments declined by 15 percent in 1993, then rose by 41 percent in 1994, were up 60 percent in 1995, and continued on an upward arc through 2001. Gerstner fully embraced microelectronics as a core business within IBM. Encouraged by Jack Kuehler and the PowerPC opportunity, he announced plans to build a $2.5 billion semiconductor plant at East Fishkill, part of a $5 billion investment in the technology.

---

## AI AND DEEP BLUE

IBM had been on the sidelines of supercomputer technology since Frank Cary canceled Gene Amdahl's ACS/360 machine in 1969. The company refocused on viable, high-end commercial 360 and 370 machines. Twenty years later, Gerstner and IBM would take the first steps to get back into this high-end sector. Interestingly, it was the game of chess and a team of graduate students at Carnegie Mellon University that hastened the move.

The idea of employing a computing machine for chess had been around for some time. Charles Babbage wanted to apply his computing engines to chess but he settled for Tic-Tac-Toe[28]. In the 19th Century, there were a number of mechanical chess machines, including "The Turk" and "Mephisto." They turned out to be hoaxes, operated by a human in an adjacent room. Konrad Zuse saw the potential to apply computing power to the game. In 1938, he predicted that a machine would beat the world's reigning chess champion within 50 years. His prediction would turn out to be very close.

It would be up to Claude Shannon to provide the working roadmap to a machine victory. Shannon had touched many different fields, including binary logic, cryptography, and even artillery fire control. Chess was another area that interested him. In his 1950 paper "Programming a Computer for Playing Chess"[29], he first worked the immense scale of the challenge. He estimated the total number of possible move combinations at ten to the hundredth power. Known as the "Shannon Number", this was apparently greater than the total number of atoms in the universe.

He next laid out two approaches to computing a chess move. His "Type A" strategy applied brute force, computing all the possible positions, then rating them before selecting a move. With the computers of his day, this would be painfully slow. He estimated a contemporary computer would take at least 16 minutes to calculate each move. His alternative strategy, "Type B", applied chess logic, limiting the search to moves that scored high on probability analysis. He also suggested the addition of a chess-opening database, a selection of proven sequences for use at the beginning of a game. Shannon concluded his analysis by saying automated chess play by either strategy, "will force us either to admit the possibility of mechanized thinking or further restrict our concept of "thinking".

Arthur Samuel was one of the early readers of Shannon's treatise. As we have seen, he opted for the much simpler game of checkers. Still, his televised checkers matches on the IBM 701 in 1956 created a stir, and was a publicity boon for IBM. Another IBMer, Alex Bernstein, took the next step. He had been captain of the chess team at Bronx Science High School. He joined IBM in 1956 in the Applied Science Department, working for Charlie DeCarlo. He secured DeCarlo's approval to work part-time on a chess program. He led a team of programmers that shared the IBM 704 machine on third shift with John Backus and his FORTRAN group. Though the 704 was one of the fastest computers at the time, his chess program still took eight minutes to calculate each move. Bernstein's program employed Shannon's Type B approach, using chess strategies, not brute speed. His system could potentially beat a chess beginner. Bernstein enjoyed moderate success, with profiles in Scientific American and Life Magazine. Tom Watson, Jr. was not entirely happy as it gave the impression that IBM employees sat around playing games[30].

A major force in Artificial Intelligence made the next move. John McCarthy studied at Caltech and Princeton. A lecture by John von Neumann set him on a career in computing. As a Dartmouth professor in 1955, he arranged the time-sharing of the MIT's 704 computer. He also directed MIT's Project MAC on timesharing. A decade later, now at Stanford, McCarthy supervised several graduate students, including Alan Kotok, in the development of a chess program. Called the Kotok-McCarthy program, it ran on the University's IBM 7090, which at 229,000 instructions per second, was nearly sixty times the performance of the IBM 704. In 1967, he set up a teleconferenced chess match with his Russian chess counterparts. Though he lost the match, the Kotok-McCarthy program computed the best chess in the 1970s[31].

Entering the next decade, the Belle computer took over as the best chess computer[32]. Ken Thompson, the co-inventor of UNIX, and Joe Condon took advantage of a high-performance version of the DEC PDP-11. Though the machine ran at only half the performance of the IBM 7090, they deployed special LSI (large-scale integration) chess-specific circuits. Belle searched 180,000 chess positions per second, compared to a comparable general-purpose computer that could only muster 5,000 positions per second[33]. Belle was unbeatable in the early 1980s.

A former Air Force fighter pilot moved the hunt for a chess machine into high gear. Edward Fredkin worked on the DEC PDP-1 at Caltech before joining the Air Force and training as a jet pilot. Instead, the Air Force assigned him to MIT's Lincoln Lab to work on the SAGE program. After leaving the service, Fredkin started a technology company that made him a fortune. He bought a Caribbean island and a seaplane to fly him there. Deciding to return to academics, MIT hired him as a full professor, despite his lack of a college degree. He spent time at Caltech and Boston University before joining the faculty at Carnegie Mellon, where he developed an early chess hardware design.

In 1980, Fredkin contributed $100,000 to fund the Fredkin Prize[34]. The prize had three tiers. The first tier awarded $5,000 to the first chess computer to achieve a chess master rating. The second tier awarded $10,000 to the first computer to achieve international master status, with a chess rating over 2500 across 25 consecutive games. The third tier provided $100,000 to the first computer that defeated a world chess champion.

The international chess federation oversaw the chess rating system. On a scale from zero to 3000 and above, it reflected the probability of one rated player beating another rated player[35]. A beginning chess player might have a rating of 100. A national chess master had a rating of 2200 or greater. An international chess master ranked 2400 or above. Chess great Bobby Fischer's rating was over 2900. The Belle computer achieved a Master-level rating of 2200[36], winning the first tier Fredkin Prize in 1981[37].

When Edward Fredkin left Carnegie Mellon, he entrusted the administration of his Prize to Hans Berliner, a resident professor and very accomplished chess player. Berliner would eventually have two separate teams at the university vying for the chess prize. His first team created the Hitech machine, the first computer to reach a chess rating of 2400.

A recently arrived graduate student proposed an alternative to Hitech. Feng-Hsiung Hsu emigrated from Taiwan, arriving at Carnegie Mellon in 1985. The Belle computer greatly influenced his thoughts on computing chess. Hsu wondered what would happen if a machine was a thousand times the "chess" speed of Belle. Could that kind of pure brute force be successful[38]? Hsu was an expert in VSLI (Very Large Scale Integration) chip design. He felt he could embed the key chess logic functions on one chip. VSLI densities in 1985 had progressed to 100,000 transistors per chip[39]. That provided enough logic room to search possible moves, evaluate and select the best move, and control play out several moves[40]. His custom chip searched 160,000 chess positions per second and could theoretically ramp up to two million positions[41]. Working with a team that included Murray Campbell and Thomas Anantharaman, Hsu integrated the chip into an existing processor, prosaically dubbed "Chip Test". The machine won the annual ACM (Association for Computing Machinery) chess tournament without a single defeat, putting the chess world on notice. Chip Test played many practice games against Hitech, though Hans Berliner squashed any publication of the results.

The success of Chip Test led to funding for a follow-on machine called Deep Thought. With a chess search speed of 700,000 positions per second, it would win the Tier 2 Fredkin prize in 1988 with a rating of 2551. The Deep Thought team realized that tackling the next step, beating a world chess champion, required a far faster and costlier system. A former graduate student from Carnegie Mellon would provide the solution. Peter Brown joined IBM Research in 1984. Aware of Carnegie Mellon's success in chess, he wondered how much publicity IBM would receive by sponsoring a match between an IBM machine and a world chess champion. He concluded the publicity would be considerable, certainly more than buying a 30-second spot during the Super Bowl[42]. It would far outdo Arthur Samuel's televised checkers game.

Brown saw the easiest path forward as simply hiring the Deep

Thought team. He pressed the idea at IBM Research. He also had some inside help. John Cocke had worked with Edward Fredkin and pushed Research to take up the challenge. In March 1989, Feng-Hsiung Hsu and Murray Campbell joined IBM. Hsu would focus on hardware. Murray would lead chess logic and programming. They added Joe Hoane, a staff programmer at Research[43] to the team.

IBM was perhaps overly anxious to move forward, deciding to start immediately with the existing Deep Thought machine. The company organized a match with the reigning world chess champion, Garry Kasparov. Kasparov had won the title in 1985, defeating Anatoly Karpov. He had since defended the title twice, both against Karpov. Garry Kasparov's chess rating was over 2800. Karpov visited IBM and played several games against Deep Thought. Robert Byrne, an American Grand Master, also played a number of games with Deep Thought. He only managed a victory when he played near perfect chess. Both masters came away impressed with the machine. Looking to establish an on-going relationship with IBM, Kasparov agreed to a discounted fee of $30,000 for a match, set for October 1989[44]. He reviewed fifty of Deep Thought's prior games[45]. It was a two game exhibition. Kasparov won both easily.

Chastened by the sound defeat, it was back to the drawing board to create an improved and faster machine. Deep Thought II sported 24 concurrent chess processors with a speed 5-8 times that of Deep Thought I[46]. Deep Thought I actually beat the new machine until the team made software changes to take full advantage of the extra performance. In addition, it was clear that the Deep Thought machines were very good on defense but needed greater capabilities on offense. Deep Thought II won the 1991 and 1994 ACM championships, playing at a 2550 rating level[47]. After several successful test matches with lesser chess champions, the team felt it was ready for a rematch with Kasparov.

It was also time for a new name and enhanced hardware. A naming contest evolved Deep Thought II into Deep Blue. Lou Gerstner had proclaimed that IBM would use its ultra-fast RISC PowerPC microprocessors on everything from entry systems to supercomputers. That included Deep Blue. IBM Research produced a high-performance computer that featured thirty-six high-end

RS/6000 PowerPC processors connected in parallel. The chess team integrated 480 of Hsu's dedicated chess chips. Deep Blue's theoretical chess search speed was 300 million positions per second, though the effective speed was somewhat less[48]. Hsu estimated that the same computing capability on a general-purpose computer would require a speed approaching one trillion instructions per second, a reflection of the impact of his custom chess chips.

The Association for Computing Machinery agreed to host the match, scheduled for February 1996 in Philadelphia. The purse was $500,000, with $400,000 going to the winner[49]. The match pitted Kasparov, rated at 2800, with Deep Blue, rated at 2700[50]. Deep Blue won game one, the first victory by a machine over a reigning world champion. Kasparov came back and won game two. Games three and four ended in draws. Kasparov rolled to victories in games five and six, winning the match. He praised the Deep Blue team, commenting, "They have succeeded in converting quantity into quality" referring to its raw speed. He also noted that "in certain kinds of positions, it sees so deeply that it plays like God.[51]"

IBM asked for a rematch and Kasparov graciously consented. In a decision that would cause untold problems, IBM took on the dual roles of host and player. The purse was $1 million with $700,000 going to the winner[52]. They agreed to conduct the match in October 1997 in Manhattan.

For the IBM team, the focus was on significantly ratcheting up the hardware. A new Deep Blue, dubbed "Deeper Blue", would emerge. Kasparov did not realize how much IBM would invest to beat him[53]. Cost was no object. The Deeper Blue hardware alone cost in the neighborhood of $20 million. Its raw speed was 11.38 gigaflops (billion floating-point operations per second).[54] This put the machine at number 259 on the Top 500 list of fastest supercomputers in the world. This was respectable supercomputer power. On top of that, the integrated chess chips gave the machine "super" chess power. Deeper Blue rated at twice the performance of Deep Blue[55].

Kasparov was somewhat dismissive of Deeper Blue's fantastical performance. His view was, "sheer power means little in chess because it is a mathematically near-infinite game.[56]" The IBM team understood this. Their focus turned to tactical chess, with beefed up evaluation function in the chess chips and deeper learning in the software. IBM added two chess Grand Masters, Joel Benjamin and Miguel Illescas to the team[57].Working from a database containing 300,000 chess games, they developed opening "books" that comprised the first thirty moves.

The rematch was a big deal. Peter Brown had made the right call in picking chess. Kasparov was the "rock star" of the chess world. Two days before match, an IBMer who had been at the 1964 System/360 press announcement called the Deep Blue press conference even bigger[58]. At the same time, the run-up to this second match was different. Kasparov commented that IBM was noticeably icier, that the gloves were off. He complained about IBM's dual roles. IBM did not help when it declined to publish Deep Blue's recent training games as was customary. In addition, IBM hired a team of Grand Masters without divulging it to Kasparov[59].

IBM held the match on the 35th floor of the Equitable Center in Manhattan. Kasparov's mother Klara attended all his major matches. She inspected the venue and confirmed her usual front row seat[60]. Lou Gerstner was also on hand for the contest. Kasparov took game one. He admitted to a bad opening and lost game two[61]. After the game, his team told him he missed a simple move that would have gotten him a draw[62]. Games three and four ended in draws. Deep Blue was lucky to escape with a draw in game five. The match was tied 2.5 to 2.5 heading into game six[63]. Kasparov made a critical error at move seven and would not recover. At one point, he turned and said something in Russian to Klara, a violation of match rules[64]. Less than an hour into the game, he resigned after only 20 moves. Deep Blue had won the match. In the Deep Blue operations room, Gerstner commented, "I just think we should look at this as a chess match between the world's greatest chess player and Garry Kasparov.[65]"

Kasparov admitted he was a sore loser[66]. He had no experience losing. He had never lost a professional chess match[67]. He was also angry that IBM hosted the meeting, suspicious about Deep Blue reboots in the middle of games, and IBM's failure to provide Deep Blue's prior matches. Joe Hoane complained, "You know he ruined the joy of it.[68]" Years later, Kasparov would comment how good Deep Blue was and how poorly he played

Kasparov and Deep Blue had played two formal matches, with one victory apiece. Kasparov talked with Gerstner about a third, rubber match. Though Gerstner was non-committal, he sensed that IBM saw no advantage in playing again[69]. The chess world was outraged. In actuality, Kasparov had already won two out of three, with the victories against Deep Thought in 1989 and Deep Blue in 1996.

The three lead Deep Blue researchers split the $100,000 Fredkin Prize. IBM reaped an avalanche of good publicity. The comments in the press were reminiscent of the "electronic brain" stories that followed the unveiling of ENIAC in 1946. This time, however, the appellation rang truer. Many considered the success of Deeper Blue as the real beginning of AI (Artificial Intelligence). Gerstner basked in the success. He brought the company back from disaster and now he had made a pivot from cost cutting to investing.

Murray Campbell summed up the lesson of Deep Blue and its predecessor machines. They illustrated the value of combining general-purpose compute power with specialized computing chips. Hsu had estimated that a general-purpose computer would have to compute at one trillion operations per second to match the combination of a high-speed computer with dedicated chess chips[70]. Custom processing chips handled the unique computation requirements (as in the chess chips) at very high speeds. This approach was applicable to many other areas of Artificial Intelligence. Hsu left IBM in 1999 and ended up working for Microsoft Research in China. Campbell was still at IBM as of 2017. Deep Blue remained in the Research lab until 2001, when IBM donated sections of it to the Smithsonian Institute and the Computer History Museum[71].

Gerstner did not stop with Deep Blue. He stayed with the color scheme and funded the Blue Pacific supercomputer. The United States had ended nuclear bomb testing in 1992 as part of larger negotiations for a comprehensive test ban treaty. Absent actual tests, the nuclear labs, principally Los Alamos and Lawrence Livermore, needed far more powerful computers to simulate nuclear explosions. IBM won a $94 million Livermore contract for Blue Pacific in 1998. Like its predecessor machines, it was RS/6000-based, with 512 eight-way 64-bit PowerPC processors. It achieved a top performance of three teraflops, or three trillion floating-point calculations per second. This represented a major computing speed milestone. Stretch computer had been the first computer to surpass the "megaflop" boundary, or a million calculations per second. The Cray-2 topped the "gigaflop" barrier, or a billion calculations per second. Blue Pacific was the first computer to exceed the "teraflop" level. By this yardstick, ENIAC or possibly Colossus were the first computers to break the "kiloflop" barrier, exceeding thousands of operations per second.

IBM's success with Blue Pacific spurred the company to announce plans to spend $100 million to develop a massively parallel supercomputer named Blue Gene, designed to do biomedical research in genetics. Using 2,048 PowerPC processors per rack, it scaled up to over 60,000 RISC processors running in parallel. Built in Rochester, Blue Gene would become a formal system product line for IBM as the "eServer Blue Gene Solution.[72]" This had been Gene Amdahl's aim with his ACS/360 design, and now it was reality but 30 years later. Blue Gene prices started at $1.5 million. The Lawrence Livermore National Lab had the largest Blue Gene installation, where it was the #1 supercomputer in the world in 2004, and stayed on top for nearly four years. In 2006, there were 27 Blue Gene computers in the Top 500 list of supercomputers. IBM sold far more during the program's duration. Iowa State in Ames installed one, a fitting milestone for a university that rediscovered John Atanasoff. IBM Rochester made a Blue Gene supercomputer available to the Mayo Clinic. The Clinic used the system to store the world's largest patient medical database.

The sale of the Federal Systems Division did not end divestments on Gerstner's watch. Many of these business units were integral to IBM's total solution for its customers. He sold the PC division to Lenovo in 2002. Perhaps more painful than the sale of the PC division was the sale of Adstar, IBM's disk business. IBM invented the disk drive and drove the industry's research and development for fifty years. The pace of advance in disk density even outpaced Moore's Law on semiconductors. In 1993, IBM Research working with San Jose increased the data density on the disk platter to 354 million bits per square inch[73]. In 2001, IBM scientists and engineers further increased that density to 35 billion bits per square inch. However, as was the case with PCs, compatible manufacturers rode the technology that IBM created until the company was over-whelmed. IBMers who carried the disk secrets to other companies aided the downfall of the division. In a final irony, IBM sold the division to Hitachi, the company that stole IBM trade secrets in the 1980s.

Nothing illustrates the failure of driving IBM technology better than networking. In 1991, this was a $5.5 billion business for IBM. The company had responded to the growth of local networks with its proprietary Token Ring technology. An upstart networking company, Cisco, with only $500 million in sales, embraced Ethernet, a simpler technology invented at Xerox's PARC lab. By 1998, IBM's networking revenues shrunk to $800 million and Cisco had ballooned to $18.9 billion, completely catching the wave of computer networking[74]. In 1999, IBM sold what was left of its networking division, mostly patents, to Cisco for $300 million and exited the space.

A direct result of Gerstner's One IBM strategy was a focus on the company's professional services capability. IBM had the products and skills to integrate a complete solution for a customer. However, this was not going to be an easy market to challenge. There were strong and entrenched competitors, including EDS and Computer Sciences Corporation. Yet, Gerstner had an outstanding executive to lean on. Dennie Welsh was head of IBM's services team, called the Integrated Systems Services Corporation. At their first meeting, Welch outlined a broad vision of what services could do. He described a palette of offerings that ran all the way to completely taking over customer IT operations. This was music to Gerstner's ears, as he had wanted to outsource all of RJR Nabisco's data centers for years.

Welch was a big man, a former Army pilot, who started with IBM in 1966 as an engineer on NASA's Apollo and Space Shuttle programs. The services unit began by accident. IBM had been out of the services business since 1973, when it sold its Service Bureau Corporation to Control Data as part of the "paper machine" lawsuit settlement. In 1988, Eastman Kodak asked IBM to take over part of its IT operation. Welch took up the challenge, and the company built and operated a data center in Rochester. Welch continued to grow the services operation, though it remained a small part of IBM. His teams even recommended solutions that included non-IBM products. In fact, he wanted IBM to be agnostic about the solution components, much to the dismay of IBM sales.

Welch's somewhat backwater life in IBM changed in 1992 when his team won the contract to take over Sears IT operations. The contract was for $8 billion[75]. That got everyone's attention. A number of similar multi-billion-dollar long-term contracts would follow. By 1995, Welch's division reached $13 billion in annual revenues, representing 18 percent of IBM's total revenue. Integrated Systems Services Corporation, now renamed IBM Global Services, was the largest services organization in the world, displacing EDS. Welch took a medical leave that year for a rare immune disorder and eventually retired. He recommended an outstanding salesman and deal closer to take his place. That person was one of his early deputies in the Services group, Sam Palmisano. A rising star within the company, Palmisano had already logged stints running the PC and mainframe divisions.

Now at the center of Gerstner's "One IBM" strategy, Palmisano continued to move Global Services forward. By 1998, it had grown it to a 33 percent share of IBM revenues. Four years later, it was 45 percent of IBM revenues. The services workforce would grow from 7,600 in 1992 to over 150,000 in 2002. That year, IBM acquired the consulting division of PricewaterhouseCoopers, adding 30,000 consultants. A significant share of the workforce growth came from offshoring, with India representing by far the greatest numbers.

Gerstner also headed the #1 software company in the world, with revenues north of $10 billion[76]. OS/2 and the almost religious fervor involved in taking back the desktop market got most of the press. However, most of IBM's software was hiding in middleware. The typical customer software "stack" had the operating system at the bottom, then middleware, and end-user applications such as ERP (Enterprise Resource Planning) programs at the top. IBM's middleware included communications control programs like CICS and database managers such as IMS and DB2. Software had outlandish gross profit margins, in the neighborhood of 80 percent. The majority of IBM's software portfolio ran on mainframes, so its destiny ran with big iron.

Gerstner had a very capable executive, John Thompson, running

IBM's software business. Thompson's immediate mission was to rationalize IBM's 4,000 software products, boiling them down to a usable set of industry-leading middleware[77]. The task required the consolidation of the many labs across the organization involved in software development. Gerstner was also comfortable with acquiring companies that complemented the overall software portfolio. Working with Thompson, he engineered the hostile takeover of Lotus Development in 1995. He acquired Tivoli Systems, a maker of systems management software in 1996 for $745 million. Three former IBMers had started the company. A key advantage was Tivoli's support for open systems, particularly UNIX. In 2001, he bought Informix, a relational database company, for $1 billion. Informix suffered in its competition with Oracle and was available. Gerstner's interest was moving its existing customers to the DB2 database.

The following year, a possible acquisition of SAP, the premier company in ERP applications, was under consideration. Gerstner backed away from it, rethinking his position on application software. He did not want to compete in the market where key software partners such as SAP, PeopleSoft, and J.D. Edwards might react by shunning IBM hardware and services. By 2001, Gerstner had IBM software rolling, reporting $13 billion revenue and $3 billion in net profit. This represented 15 percent of IBM total revenues.

A central software initiative involved answering Gerstner's question about the "buy button." Early web pages featured simple content management and personalization. Gerstner wanted to move core business functions to the Internet, including e-commerce. The company would pour substantial development resources into a middleware framework to support such a transition. A family of enabling products, under the umbrella name of WebSphere debuted in 1998. Earl Wheeler aided the WebSphere initiative by retiring in 1993, replaced by Steve Mills. Don Ferguson, an engineer who had joined Research in 1985 would emerge as the "Father of WebSphere," He led a small team in producing the initial WebSphere elements. He would become the chief architect of the IBM Software Group, defining the core architecture of WebSphere as well as the integration with middleware products such as DB2, Tivoli, Rational, and Lotus. The team responsible for WebSphere would grow to 6,000 programmers.

Ferguson would be named an IBM Fellow in 2001. He followed Glenn Henry's example, leaving IBM and joining Dell Computer as its chief technical officer. He would not stay long but left the position warm for yet another IBM Fellow, Jai Menon. Menon had earned his reputation in the San Jose lab developing the concept of RAID disk clusters. I am sure that Tom Watson did not have the rotating technical chair at Dell in mind when he created the IBM Fellow program in 1962.

# GERSTNER TAKES ON THE PC

Though addressing the mainframe business was the highest priority, the greatest number of challenges swirled around the PC division. Eight years had passed since the IBM PC business reached its peak and began a slow descent. By 1994, IBM's market share had plummeted to 8 percent[1]. The year before the division reported a $1 billion loss. In addition, Boca had $3 billion tied up in parts inventory[2], a reflection of its lumbering time to market. This was a year when the worldwide PC market grew by 16 percent to $66 billion. Microsoft rolled out Windows 95, spending upwards of $200 million on the launch alone[3].

Lou Gerstner's brother Dick had warned him about the PC group. Besides the division's bureaucracy, there were at least three other seemingly intractable challenges. IBM was stuck with Microsoft and Intel agreements that severely limited the options. The company's mainframe group had hijacked the PC strategy. And, Gates and Ballmer were now active and duplicitous competitors. Gerstner looked for any path to a win. He would rail against Intel and Microsoft, saying, "Those two companies had ridden this gift from IBM right to the top of the industry.[4]" He understood that they would continue to climb up the IT food chain and endanger IBM's core server businesses. On top of all the issues, the commoditization of PC hardware left little room for growth and scant margin for error. It was time to cut loose and focus on retrenching. Most of the projects aimed at attacking the Intel and Microsoft duopoly were still hanging around. They included OS/2, Systems Application Architecture (SAA), Workplace OS, the Taligent venture, and the PS/2 and PowerPC hardware programs. The dire state of the company allowed Gerstner to take long overdue action.

IBM's big bet to address Intel and Microsoft was the PS/2. It was far more successful in corporate markets, where build quality and support were more important. Bill Lowe returned from Rochester to lead the rollout of the PS/2 line and OS/2. He took the heat for the rapid failure of the PS/2. Passed over for promotion to senior vice president by Dick Gerstner, he left the company for General Electric. Jim Cannavino took over but soon departed for Perot Systems when he saw no path to the top.

IBM reentered the home market in 1990 with the PS/1 line, based on the PC AT architecture. It followed in the disastrous footsteps of the PC Jr., whose chiclet keyboard and relatively high price led to its withdrawal just a year after its announcement. Bill Sydnes, who had so much success with prior projects, was the system manager for that star-crossed machine. Boca replaced the PS/1 in 1994 with Aptiva, a far more polished system. It received great reviews. The PC division was apparently surprised by this and found it unprepared to fill the demand. It also arrived as clone companies had just announced the first PCs under $1,000, putting price pressure on the already thin margins.

IBM also tried "if you can't beat them, join them," with two separate efforts at a PC clone. Several ex-IBM executives formed Individual Computer Products International (ICPI)[5] in Europe, a wholly owned subsidiary of IBM. With perhaps 35 employees and a host of subcontractors, it sourced parts from China and produced the Ambra line. It was a third of the price of a PS/2 and only available via mail order. Brought to the United States in 1993, it competed with the other IBM clone, the PC division's ValuePoint line. The ValuePoint returned to the AT bus, eliminating the microchannel. The Aptiva would replace both ValuePoint and Ambra.

The most promising pathway to a successful personal computer for IBM was to go high, not low, by developing a state-of-the-art machine at a healthy price. That machine would be the ThinkPad. Compaq cloned the IBM PC and released a portable machine in 1982. By that time, Bill Lowe had managed several "luggable" computer projects, including the SCAMP APL machine and the 5100 series. The first IBM portable was the IBM Portable Computer, in 1984. Its high price and 30-pound heft led to an early exit. The PC Convertible arrived two years later. It was far more successful, but its price, hard-to-read display, and 13-pound weight made it ultimately uncompetitive.

It was time to try a new approach. The new machine, code named "Aloha," enlisted many parts of the company. Richard Sapper, an industrial designer based in Milan, provided the design[6]. IBM's Yamato Japan lab handled the engineering. IBM Research contributed an advanced thin-film color display. The Microelectronics Division reengineered an Intel 486 processor to provide higher performance at reduced heat and weight. Then there was the small, embedded, red TrackPoint replacement for the mouse. Ted Selker came up with the idea while working for Xerox at PARC, then refined it after joining IBM. Finally, the machine had to have an outstanding keyboard. This had been an IBM focus since Watson Sr. had purchased the Electromatic Typewriter Company in 1933.

All of these elements came together with the IBM ThinkPad 700, announced in 1992. "Think" was of course the legacy of Watson Sr., and "Pad" reflected its original design as a notepad coupled with a writing stylus. Weighing in at just 5.7 pounds, the ThinkPad showed off what the company did best. It received rave reviews and numerous design and performance awards. *PC Magazine*'s review called it a "bold and a great success. . . After years of designing undistinguished portables, IBM has finally gotten it right.[7]" By 1993, the ThinkPad was a $1 billion business for the company. A smaller version code-named "Butterfly" with a unique expanding keyboard debuted the as the 701C. Tim Cook, later to become Apple's CEO, shepherded it to market.

This author can speak to personal experience with the ThinkPad. The arrival of the first ThinkPad coincided with IBM's tailspin under John Akers. Gerstner and York looked for any way to reduce costs, including the vast real estate owned for IBM branch offices. The new laptop opened the door to a work-from-home mobility program within the company. I was responsible for the implementation of mobility within the Rocky Mountain region of the United States. We issued ThinkPads to our field personnel, enabling remote work at a customer site or at home. With space in the offices scaled back, users could log into one of the remaining desks for the day.

The success of the ThinkPad demonstrated that the farther IBM moved away from the consumer market, the better it did. Gerstner could not be faulted for trying. He would move on from the PS/1, Ambra, and ValuePoint while keeping the Aptiva. He dropped the PS/2 in 1995. The larger Netfinity servers had good margins. Mail order fulfillment replaced retail sales channels. Gerstner came to the view that IBM should have abandoned the direct sales model far earlier. It would have reduced costs and might have led to a better outcome.

These many changes failed to make much of a dent in IBM's bottom line. The PC Company business grew to $14.4 billion by 1997 with small but recurring losses. In 1998, revenues slumped to $12.8 billion, and the PC operation recorded another $1 billion loss. In the 1998 annual report, Gerstner apparently telegraphed his conclusion. He wrote, "the PC era is over,[8]" certainly a view not held by Microsoft, Intel, and the many clones. Gerstner outsourced PC manufacturing in 2002. The losses continued, but Gerstner resisted the impulse to cut the PC Company loose, given its importance to IBM's corporate accounts.

Without a doubt, the biggest disappointment and financial drain was OS/2. Jerry York was aghast at the ongoing cost of the program, reportedly $600-$800 million annually[9]. This investment would garner a mere five percent share of the personal computer operating system market, with DOS and Windows accounting for 90 percent. As Dick Gerstner advised, there remained a broad constituency in Boca arguing OS/2 was technically superior and would eventually win. However, it was also far more expensive, less user friendly, and supported very few user applications. York urged Gerstner to reduce or eliminate the OS/2 expense. Gerstner came to the same conclusion, calling OS/2 the "troubled child.[10]" However, a number of IBM's larger customers had committed to OS/2 and pressed him to continue to enhance and support it.

One major developer fully embraced OS/2 -- Lotus Development. Mitch Kapor started Lotus in 1982 and the company rapidly expanded with its "killer application," the 1-2-3 spreadsheet. Ray Ozzie, who started programming with VisiCalc, was the gifted programmer who ran its development operation. Ozzie led a programming team of just twenty to create Lotus Notes, an application doing what OfficeVision struggled to do. John Thompson, Gerstner's senior executive for software, broached the idea of acquiring Lotus. Lotus CEO Jim Manzi rejected any discussion. Gerstner decided to proceed with a hostile takeover, something definitely not part of IBM's DNA. He called Manzi minutes before the formal release of the unsolicited bid[11]. Manzi and Gerstner both got their starts at McKinsey and actually developed a rapport. Gerstner also met with Ray Ozzie. Negotiations followed, and the acquisition went through in 1995 for $3.2 billion. Most of the Lotus senior executives eventually left, but Ozzie promised to stay on until the next release of Lotus Notes was completed. The odd man out in the deal was OfficeVision. IBM reconstituted the Lotus applications into the SmartSuite product. It started fast, but continued development in OS/2 sapped programming resources that could have developed 32-bit code for Windows 95.

IBM's continued push on OS/2 and now the purchase of Lotus incensed Bill Gates. His empire consisted of Windows and Microsoft Office, and IBM's SmartSuite competed directly. He had focused on killing OS/2, and now he faced another challenge. Gates became unresponsive when Boca executives tried to call. He finally decided to vent his anger on the Aptiva line. He waited until the last possible moment to negotiate royalties for the coming year. IBM had been paying Microsoft $40 million per year for Windows licensing. In 1996, Gates raised the fee to $220 million. The following year, with Windows NT licensing added in, he raised the payment to $440 million[12]. The royalty payments were far higher than what Microsoft charged other PC manufacturers. Gates went further and limited the level of support he provided to IBM. Microsoft also threatened a number of PC companies, including Compaq, if they sold OS/2.

Microsoft's actions found their way into the Justice Department's antitrust lawsuit against the company in 1998. Seeking to protect its monopoly, Microsoft clearly sought to "punish" IBM for simply competing. Bill Gates answers to Justice Department questions were characterized as "quibbling, unresponsive, and forgetful."[13] He purported not to grasp the concept of "market share." The suit went to trial, and Microsoft was ruled guilty of monopolistic practices. Judge Thomas Penfield Tate recommended that the company be split in two. Microsoft would appeal the verdict and escape intact with minor penalties. The company found itself on the other side of a Justice Department antitrust suit in 2023. They were called to testify against Google's monopolistic power. Satya Nadella, Micosoft's CEO, bemoaned that "Microsoft could not overcome Google's use of multi-billion deals" and competing with Google was a "nightmare.[14]" Those comments ring hollow considering Microsoft's tawdry monopolistic history. It should be noted that Google is also in the crosshairs of the Justice Department. It search business ($26 billion in 2021) enabled it to easily pay out $26 billion to several companies, most prominently Apple, to ensure that Google was the default search engine[15].

OS/2 Version 3.0 was already nearing completion in 1993 and targeted for release in early 1994. Since performance was a key criti-cism of OS/2, Abby Kohnstamm and IBM's marketing agency, Ogilvy & Mather, decided to address it head-on. The consumer version of OS/2 would have a new name, OS/2 Warp. As with so many initia-tives with the product, there was a slight problem. The name meant to connote speed, as in Captain Kirk of the *Enterprise* asking Scottie to accelerate to "warp factor three." IBM signed Patrick Stewart, the actor who helmed the current Star Trek series, for the product launch. However, IBM Legal was apparently asleep at the switch and had failed to secure the right to use the name "warp" from Paramount Pictures, the owners of the franchise[16]. IBM had to take the position that they really meant a different definition for "warp", something that is bent or twisted. This was hardly ideal, though perhaps more appropriate.

By late 1995, Gerstner decided to fund one final release of OS/2. Many of the programmers were already redeployed to other projects. OS/2 Warp 4.0 was completed in 1996. In a fitting end, the last Warp version did not run properly on Aptiva machines. At its sunset, OS/2 had just a 3.3 percent share of the desktop market. Gerstner commissioned a confidential study of the PC Division, with a report produced only in numbered copies. The report led Gerstner to end all OS/2 development and cut funding by 95 percent. He eliminated the Boca lab, and laid off 1,300 employees[17]. Gerstner could have withdrawn OS/2 immediately. However, as a former customer, he agreed to continue support, even at a cost in excess of $100 million. Full support expired in 2006, and even then, IBM offered it on a fee basis. Gerstner would comment on OS/2, "It looked like we had lost, so why don't we get on with doing something else?[18]" He simply wanted this program behind him. IBM had spent at least $2 billion on OS/2 after it was clear that the desktop race was over[19].

For Boca Raton, the loss of OS/2 and the programming lab was a gut punch. The IBM PC era, which started so triumphantly in 1981, came crashing down. The division was caught between two forces that it could not control. On the one hand, it had to deal with Bill Gates and Microsoft. Starting with the flawed agreement giving Microsoft control of the operating system, the lab endured an uneasy working relationship while Gates worked behind their backs on Windows and NT. Clearly, everyone from Opel, Lowe, and on down the organization misread the ruthlessness and ambition of Gates and Ballmer. Then, Boca's independence ended with corporate and mainframe ascendency. Those early, giddy days working under Frank Cary's wing evaporated. The fences erected by the mainframe business and the grand ambitions of Earl Wheeler limited the lab's choices.

There were certainly options along the way. Watts Humphrey pressed to develop the basic PC software in-house, but was overruled. Early on, Gates had suggested a more measured approach to fixing DOS by simply adding a front-end and implementing multi-tasking. When IBM forged ahead with OS/2, Gates would adopt this approach as Windows. IBM Research and its CP/88 operating system offered another avenue out, but IBM caved in response to Gates's threats. Later on, Gates would offer to take over OS/2 development, or even sell part of his company. As time passed and Microsoft got stronger, the options for IBM got fewer and less attractive. The final indignity was Gates's offer to lower the royalty for Windows if IBM cancelled the release of OS/2. IBM dismissed each of the options and the OS/2 team soldiered on to the end.

It was also clear that IBM was no match for the marketing and business savvy of Gates and Ballmer. They recognized the market potential early and focused on locking up application developers and licensing MS-DOS to every PC manufacturer. They also saw the need for two different operating systems. OS/2 was trying to serve two masters, the consumer desktop and the office server. It is telling that Dave Cutler chose to shy away from OS/2 and develop a corporate-grade operating system. NT looked a lot like the VMS he designed at DEC. Likewise, IBM had a wide variety of operating systems to choose from other than building OS/2 from scratch, as Watts Humphrey argued.

With decision on OS/2 behind him, Gerstner still had work to do. He needed to address another piece of the desktop strategy, the Taligent joint venture with Apple and HP. It started with great promise in 1991. There was significant buzz in the industry in the early years, seeing the effort as a serious challenge to Microsoft's hegemony. Even Bill Gates admitted that he was concerned. As the years passed with nothing to show for the joint effort, there was speculation that IBM and Apple were having second thoughts about a new operating system's impact on OS/2 and Mac OS. The truth was progress was slow and getting to a formal release proved elusive. In 1995, Apple and HP dropped out, leaving IBM to carry on alone. Gerstner folded the Taligent operation in 1997, after IBM programmers incorporated valuable components of the Taligent code into other IBM software. In the end, the Taligent venture represented an investment in excess of $100 million by the three parties, with little to show for it.

Apple continued to work with IBM on the PowerPC program. Despite the struggles with Taligent, there was a sense in IBM that some kind of combination with Apple would work. In 1994, Gerstner was early into his acquisition mode. Apple was having financial problems and looking for a partner. Michael Spindler, who had replaced John Sculley as CEO, considered IBM and Sun. Gerstner offered to acquire Apple for $40 a share, a $5 premium over the current share price[20]. Gerstner met with Spindler and his top executives in September 1994 in Chicago. Spindler was looking for at least $60 a share. Gerstner was unwilling to go that high and withdrew because the Apple executives seemed mainly interested in golden parachutes. Two years later, Apple acquired neXT, the computer company that Steve Jobs started after being forced out of Apple. As Apple CEO, he replaced Apple's existing Mac operating system with the Unix-based OS his team developed at NeXT. Apple was soon on a sharp upward trajectory.

The failure of OS/2 sealed the fate of another software program— Systems Application Architecture (SAA). Few mainframe System/370 customers jumped on SAA. Most waited on the sidelines to see how it developed. It was the same story with software developers. In 1989 and 1990, IBM paid an estimated $500 million for stakes in smaller PC software companies in exchange for their pledges to develop SAA applications[21]. This sometimes swamped the companies with IBM bureaucratic requirements. IBM had a successful beginning in PC software. Joyce Wrenn of Application Customizer Service (ACS) fame grew a small software group in Boca into a $100 million business[22]. When IBM failed to support Wrenn's initiatives to grow the volume, she left the company. Her replacement essentially wound down the entire operation to zero, the same fate as ACS. IBM would take a $2 billion loss to leave the PC software business entirely.

SAA did not go anywhere. SAA could have worked, but was doomed by several factors. Foremost was the failure of OS/2, which was central to the concept. In addition, the slow progress of SAA development came at a time of rapid computing change and innovation, including the emergence of open systems. They stood in stark contrast to the closed and proprietary nature of SAA.

Work continued on Workplace OS, the microkernel-based scheme to provide interoperability between desktop operating systems and IBM systems. David Schleicher led the massive effort, working under Earl Wheeler. Schleicher had the Boca Raton programming center and parts of the Austin and Rochester labs reporting to him. This added up to roughly 2,000 programmers. Austin and AIX opted out of the program, soon followed by Rochester and the AS/400 operating system.

Gerstner and York reduced the scope of the program in late 1994 due to expense. Gerstner cancelled the entire project in March 1996.

He also shut down the Personal PowerPC division, the group building workstations to run Workplace OS. IBM spent nearly $2 billion on the Workplace OS program over four years[23]. Wheeler retired in 1993, before his grand visions of SAA, Workplace OS, and a mainframe-centric OS/2 all collapsed.

Workplace was as ambitious as OS/360, FS, or Fort Knox. The microkernel was no problem, but adding the OS personalities dramatically increased the degree of difficulty. The personality programs suffered with what Fred Brooks had identified as the "second system effect," where additional function is piled on. David Schleicher might have recognized a different path for Workplace. A microkernel with multiple "personalities" had already been done within IBM. In fact, Schleicher had led the effort, with the FS implementations on the System/38 and AS/400. The two layers of microcode of the high-level machine interface were the microkernel. This structure facilitated the consolidation of the System/36 and System/38, two very different operating systems.

## LINUX

Gerstner and his team decided to stop trying to compete with Microsoft in the PC-centric client-server market and position the company for the next big technological shift, cloud computing. Led by Irving Wladawsky-Berger, the company embraced an open-source variant of UNIX called Linux. It was the right play for the future. It also offered the perfect anecdote to Windows.

Linus Torvalds announced Linux in 1991. He was already experienced with Intel computers when he took a UNIX course at the University of Helsinki in 1990. Intrigued, he developed a kernel for the MIMIX version of UNIX licensed by the university. Unhappy with the licensing arrangement, Torvalds turned to a free open-source version of UNIX created by the GNU project. Started by Richard Stallman at MIT, GNU stood for "GNU's Not Unix" with the goal of a free, collaborative version of UNIX.

Torvalds was not willing to wait for the GNU kernel and decided to develop his own[24]. He replaced all of the Mimix programs with GNU code and announced his UNIX version in 1991. He originally tackled the project as a hobby but ended up releasing a complete operating system. He based his master's thesis on the software. Associates convinced him to name the system "Linux." Penguins fascinated Torvalds and they became the avatar for Linux.

There were other versions of UNIX with a microkernel, but Torvald's version took off. Torvalds believes that getting his version out first was a key advantage. In addition, the Linux kernel was the result of one smart individual crafting the design. His success underscores the conclusion that Fred Brooks came to regarding OS/360 development. He should have waited for one individual or small team to design the architecture before turning thousands of programmers loose.

UNIX was a significant revenue source for IBM, with its AIX variant. In 1997, Project Monterey began with the mission to develop an enterprise-class version of UNIX[25]. It would bridge several versions of UNIX and support Intel's upcoming Itanium architecture (soon to be derided as "Itanic"). As this program proceeded, IBM was hearing more about Linux. Sam Palmisano met with various Internet companies, and the subject of Linux kept coming up. Irving Wladawsky-Berger heard the same thing from academia. Some universities clustered Intel servers and added Linux on top, achieving a poor-man's supercomputer. Gerstner approved two studies to assess the use of Linux[26].

In a way, Linux did what Workplace OS purported to do. It ran on all the major platforms. IBM embraced multi-vendor approaches to computing with IBM Global Services. The company had significant experience with open systems. Linux provided a path around the Wintel duopoly. It also gave the company a head start in Internet-based computing. Palmisano reviewed the proposal to adopt Linux in 2000. He supported the move and convinced Gerstner to buy in. Gerstner cancelled the Monterey program and asked Wladawsky-Berger to lead the Linux effort. Wladawsky-Berger noted how alien this new direction was for IBM. The company had thousands of inventions and patents, and now it was going outside for a central piece of technology, one that was publicly controlled.

Once the new direction was set, the company moved fast. It was one thing to announce the Linux plan and quite another to convince the computing world and the company's commercial customers that proprietary IBM was serious about it. Marketing switched into high gear with the typical mix of ad buys in papers, magazines, and even billboards. There was even a campaign to paint the Linux penguin on sidewalks in major cities. This did not work out as well. San Francisco charged the company as a graffiti scofflaw, fining the company $100,000 and ordering a clean-up[27]. Press coverage of this fiasco actually helped with the visibility of the campaign. Marketing rebounded with a Super Bowl ad campaign that featured Muhammad Ali, John Wooden (the famed UCLA basketball coach), and Penny Marshall (of the TV series *Laverne and Shirley*).

Gerstner demonstrated that IBM was indeed serious about Linux by the company's investments. He announced in December 2000 that IBM would be investing $1 billion during the following year on Linux. He added that an additional $4 billion would be spent in building the infrastructure for Linux-based applications. This included building new data centers, surrounding Linux with middleware, and incentivizing business partners to support Linux-based Internet applications. IBM joined with HP, Intel, and others to foster the continued development of Linux for commercial applications. This evolved into the Linux Foundation. Over time, the organization estimated a billion lines of open-source Linux code were produced, the result of 750,000 programmers around the world working in 18,000 companies[28].

# "THE MAN WHO SAVED IBM"

L ou Gerstner led the company for nine years, from 1993 to 2002. During this period, revenue grew from $64.5 billion to $85.9 billion, an increase of 33 percent, or an average of about 3 percent per year. IBM rebounded from the $5 billion loss in 1992 to income of $7.7 billion his last year, representing an increase of 255 percent. This put IBM back in the top ten companies of the Fortune 500. Earnings per share increased from a loss of $2.17 to earnings of $4.35 per share. In his first three years, he reduced headcount from 302,000 to 225,000. When he left in 2002, the workforce was up to 320,000.

Gerstner and York took much of the pain of restructuring the company fiscal 1993, resulting in the loss of $8.1 billion. IBM turned in a $3 billion profit the following year. York reduced expenses from 40 percent of sales to 33 percent in the first two years, though this ratio remained higher than industry averages. By 1995, sales reached $72 billion, up 12 percent. Gerstner declared the turnaround complete by 1997.

IBM's stock price and market value reflected these results. When Gerstner took over, the stock was trading at $12 and the market value was at $29 billion, a figure certainly less than the book value of the company. He left with the stock at $118 and the market value at $180 billion. The rising stock price during his tenure required a couple of two-for-one stock splits, one in 1997 and the second in 1999. The revenue mix changed significantly. In 1992, hardware was $33.8 billion (57 percent), software $11.1 billion (18 percent), and services $15 billion (25 percent). In 2001, hardware was $25.7 billion (31 percent), software $12.9 billion (16 percent), services $35 billion (43 percent), and technology including patent sales $8.0 billion (10 percent)[1].

Though Lou Gerstner clearly rescued IBM, there were other more measured views of his overall performance. Peter Greulich credits Gerstner with saving IBM but leaving a company that was no longer the technology juggernaut it had been. The failures in personal computers and the focus on services resulted in missing major emerging technologies. Gerstner ran the company during a period that saw one of the longest spans of business expansion in history. With such a tailwind, the company should have produced more than 3 percent revenue growth. There was a also conflict of interest between the stock price and the executive options that focused on it, and real strategies for growth.

Gerstner ostracized *Fortune* magazine, but that did not deter *Fortune*'s coverage of IBM in the least. Bethany McLean, a *Fortune* journalist who had been the first to question Enron's finances, also wrote about the financial angles employed by IBM[2]. Her analysis starts with the question, "How do you grow earnings five times faster than revenues?" Given the singular focus on the IBM share price as the measure of performance, she examined elements of the financial arsenal that IBM used to stimulate that price. The key indicator was the mismatch between revenues and earnings. Revenues grew at a 4 percent pace during Gerstner's time as CEO, but earnings grew at close to a 20 percent pace. The core underlying tactic was the stock buyback. As McLean observes, "from 1995 through 2001, IBM spent around $44 billion buying back shares." That was an amount just slightly less than the $45.5 billion in income earned over that span.

Gerstner incented the performance of his managers with plentiful stock options. There was no need to account for stock options in the P&L. At the same time, deductions for options reduced corporate taxes. However, exercised stock options increased the number of outstanding shares and diluted earnings per share. The stock buyback was used to offset this effect. Buybacks got an ironic boost in 1993 when a change in the tax code exacted a corporate penalty on any executive whose salary was in excess of $1 million. Most companies ignored the additional tax and continued to pay big salaries. However, it also led to more companies using stock options as a way to pay their executives. As the use of stock options increased, the tactic of stock buybacks surged.

John Opel stopped Cary's modest buyback program, as he needed all available funds to support his expansion of the company. John Akers had authorization by the board to restart buybacks, but IBM's deteriorating condition put that plan aside. Gerstner embraced, restarted, and institutionalized the practice at IBM. He increased the level of buybacks to $6 billion a year. It sent a stark signal that IBM cash was better deployed in financial maneuvering than developing new technology for growth. York started the program in 1995, shortly after the first two years of brutal job cuts and the $8 billion loss in 1993. IBM and to a lesser extent Cisco would emerge as the poster children for stock buybacks. The even greater problem with Gerstner's program of stock buybacks was the use by his successors, who would fully adopt and expand the practice.

The financial press hailed Gerstner's turnaround of IBM. To paraphrase Joshua Lederberg, the Nobel laureate who cautioned Gerstner about his stewardship of a "national treasure," he certainly did not screw it up. The company went through a near-death experience and emerged on a strong financial footing. Gerstner's "One IBM" was indeed the right strategy. Though it was painful and sacrificed some of the fabric of the Watsons' basic beliefs, the alternative was too dark to consider. He reversed the expansion set in motion by John Opel and achieved a better balance between the company's revenues and its expense structure. He reduced layers of management and worked to reduce the sheer bureaucracy. His successor, Sam Palmisano, would say Gerstner's central legacy was his strategic approach, where all actions started and ended with the customer.

Gerstner was big on financial performance but less so on technology. Though he wanted to get good ideas out of IBM Research and into IBM products, he had a hard time justifying the huge expense while the company was hemorrhaging. He also struggled to see how spending big on R&D would help him in the early years. In 1992, IBM's R&D expenditure was $6 billion. Against revenue of $64.5 billion, this represented 9 percent. Starting in 1994, he cut the R&D rate to 6 percent, where it remained. At the same time, Microsoft was spending upwards of 60 percent of revenue on R&D. On a per-employee basis, IBM was far down the list at $16,000 while Microsoft was #1 at $141,000 per employee. Of course, that is not a completely valid comparison, but the gulf between the two companies is still striking[3].

Gerstner embraced strategies that would inhibit the company's future growth. The focus on services came at the expense of IBM's traditional role as the technology leader. He apparently did not trust IBM to develop market-leading products, so he acquired companies instead. In spite of the success of IBM's push into "e-business", the IBM that Gerstner left was not well positioned to take advantage of new technologies as they emerged, most prominently cloud computing.

Lou Gerstner pulled IBM back from the abyss and set it on a growth path going forward. A great deal of the fruits of IBM's turnaround went to the senior executives who benefited by significant stock options. A considerable amount went to one individual in particular, Lou Gerstner. He took IBM executive pay to uncharted heights. He began fast with the signing bonus of $4 million, $2 million salary, $1.5 million in performance incentives, and 500,000 stock options. The next year the company granted two million additional share options as he steered IBM back to growth. He collected $35 million from those options as the stock rebounded. In 1997, he agreed to stay on for five more years in exchange for two million additional share options. In 1998, his earnings rose to $110 million, with $8 million in salary and $96 million in exercised options. The IBM stock was $66 a share when he left. He would cash out his remaining options for a reported $311 million[4]. A Forbes analysis of the "Top 10 Largest CEO Severance Packages of the Past Decade" put Gerstner's package at $189 million, good enough for eighth place on the list. Jack Welch of General Electric led the list with severance of $417 million[5]. A 2021 estimate of Gerstner's net worth was at least $600 million. Much of that came from his IBM tenure. He set the bar for IBM executive pay at a stratospheric level. It would be a level enjoyed by his successors.

Gerstner also invested in his executive offices. He had never liked the closed, sterile, 1960s look of the Armonk headquarters. In 1997, he moved into a new headquarters building less than a mile from the old structure. It had an open floor plan, though Gerstner and his senior executives had private offices on the third floor. It was fitted out with a cafeteria, fitness room, and jogging path through the woods. There were IBM historical exhibits throughout, including a Selectric typewriter, an 1855 Thomas de Colmar upright piano-size calculator, and the main console for the SSEC (Selective Sequence Electronic Calculator), the one that was featured in the film noir movie *Walk East on Beacon*.

Gerstner retired from IBM at age 60 in March of 2002. He offered his own assessment of his work at IBM in the memoir "Who Says Elephants Can't Dance?" As he might say, he "made the elephant dance." He agreed that despite his many other roles and accomplishments, he will always be known as "the man who saved IBM.[6]"

Gerstner spent nine years at the company. He considered his work at IBM as his most important. He would teach occasionally at IBM management training schools. He became a "true Blue" IBMer, though he would say he would always be considered an outsider. In his final communication to IBMers, he said, "I fell in love with IBM" and "I am an IBMer for life.[7]"

# PALMISANO AND THE ROADMAP ERA

L ou Gerstner felt key part of a CEO's job was to find and mentor a successor. Three potential candidates sat on the third floor of the new IBM headquarters building in Armonk. John Thompson headed IBM's software division. Nick Donofrio was the company's chief technology officer, a future IBM Fellow, with a storied history of engineering developments. However, Gerstner gravitated to the third senior vice president, Sam Palmisano. Services played a key role in IBM's reemergence, and Palmisano logged two tours in that business. Gerstner promoted him to president of IBM Global Services in 1998 and then moved him to Enterprise Systems to learn the mainframe side of the company. In 2000, he elevated Palmisano to President and Chief Operating Officer of the company. In Palmisano, Gerstner found someone who shared his focus on the customer and was a stickler for financial commitments and performance. The two would occasionally enjoy a glass of cognac and a cigar together after a challenging day[1].

Gerstner notified the IBM board that he would cede CEO duties to Palmisano effective March 1, 2002. Gerstner would stay on as board chairman until January 1, 2003. Growing up in Baltimore, Palmisano came from a humble background. His father was an auto mechanic. Sam played saxophone in the school band. He graduated from Johns Hopkins in 1972, where he played football. Solidly built, his fraternity brothers christened him "Baloo" after the big bear in *The Jungle Book*[2]. He had an opportunity to try out for the NFL Oakland Raiders but realized that a 225-pound interior lineman from a Division III school would be a long shot for the pros. IBM recruiters visited the Johns Hopkins campus, where they interviewed and hired him.

Palmisano started as a data entry clerk but soon switched to sales. He quickly established himself selling mainframes. Once on the executive fast track list, he rose rapidly through the ranks before John Akers tapped him as his AA in 1989. He recognized the AA job as a "flunkie" position but one that could make or break one's career. He had lunch with Tom Watson once a month, a sure indicator he was being groomed for greater things. Akers sent him to run IBM Japan to see how he would do working in a different culture. Palmisano and his Tokyo CFO were the only Americans among the 23,000 employees in Japan[3].

Palmisano and Gerstner were quite different. Gerstner was focused, impatient, and quite private at times. Palmisano was very personable, able to strike up a conversation with anyone. As CEO, he moved around the company asking, "How do you think we're doing?[4]" His trademark horn-rimmed glasses softened his imposing football lineman presence. Everyone in and out of IBM called him Sam. Until he was elevated to CEO, he drove himself to work each day. He was well connected outside of IBM, in part a result of his wife's relationships. He was friends with both Bush presidents and, in fact, purchased the Bush family's Maine property at Kennebunkport in 1996.

Given his extensive experience with the services business and long working relationship with Lou Gerstner, it was no surprise that Palmisano's strategies were similar to his predecessor's. Like Gerstner, Palmisano aimed to speak to a customer every day. Neither was a visionary. They executed based on defined strategies and financial goals. However, the IBM Palmisano faced had changed. Gerstner deployed an arsenal of financial techniques to pull IBM back from the brink. The company was now financially secure, and the focus needed to shift to growth. Gerstner measured progress by the company's share price, a number that would reflect growth. And, growth in IBM was a result of continuous innovation. However, Palmisano chose maximizing shareholder value through increased earnings per share as his core focus. This was an odd objective for an IBMer ingrained with the traditions of the Watsons. Jack Welch of General Electric commented about this strategy:

> "Shareholder value is the dumbest idea in the world. Shareholder value is a result, not a strategy—your main constituencies are your employees, your customers, and your products.[5]"

Palmisano's mentor, Lou Gerstner, was even more specific. He would say, "There is only one financial scoreboard, and it was the stock price.[6]" In contrast, Palmisano's viewed earning per share as the central objective.

Moreover, he was entirely upfront about this single-minded approach. He laid out his plan in a five-year road map, adopted in 2005. The goal was doubling earnings per share by 2010 to $10. In 2010, he would release another five-year road map, aiming to double earning per share again by 2015, this time to $20. He maintained his roadmaps signaled that IBM was not focused on quarter-to-quarter results. However, the financial roadmap also signaled that IBM was no longer the high-growth technology leader that it had been for nearly a century.

Palmisano translated his goal of maximizing shareholder value by divesting unprofitable businesses, reducing labor costs, outsourcing, and growing by acquisition. More importantly, he would embrace a range of financial techniques that prominently included stock buybacks. A year after he assumed both the CEO and chairman's positions, Palmisano disbanded the Corporate Executive Council, the group of senior executives that Gerstner had set up to review key strategies and actions. Palmisano now had free, unfettered reins to run the company.

Palmisano was serious about divesting the least profitable units and focusing on high-margin businesses. It was essential to prop up earnings per share. Nearly all of his divestments involved hardware technology, the critical components in Gerstner's "One IBM' strategy.

The iconic disk drive business, the pride of San Jose, would be one of the first units cut loose. He sold the division, called ADSTAR, to Hitachi in 2002 for $2 billion. In 2004, he sold the consumer lines of the PC Company to Lenovo for $1.75 billion while maintaining a small stake in the business. Despite the widespread emotional attachment to the PC division, Palmisano characterized the decision as one of his easiest[7]. The PC market had become consumer-centric, a market that was too far afield from IBM's core competencies. IBM's market share had dropped to roughly 5 percent, and margins were thin at 3-4 percent. However, revenues remained substantial at $9.5 billion per year.

Palmisano could have sold the division to Dell or HP, or even a private equity firm. Dell had expressed an interest in IBM's PC business, and HP had paid $25 billion for Compaq just three years earlier. Yet, Palmisano wanted a partner that provided entry into the Chinese market. He concluded that continued international growth would not come from the G7 countries but from emerging economies, especially China. He passed up more lucrative offers from HP and Dell in order to establish a foothold and a partner in China. Palmisano would later sell the ThinkPad unit to Lenovo for $1.75 billion[8].

Palmisano cut loose the printing division. In 2007, he sold the IBM Printing Systems Company to Ricoh for $725 million. IBM and Ricoh ran the business as a joint venture for three years before it became a wholly owned part of Ricoh. The printing group's revenue had declined to $1 billion. However, the maintenance, supplies, and especially software revenue elements had high margins, and the unit was very profitable. These three hardware divisions generated $20 billion in sales annually. They simply did not fit into Palmisano's earning per share road map.

At the same time, Palmisano acquired in a wide range of companies to fill gaps in the company's software and services offerings. He acquired roughly 120 companies during his tenure. The major purchases were Rational Software (application development), FileNet (content management), Cognos (business intelligence), and SPSS (statistics). Many fit into the targeted growth segments of data mining and data analytics. Palmisano continued in the tradition set by his predecessor, looking outside for innovation instead of developing it in-house.

The days of thousands of programmers developing IBM software were gone.

Palmisano oversaw a seismic shift in IBM revenues from hardware to software and services. The key piece was the acquisition of Price-waterhouseCoopers (PWC) in 2002 for $3.5 billion. HP apparently bid $18 billion for PWC just two years previously.[9] PWC added $5 billion in revenue and roughly 25,000 new consultants to IBM Global Services. They joined a like number of consultants already in the IBM services division. PWC business skills complemented IBM's technology skills. Ginni Rometty, general manager of IBM Global Services, championed the acquisition and managed the successful integration of the PWC teams.

By 2005, services revenue crossed the 50 percent threshold of IBM's business. Now more than half of the company, it was near zero when the Watsons were in charge. The pressure to grow services led to bidding for business at low margins. Services revenue would plateau and margins declined. Typical margins were 25 percent compared to software margins of 80-90 percent and 60-80 percent margins that IBM had previously enjoyed with hardware. Palmisano and his successor tried to reduce services cost by going offshore. Offshore workers at $10-$12 per hour replaced IBM services personnel at an average salary of $56 per hour. The offshore employee also did not incur IBM's substantial benefit costs. IBM grew its workforce in India from less than 3,000 when Palmisano took over to an estimated 130,000 by 2017[10]. A parallel process, the Lean Initiative, identified positions that were candidates for offshoring. IBM's workforce increased from 316,000 in 2002 to 433,000 in 2012. At the same time, the U.S. population decreased through layoffs and attrition to 140,000 by 1999 and continued to go down, though IBM stopped breaking out the numbers. The offshore workforce would reach 200,000.

Palmisano made another decision that affected the existing workforce: He eliminated the IBM pension in 2005. The company implemented a two-tier arrangement that preserved the pension for older employees and switched to a 401K-based system for everyone else. He maintained that this had to be done because of the growing pension liability. Though other companies were making the same change, it was a tougher sell to IBM employees when the company was spending up to $19 billion a year on stock buybacks[11]. And, it came in the midst of domestic layoffs to offshore their jobs.

In June 2011, Palmisano had the opportunity to temporarily get away from IBM's current struggles and celebrate the company's 100th anniversary[12]. It marked the founding at C-T-R in 1911. Attendees included Lou Gerstner, John Akers, the IBM board members, top IBM executives and engineers, and a number of CEOs from major IBM customers. IBM hosted the celebration at the IBM Research building in Yorktown Heights, enabling 2,000 IBM researchers to join. Thirty members of the Watson family were also there. Ken Chenault, an IBM board member and CEO of American Express, described how IBM had stepped in when his company's headquarters were damaged during the 9/11 attack in New York City. He mentioned that his company's first acquisition from IBM was a time clock. He said that IBM's greatest invention was the IBMer. The celebration ended with all present singing the IBM song "Ever Onward.[13]" Watson would have been proud.

> There's a thrill in store for all,
> For we're about to toast,
> The corporation known in every land.
> We're here to cheer each pioneer
> And also proudly boast
> Of that "man of men," our friend and guiding hand.
> The name of T. J. Watson means a courage none can stem;
> And we feel honored to be here to toast the "IBM."
> EVER ONWARD—EVER ONWARD!
> That's the spirit that has brought us fame!
> We're big, but bigger we will be
> We can't fail for all can see
> That to serve humanity has been our aim!
> Our products now are known, in every zone,
> Our reputation sparkles like a gem!
> We've fought our way through—and new
> Fields we're sure to conquer too
> For the EVER ONWARD I.B.M.

EVER ONWARD—EVER ONWARD!
We're bound for the top to never fall!
Right here and now we thankfully
Pledge sincerest loyalty
To the corporation that's the best of all!
Our leaders we revere, and while we're here,
Let's show the world just what we think of them!
So let us sing, men! SING, MEN!
Once or twice then sing again
For the EVER ONWARD I.B.M.

Palmisano led IBM for ten years, from 2002 to 2012. He increased revenue thirty percent from $81.9 billion to $106.9 billion, a rate just under 3 percent per year. Revenues from the divested units, such as the PC Company, would not have significantly changed the outcome. He grew income from $3.6 billion to $15.9 billion, an increase of 106 percent, at compound rate of 16 percent per year. The difference between revenue and income growth points to the road-map program to prop up earnings per share. In 2004, Microsoft was one-third the size of IBM but its profits outpaced IBM's by over 30 percent. Palmisano made his first road-map target with $11.52 per share in 2010, up from $3.00 per share in 2002. IBM's stock price rose from 118 to 177, an increase of 50 percent. Market capitalization was slightly up, from $208 billion to $214 billion.

Stock buybacks were center stage, with a dramatic effect on both earnings per share and the stock price. Palmisano averaged $10 billion in buybacks per year, racking up a total expenditure of $99 billion over his tenure[14]. His peak year for buybacks was $19 billion in 2007, a year in which the company only recorded $10 billion in net income. This high-water mark involved using funds borrowed from a foreign subsidiary to avoid U.S. taxes. The technique was the "Killer B" tax shelter[15]. Due in part to IBM's prominence in using it, the Internal Revenue Service soon outlawed it. It does illustrate the lengths that the financially driven IBM would go to meet the road-map goal. As Palmisano approached the end of his first road map in 2010, he was working on the next one by announcing the second $15 billion stock buyback in less than a year.

Lou Gerstner used a range of financial measures to pull IBM back from the brink of collapse. He did start buybacks, but as a technique to offset the high volume of stock options he awarded. He focused on the stock price, the real indicator of market growth and industry leadership. IBM issued stock splits in 1997 and 1999 because the company was doing so well. After getting IBM's house in order, he then focused on growth. As IBM's traditional market segments, including mainframes, were gradually declining, technology growth through outstanding new products should have been the order of the day. Instead, Palmisano embraced the financial approaches and, in fact, doubled down on them. He divested low-margin businesses. There was nothing inherently wrong in this. The problem was that many of those businesses were central to the "One IBM: strategy, including personal computers, servers, disk drives, and printers. His strategy leaned heavily on moving the company to services. Palmisano increased the workforce from 320,000 to 433,000. The growth came mainly outside the United States and Europe, principally in India. The ultimate irony is the push to services resulted in another low-margin business – and that was services. Palmisano would pass his financial strategies, the second road map, and the declining state of the company to his successor.

Investors and the press had a far more positive view of Palmisano's tenure. IBM moved to #12 on *Fortune*'s list of Most Admired Companies[16]. Palmisano's aim of consistent and rising earnings per share was particularly attractive to large investors. The largest of large investors, Warren Buffet, became a fan, despite his disdain for tech stocks. The announcement of the second road map prompted him, the famed "Oracle of Omaha," to invest $10 billion in IBM[17]. He was impressed by the company's "methodical way" of making money for shareholders. Palmisano had turned a technology company into a financial company. Palmisano benefited from the extraordinary high compensation bar set by his predecessor. He assumed Gerstner-class salary numbers and left with a retirement package estimated at $271 million. The Wall Street Journal placed that in the top ten for severance packages, far eclipsing Gerstner's $189 million mark[18].

# ROMETTY STAYS THE COURSE

Palmisano turned 60 in 2011, and acting on his recommendation, the board selected Ginny Rometty as the next CEO. She beat out Mike Daniels, her peer in Global Services, in part because Daniels was much closer to the IBM retirement age. Steve Mills, the head of the software division, was also a strong candidate but again, deemed too close to retirement age. She took over the reins in January 2012. Her career in IBM services, orchestration of the PricewaterhouseCoopers acquisition, and close working relationship with Palmisano were deciding factors.

**Figure 58.** Sam Palmisano (CEO 2002-2012) and Ginni Rometty (CEO 2012-2020). Reprint Courtesy of IBM Corporation ©.

Rometty grew up in a single mom household. She graduated from Northwestern in 1978 with a computer science degree, attending via a scholarship from General Motors. She interned at GM between her junior and senior years, and found time to be president of her sorority. She attended the GM Institute to develop her programming skills and in 1981, she joined IBM as a systems engineer in Detroit. She was just one of 355,000 employees worldwide. There were ten to fifteen levels of management between her and CEO John Opel. IBMers with sales and marketing backgrounds, mostly men, dominated most of the upper levels. Never holding a territory as a salesperson was a real inhibitor. A star salesman could climb fast. He might be plucked by an upper executive as an AA and skip several rungs. Getting to the very top of IBM required a fast ascent. Though it took thirty years for Rometty to make it to the top, there was not much time to spend in any one position.

Rometty stayed in technical positions for ten years and then moved to the consulting group in 1991. She benefited immensely by moving into Services early, catching the wave started by Dennie Welch and then amplified by Gerstner's "One IBM" focus on solutions. She rose fast, becoming head of Global Services in 2002. Palmisano named her senior vice president for Sales, Marketing, and Strategy in 2010, a part of the inner circle.

Rometty was not as approachable and down to earth as Palmisano. She was ambitious to a fault and quick to claim credit. She tended to view employees as "resources" that could be molded to power her ascent[1]. The higher you went in IBM, the less importance that attached to how you treated your employees. Vin Learson was certainly a case in point and Sam Palmisano was the exception. A focused and unrelenting ambition was characteristic of most IBM CEOs. However, most came from sales where you had to be personable to close business. Rometty ironically did make it easier for those that followed in her footsteps. The shrinking of the company during her watch, with layoffs, outsourcing, and offshoring, reduced the number of rungs one had to climb to get to the CEO office.

Palmisano stepped down as the books closed for 2011. For the year, IBM produced $107 billion in revenue and had a market capitalization of $235 billion. The second road map, to hit $20 per share by 2015, was already in place. When taking over, Rometty had the opportunity to set her own course. The lesson of John Akers, accepting Opel's ambitious goals and the extensive expansion that went with them, was still fresh. Instead, she decided to adopt Palmisano's road-map and focus on earnings per share. She faced challenges immediately at the start. Revenues declined in the second quarter of 2012.

Her strategies, driven by the EPS targets, closely mirrored those of her predecessor. She would continue to divest what she viewed as unprofitable parts of the company. The transition from hardware to software and services continued. The role of the mainframe would diminish. Where the strategies fell short to increase earnings per share, financial engineering took over. The game plan included vast stock buybacks coupled with continued reductions in costs through layoffs and offshoring.

## ROMETTY'S "MOON SHOT"

Lou Gerstner knew the odds were against him when he agreed to take the reins of IBM. Despite his best efforts and the financial magic of Jerry York, mainframe revenues could have collapsed, sending the company into the financial abyss. He was able ride the Internet and e-business waves back to prominence. Ginni Rometty was in a similar situation as she advanced to the CEO office, with potentially years of declining revenues ahead of her. Work was underway in Yorktown Heights on a project that might come to her aid. IBM researchers had been on the lookout for a new challenge after the success of Deep Blue in chess. The incredible run of Ken Jennings on the *Jeopardy* game show in 1997 planted the seed. As Jennings extended his winning streak, interest skyrocketed. One IBM Research manager, Charles Lickel, watched as one of Jennings' late appearances transfixed patrons at a local bar. Lickel contemplated if the Deep Blue super-computer could beat Gary Kasparov in chess, could a different type of computer beat someone like Jennings?[2]

As he pursued the idea, he found a ready acolyte in David Ferrucci. With the tenth anniversary of the Deep Blue success in chess approaching, IBM was looking for something similar that could demonstrate its technology. Discussion of a Jeopardy challenge continued through 2004 and 2005. Ferrucci would comment, "I was the only one in IBM Research, even in the academic community, who thought it could be done.[3]" IBM Research was supposed to tackle projects of this sort. However, it was an area of special interest for Ferrucci. Finally securing a commitment, he assembled a team of over 20 researchers with the mandate to come up with something in three to five years[4]. The mission was to develop an Artificial Intelligence (AI) system to beat the best at Jeopardy. They dubbed the project "DeepJ!", a homage to the Deep Blue program. Given the lofty goals of the effort, the name changed to "Watson", after the founder of the company[5].

The challenge was far different in scope from the IBM machines that played checkers and chess. The system needed to access much of the world's knowledge as could be committed to storage bits. It needed to have natural language skills to parse the *Jeopardy!* clues into a search, scan the database, and put together the winning question. Watson hardware and software needed natural language processing, massive data analysis, automated reasoning, and machine learning functions. Even the first step of separating the Jeopardy clue into subject, verb, and object was challenging. Over time, Jeopardy clues became more witty and tricky, complicating the work[6]. An exhaustive analysis of prior Jeopardy questions concluded that Watson could realistically only answer 85%. Ferrucci's team felt that Watson had to solve at least 70% of the questions to have a chance to win[7]. The team decided to pursue multiple avenues in parallel for possible answers. Then, Watson would score the selected answers, choosing the one with the highest probability, and only then, ring the buzzer[8]. Ferrucci would comment on any hesitation by the machine, "It sort of wants to get beaten when it doesn't have high confidence. It doesn't want to look stupid."[9]

Watson required significant computer horsepower. An early test with a single core Power7 server took two hours or longer to generate an answer. Upgraded hardware included ninety Power7 servers with nearly 2,880 individual processors. The assembled machine processed up to 80 teraflops (80 trillion floating-point operations per second)[10]. It had sixteen terabytes of random-access memory. The team populated that memory with as much of the world's knowledge as could be identified and loaded. They included Wikipedia, encyclopedias, dictionaries, and news publications. Accessing information directly from the Web would be far too slow. This configuration produced answers in about three seconds.

In early tests of the system, Watson came up with the correct answer about 15 percent of the time. More work brought the success rate up to the point where the team looked for live contests. Past *Jeopardy!* champions were invited to Yorktown Heights for practice games. There were perhaps a hundred of these contests, with Watson winning 65 percent. The *Jeopardy!* producers were contacted in 2008 to gauge their interest in a challenge match with all-time champions Ken Jennings and Brad Rutter.

After negotiation on ground rules, plans for a match came together in January 2011. They installed Watson in a back room at the IBM Research studio in Yorktown Heights. The team outfitted Watson with an avatar and buzzer light. The buzzer was the source of some controversy. Jennings knew that the first to hit the buzzer had a significant advantage. Watson could activate the buzzer at electronic speed. IBM and Jeopardy reached a compromise process to level the playing field.

Three games were held, starting on January 14. After the first day, Watson and Rutter were tied with Jennings slightly behind[11]. There were several instances where Watson stumbled in "understanding" the context of the clue, revealing the challenges facing language processing in AI. Watson powered ahead on the second day with $35,734 accumulated earnings, triple that of second-place Rutter. Watson continued to excel on the third and final day, though it benefited by hitting a Daily Double. Watson bet big on Final Jeopardy. The clue was "William Wilkinson's "An Account of the Principalities of Wallachia and Moldavia" inspired this author's most famous novel." The correct response was "Who is Bram Stoker." All three got the right answer but Watson retained the lead and the victory. Jennings took the loss with good humor, writing a tongue-in-cheek comment on his game pad, "I for one welcome our new computer overlords.[12]" Watson took home the $1 million prize, which the company donated to charity. Watson was featured in advertisements talking with director Ridley Scott, tennis star Serena Williams, and even Bob Dylan[13]. The victory was a high point for the company, a moment of pride among IBMers.

Rometty did not lose much time after Watson's high-profile victory to move on commercializing the technology. She chose healthcare as the new focus for Watson's AI capabilities. It was a rapidly growing sector of the economy with a focus on computerization. IBM set up several test beds for Watson, including Memorial Sloan-Kettering Cancer Center, the Cleveland Clinic, and the MD Anderson Cancer Center. Sloan-Kettering initiated a massive program to "train" Watson for cancer diagnosis. Their test team loaded 600,000 pieces of medical evidence and two million pages of journals and trials[14]. The goal was to compare patient data with known treatments and outcomes in order to recommend the best course of action for new patients. The trial at MD Anderson was unusual in that the customer paid IBM for the program, a reflection of the high interest in the technology.

In April 2015, Rometty announced plans to spend $1 billion on a new subsidiary called Watson Health. She called the new venture her "moon shot.[15]" The new group rapidly grew to 600 and located in New York City's "Silicon Alley" next to the offices of Facebook, Google, and Twitter. Rometty moved 2,000 employees out of IBM Research to work on the new venture[16]. The subsidiary crested at 7,000 employees. The complete offering required an ecosystem of software and services, supplied mainly by acquisition. The high-profile additions were Truven Health for $2.6 billion and Merge Healthcare for $1 billion. The acquisitions alone totaled more than $5 billion[17]. David Ferrucci, the lead researcher on Watson, sought to tamp down IBM marketing's claims about Watson's capabilities, with limited success.

With the cancer pilots, it soon became clear that the assigned task of diagnosis was more complex than anticipated. Despite mammoth machine training efforts, the results at Memorial Sloan-Kettering were underwhelming. They expected Watson to review all of the stored information and produce a recommendation within three seconds. However, this required too much time on the part of staff researchers to keep the databases loaded and up to date. The MD Anderson Cancer Center spent upwards of $62 million with IBM before missed deadlines and excessive costs doomed the project[18]. AI was the right bet for the Watson technology but health care, and especially cancer diagnosis, was the wrong choice. However, with the money invested in the pilot projects and the number of employees working on it, the Watson unit was too big and too important for Rometty to pull the plug, a situation that sounds a lot like FS. IBM did broaden the use of Watson AI technology to other fields where it did better.

Watson Health was only the first foray into expanding software portfolio. In 2009, "75% of all profit in the software sector [was] earned by just four companies: Microsoft, IBM, Oracle, and SAP.[19]" This was a good base to build on. IBM's targeted technology segments of cloud computing, mobile support, and AI required many gaps to be filled. Rometty acquired 73 companies on her watch. When combined with the companies acquired by Palmisano, the total ran to over 200.

The perennial question was whether IBM could absorb those firms without damaging its culture and innovation. Rometty made one software divestment. She sold what was left of the Lotus division, Gerstner's big software bet, to an Indian company in 2018 for $1.8 billion.

Meanwhile, Oracle schooled IBM on what was possible with software, growing from the foundation of a database manager to a full-blown enterprise software company. IBM's Ted Codd invented the relational database, but Larry Ellison and Oracle were first to market. In 2001, Gerstner had acquired the Informix database company. This propelled IBM to the top spot in databases with a 35 percent share. Microsoft entered the market with SQL Server and rose quickly to the third slot. Oracle surged back by 2005 with a nearly 50 percent share. He built an ERP (Enterprise Resource Planning) stack on top of Oracle's databases. The ERP offering included traditional business accounting applications as well as Human Resources, Supply Chain Management, and Customer Relationship Management (CRM). Ellison acquired PeopleSoft in 2005, shortly after PeopleSoft had purchased J.D. Edwards. This positioned Oracle, along with SAP, as the two major forces in the ERP market. IBM continued to dominate in mainframe databases, but its overall market share would plummet to less than 6 percent by 2021, according to Gartner analysis.

Watson was not the only supercomputer IBM produced during the Palmisano and Rometty tenures. Powered by the RISC-based POWER architecture and ever more dense microprocessors, the company developed a series of large supercomputers. On the heels of Deep Blue and Blue Gene, IBM won a $244 million contract to develop a supercomputer capable of two to four petaflops performance[20]. A petaflop is 1,000 trillion floating-point operations per second. At the time, the fastest supercomputer was the IBM Blue Gene, running at 360 trillion calculations per second. The design and build was a joint effort by IBM sites at Rochester, Austin, Essex Junction, and Boblingen as well as IBM Research. The resulting supercomputer, called Roadrunner, cost $100 million to develop. Los Alamos installed the machine and ramped it up to one petaflop on March 25, 2008, securing the #1 spot on the Top 500 list.

In 2014, IBM was selected by the Department of Energy to build two new supercomputers, to be called Summit and Sierra. The Oak

Ridge National Lab installed Summit for civilian AI-based research, replacing a Cray system called Titan that was only two years old. Lawrence Livermore National Lab installed Sierra for nuclear bomb modeling, replacing an IBM Blue Gene-based supercomputer called Sequoia. Summit and Sierra combined IBM POWER processors with a massive number of Graphic Processing Units (GPUs) supplied by NVIDIA. Summit contained nearly 10,000 POWER7 processors and 27,648 NVIDIA graphic processors[21]. The Summit footprint was the size of two tennis courts. It was the first exascale supercomputer, performing at 3.3 exaops. An exaop is 1,000 times the speed of a petaflop, at one quintillion operations per second. Summit and Sierra were ranked #1 and #2 on the Top 500 supercomputer list in 2018.

IBM failed to win the next big supercomputer contract, the Frontier machine destined for the Oak Ridge lab. Cray, now an HP subsidiary, won the contract for $600 million. IBM had a great run at supercomputers[22]. Unlike the Blue Gene series of machines, the new environment was one of expensive, complex, one-off systems with high risk to break even on them. Years earlier, that was the impetus for the IBM 701. Now, the urgency at IBM had shifted to the cloud.

# THE CLOUD COMPUTING WAKE-UP CALL

I n March 2013, the Central Intelligence Agency (CIA) awarded a $600 million cloud-computing contract to Amazon Web Services (AWS), beating out IBM and one other bidder. The loss was a massive wake-up call for IBM. Though the federal government declared that cloud-first would be its ongoing strategy, it came as a big surprise that the CIA was the first to take plunge. The agency pursued a program called the Cloud Computing System (C2S) designed to provide flexible cloud services to the CIA as well as sixteen other intelligence agencies. That included the highly secret NSA. C2S would use a public cloud service despite the sensitive nature of intelligence operations. It was a stunning endorsement of cloud computing. It served to alleviate the concerns of many companies looking to move their IT operations to the cloud.

IBM protested the award, arguing that the company's bid was $54 million lower than AWS[1]. The challenge dragged on for months and exacted additional damage on IBM's cloud reputation. The CIA responded that Amazon simply had the better technical solution. It cited a data analytics test run that processed 100 terabytes of raw data looking for patterns. It was a latter-day version of the NSA processing of 40 gigabytes of Tractor tape data on the Stretch/Harvest machine. The test ran far faster on Amazon's cloud. After eight months, IBM finally gave up and withdrew its protest. The loss would echo the company's earlier loss to UNIVAC at the Census Bureau. However, this time there was no Tom Watson or Vin Learson to energize an IBM response.

The CIA and intelligence agencies bid a follow-on cloud implementation in 2020 called C2E (Cloud Computing Enterprise)[2]. This time, the award was spread between five companies—Amazon, Microsoft, Google, Oracle, and IBM. The NSA awarded its Wild and Stormy cloud project to Amazon in 2021. The program aimed to modernize the agency's intelligence operations. Microsoft protested, but the award was upheld the following year. IBM did not even make the cut. The company that supported the NSA through generations of computers was unable to take them to the next level.

Irving Wladawsky-Berger viewed cloud computing as the third major technology shift in computing, after centralized systems and client-server[3]. The term "cloud computing" took hold as early as 2000. Cloud companies provided computational services directly over the Internet, or via a private cloud. It replaced the need for companies to purchase, construct, and manage their own data centers. It also provided the flexibility to scale up or down as workloads fluctuated. The workloads placed in the cloud included enterprise applications, storage of big data, hosting of web services, and even AI-focused machine learning. The cloud providers maintained massive data centers with vast blocks of servers. Amazon Web Services, for instance, maintained large data centers in 24 regions around the world. In principle, this provided the ability to scale up, even to supercomputer demands, if needed.

Cloud services came naturally to Amazon. Its massive, far-flung consumer data centers already provided a flexible infrastructure to get started. The company benefited by being the first mover. Andy Jassy had led Amazon's entry into music sales[4], then developed a proposal to enter cloud computing. Per Amazon business rules, his project plan could be no more than six pages. The offering went live in 2006. AWS experienced a big bump in 2009 when Netflix signed on, and it secured the big win at CIA in 2013, becoming the market leader. In the beginning, AWS was such a dominant force that the market analysis firm Gartner had to revise its "Magic Quadrant" positioning for cloud computing vendors. Otherwise, Amazon position on the graph would have been off the chart altogether[5]. Since 2008, its cloud revenues have grown at 40 percent a year. In 2021, AWS was 13 percent of Amazon's $469 billion in total revenue. However, with its 35 percent margins, it represented 75 percent of the company's profit[6].

Microsoft did not have the built-in advantage of Amazon and started late in the game. Its entry into cloud computing was helped by the work of former Lotus Notes developer Ray Ozzie. After leaving IBM, he joined Microsoft and took over as chief software architect from Bill Gates in 2006. Microsoft launched the Azure cloud two years later. Under the direction of CEO Satya Nadella, the company expanded its offering rapidly, adding a public cloud option in 2014. The Azure strategy centered on Fortune 500 customers, where Microsoft already had a strong presence due to Windows and Office applications. Microsoft saw growth at a 40 percent rate with 70 percent gross margins. It also pursued a hybrid cloud approach, splitting customer applications between its public cloud service and the client's in-house systems.

Google was also late to the cloud market. Like Amazon, it had a ready infrastructure to build on, though not as extensive as Amazon's. It had vast experience in Internet applications and a robust library of cloud-ready middleware. Google already used many of the same components for their end-user services. Google went live with their cloud in late 2010. It would struggle with enterprise customers because it had neither prior relationships nor support for legacy applications. Google also lacked the security classifications to work with federal agencies. Its key differentiator was flexible terms and low price. The rise in importance of AI and machine learning in cloud support would play to its strengths. The Google Cloud service grew from $4 billion in 2017 to $19 billion in 2021, a 47 percent compound annual growth rate (CAGR).

Google's entry into cloud computing completed the Big Three. In 2021 Amazon's share of the market was 33 percent, Microsoft had 22 percent, and Google's share was 10 percent but rising[7]. The overall market had expanded from $131 billion in 2013 to $405 billion in

2021, a CAGR of 15 percent. The market was expected to grow to $1.7 trillion by 2028, representing a 20 percent CAGR.

A very late participant in the cloud segment was Oracle, which had a cloud offering on the market in 2016. Larry Ellison had stated his desire to be the 21st-century IBM so he could not miss the cloud transition[8]. The Oracle cloud push would involve hardware as well as software. Oracle would focus on developing and manufacturing high-performance servers to populate its cloud data centers around the globe. It started with Intel processors in 2016 but moved to AMD processors two years later and then to ARM processors in 2021. The Oracle cloud network reached 37 regions by 2022.

IBM could have blazed a similar trail in developing a cloud infrastructure. The company was a leader in high-performance microprocessors, disk drives, and PC servers. This advantage evaporated as Palmisano and Rometty divested those units in their campaign to make the earnings per share roadmap goal. IBM had been early to the cloud game but failed to capitalize. The takeover of Eastman Kodak's IT operations in 1988 led to Gerstner's focus on "on-demand" computing. However, the shift to services and the shedding of core hardware businesses under Palmisano and Rometty reduced IBM's ability to tackle the cloud.

Then came the crushing loss at the Central Intelligence Agency. Now, it was a scramble to do damage control. Several months after the CIA award, Rometty announced the acquisition of the cloud-hosting firm SoftLayer for $2 billion. A key objective of the purchase was SoftLayer's existing network of data centers. Rometty would subsequently announce plans to spend $1.2 billion to add forty data centers to SoftLayer's existing 15 cloud-hosting centers[9]. SoftLayer built its business on small- and medium-size companies that did not require industrial strength facilities. It did not offer the performance, reliability, and security that IBM enterprise customers would demand. Rometty announced that key cloud elements such as the Watson AI system would run on POWER-based microprocessors. However, SoftLayer only supported Intel architecture. With the PC division gone, Rometty had to look elsewhere for servers to fit out the new data centers. The company with "machines" in its name had to turn to Dell.

Aside from the hardware challenges, IBM was far behind the leaders in software support. Google, Amazon, and Microsoft had made significant investments in developer toolkits and support middleware for cloud applications. IBM based its first cloud software stack on the open-source Cloud Foundry, released in 2014 as IBM Bluemix. The company later rebranded both SoftLayer and Bluemix as the IBM Cloud. In parallel, work was underway on two separate initiatives to develop IBM's own cloud architecture. The first effort, called Genesis[10], attempted to build a software infrastructure from the ground up. IBM hired several executives from Verizon's cloud operation to run the program. Genesis lasted until 2017, when a review by IBM Research concluded that it would not scale up sufficiently to be competitive. The cancellation of Genesis included a $250 million write-off of the Dell servers.

The Genesis team resurfaced with the NG project (apparently for "New Genesis"). IBM Research was working in parallel on the "GC" program, for "Global Cloud"[11]. IBM announced both NG and GC in 2019, representing two completely different cloud architectures. It did not help that IBM spent critical time on one-off cloud projects for key customers rather than seeing its cloud strategy and architecture through to the finish line. Arvind Krishna forced the NG and GC dual tracks into one. Krishna had risen to the top at IBM Research and then to head of the cloud division. He saw the window of opportunity for cloud rapidly closing and decided IBM needed a new approach. He engineered the acquisition of the cloud software company Red Hat in 2019.

Red Hat produced a superior stack of cloud-enabling software sitting above an enterprise version of Linux. It was the first open-source company to exceed $1 billion in revenue. The company's founder, Mark Ewing, was the son of an IBM programmer[12]. He grew up in Poughkeepsie and learned programming on Commodore and Apollo personal computers before leaving for Carnegie Mellon University. At college, he was known for his computer skills and for the red lacrosse hat he always wore. Fellow students needing technical support were told to look for the guy with the red hat.

After graduation, Ewing joined IBM as an engineer. He worked tailoring Linux installations for customers, a job he found less than exciting. He left IBM and started the Red Hat Linux project. He would sell copies of his custom Linux version out of his apartment in Durham, North Carolina[13]. In 1994, he joined with Bob Young to found the Red Hat company. Their focus was a highly enterprise version of Linux with all the supporting software. The business model provided the base custom Linux at no charge, then added software and support for a fee. Red Hat grew rapidly to 13,000 employees based in Raleigh. That actually offset the many layoffs made in the same area by IBM. Ewing would leave Red Hat a very wealthy man in 2002.

If acquisition was the right move, then Red Hat was certainly the right partner. It focused on the enterprise and had worked with IBM for nearly 20 years. The big issue was the price IBM paid. It offered $190 a share, a 63 percent premium over the current share price, resulting in a purchase price of $34 billion. By way of contrast, VMWare purchased Pivotal Software to jump-start its cloud offering and received similar assets for less than $3 billion. The IBM deal enriched the bank accounts of Red Hat CEO Jim Whitehurst and other senior Red Hat executives. Whitehurst's 382,000 shares became $73 million with the purchase. This did not sit well with Red Hat employees.

IBM finally had a competitive cloud software offering, but the costs were staggering. Besides the Red Hat expense, there were the many lost years in development of internal cloud projects. At the same time, IBM spent billions on share buybacks and Watson Health with nothing to show for its mammoth investment. Looking forward, Krishna recognized the challenges but saw the strengths that IBM could bring to bear. The company had an extensive worldwide network of data centers and the expertise to run customers' enterprise applications. With the push of e-business, IBM had become a leader in the application server market. There was market-leading expertise in AI, machine learning, and data analytics. Then there was the cloud market itself, one that projected to be so large that surely more than three major players could fit in. For Krishna, the spot for IBM was in the hybrid cloud, an integration of public and on premise applications.

IBM faced one fundamental challenge regardless of how good IBM expertise and Red Hat software stack were. That was the huge cost to build, maintain, and grow a competitive cloud infrastructure. In a situation of some irony for IBM, the Big Three had constructed huge barriers to entry for potential cloud serving players. Exhibit A was Amazon's infrastructure build-up. Since the inception of its service in 2006, Amazon's AWS unit spent an estimated $100 billion on physical infrastructure. And, their annual spend was accelerating. The tab for the two years ending with 2021 was an estimated $50 billion[14]. The 15-year estimated cost for just the new Canadian region datacenters was $17 billion. A new region in Thailand cost $5 billion. AWS could fund this expansion as its cloud revenues continue to grow at a 40 percent pace and income is expanding at an even higher rate. For example, in the fourth quarter of 2021 alone, AWS had sales of $18 billion and income of $5.3 billion. Microsoft and Google had to keep up with Amazon or risk losing market share. Microsoft spent $15 billion on cloud in 2021 and planned another $15 billion in 2022. Google struggled to keep pace. As the low-price provider and vendor with the smallest market share at 10 percent, the Google Cloud unit is still reporting losses, let alone the huge profits thrown off by its two major competitors.

How can IBM compete when the ante to play is so high? And, there are other challenges. All three major competitors are fast-moving organizations with continuous innovation, enhancing their offerings in months, not years. The competition between the Big Three will quite likely drive prices further down in the quest for market share. For IBM, this could mean huge investments to get into the game, only to discover that it was in another low-margin business.

# BYE BYE ROAD MAP

In IBM's quarterly earnings conference call with financial analysts on October 24, 2014, CFO Martin Schroeter reported a revenue decline of 4 percent for the quarter and a projected earnings shortfall. This was the tenth straight quarter of revenue decline. He added that IBM did not expect to achieve the $20 EPS (earning per share) goal of the 2015 road map, effectively abandoning the program[1]. In an unusual move, Ginni Rometty joined the call. She wanted to stress IBM's continued transition to high-margin businesses of cloud, AI, and data analytics. She cited $16 billion in revenue for data analytics and $3 billion for IBM's emerging cloud solution. At the same time, the company was opaque on the details of what kind of revenues went into those measurements. Internally, there was significant pressure to put newly booked deals into one of those target growth categories. It is instructive that the end of the road map era and new technologies areas were addressed in the same call but not connected. The architect of the road map program, James Kavanaugh, would comment on this years later, stating, "I found out, when this industry moves so fast, you can't get locked into something that's long-term that might prevent you from identifying new, emerging areas of opportunity.[2]"

In the same conference call, Martin Schroeter announced the sale of two hardware divisions. They represented $7 billion in revenue but

$500 million in losses[3]. IBM sold the Netfinity server division to
Lenovo. They had asked $4 billion for the unit but settled for $2.3
billion. The xSeries line lost money but it was a key building block for
"One IBM" solutions. It was also a key building block for cloud
computing, where Intel-based servers were the standard. IBM would
have been a significant customer of its own operation as it built out
new cloud data centers. At the same time, the Big Three cloud
vendors were themselves getting into both Intel servers and specialized microprocessor chips to enhance the function, performance, and
reliability of their cloud installations. The timing to jettison the PC
division, and at fire-sale prices, was unfortunate. With the Lenovo
sale, IBM was completely out of the PC business that it created.

The other sale that Martin Schroeter announced was IBM Microelectronics. It represented a fundamental part of IBM technology.
Furthermore, it was not even a sale. IBM would be paying a company
called Global Foundries $1.5 billion to take over the division. On top
of that, IBM would be taking a $4.7 billion charge to earnings to
reflect costs related to the divestiture. IBM had a long and important
history in component and semiconductor technology. This extended
from Ralph Palmer's pluggable units to SMS and SLT cards, integrated circuits, and now RISC-based POWER-based microprocessors.
It included manufacturing as well, illustrated by the innovation with
core and SLT production that provided a competitive edge in cost.
The division had lost $700 million on sales of $1.4 billion in 2013[4],
but it was so much more than a freestanding profit center. Jack
Kuehler, the lead technology executive and IBM president in 1991,
argued that the best and fastest chips were central to IBM and could
not be outsourced. Lou Gerstner had invested $2.5 billion in Micro-
electronics in 2000 to build an advanced semiconductor plan in East
Fishkill.

In spite of all that history, IBM paid to get rid of it. Global
Foundries, a company created from Advanced Micro Devices and
owned by the Emirates of Abu Dhabi, would take over IBM's entire
microprocessor business, including the Essex Junction, Vermont, and
East Fishkill fabrication plants, as well as some parts of IBM Research.
Global Foundries would also get 16,000 patents covering microelec-
tronic processes. On top of that, IBM committed to invest an addi-

tional $3 billion in semiconductor R&D and share those developments with Global Foundries. IBM would continue to do basic research in high-density chips, and work with Global Foundries, Samsung, and Intel on production. IBM structured the deal financially so that it did not affect current business results.

Prior to the sale, IBM Research had driven circuit densities down to 32 nanometers with the POWER7 line in 2010 and to 22 nanometers with POWER8 in 2014. Densities would continue down to 14 nanometers with POWER9 in 2017. These were all fundamental building blocks of IBM's Power Systems, mainframes, and supercomputers. A core part of the agreement required Global Foundries to produce high-end chips for a period of ten years, starting with 10-nanometer production.

Within months of the agreement, Global Foundries announced that it would not be making the 10-nanometer chips[5]. At first, Global Foundries promised to develop a more advanced chip, at the seven-nanometer density. Then in 2018, it revealed that it would only produce the more advanced chip if IBM provided it with an additional $2.5 billion. While this disagreement was still up in the air, Global sold the East Fishkill plant to ON Semiconductor in 2019. Employment at Essex Junction plummeted from 8,000 to 2,600.

In 2021, IBM sued Global Foundries for breach of contract, contending that Global Foundries did not supply the agreed-to chips. Furthermore, IBM alleged that Global Foundries had used the original $1.5 billion to renovate the Microelectronics Division plants for its other businesses. Global Foundries countersued, claiming that $1.5 billion was inadequate and that billions more would be required to produce the chips IBM wanted. In the meantime, IBM had to secure its microchips from Samsung, at the going market rate. Samsung already had the manufacturing process for seven nanometers and then five nanometers in its production for smartphones and other devices. It had plans in place for two nanometers by 2025 and 1.4 nanometers by 2027. Samsung wanted to get into server chip production.

IBM Microelectronics had been the driving force behind the PowerPC microprocessor chips of the AIM joint venture between Apple, IBM, and Motorola. The AIM group (known internally as the Somerset program) spent at least $1 billion of development during the program's life. Apple would use PowerPC processors to power its Mac line from 1994 to 2006. Steve Jobs would announce in 2006 that Apple would return to Intel for chip designs and fabrication, declaring that IBM was too slow in developing new designs. Apple designed its own chips for the iPhone and iPad. It would move on from Intel in 2020, embracing a British company, Advanced RISC Machines (ARM), which enabled Apple to design its own microprocessors and then use Taiwan Semiconductor Manufacturing Company (TSMC) to produce them. IBM Research continued on the leading edge of microprocessor design.

Not long after IBM's sale to Global Foundries, microprocessor chips became one of the most important technologies on the planet, with fierce competition at the cutting edge of the technology and a widespread shortage of fabrication plants. The Biden administration's CHIPS (Creating Helpful Incentives to Product Semiconductors and Science) act aimed to underwrite competitiveness in this crucial area. Intel was considering purchasing Global Foundries for $30 billion. Rometty paid $1.5 billion to get rid of it. Rometty commented on the outgoing division, "That is someone else's business," adding, "We need to deploy that capital to other things.[6]" By this, we could assume she meant stock buybacks.

When Ginni Rometty finally announced she was stepping down, IBM stock gained 5 percent. That was a good overall indicator of her performance. She took over an IBM from Sam Palmisano that had just recorded $107 billion in revenue and reached a market capitalization of $214 billion. After eight years, Rometty oversaw revenue spiraling down 25 percent, to $79.6 billion. This included a string of 22 straight quarters of revenue decline[7]. IBM's market value plummeted to $119 billion, an average drop of $15 billion for each of her eight years as CEO[8]. The company's market value in 2012 was roughly equal to Microsoft. In 2020, the market capitalization of Microsoft had risen to $1.3 trillion[9].

Rometty contended that this decline was the result of divesting business units and currency fluctuations. However, IBM's low sales per employee disputes this view. In 2022, Apple revenue per employee was $2.4 million, Alphabet (Google) $1.5 million, and Microsoft $1.0 million[10]. In 2013, IBM revenue per employee stood at

$213,000[11]. IBM's low numbers reflected the transformation of the company into a services-led enterprise. A pure consulting competitor like Accenture had revenue per employee at only $110,525. The low production also hints that there were no new market-leading products available to sell at any time during her tenure[12]. Earnings plummeted 41 percent, from $16 billion to $9 billion. This occurred despite all the financial tricks deployed to boost earnings, including divestments and layoffs. Market capitalization declined nearly 50 percent to $120 billion. IBM's share price slid from $184 to $112, a decline of 25 percent during a period when the S&P 500 index was up 208 percent. It was also a time when technology stocks were king, but just not IBM. Microsoft was up 500 percent during this period. If IBM shares had kept pace with inflation, then the price would have been $335. Warren Buffett sold his IBM investment in 2018 and took his losses.

In 2022, a new acronym surfaced for Big Tech -- "FAANG" -- which stood for Facebook, Amazon, Apple, Netflix, and Google. The combined market capitalization of those five companies was $4 trillion. Where IBM once stood atop Big Tech with the Seven Dwarfs far below, the company was now far down the line behind Microsoft, Apple, Dell, Amazon, and Google. IBM was #49 on the Fortune 500 list.[13] In 2021, revenues were stated in three major segments—software, consulting services, and infrastructure, all of which were services-related. Software was 42 percent of revenue, consulting services 31 percent, and infrastructure 25 percent. Missing was hardware, the "machines" in the IBM name. What little hardware remained was buried in the services categories. Disk drives, PCs, printers, and microprocessor chips were all gone.

The central feature of the Rometty years was the revenue decline. She ran the company for nine years. Only two of those years showed a revenue increase, and the increases were less than 1 percent. Tom Watson never had a down year. His father had the shocker decline in 1921, two modest declines in the depth of the Depression (1932 and 1933), and a decline in 1946 in the transition back from the war economy. He also had stratospheric revenue increase years of 48 percent, 49 percent, and 50 percent. Watson Jr. had consistently high revenue increases, year after year. With Learson, Cary, and Opel, the revenue increases settled down into the mid-teens. Akers was generally under

10 percent growth until disaster struck in 1991. Gerstner had the financially calculated down year in 1993, then modest growth thereafter. Palmisano had three down years out of ten, and modest growth in the other years.

The push of the road maps to reduce costs and boost earnings per share resulted in more layoffs and increased offshoring of skills. At the end of 2012, IBM reported 434,246 employees worldwide and stopped reporting workforce numbers. The company did not want to highlight the dual trends of layoffs and the offshoring of jobs.

Rometty's core focus was expanding the services model. She spent 20 years in various IBM services positions. It was the source of her reputation. The offshoring started by Palmisano continued under Rometty. By 2015, U.S. employment by IBM was down to 75,000, from a peak of 200,000 in the 1980s. Staffing in India had risen to 130,000 by 2017. This represented one-third of the total IBM worldwide population. Many IBMers called the Rometty road map "Roadkill 2015.[14]" There were three rounds of layoffs in 2016. IBM managers needed to assess a certain percentage of their employees to be fired, even if their staff consisted of all strong performers. Employees to be laid off could search for other positions within the company, but IBM discouraged managers from making offers to them. In some cases, laid-off workers had to train their offshore replacements. Though labor costs declined, the quality of IBM's services solution, including technical and business competence, suffered.

The selection of who should go in "resource actions" was sometimes less a factor of an employee's rating than where they worked. The United States and United Kingdom were major centers of layoffs, as layoffs were easier in those countries. Those pushed out left with far smaller exit packages. In 2015, the severance was just one month's pay. Palmisano dropped the company-funded pension in 2005, though he replaced it with a 401K match. Rometty eliminated the match.

In 2016, Rometty announced IBM's plans to hire 25,000 professionals in the United States over the next four years. It appears that a driving factor was replacing older workers with younger millennial ones. Some executive correspondence termed older workers "dinobabies"[15]. The argument was that "Accenture is 72 percent millennial we are at 42 percent." In 2014, IBM asked employees receiving sever-ance pay to waive their right to sue IBM, a reaction to the many age discrimination lawsuits the company received. Most telling was IBM new place in the *Fortune* rankings of the most admired companies. IBM sunk to the 46th position, after being as high as 19th in 2006 under Palmisano. It had been as low as 45th in the depths of Akers' time as CEO[16]. All of this had an impact on morale and on individual productivity.

Rometty continued her predecessor's financial approach with stock buybacks. She would spend $58 billion over her tenure. This effort continued even as the company's market capitalization cratered. Rometty started reducing buybacks in 2015, with only $9 billion spent. This decreased to $4 billion in the following three years, before she suspended buybacks altogether in 2019. Even with this dialing back of the program, she still averaged $7.2 billion a year over her tenure. Palmisano never slowed and finished with an average of $10 billion per year. Gerstner averaged $6.3 billion per year[17], but he was issuing stock splits in 1997 and 1999 because IBM doing so well.

The result of this financial maneuvering was meeting the road-map target of earnings per share and continuing dividend payouts to shareholders. At the same time, the actions crippled IBM during a time of technological upheaval. The value of shareholder stock cratered 44 percent as IBM performance continued in a downward spiral. The $58 billion that Rometty spent on stock buybacks would have come in very handy in competing with the Big Three in the next technological shift, cloud computing. It would have purchased consid-erable R&D, internal stack development, acquisitions, and the build-out of data centers.

Despite her record, Rometty benefited from the CEO pay stan-dard set by Gerstner and continued by Palmisano. Over the first seven years, she was paid $137 million in salary, not counting stock options[18]. There was no apparent linkage to performance, despite quarter after quarter of declining revenues. In 2012, her first year, Rometty earned $15.4 million in salary plus nearly $10 million in stock[19]. In 2014, she was awarded a $3.6 million bonus that grew to $6 million with stock sales. In 2015, she recorded her 15th straight quarter of declining revenue while income was off by 15 percent and the stock price down 36 percent. She still received a $4.5 million bonus plus additional stock options. In 2017, the board raised her salary to $33 million, making her the eighth-highest-paid CEO in the United States. Satya Nadella, the CEO of Microsoft, earned $18 million that year, along with $84 million in stock granted in his first year. Tim Cook, the CEO at Apple, earned $9 million in salary though he would receive $378 million in stock his first year. Rometty's performance paled beside those two companies.

Rometty's cumulative earnings as CEO were an estimated $158 million. She sold her stock options quickly as the share price continued to sink despite the buybacks. It is interesting that the IBM board never chose to act, giving Rometty free rein. The modern IBM board appeared to completely lapse into window dressing, able to attract top-flight executives to fill the board but without any inclination or mandate to question the CEO. With Rometty, the board missed a final opportunity to chart a new path. Tom Watson had set the standard for IBM CEOs to retire at age 60. Those who followed honored that precedent. Even a giant like Vin Learson, who was in the corner office for just two years, left at age 60. Ginny Rometty stayed as IBM's leader until she was almost 63. That's three years that could have been given over to a fresh start.

There were 12 companies in the Dow Jones Industrial Average when it debuted in 1896[20]. General Electric (from Thomas Edison) was for years the only surviving member before being dropped in 2019 after 122 years. American Cotton Oil and Distilling & Cattle Feeding were among the original companies that did not hang around long. The Dow went to 30 companies in 1928. The index dropped Sears after 75 years. Tech companies Cisco, Honeywell, Intel, and Microsoft were relatively recent additions. Surprisingly, IBM was not added until 1979. Amazon and Alphabet appear to be waiting in the wings for a possible selection. If either of those were added, IBM would quite likely be the prime candidate to be dropped. IBM has underperformed since 2010, when it was down 1 percent while the Dow was up 216 percent[21].

Combining the records of Palmisano and Rometty presents an even bleaker picture. It appears that both CEOs, in their rush to embrace a financial road map, missed the core of Gerstner's "One

IBM" concept. Not all of the various divisions were high-margin enterprises, but together they comprised a consistent and integrated whole. However, Palmisano and Rometty cut many of them loose. Gone were personal computers, Intel servers, disk drives, printers, and even microelectronics. Gerstner had cut loose networking gear, and Akers had spun-off low-end printers to Lexmark. Palmisano and Rometty jettisoned the rest. Across the tenures of three CEOs, IBM essentially lost nearly all the "machines" that were part of its name.

With the exit of the machines went the plants and labs that produced them. This compounded the original error of divestment. Only a portion of IBM's technological advantage came from IBM Research. As Glenn Henry found out as he moved forward with System/38 development, he received no assistance from the researchers in Yorktown Heights. This was not a problem, as the labs associated with the plants had the engineering and software expertise to go it alone. Those resources disappeared as the various hardware divisions were sold. IBM Research could not pick up those missing pieces.

Palmisano and Rometty had their sights set on new areas of growth, mainly anchored in software and services. Their emerging IBM was dominated by the vast, low-margin services enterprise they created and a software business that owed much to one acquisition -- Red Hat. Their strategy might have worked if they were able to establish primacy in the new technological waves of cloud computing, AI, and machine learning. However, progress was slow and, especially for Rometty, not fast enough to offset the declines of legacy existing businesses, the few that survived divestment.

The core element of the Palmisano and Rometty strategy was the stock buyback. Palmisano spent $99 billion on buybacks, and Rometty spent $58 billion. Combining their records —probably unfair to Palmisano—you get a picture of how IBM fared in the roughly 20 years that passed since Lou Gerstner turned over the reins. Revenue growth actually declined at a rate of 0.4 percent per year while income grew at rate of only 1.1 percent annually. The IBM workforce saw a massive shift from North America and Europe to offshore India and South America. During this period, the S&P 500 grew by 144 percent, at a rate of 5 percent per year. IBM stock increased 11 percent, at a rate of 0.6 percent per year. In 2002, Microsoft was already a large, monopolistic force. Yet its stock increased by 858 percent over the period, a compound growth rate of 13.3 percent.

A broader perspective of IBM's share buybacks would include Lou Gerstner and survey the 25 years beginning in 1995. According to the analysis done by Peter Greulich[22], IBM spent $201 billion on share buybacks during this time, an overall average of $8.8 billion per year. The program helped meet Palmisano's EPS target but did not appreciably move the stock price. From 2000 to 2016, for example, IBM stock was up only 30 percent, while the overall market was up over 300 percent. The buyback expense propped up earning per share but did nothing to increase the overall value and position of IBM.

IBM's investment in research and development had remained remarkably consistent during the buyback years, in the neighborhood of $6 billion annually. This budget supported 12 labs and 3,000 researchers. During this period, IBM spent $90 billion on R&D, which represented 6 percent of revenues. This was significantly lower that other companies in the technology market, which averaged 12-14 percent of revenues. In the years from 2008 to 2014, IBM was still in the top 20 by R&D expense. It slipped to #31 in 2018. The leaders in 2020 were Amazon ($43 billion), Google ($28 billion), Microsoft ($19 billion), and Apple ($19 billion)[23]. IBM actually got a rebate for its R&D investment, in the form of intellectual property revenues. The R&D community generated patents, lots of them. In 1990, IBM was #9 in the number of patents produced for the year. It moved to the #1 spot in 1993 and stayed there. The year 2021 was the 29th consecutive year at the top. Intellectual property revenue averaged over $1 billion annually, a significant amount as the company's revenues declined. However, the many patents did not translate into market-leading products, as Lou Gerstner would note and Jack Welch would certainly second.

This is quite a change from the IBM of the 1950s and 1960s, when its R&D expense dominated the industry. The lessened reliance on R&D to move the company forward would certainly be anathema to either of the Watsons. Their mantra was to invest in making better products and everything good would flow from that. Watson Sr. estab-lished that practice from the day he borrowed the $40,000 from Guarantee Trust Bank to enhance the Hollerith machine. IBM became the preeminent R&D company, the successor to Bell Labs. Watson Sr. had won the battle with George Fairchild, that investment and continual innovation was more important than short-term financial techniques. This was a lesson lost on Palmisano and Rometty.

By far, the biggest technology miss across the tenures of Palmisano and Rometty was cloud computing. For Gerstner, it was clear from the start that the mainframe was "do or die" for the company. Cloud computing was, in essence, the replacement for the traditional in-house mainframe. IBM was mostly missing in this sea change. The three companies that did not miss the transition – Apple, Google, and Amazon – now dominate computing.

The IBM Company essentially invented the computing industry. Just a partial list of breakthrough inventions is very long:

- Punched card tabulation and the 80-column card
- The disk drive
- The floppy disk drive
- Computer tape and tape drives
- Relational databases
- RISC technology
- Selectric typewriter
- DRAM memory
- POWER microprocessors
- FORTRAN
- Supercomputers – Blue Gene, Sierra, NORC, Roadrunner, Stretch
- The mainframe computer
- Watson, Jeopardy and Artificial Intelligence
- First computer to sell more than 1,000 (IBM 650)
- First computer to sell more than 10,000 (1401)
- First computer to sell more than 100,000 (AS/400)
- System/360, deemed one of top three inventions in the 20$^{th}$ Century
- IBM PC
- IBM ThinkPad
- The scanning tunneling microscope

- Nobel Prize for superconductivity
- UPC (Universal Product Code)
- SABRE airline reservation system
- SAGE early warning system with first displays and fault tolerance

# CAN THIS MAN SAVE IBM, PART TWO

With the acquisition of Red Hat, Jim Whitehurst became IBM President. Ginni Rometty had already overstayed Watson's retirement guideline by almost three years. With the president usually the heir apparent in IBM, many expected that Whitehurst would take over. He had an outstanding record of accomplishment, reminiscent of Lou Gerstner. He attended the London School of Economics, earned an MBA at Harvard, rose to vice president at Boston Consulting Group, and was chief operating officer of Delta Air Lines, where he oversaw the company's comeback from bankruptcy. Red Hat hired him as CEO in 2007. In March 2020, IBM awarded him $22.4 million in IBM shares as part of a $27 million compensation package. Despite all this groundwork for the top job, Rometty would recommend her senior vice president for Cloud and Cognitive Software, Arvind Krishna, to be the next IBM CEO. Whitehurst left IBM, in search of another CEO position. IBM shares dropped 5 percent on the news that Whitehurst was leaving[1].

Arvind Krishna took over in March 2020. He grew up in India, where his father was a general in the Indian Army. He earned an electrical engineering degree at the Indian Institute of Technology before moving to the United States and completing a doctorate in electrical engineering at the University of Illinois. He joined IBM at Watson

Research in 1990, doing fundamental research work on block chain technology. It provided for the deployment and security of distributed cloud databases. As he rose in the company, he gained experience in sales, managing the Tivoli and Informix products under Bob LaBant. Krishna rose to general manager of the technology group in 2009 and senior vice president and head of IBM Research in 2015. He moved over to the cloud and AI divisions, where he led the acquisition of Red Hat. With cloud computing an existential threat, Rometty and the IBM board selected him as the next CEO. Microsoft had done the same thing in 2014, choosing Satya Nadella, the head of its cloud division, as CEO. Krishna started with a $17 million compensation package in 2020. He was two years shy of 60 but Rometty had already broken the long-standing tradition of the CEO retiring at that age. Krishna stayed.

Krishna wasted no time in making several major decisions. He split IBM into two separate companies. The legacy infrastructure business became Kyndryl, apparently a combination of "kinship" and "tendrils.[2"] The cleaved business was a global operation with 4,600 customers and 90,000 employees[3]. Three-quarters of the Fortune 100 used its hosting services. Revenues were declining at 5-6 percent per year, going from $40 billion in 2011 to $26 billion in 2020. Margins were in the 20 percent range. Working from the Palmisano and Rometty playbook, Krishna viewed the business as keeping down the revenue and income growth of the rest of IBM. The cost of orchestrating the split was an estimated $5 billion.

Martin Schroeter, the former IBM CFO, was the new CEO of Kyndryl. He viewed the declining performance of Kyndryl a result of very limited investment over time. IBM prevented the unit from using competitive cloud services as they conflicted with IBM's cloud strategies. Released from that restriction, Schroeter focused on building relationships with Amazon, Microsoft, Google, and VMWare. Kyndryl's biggest customer remained IBM, with the mission of maintaining IBM's internal mainframes. At the same time, Schroeter faced the challenge of getting out from under IBM's legacy business and financial systems.

Krishna's second major decision was to sell Watson Health. Though AI was a core strategy going forward, the use of AI in health care had proven to be a money pit. Much in the manner of Lou Gerstner moving quickly to get out from under the mess left by John Akers, he sold Watson Health to the private equity firm Francisco Partners in January 2022 for just $1 billion. Rometty had invested far in excess of $5 billion in the venture with little to show for it. In 2023, Krishna also sold The Weather Channel to Francisco Partners, reportedly for $2 billion. The Weather Channel had been an application showcase for the supercomputer technology used in the Summit and Sierra systems. Rometty purchased the company in 2016 for $2 billion[4].

Krishna now led a much smaller IBM[5]. As of 2023, IBM was now a $57.35 billion company that was split between software at 42 percent, consulting services at 31 percent, and infrastructure (mainframes and Power Systems) at 25 percent. The workforce at the end of 2022 was just below 300,000. With the reduction in size, Krishna announced a reorganization of the sales force. One group would manage IBM's top 500 customers and he organized the rest by product. The twin technologies of cloud computing and AI would form the core of his strategy going forward. He has a very steep hill to climb in competing in cloud computing. The three major players all had market capitalizations north of $1 trillion. The new slimmed-down IBM had a market cap of just $130 billion.

Cutting loose Kyndryl and Watson Health provided a far greater focus on cloud computing. IBM aimed to be the leader in the hybrid cloud segment, where an enterprise's computing was split between the cloud and its own data center. This emphasis on the hybrid cloud hurt the company in competition with the Big Three cloud companies, as they would press the cost advantage of their cloud implementation. One analyst commented about companies selecting AWS for cloud computing, "You don't get fired for going with Amazon Web Services.[6]" IBM once had that vaunted position. Krishna believes the cloud market has room for IBM if the company makes the right moves. He notes that, "cloud is not going away, but the days of 50% and 100% growth are over.[7]" And, IBM has missed all those go-go years chasing earning per share with endless stock buybacks.

A promising avenue of attack for IBM is the loss of data control with the public cloud. Antonio Neri, the president of Hewlett

Packard Enterprise, describes Amazon Web Services as not unlike the "Hotel California" in the Eagles song: Once you check in, you can't leave. He means that can be prohibitively costly to get your data back[8]. An early adopter of the cloud, Dropbox, moved its consumer repository service to AWS. As its volume of data hit 600 petabytes (each petabyte equaling one million gigabytes), it hit a cost wall with Amazon and elected to return to in-house data centers. The move was termed "repatriation." Dropbox documented savings of $75 million over two years with the change. Amazon reacted with an offering called Outposts that put servers into a customer's location. However, IBM is far better positioned to combine in-house installations with cloud applications.

At the same time, the clock is ticking. IBM's traditional server business continues to decline faster than the new technologies are gaining steam. Though $34 billion was far too much to pay for Red Hat, the acquisition continues to work wonders for IBM's financials. Red Hat revenues have grown at 18-20 percent annually since joining IBM while net income has risen at a 15 percent per year clip. Red Hat finished 2021 with $5.6 billion in revenue, representing a full 10 percent of the new IBM's $57.4 billion total. Red Hat was 15 percent of IBM's net income. It has been the one bright spot for the company.

Meanwhile, the daily battle for cloud business continues to be a struggle for IBM. In 2018, the Department of Defense issued a request for proposals for the Pentagon's AI initiative. The $10 billion project was the Joint Enterprise Defense Initiative (JEDI). Google dropped out because of employee pressure, citing the company's motto of "do no evil." Microsoft won the contract in October 2019. Amazon, Oracle, and IBM all protested. Facing interminable delays in the courts, the Defense Department canceled JEDI in July 2021. The Defense Department started over with the Joint Warfighting Cloud Capability (JWCC), a similar multi-cloud AI project worth $9 billion[9]. They asked two companies, Microsoft and Amazon, for proposals. They added Google and Oracle later. In December 2022, the Pentagon announced that it would award the contract to all four companies. IBM did not even receive an invitation to submit a proposal. The Defense Department concluded that IBM's cloud offering did not provide hyper scale cloud capability or the flexibility to add resources on demand.

---

## ANOTHER WAKE-UP CALL

In December 2022, the Artificial Intelligence (AI) start-up company OpenAI released an Internet chatbot called ChatGPT. It provided a natural language conversation with a massive Internet database to produce an "intelligent" response. ChatGPT scoured its accumulated data for text matches, rated the hits for degree of relevance, and then produced a response in complete sentences. This represented a significant step beyond the capabilities of Google Search, which located relevant sites and simply presented a list. The reaction to ChatGPT was swift and explosive. There were over a million enrollments in the first five days[10], with 30 million users and five million daily visits within a couple of months. It was a global phenomenon. Early use of ChapGPT gravitated to applications such as writing essays or scripts. Some colleges moved to ban its use. The press reaction mirrored the coverage of ENIAC in 1946 where the flashing ping-pong balls indicated the machine as "thinking." There were corrective missives posted that reiterated, "AI can learn, it cannot think"[11]. ChatGPT was certainly a milestone in AI but its focus was not where the core AI opportunity lies. It provided stilted, formulaic text responses to basic inquiries. The future is in targeted AI where machine learning occurs through high volumes of use cases. IBM's Jeopardy and chess programs demonstrated the power of AI when the machine learns to perform specific tasks.

A group of Silicon Valley investors including Elon Musk, Peter Theil, and Reid Hoffman, founded OpenAI in 2015 as a nonprofit. Armed with substantial initial funding, the new company sought and hired the best AI researchers. Their early focus was reimagining the way they trained raw data for use. Google researchers invented a technique called "generative pre-trained" (GP) learning in 2017. It provided a path to preparing AI databases with far less manual supervision. OpenAI expanded the concept to "GPT", for "generative pre-trained transformer" and released its first version as GPT-1 in 2018. ChatGPT is based on GPT-3, the 2020 release of this training technology. Even with the new approach, OpenAI would spend enormous amounts of cloud computing time to train their data models. One project used 128,000 Google CPUs for several weeks. When Microsoft invested in the company in 2019, OpenAI moved to the Azure supercomputing platform for such tasks. The OpenAI approach does not involve any real-time Internet "crawling" to add new data as Google Search does. The OpenAI database comes from a one-time retrieval done prior to 2021.

With the release and frenzied reaction to ChatGPT, the existing AI players scrambled to contain the damage. Microsoft was already an investor in OpenAI, having provided $1 billion in 2019 and an additional $2 billion a year later. After the ChatGPT debut, the two companies came to a further agreement, with Microsoft providing $10 billion in exchange for substantial stake in the company. Microsoft's Bing and Google's Bard added support for ChatGPT. Google has been focused on buying AI and robotics companies. It planned to build a huge cloud-computing farm that could contain up to 100 million server nodes[12]. Amazon announced investment of $4 billion in the AI startup Anthropic.

At the outset, OpenAI did not have a revenue model. Then, they announced premium subscription access to ChatGPT. Revenue is projected to grow to $1 billion by 2024. As of April 2023, OpenAI had a market valuation of $27-29 billion. That market value ballooned to $157 billion. The valuations reflect the massive growth opportunity of AI. IBM projects AI could add $16 trillion to the global economy by 2030. The AI market was $8.2 billion in 2021, $11 billion in 2022, and projected to be $126.5 billion by 2031[13]. IBM's latest earnings show 3% year on year growth. In contrast, AI is expected to grow at a rate of 37% CAGR (compound annual growth rate), or even higher[14].

With the deafening AI buzz created by ChatGPT, where was IBM? Besides the focus on OpenAI, the talk was mostly about Microsoft, Google, and Amazon Web Services. Yet, IBM has decades of experience in AI, starting when Arthur Samuel trained the IBM 701 to play checkers in 1952. The company was there at the birth of AI. John McCarthy, a mathematics professor at Dartmouth College and the inventor of timesharing at MIT, organized a two-month study of Artificial Intelligence in the summer of 1956[15]. He invited leading experts in AI, including Nathaniel Rochester, Claude Shannon, and Arthur Samuel. He asked Alex Bernstein to talk about his chess program. There were arguments about the name "Artificial Intelligence" with Samuel commenting, "you think there's something kind of phony about this." McCarthy hoped that the summer work would move AI forward and was disappointed at the lackluster results.

Then came the huge progress fueled by supercomputers and AI at IBM Research. In 1997, the Deep Blue team trained their machine to beat the best chess masters. Researchers developed Watson from 2007 to 2011 to compete on Jeopardy! In fact, the Watson system was not all that much different from ChatGPT. To vie against the Jeopardy champions, IBM needed to create an enormous database of knowledge, develop a natural language interface to access that information, retrieve many different possible answers to the Jeopardy clue, determine which answer with the highest probability, and ring the buzzer. However, following this success, IBM took a wrong turn and tried to forge a path forward with Watson Health, tackling the most difficult AI task imaginable. Fast forward to today and AI presents another major technological wake-up call for the company.

IBM has to contend with much-heralded AI players such as Google, Microsoft, and Apple. However, there are many others. One emerging competitor is David Ferrucci, who led the IBM Watson program. IBM named him an IBM Fellow in 2011. Ferrucci knew better than anyone that Rometty's "moon shot" was a bridge too far. He left IBM the following year after spending 18 years at IBM Research. Ferrucci had a noncompete contract with IBM that prevented working for another technology company[16]. He joined Bridgewater Associates, a hedge fund, enabling him to continue his work on AI as an incubation program. Chess Grandmaster Garry Kasparov gave a presentation on AI at Bridgewater in 2014.He talked with Ferrucci, who revealed that he left IBM because he wanted to do deeper analysis AI and thought IBM's Watson Health was the wrong direction. He wanted to know what questions were the right ones for AI as opposed to simply looking up facts[17]. Bridgewater would spin off his operation as an AI startup in 2015, called Elemental Cognition.

Unlike OpenAI, Ferrucci's focus was on enterprise AI applications. His company has released two chatbots (Cogent and Cora) that were set up and trained for specific industries (financial services and travel planning). An early commercial project created an AI travel assistant for the OneWorld consortium of airlines, enabling customers to plan their around-the-world trips. One of Elemental Cognition's early investors was Sam Palmisano[18]. With IBM, he had invested in stock buybacks and divested in IBM technology. With his own money, he is investing in leading edge computing technology.

As of 2023, IBM remained in the "Leaders" section of Gartner's Cloud AI Developer Magic Quadrant, along with Microsoft, Google, and Amazon Web Services. It is important to remember that the success of OpenAI had temporarily shifted the focus from business use of AI to individual applications. With the divestment of Watson Health, IBM focused on usable and affordable artificial intelligence applications. IBM embraced the GPT approach to training AI data. It replaces the brute-force method unsuccessfully tried in health care. To illustrate, the GPT-2 language model trained with 40 gigabytes of data while its follow-up GPT-3 model was able to effectively use 570 gigabytes of data. It "ingested" large volumes of recorded information (Wikipedia, digital books, large sections of the Internet) much in the way that the original Watson machine was trained to play Jeopardy! IBM has 400-500 people working on training models and machine learning techniques with plans to have over 1,000 AI consultants[19]. IBM has also sought to supercharge AI deep learning with hardware. IBM Research produced an AI-geared supercomputer called Vela. It uses a purpose-built processor chip for AI, the Artificial Intelligence Unit (AIU). The AIU is based on 5-nanometer circuit density and contains 23 billion transistors on a single chip.

In May 2023, IBM announced a re-launch of IBM's AI solutions as WatsonX. It includes an enterprise toolkit geared to train, test, and deploy AI applications. The focus will be on models trained for business applications such as supply chain, CRM (Customer Relationship Management), and cyber security. IBM CEO Arvind Krishna said that the new IBM AI is 100 times less costly than the original Watson model. Krishna went on to describe IBM's internal plans for AI. There will be a focus on replacing some back office roles with AI. IBM has 26,000 employees in such jobs and AI could handle up to 30% of those roles. That equates to 7,800 employees, about 3% of current the IBM workforce (this puts IBM's workforce at 260,000)[20]. IBM has also announced Code Assistant for IBM Z, a program for converting COBOL programs to Java. There are an estimated 800 billion lines of COBOL remaining, with 84% of mainframe accounts still running on it[21].

Can IBM still catch the AI wave? IBM has refocused almost entirely on AI and the hybrid cloud. This is clearly the right focus. The company has significant experience and technology to bring to bear. They have switched to cutting edge AI model development and targeted business areas that are ripe for AI. Watson Health is in the rear view mirror. The company has an enterprise focus, strong relationships with the largest companies, and more than enough consulting services to aid customers in the implementation. Krishna is realistic about the 2023 mania surrounding AI, commenting, "AI is nowhere as good as a 9-month human baby. You show them a dog three times, the fourth time they'll say dog.[22]"

On the other hand, there is a raft of competitors, small and large, that are also gearing up. Many have greater mindshare with AI and can move more rapidly than IBM. Demis Hassabis, a British chess prodigy, started Deepmind in 2010[23]. Following the lead set by IBM's Watson, the British company trained their systems on simpler Atari games before moving to chess. In 2016, they developed the AI learning functions to master the game of "Go" which has far more combinations than chess. They proceeded to attack and master the even more complex game StarCraft II[24]. Google acquired Deepmind in 2014. It is interesting to note the Google and Deepmind ventured into the hospitals of Britain's National Health Service. No longer addressing a specific mathematical game, they are struggling with challenges there.

Then, there is IBM's lack of leading edge cloud data centers. The Big Three players are adding supercomputers to their cloud centers to handle AI. IBM dropped out of supercomputers after Summit and Sierra. The last two IBM CEOs have divested key parts of the company that would have been very useful in this burgeoning market. That includes the Microelectronics and PC Server divisions.

---

## QUANTUM COMPUTING

IBM remains a leader in quantum computing. The technology is still in the future, perhaps far in the future. It would be unlikely to result in any significant revenue during Krishna's time at the IBM helm (though he has already breached the age 60 limit, so who knows?). The concept of a quantum computer is as difficult to explain as is the underlying theory of quantum mechanics. It is perhaps fitting that Richard Feynman provided some of the early guideposts to the idea. Like John von Neumann, he was always on the lookout for faster and faster computers to simulate the behavior of physics at the subatomic level. Quantum computers seek to harness the strange behavior at this level to provide a vastly more powerful computing machine. A traditional computer works in binary, with information bits having a state of zero or one. The corresponding unit of information in a quantum computer is the qubit, with has the probability of being in the one or zero state, as well as any state in between. This characteristic of the qubit offers near limitless states. For example, a quantum computer with just the capacity of ten qubits would be roughly equal to a traditional computer with 16,000 bits. A quantum computer with 500 qubits might have the capacity of a traditional computer with more bits than there are atoms in the universe.

The promise of such a computing device does not end with the stark difference between a bit and a qubit. The quantum computer also has the potential to run vast numbers of computations in parallel. This introduces the concept of "quantum supremacy," the idea that such a processor will be able to solve problems that a traditional computer could not solve, regardless of the time allotted. A quantum computer could address problems that existing computers, even the fastest supercomputers, could not touch. Those involve deep scientific problems in biology, weather forecasting, advanced machine learning, and certainly Feynman's calculations of quantum mechanics. In addition, a quantum computer would likely be able to break any cryptographic scheme, a significant security exposure.

The hitch in the concept of the quantum computer is the sheer difficulty in building and operating such a device. To get usable qubits, temperatures need to be close to absolute zero. The device needs to be completely isolated from any external effects that might alter qubit probabilities and render the computation useless. Skeptics argue that the engineering problems might never be solved. Nevertheless, extensive research continues. The major participants are the same companies vying for the lead in cloud computing—Google, Microsoft, Amazon, and IBM. In 2019, Google finished its Sycamore quantum computer, a device with 53 qubits. Google claimed Sycamore could perform computations three million times faster than the Summit supercomputer, at the time the world's fastest. Google also claimed that a task that ran in 200 seconds on Sycamore would take 10,000 years on a supercomputer. Quantum experts have questioned this conclusion.

Krishna remains confident that IBM's quantum technology will make it out of IBM Research. The company is following an accelerated roadmap that is cycling through a host of bird names[25]. The first quantum device, Canary, contained five qubits. IBM Research progressed with rising qubit capacity through Albatross (16), Penguin (20), Falcon (27), Eagle (127), and Osprey (433). In 2023, IBM announced the first 1,000-qubit quantum machine, the Condor at 1,121 qubits, with Kookaburra to follow in 2025 with 4,158 qubits. Quantum System One was the company's first commercial quantum computer. Based on the Eagle prototype, it arrived in 2019. A number of research labs (CERN, Fermilab, Argonne National Lab, and Lawrence Livermore Lab) accessed the system remotely. The Cleveland Clinic installed Quantum System One for biomedical research. IBM plans to have eight Quantum computing centers operational by the end of 2024. Meanwhile, Research continues to increase quantum capacities with the Blue Jay program expected to top the 2,000-qubit plateau in the 2033 time frame.

## THE FUTURE OF THE MAINFRAME

The classic computer, where the bit is represented by a transistor, remains central to IBM. The System/360 was wildly successful and its successors continue to this day. Though a much smaller part of the new IBM, the mainframe is central to the company's most important accounts and key to its hybrid cloud strategy. New generations of IBM mainframes have followed the POWER curve and have been debuting roughly every three years. IBM's mainframes switched from the zSeries branding in 2005 to simply "Z", standing for "zero downtime." The Z13, arriving in 2015, was particularly transformative. It featured circuits at the 22-nanometer density with four billion transistors on a single chip. Memory had grown to ten terabytes (trillion bytes), a factor of nearly 80 million times the memory on the original System/360. Taking advantage of Denard's Law of Scaling, the Z13 had the same power requirements as its immediate predecessor.

In April 2022, IBM announced the Z16 mainframe[26]. It featured the Telum microprocessor designed by IBM engineers at seven-nanometer density and produced by Samsung. With up to 9.2 billion transistors per chip, it provided speeds up to 200 teraflops (each teraflop runs at a trillion floating-point operations per second). This represented a 77 percent performance boost over the Z15. The system remains backwards compatible with the 1964 System/360. The Z16 design focuses on handling complex AI application workloads. IBM believes that cloud and AI workloads will force more supercomputer technologies onto the Z platform. IBM Research developed the first two-nanometer chips in 2021, containing 50 billion transistors[27]. It featured a layered "nanosheet" three-dimensional structure that could potentially keep Moore's Law going for some time

Over the past several years, the market for semiconductors completely changed, with the twin challenges of worldwide shortages and the erosion of U.S. technical supremacy. The CHIPS act provides $280 billion to rebuild U.S. semiconductor capabilities. The legislation aims to increase U.S. competitiveness in semiconductor research, design, and production. The legislation provides $170 billion to fund research and $53 billion to underwrite increased chip production.

In parallel with CHIPS' passage, leading semiconductor companies announced significant expansions. Micron announced a $100 billion investment in fabrication plants in New York State. The Taiwan Semiconductor Manufacturing Company (TSMC), the world leader, plans to invest $40 billion in Phoenix. Samsung will build a $17 billion chip foundry in Texas. Intel will invest $20 billion in a new facility in Columbus, Ohio. As for IBM, the company announced plans to commit $20 billion in semiconductors, quantum computers, and AI.

Arvind Krishna virtually eliminated the practice of stock buybacks. The company limited its share purchases to a total of under $1 billion over his first three years. There was growing clamor with the Biden administration to curb the practice. The Inflation Reduction Act of 2022 exacted a 1 percent tax on corporate stock buybacks. In addition, the Biden administration indicated that it favors companies that eschew buybacks for investment funds from the CHIPS act. More action on buybacks is likely. Yet, the volume of stock buybacks continued unabated in 2022. It was projected to surpass one trillion, compared to $500 billion in dividends[28]. The recent leaders in buybacks were Apple, Google, Facebook, and Microsoft. These were all companies with far more cash than they knew what to do with.

The heightened importance of semiconductor technology casts doubt on IBM's 2015 decision to sell its Microelectronics division to Global Foundries. It demonstrates the risks of ceding critical, core technology to another company. The ensuing legal struggles with Global Foundries simply made a bad decision worse. IBM has far less leverage over the nanometer technology that is central to the company's future ambitions. Its cloud computing competitors are investing, not divesting, in chip technology. Lou Gerstner viewed the company's microtechnology as essential and invested billions in it. From the days of Ralph Palmer, memory and logic technology was a driving technological force.

Moore's Law, the near doubling of transistor densities every two years, is increasingly exposed. By 2018, a gap developed, with density starting to fall short of doubling. Three factors drove this change. First, the market for semiconductors broadened, with vastly more applications that do not require leading-edge densities. Second, as chip densities increased, the cost to build and operate fabrication plants grew dramatically. At present, there remain only a handful of foundries still making leading-edge chips. Global Foundries sought to abdicate its commitment to IBM on the densest chips and focus on the consumer chip market. As to transistor density, sheer physics is beginning to come into play. The current width between transistors on the leading chips is just two nanometers. The silicon atom is only 0.2 nanometers wide, so we are now at the point where the transistor gap is just ten times the width of the substrate atoms[29].

The potential end of the Moore and Dennard laws will pose a challenge for continued growth in computer performance. At the same time, there have been reduced performance gains with other techniques such as parallel processors, pipelining, and instruction branch prediction. When architects combined both hardware and software techniques, it was not unusual to see 50 percent performance gains year in and year out. Hennessy and Patterson (the Stanford RISC researchers) in *A New Golden Age for Computer Architecture* argue that this level of annual gain is unlikely in the future unless completely new computer architectures come to the rescue[30]. The potential candidates include quantum computers or greater use of graphics accelerator chips that have found a place in supercomputers.

# EPILOGUE, A SMALLER IBM

As of this writing, IBM is a much smaller company, one that is decidedly less physical in nature. Most of the manufacturing plants and labs are gone, along with the people who staffed them. The switch to services and, lately, work from home has resulted in many of the buildings that the Watsons created being sold or razed.

IBM Atlanta faced considerable "rearranging of the deck chairs" over the years. The company whittled down the original mandate of the General Systems Division with each reorganization. Lew Gray and the Southwest Marketing division moved to New York in 1988. This marked the end of the "Big Chicken" era in Atlanta after a 20-year run. As for the Big Chicken itself, it is still there and as popular as ever. For a time, IBM in Atlanta continued to grow. The Hillside headquarters building was added next to the Lakeside building in 1987. The 50-story IBM Tower in midtown opened in 1990. The year was the high point for IBM presence in Atlanta, with the workforce peaking at 6,300. It was a fleeting moment as the Akers downturn led to significant reductions. By 1995, the Atlanta IBM population had dropped to 3,500.

IBM moved out of the Hillside and Lakeside buildings in 2012. The 5,000 employees working there dispersed to working from home or reductions in force. The Atlanta Public Schools purchased the headquarters property. They needed a new high school in the north part of town. They demolished the entire Hillside building. The 11-story structure, including my former office, came down in a matter of seconds after explosive charges were detonated. The companion 11-story Lakeside building, suspended over a small lake, became North Atlanta High School. It opened in the fall of 2013 with 2,400 students. At $147 million, it was one of the most expensive high school construction projects ever. The average U.S. high school cost $38 million[1]. The 50-story IBM Tower, where IBM once occupied half of the floors, reverted to its original name of One Atlantic Center.

In 2016, IBM announced its intention to sell the entire Rochester, Minnesota site. It was the end of an era that began with the production of punched-card sorters and collators in 1956. The "Big Blue Zoo" facility had grown through the System/3 era. It had survived the loss of the integrated General Systems Division and soared to new heights with the AS/400. Production peaked in 1991 with 8,100 IBMers in 34 buildings[2], contributing $14 billion to the IBM Company. Rochester managed through the Akers down years though 2,100 employees did not make it. Rochester shifted gears with the RISC era in 1995 and by 1998 was shipping an AS/400 every 12 minutes. It achieved this record despite the further loss of autonomy to Austin, Poughkeepsie, and IBM Corporate. The site's experience with RISC microprocessors and rack-mounted systems led to a supercomputer mission, with roles in Deep Blue, Blue Pacific, Blue Gene, Roadrunner, Summit, and Sierra.

In 2008, the last time IBM reported official numbers Rochester's workforce was down to 4,200. The AS/400 had been through eServer iSeries, "Server i5, System i, and IBM i before becoming part of the Power Systems family. In 2013, manufacturing moved to Mexico and that space was mothballed. The workforce plunged to 2,800. IBM completed the sale of the 500-acre site in 2018. They leased eight of the 34 buildings for continuing operations. Those included Power systems (the AS/400), cloud computing, a large data center that became part of the Kyndryl spin-off, and a section of Watson Health, soon to be sold. In 2004, IBM provided a Blue Gene supercomputer to the Mayo Clinic for patient data analysis[3]. In 2014, the Mayo Clinic began piloting the Watson Health AI solution. In 2019, the clinic selected the Google Cloud, signing a ten-year agreement for cloud computing, data storage, machine learning, and AI. It was the end of a long relationship between IBM and Mayo[4]. In 1966, IBM and the Mayo Clinic each had 3,600 employees. By 2020, IBM had 2,500 while the clinic had grown to 35,000.

The AS/400 lives on though the name has changed many times. The system is now running on Power Systems hardware. Releases of POWER9 architecture (at 14-nanometer density) and POWER10 (at 7-nanometer density) were delayed due to the sale of IBM Microelectronics to Global Foundries and IBM's subsequent move to Samsung for chip production. The FS architecture, with its high-level machine interface, ease of application development, integrated databases, and object orientation, remains. Over the years, Rochester has added a host of new technologies, including Java, C++, dot Net, and SQL. And, what about RPG? With well over 750,000 systems[5] shipped using RPG, it should be high on the list of programming languages. Alas, according to the TIOBE ("The Importance of Being Earnest") index that rates programming languages by Internet search hits, RPG does not break the top 50[6]. This list is predictably headed by the likes of Python, C++, Java, and Visual Basic, though Assembler sneaks in at #8. The platform and its funny programming language are still secrets after all these years.

A sizeable inventory of AS/400 persists and continues to be upgraded by faster Power System processors. Timothy Prickett Morgan estimates that there are 160,000 customers running on Power Systems, the vast majority of them running as AS/400s. John Rockwell identified at least 293 companies of the Fortune 500 that still operate on the platform. These include AmerisourceBergen, Baxter, Best Buy, Cisco, Costco, DR Horton, Dick's, eBay, Ford, Harley, Home Depot, Kohl's, Lennar, Levi Strauss, Lowe's, McDonald's, McKesson, Nordstrom, Office Depot, Rite Aid, Ryder, Sysco, Target, Tesla, Texas Instruments, TJX, UnitedHealth, Wal-Mart, Whirlpool, Williams-Sonoma, Xcel Energy, and Hertz[7].

Costco has green-screen terminals throughout its stores that are used for inventory inquiries. State Farm and American General Finance continue to run large networks of AS/400 systems in their local offices. McKesson has had a long history with IBM. It installed

IBM unit record equipment in the 1930s for BICARSA applications. E.D. Smith, the jelly and jam manufacturer in Winona, Ontario, and one of the original customers of the System/38 Remote Test Service, runs its enterprise on a far more powerful Power Systems model. Its U.S. competitor, the J.M. Smucker Company, is also a long-time AS/400 user.

Microsoft was finally able to replace its corporate AS/400 systems in 1999, migrating to a SAP R/3 enterprise suite running on Windows NT servers. Bill Gates and Steve Ballmer wanted to move NT up the food chain and into IBM's midrange customer base. In the early 2000s, Microsoft tried to entice a move to their .NET platform running on NT. It was less than successful. As Microsoft had been an AS/400 customer for many years, it should have known how loyal the install base was. There was no need to change computers when you already had the most advanced architecture there was. Falling back to a "if you can't beat them, join them" strategy, Microsoft added POWER9 AS/400 servers to its Azure data centers in 2020, hoping to skim off accounts by moving applications to the cloud.

Endicott was the birthplace of C-T-R and IBM. It was the core of the company for many years. Starting with the original Bundy building in 1906, the sprawling site of Plant No. 1 grew to an enormous complex of 29 buildings and four million square feet of floor space. A long central walkway, called the Ivan Fredin Expressway after the Endicott site manager of the 1960s, connected the buildings. IBM's birthplace in Endicott and the surrounding plants and labs underwent drastic reductions or outright sale. At its height, the Endicott area, including the Federal Systems Division in Owego, and the Glendale Federal Systems lab had more than 17,000 IBM employees[8]. The long down slide began in 1993 when Gerstner sold the Owego and Glendale Federal Systems locations to Lockheed Martin.

IBM sold the entire 130-acre campus in 2002 to Huron Associates, a New York State-sponsored real estate company, for $65 million. At that time, fewer than 1,000 IBMers still worked at the site. In 2021, Huron sold the site to Phoenix Investors for $41 million[9]. By the following year, IBM reported only several hundred employees still in Endicott. The unoccupied buildings deteriorated without regular maintenance. Five of the buildings were slated for demolition in 2022. Watson's beloved Homestead was saved, becoming Traditions at the Glen, a wedding venue. As of 2023, there were perhaps 50 remaining employees in Endicott, all working from home[10].

The Boca Raton complex, the hurricane-proof design by Marcel Breuer, grew to four million square feet of space and peaked at 10,000 employees during the PC era. The workforce declined to 5,500 by 1988. The failure of OS/2 and loss of leadership in the PC market led Lou Gerstner to pull the plug on OS/2 in 1996, eliminate the lab, and lay off 1,300 people. Major operations moved to Raleigh, which had free space when IBM sold its networking division to Cisco. The 1,500 remaining employees in Boca moved to different locations in the area, and the site sold for $46 million. It sold again in 2018 for $170 million and once more in 2021 for $320 million. By this time, extensive office and residential development surrounded the massive main concrete buildings.[11].

The Boulder site peaked at 5,000 employees working in 24 buildings. The tape mission moved to Tucson in 1978 in part because of the continued loss of engineers to nearby Storage Technology and its many high-tech offspring companies. The three-year joint venture between IBM's Pennant printing division and Ricoh ended in 2010. IBM's infrastructure services continued to grow at the site, including the addition of the world's largest data center. With the 2022 split-up of the IBM, the spin-off company Kyndryl took over the campus, occupying 75 percent of the buildings, with IBM, Ricoh, and Lexmark using the rest.

San Jose, the birthplace of the disk drive, became the headquarters of ADSTAR, one of John Akers' semi-independent divisions. The rebadged operation struggled. The defection of so many engineers did not help, but some problems were self-inflicted. A series of general managers came and went. The sheer explosiveness and price pressure of the drive business, starting at the low end and moving up the value chain, was unrelenting. Hitachi bought the disk division in 2003. As various buildings on the site were either razed or re-missioned, a fight developed over the iconic Building 25 and its reflecting pool. Lowe's wanted to raze the site for an expanded building supply store and parking lot. A compromise had Lowe's building a smaller store while preserving Building 25 as a historic landmark. However, an unoccupied building is always a risk. In 2008, Building 25 burned to the ground[12]. Arson was suspected. Lowe's went ahead and built its big store but consented to one small change in recognition of the historical importance of the site. They decorated the store façade with the same punched-card pattern that had faced the IBM building.

The remainder of the site remained neglected for many years. It is now a small park surrounded by Target, Lowe's, apartments, and an In-n-Out restaurant. The remnant of the reflecting pool continues to deteriorate. IBM had no interest in the site but wanted to see some recognition of the original lab site at 99 Notre Dame. That nondescript low-slung building still sits there to this day, completely boarded up.

The Hudson Valley north of New York City had at one time 30,000 employees working at the Poughkeepsie, Kingston, and East Fishkill sites. The Kingston site, which went through a number of missions and never had a real identity, closed in 1995. East Fishkill became part of Global Foundries.

As for Poughkeepsie, though diminished from the glory years, it remains the mainframe center of operations for IBM. It was IBM's largest plant in 1985, with 12,500 employees. As of 2022, about 3,000 employees still worked at the site. In April of that year, IBM announced a renovation within one of the manufacturing buildings to produce mainframes and Power Systems servers. In 2019, Poughkeepsie opened the Quantum Computation Center, a first step in eventually building quantum computers[13]. The famous Building 701, built in 1953, is now Our Lady of Lourdes High School. This could change. The CHIPS act has the potential to breathe new semiconductor life into the Hudson Valley.

# FINAL COMMENTS

I BM is no longer the "national treasure" it once was. It might be a "national treasure emeritus" along the lines of Bell Labs. How did IBM get to this point? It was years of slow decline, chasing faulty strategies. Hardware became less important to the company. This is tough to digest for a company with "machines" in its name. Those "machine" businesses deemed too marginal included disk drives, PCs, PC servers, ThinkPads, low-end and high-end printers, and networking gear. The all-out switch to services, initially a savior, proved to be just another low-margin business. This would have mattered less if the company had not missed the seismic shift to cloud computing.

There are many "what ifs" in the company's history. What if Charles Flint had not supported Watson Sr. with his early and on-going investment in R & D? What if Tom Watson Jr. became an airline pilot instead of returning to IBM after the war? Would it have been IBM and the Seven Dwarfs, or would the company under Watson Sr. ridden punched cards to an early demise? What if Watson Jr. decided against betting the company on System/360? With Frank Cary's foresight, the company had an early and commanding lead in personal computers. What if Watts Humphrey won the argument to develop the essential PC software in-house? What if John Opel had taken a more measured approach to moving the company forward instead of abandoning all rental agreements and committing to a massive expansion? What if the IBM board of directors had acted sooner, perhaps replacing John Akers with George Conrades?

Looking to the future, there are more questions than answers for IBM. Does IBM have the resources to compete with deep-pocketed competitors in cloud computing? Will cloud computing turn out to be just another low-margin business? Is there a viable high-margin play for IBM? After so many rounds of layoffs and offshoring, is the IBM culture up to the today's challenges? Most critically, is Arvind Krishna the right person to lead the charge? The future requires dynamic leadership with winning strategies. Watson Jr. lived by the creed, "solve it, solve it quickly, solve it right or wrong." If it turned out to be wrong, you could go back and adjust, now armed with the knowledge of what didn't work. Of course, he was aided by an executive who would take the plan and run with it. Vin Learson would simply not accept failure.

The slow death by years of quarterly declines obscured IBM's need for drastic action. The current problems are equally as existential as those faced by John Akers, but there may not be the same pressure to act. Some actions seem obvious. There should be fewer financial tricks. IBM must focus on innovation, driven by IBM Research and its few remaining labs. The proof will be in market-leading products.

It is also possible that IBM is no longer a growth business. It may finally catch the cloud shift or the next one. Or, it may not. Such an outcome would not dim the brilliance of the company. Its history is storied and unparalleled. It is this past excellence that makes the current IBM shrink in comparison. The company is well over a century old. Watson Sr. felt that his company would go on forever[1]. Watson Jr. was probably more realistic. He would die not knowing if Lou Gerstner managed to turn the company around. Nothing lasts forever. Or, does it? The Big Chicken is 59 years old and still has legs.

# ACKNOWLEDGMENTS

Through the course of putting this book together, I was indebted to the many books and articles on IBM that came before. A special thanks to James Cortada, who suggested I tackle this project. . His exhaustive history, *IBM: The Rise and Fall and Reinvention of a Global Icon*, was a key source for this project.

A primary and trusted resource was the oral histories of those who were there, the prime movers of the IBM Company over its long history. Invaluable were the recollections of Glenn Henry, David Schleicher, Steve Dunwell, Fred Brooks, Ralph Mork, Chuck Branscomb, John Atanasoff, Karl Ganzhorn, Max Paley, and many, many others. Glenn Henry reviewed the initial manuscript and made a number of corrections. A giant at the center of IBM's success, T. Vincent Learson, left very little. I relied on the many who worked with him to tell his story.

In my coverage of early computing, I leaned on several excellent sources. Walter Isaacson's *The Innovators* provided extensive portraits of computing pioneers such as Charles Babbage, Ada Lovelace, and the Bell Labs trio of Shockley, Bardeen, and Brattain. Kenneth Flamm's *Creating the Computer* was another comprehensive history that provided significant coverage and insights into computing history.

For the Watsons, I used Kevin Maney's *The Maverick and His Machine* as well as the Thomas Watson Jr. biography *Father and Son: My Life at IBM*. An insider's look that focused on IBM during the critical 1950s was *Computer: Bit Slices from a Life*, by Herb Grosch. Frank Da Cruz documented the long and fruitful collaboration between IBM and Columbia University.

A group of IBM engineers who were there wrote the definitive guides to IBM's early computer development. *IBM's Early Computers*, by Charles Bashe, Lyle Johnson, John Palmer, and Emerson Pugh, covers IBM's punched-card machines, the 700 and 7000 series, and key additional technologies such as disk, tape, and memories. Pugh, Johnson, and Palmer then tackled IBM's "bet your company" project with *IBM's 360 and Early 370 Systems*. Speaking of firsthand sources, it does not get more firsthand than Bob Evans's *The Genesis of the Mainframe*. He was at the absolute center of the System/360 revolution, and so much else.

I had a number of excellent sources in examining the Stretch, FS, and Projects X and Y supercomputer programs. Werner Buchholtz's *Planning a Computer System: Project Stretch* was the original blueprint for Stretch. The academic work by Mark Smotherman was also essential. I found the famous Memorandum No. 125 by John Sowa essential in understanding what went wrong with FS.

A main goal of this book was to shine a light on that lesser-known part of IBM—its small and midrange systems. While there is a complete library of books on "big iron" IBM, the pickings are slim in the IBM outside the New York corridor. Frank Soltis wrote a firsthand account in *Fortress Rochester*. Another key source was *The Silverlake Project*, by Roy Bauer, Emilio Collar, and Victor Tang. Two journalists, Timothy Prickett Morgan and Alex Woodie, provided in-depth coverage of all things Rochester for many years.

Looking at more recent IBM history, I found Paul Carroll's *Big Blues: The Unmaking of IBM* instructive in the PC wars and the lead-up to the collapse under John Akers. I am indebted to Peter Greulich for his insightful analysis of the financial manipulations of recent IBM. Steve Lohr, the New York Times technology and business reporter, provided excellent coverage of the years after Watson, Jr.

Finally and most prominently, I am very much indebted to IBM and their Archive team who assisted in the research and graciously supplied a multitude of wonderful photographs.

# ABOUT THE AUTHOR

After graduating from Dartmouth College, Shaffer spent 42 years with IBM as a systems engineer, technical "Top Gun", systems engineering and marketing manager, and AS/400 product manager. He retired from the company in 2015.

Bill lives in Boulder, Colorado, with his wife, Cynthia. Their two sons live in New York City and Seattle. Both have computing careers, with one spending ten years with IBM and the other working at SpaceX. This is Shaffer's second book. Hi first book -- *Shifting Gears: One Family's Journey through the Automobile Age* -- is an automotive history as seen through the perspective of a small farming town in the Midwest.

# NOTES

## COUNTING BEADS AND SPINNING WHEELS

1. Richard P. Feynman and Ralph Leighton, *"Surely You're Joking, Mr. Feynman"* (W. W. Norton, 2018), 226-227.
2. "The Abacus vs. the Electric Calculator", Toronto Metropolitan University, retrieved February 6, 2023.
3. Motoko Rich, "The Right Answer? 8,186,699,633,530,061 (An Abacus Makes It Look Almost Easy", New York Times, August 21, 2019.
4. Guy Mourlevat, *Les machines arithmétiques de Blaise Pascal* (Clermont-Ferrand: La Française d'Edition et d'Imprimerie, 1988).
5. Mourlevat, *Les machines arithmétiques de Blaise Pascal*.
6. James Tam, "Early Mechanical Computers", A History of Computing, slides from course CPSC409, University of Calgary, Fall 2020.
7. Kathick Nambi, "The Inventor of the First Mechanical Calculator", (Medium, June 13, 2020).
8. Herman Goldstine, *The Computer: From Pascal to Von Neumann* (Princeton University Press, 1972), 9.
9. William Aspray, editor, *Computing Before Computers* (Iowa State University Press, 1990), 28-30.
10. Martin Campbell-Kelly, William Aspray, Nathan Ensmenger, and Jeffrey Yost *Computer: A History of the Information Machine* (Westview Press, 2014), 27.
11. Valery Monnier, "Les Arithmomètres de Thomas de Colmar & Payen", arithmome-tre.org, 2012.
12. "Calculator's Calculators", Time Magazine, October 25, 1943.

## BABBAGE AND "STEAM" COMPUTING

1. Denis Roegel, "The great logarithmic and trigonometric tables of the French Cadastre: a preliminary investigation", *Research Report*, 2010.
2. Raymond Flood, "Babbage and Lovelace", Lecture at Gresham College, January 19, 2016.
3. Walter Isaacson, *The Innovators* (Simon and Shuster, 2014), 20.
4. Roegel, "The great logarithmic and trigonometric tables of the French Cadastre: a preliminary investigation".
5. Lorraine Daston, "Enlightenment Calculations", *Critical Inquiry* (University of Chicago Press, Autumn 1994), 182-202.
6. Doron D. Swade, "Calculation and Tabulation in the Nineteenth Century" (PhD thesis, University College London, 2003).
7. Charles Babbage, *Passages from the Life of a Philosopher* (Longman, Green, Longman, Roberts & Green, 1864), 42.
8. Worth approximately 280,000 British pounds in 2024.
9. Stan Augarten, *Bit by Bit: An Illustrated History of Computers* (Ticknor and Fields, 1984), 50.

10. Ernest Goodall, "The Jacquard Loom in 1820", (*Journal of the Royal Society of Arts*, vol. 108, no. 545, April, 1960), 374-376..

11. Pamela McCorduck, *Machines Who Think* (A. K. Peters, 2004), p. 29.

12. James Gleick, *The Information: A History, A Theory, A Flood* (Pantheon Books, 2011), 163.

13. Swade, "Calculation and Tabulation in the Nineteenth Century."

14. Uta C. Merzbach, "Georg Scheutz and the First Printing Calculator" (Smithsonian Studies in History and Technology, Number 36, Smithsonian Institution Press, 1977).

15. Alexander Arbel, editor, *Routes to the Information Revolution* (Cambridge Scholars Publishing, 2019), 47.

16. Babbage, *Passages From the Life*, 67.

## HOLLERITH AND DATA PROCESSING

1. Keith S. Reid-Green, "The History of Census Tabulation" (Scientific American, February, 1989).

2. Martin Campbell-Kelly, William Aspray, Nathan Ensmenger, and Jeffrey Yost, *Computer: A History of the Information Machine* (Westview Press, 2014), 13.

3. Leon E. Truesdell, *The Development of Punch Card Tabulation in the Bureau of the Census* (Bureau of the Census, 1965), 10.

4. Truesdale, *The Development of Punched Card Tabulation*, 13.

5. Virginia Hollerith, "Biographical Sketch of Herman Hollerith", (University of Chicago Press, Vol. 62, No. 1, Spring 1971).

6. Truesdale, *The Development of Punched Card Tabulation*, 39.

7. Truesdale, *The Development of Punched Card Tabulation*, 46.

8. Truesdale, *The Development of Punched Card Tabulation*, 40-43.

9. Truesdale, *The Development of Punched Card Tabulation*, 56.

10. Lars Heide, *Punched Card Systems and the Early Information Explosion* (Johns Hopkins University Press, 2009), 24.

11. George E. Biles, Alfred A. Bolton, and Bernadette M. Dire, "Herman Hollerith: His 100 Year Legacy" (*Academy of Management Proceedings*, Vol. 1998, No. 1).

12. Heide, Punched Card Systems, 58.

13. Arthur L. Norberg, "High Technology Calculation in the Early 20th Century: Punched Card Machinery in Business and Government", *Technology and Culture* (Johns Hopkins University Press, October, 1990).

14. Byron Paul Mobley, "The ingenuity of common workmen: and the invention of the computer" (Iowa State University Capstones, Theses, and Dissertations, 2001), 88.

15. Truesdale, *The Development of Punched Card Tabulation*, 118.

16. Norberg, "High Technology Calculation in the Early 20th Century: Punched Card Machinery in Business and Government".

## CHARLES FLINT AND HIS TRUSTS

1. "Antonio Lopez de Santa Ana", Wikipedia, retrieved February 6, 2023.

2. Cheril Vernon, "Chewing Gum", Ever Wonder, August 30, 2017.

3. Robert Sobel, *IBM: Colossus in Transition* (Bantam Books, 1983), 4-12.

4. Charles R. Flint, *Memories of an Active Life* (G. P. Putnam and Sons, 1923), 242.

5. Flint, *Memories of an Active Life*, 289-296.

6. *James T. Cortada, IBM: The Rise and Fall and Reinvention of a Global Icon* (The MIT Press, 2019), 10.
7. Flint, *Memories of an Active Life*, 308.
8. Martin Campbell-Kelly, William Aspray, Nathan Ensmenger, and Jeffrey Yost, *Computer: A History of the Information Machine* (Westview Press, 2014), 36.
9. Flint, *Memories of an Active Life*, 312.
10. "Tabulating Concerns Unite", The New York Times, June 10, 1911.

## THOMAS WATSON, SR. AND NCR

1. Kevin Maney, *The Maverick and His Machine* (John Wiley and Sons, 2003), 16.
2. Thomas J. Watson, Jr. and Peter Petre, *Father, Son, and Co*, (Bantam, 1990), 49.
3. Gerald Carson, "The Machine that kept them honest", American Heritage, Vol. 17, Issue 5, August, 1966.
4. Robert Sobel, *IBM: Colossus in Transition* (Bantam Books, 1981), 35.
5. William Rodgers, *Think: A Biography of the Watsons and IBM* (Stein and Day, 1969), 41.
6. Martin Campbell-Kelly, William Aspray, Nathan Ensmenger, and Jeffrey Yost, *Computer: A History of the Information Machine* (Westview Press, 2014), 31.
7. Rodgers, *Think*, 62.

## THE ODD TRUST OF COMPUTING-TABULATING-RECORDING

1. William Rodgers, *Think :A Biography of the Watsons and* IBM (New American Library, 1969), 70.
2. Mobley, The Ingenuity of Common Workmen, 89.
3. Thomas J. Watson, Jr., A Business and its Beliefs (McGraw-Hill, 2003), 36.
4. Mark Bernstein , *Grand Eccentrics: Turning the Century: Dayton and the Inventing of America* (Orange Frazer, Press, 1996), 87.
5. William C. Shaffer, *Shifting Gears: One Family's Journey Through the Automobile Age* (self-published 2021), 120.
6. J. M. Hiznay, "Interview with Clair Lake", 1954, Box A-25-3, IBM Archives.
7. Kevin Maney, *The Maverick and His Machine* (John Wiley and Sons, 2003). 85.
8. Lars Heide, *Punched Card Systems and the Early Information Explosion* (Johns Hopkins University Press, 2009), 109.
9. Figures on IBM revenue, income, and workforce are pulled directly from the IBM archives at ibm.com.
10. Georgi Dalakov, "Gustav Tauschek", Computer Timeline, retrieved from http://www.computer-timeline.com/timeline/gustav-tauschek on February 6, 2023. .
11. Walter D. Jones, "Personal Observations and Comments Concerning the History and Development of IBM: The Perspective of Walter D. Jones", from the compilation by Jeffrey R. Yost, *The IBM Century: Creating the IT Revolution* (IEEE Computer Society, 2011), 39-52. Jones was IBM's #1 salesman in 1915. He moved to Paris in 1929 as IBM's European General Manager.
12. "Songs of the IBM", IBM Company, 1937, 7.

# IBM AND THE DEPRESSION

1. "The Dirty Secret About the 1929 Stock Market Crash" (The Conservative Investor, Seeking Alpha, April 14, 2013).

2. Jeffrey R. Yost, editor, *The IBM Century: Creating the IT Revolution* (IEEE Computer Society, 2011), 95. This observation from the essay "Pioneering: On the Frontier of Electronic Data Processing, a Personal Memoir" by James W. Birkenstock.

3. Gerald Breckenridge, "Salesman No. 1", Saturday Evening Post, May 24, 1941.

4. Martin Campbell-Kelly, William Aspray, Nathan Ensmenger, and Jeffrey Yost, *Computer: A History of the Information Machine* (Westview Press, 2014), 38.

5. "Ken Shirriff, "1950's tax preparation: plugboard programming with an IBM 403 Accounting Machine", retrieved at righto.com on February 6, 2023.

6. Shirriff, "1950's tax preparation: plugboard programming with an IBM 403 Accounting Machine".

7. Paul Breedveld, "Rise and Fall of the Mechanical Calculator", Delft University of Technology.

8. Bernard Cohen, *Howard Aiken, Portrait of a Computer Pioneer* (The MIT Press, 1999), 88-89.

9. Charles J. Bashe, Lyle R. Johnson, John H. Palmer, and Emerson W. Pugh, IBM's Early Computers (The MIT Press, 1986) p. 15.

10. Henry S. Tropp, "Interview of Arthur Halsey Dickinson", March 8, 1973, Smithsonian Museum of American History.

11. William Rodgers, *Think: A Biography of the Watsons and IBM* (Stein and Day, 1969), 135.

12. Frank Da Cruz, "Columbia University Computing History", retrieved from Columbia.edu/computinghistory on February 6, 2023.

13. Ralph Watson MeElvenny and Marc Wortman, *The Greatest Capitalist Who Ever Lived: Tom Watson, Jr. and the Epic Story of How IBM Created the Digital Age* (Public Affairs, 2023), 175.

14. Rodgers, *Think: A Biography of the Watsons and IBM* (Stein and Day, 1969), 146-148.

15. William Aspray, "Oral History of John McPherson", conducted on April 29 and May 12, 1992 in Short Hills, New Jersey.

16. Rodgers, *Think: A Biography of the Watsons and IBM* (Stein and Day, 1969), 148-149.

17. "Columbia Alumni News", vol. XXIII, no. 11, December 11, 1931.

18. Larry Saphire, "Wallace Eckert interview", for the Technical History Project, interviewed in July, 1967 at the New York City Watson Lab.

19. Martin C. Gutzwiller, "Wallace Eckert, Computers, and the Nautical Almanac Office", U. S. Naval Observatory, 1999.

20. Frank da Cruz, "Interconnected Punched Card Equipment: The First Automated Scientific Calculations", Columbia University History, 2004.

21. Wallace Eckert, *Punched Card Methods in Scientific Computing*, (Thomas J. Watson Astronomical Computing Bureau, Columbia University, 1940).

# IBM ACCELERATES TO THE LEAD

1. Kevin Maney, *The Maverick and His Machin,* (John Wiley and Sons, 2003), 159.
2. William Rodgers, *Think: A Biography of the Watsons and IBM* (New American Library, 1969), 138.
3. Rodgers, *Think: A Biography of the Watsons and IBM,* 127-128.
4. "People", Time Magazine, July 15, 1946.
5. Ralph Watson MeElvenny and Marc Wortman, *The Greatest Capitalist Who Ever Lived: Tom Watson, Jr. and the Epic Story of How IBM Created the Digital Age* (Public Affairs, 2023), 88.
6. Ralph Watson MeElvenny and Marc Wortman, *The Greatest Capitalist Who Ever Lived,* 85.
7. Richard Tedlow, *The Watson Dynasty: The Fiery and Troubled Reign of IBM's Founding Father and Son* (Harper Business, 2003), 5.
8. James T. Cortada, *IBM: The Rise and Fall and Reinvention of a Global Icon* (The MIT Press, 2019), 101.
9. Cortada, 101.
10. "International Business Machines v. United States", from caselaw.findlaw.com, retrieved October 18, 2021.
11. Martin Campbell-Kelly, William Aspray, Nathan Ensmenger, and Jeffrey Yost, *Computer: A History of the Information Machine* (Westview Press, 2014), 21.
12. Saul Engelbourg, *International Business Machines: A Business History* (Arno Press, 1970), 316.
13. William Rodgers, *Think: A Biography of the Watsons and IBM* (Stein and Day, 1969), 84.

# VANNEVAR BUSH AND AN ANALOG INCURSION

1. Walter Isaacson, *The Innovators* (Simon and Shuster, 2014), 37.
2. Jimmy Soni and Rob Goodman, *A Mind At Play* (Simon and Schuster, 2017), 83.
3. "The Modern History of Computing", Stanford Encyclopedia of Philosophy, retrieved from plato,Stanford.edu on October 5, 2023.
4. Jerome B. Wiesner, "Vannevar Bush: A Biographical Memoir", National Academy of Sciences, 1979.
5. Wiesner, "Vannevar Bush".
6. Soni and Goodman, *A Mind At Play,* 27-28.
7. G. Pascal Zachary, *Endless Frontier: Vannevar Bush, Engineer of the American Century* (Free Press, 1997), 49-55.
8. Jimmy Soni and Rob Goodman, *A Mind At Play,* 30.
9. John G. Brainerd, "Genesis of the ENIAC, Technology and Culture, Vol. 17, No. 3 (July, 1976), 482-488.
10. "UCLA's Bush Analyzer Retires to Smithsonian", Computerworld, January 9, 1978.
11. "Electromechanical and Analog Computing ", Carnegie Mellon University, School of Computer Science, 2017.
12. Tim Robinson, "The Meccano Set Computers", IEEE Control Systems Magazine, June, 2005.
13. Maurice Wilkes, *Memoirs of a Computer Pioneer* (MIT Press, 1985), 25.
14. Wiesner, "Vannevar Bush".

15. Claude Shannon, "A Symbolic Analysis of Relay and Switching Circuits", MS thesis, MIT, 1940.
16. Paul E. Ceruzzi, *Reckoners: The Prehistory of the Digital Computer, From Relays to the Store Program Computer* (Greenwood Press, 1983),78-98.
17. Paul Ceruzzi, "Crossing the Divide: Architectural Issues and the Emergence of the Stored Program Computer, 1935-1955" (*IEEE Annals of the History of Computing*, vol.19, no. 1, 1997).

# THE FIRST COMPUTER

1. John V. Atanasoff, "The Forces That Led to the Design of the Atanasoff-Berry Computer", Lecture at Iowa State University at Ames on November 11, 1980.)
2. John V. Atanasoff, "Advent of Electronic Digital Computing", Annals of the History of Computing, 1983.
3. Clark R. Mollenhoff, *Atanasoff: Forgotten Father of the Computer* (Iowa State University Press, 1988), 21.
4. Mollenhoff, *Atanasoff*, 21.
5. Byron Paul Mobley, "The ingenuity of common workmen: and the invention of the computer", (Iowa State University Capstones, Theses, and Dissertations, 2001), 49.
6. Jimmy Soni and Rob Goodman, *A Mind At Play* (Simon and Schuster, 2017), 83.
7. Mollenhoff, *Atanasoff*, 26.
8. Mobley, *The Ingenuity*, 56.
9. Mollenhoff, *Atanasoff*, 29.
10. Mollenhoff, *Atanasoff*, 27-28.
11. Jane Smiley, *The Man Who Invented the Computer* (Doubleday, 2010), 64.
12. John V. Atanasoff, "Computing Machine for the Solution of Large Systems of Linear Algebraic Equations", Iowa State Digital Collections, retrieved October 31, 2021.
13. Mollenhoff, Atanasoff, 42.
14. Alice R. Burks and Arthur W. Burks, *The First Electronic Computer: The Atanasoff Story* (University of Michigan Press, 1988), 12.
15. Burks and Burks, *The First Electronic Computer: The Atanasoff Story*, 14.
16. Atanasoff, "Computing Machine".
17. John L. Gustafson, "An FPS Forerunner: The Atanasoff-Berry Computer", Ames Laboratory, 2002.
18. "Atanasoff: Father of the Computer", documentary film directed by Mila Aung-Thwin and Daniel Cross, Shaw Media, 2012.
19. Mobley, *The Ingenuity*, 168.
20. Mobley, *The Ingenuity*, 154.
21. Mollenhoff, *Atanasoff*, 52.
22. Smiley, *The Man Who Invented the Computer*, 77.
23. Paul E. Ceruzzi, *Reckoners: The Prehistory of the Digital Computer, From Relays to the Store Program Computer* (Greenwood Press, 1983), 92-93.
24. Nancy Stern, "Interview of John Mauchly", conducted on May 6, 1977 at Ambler, PA, Niels Bohr Library & Archives, American Institute of Physics, College Park, MD.
25. Stern, "Interview of John Mauchly".
26. Nancy Stern, "Interview of John Mauchly."
27. Smiley, *The Man Who Invented the Computer*, 79.

# WORLD WAR

1. William C. Shaffer, *Shifting Gears: One Family's Journey Through the Automobile Age* (self-published, 2021), 158.
2. James T. Cortada, *IBM: The Rise and Fall and Reinvention of a Global Icon* (The MIT Press, 2019), 132.
3. Cortada, 125.
4. William Aspray, "Oral History of John McPherson", conducted on April 29 and May 12, 1992 in Short Hills, New Jersey.
5. Emerson W. Pugh, *Building IBM: Shaping An Industry and Its Technology* (MIT Press, 1995), 93.
6. Larry Saphire, "Wallace Eckert interview", for the Technical History Project, interviewed in July, 1967 at the New York City Watson Lab.

# ENIAC: "THE GIANT BRAIN"

1. Herman Goldstine, *The Computer: From Pascal to Von Neumann* (Princeton University Press, 1972), 137.
2. Uta C. Merzbach, "Oral History: Irven Travis", interviewed May 9, 1969, Smithsonian Institute.
3. John G. Brainerd, "Genesis of the ENIAC", Technology and Culture, Vol. 17, No. 3 (July, 1976), pp. 482-488.
4. V. R. Cardozier, Colleges and Universities in World War II, (Praeger Publishers, 1993), 178.
5. Jane Smiley, *The Man Who Invented the Computer*, (Doubleday, 2010), 81.
6. Clark R. Mollenhoff, *Atanasoff: Forgotten Father of the Computer*, (Iowa State University Press, 1988), 255.
7. Scott McCartney, *ENIAC: The Triumphs and Tragedies of the World's First Computer*, (Walker and Company, 1999), 49.
8. McCartney, ENIAC, 61.
9. Byron Paul Mobley, "The ingenuity of common workmen: and the invention of the computer", (Iowa State University Capstones, Theses, and Dissertations, 2001), 177.
10. Nancy Stern, "An Interview with Arthur W. and Alice R. Burks", June 20, 1980, at Ann Arbor, Michigan
11. T.A. Heppenhelmer, "How von Neumann Showed the Way" (American Heritage of Invention and Technology, 1990).
12. Thomas J. Bergin, editor, "50 Years of Army Computing: From ENIAC to MSRC", Army Research Laboratory, 2000.
13. Mollenhoff, *Atanasoff*, 65.
14. Heppenhelmer, "How von Neumann Showed the Way".
15. Goldstine, *The Computer*, 170.
16. "ENIAC Display", Computer Science and Engineering Department, University of Michigan, retrieved from cse.engin.umich.edu on February 5, 2023.
17. E.C. Stevenson and I. Getting, "A Vacuum Tube Circuit for Scaling Down Counting Rates", Review of Scientific Instruments, November, 1937.
18. Burks, Arthur W., "From ENIAC to the Stored Program Computer: Two Revolutions in Computers", Logic of Computers Group Technical Report No. 210, January, 1978.

19. William Lokke, "Early Computing and Its Impact on Lawrence Livermore National Laboratory", Department of Energy, March 19, 2007.
20. Paul Atkinson, *ENIAC versus Colossus and the early presentation of electronic computer* (University of Oxford, 2014).
21. Bergin,"50 Years of Army Computing: From ENIAC to MSRC".
22. Goldstine, *The Computer*, 226.
23. Goldstine, *The Computer*, 198.
24. Stan Augarten, *Bit by Bit: An Illustrated History of Computers*, (Ticknor and Fields, 1984), 99.
25. Kenneth Flamm, *Creating the Computer* (The Brookings Institute, 1988), 252.

# A MUCH BETTER ENIAC

1. Karl Kempf, "Electronic Computers within the Ordnance Corps: Historical Monograph from 1961" Aberdeen Proving Ground, November, 1961.
2. John von Neumann, "First Draft of a Report on the EDVAC", draft completed and distributed by Herman Goldstine starting June 30, 1945.
3. Herman Goldstine, *The Computer: From Pascal to Von Neumann* (Princeton University Press, 1972), 191.
4. Goldstine, *The Computer*, 198.
5. Nancy Stern, "An Interview with Arthur W. and Alice R. Burks", June 20, 1980, at Ann Arbor, Michigan.
6. Martin Campbell-Kelly and Michael Roy Williams, editors, *The Moore School Lectures: Theory and Techniques for Design of Electronic Digital Computers* (MIT Press, 1985), xvii.
7. Maurice Wilkes, *Memoirs of a Computer Pioneer* (MIT Press, 1985), 116-126.
8. Martin H. Weik, "A Survey of Domestic Electronic Digital Computing Systems", Aberdeen Proving Ground, December, 1955.
9. Maurice Wilkes, "Computers Then and Now" (Journal of the ACM, January, 1968).
10. Scott McCartney, *ENIAC: The Triumphs and Tragedies of the World's First Computer* (Walker and Company, 1999),149.
11. Richard R. Mertz, "An Interview with Julian Bigelow", conducted on January 20, 1971, Smithsonian National Museum of American History.

# THE MARK I

1. Bernard Cohen, *Howard Aiken, Portrait of a Computer Pioneer* (The MIT Press, 1999), 11-13.
2. Cohen, *Howard Aiken*, 66.
3. Cohen, *Howard Aiken*, 53.
4. Paul E. Ceruzzi, *Reckoners: The Prehistory of the Digital Computer, From Relays to the Store Program Computer* (Greenwood Press, 1983), 48.
5. Frank Da Cruz, "Columbia University Computing History", retrieved from Columbia.edu/computinghistory on February 6, 2023.
6. Cohen, *Howard Aiken*, 46.
7. Ceruzzi, *Reckoners*, 57.
8. Ceruzzi, *Reckoners*, 49.

9. Stan Augarten, *Bit by Bit: An Illustrated History of Computers*, (Ticknor and Fields, 1984), 104.
10. Ceruzzi, *Reckoners*, 60
11. J. A. N. Lee, "Howard Hathaway Aiken", IEEE Computer Society, 1995.
12. Bernard Cohen and Gregory W. Welch, editors, *Makin' Numbers: Howard Aiken and the Computer* (MIT Press, 1999), 59.
13. Ralph Watson McElvenny and Marc Wortman, *The Greatest Capitalist Who Ever Lived: Tom Watson Jr. and the Epic Story of How IBM Created the Digital Age* (Public Affairs, 2023), 358.
14. Cohen, *Makin' Numbers*, 60.
15. Kathleen Broome Williams, "Scientists in Uniform: The Harvard Computation Laboratory in World War II" (U.S. Naval War College Press, Vol. 52, No. 3, Summer 1999).
16. Maurice Wilkes, *Memoirs of a Computer Pioneer* (MIT Press, 1985), 174.
17. Wilkes, *Memoirs of a Computer Pioneer,* 175.
18. Broome, "Scientists in Uniform: The Harvard Computation Laboratory in World War II."
19. Ceruzzi, *Reckoners*, 66.
20. Herbert R. J. Grosch, *Computer: Bit Slices from a Life* (Underwood Books, 1991), 79.
21. Cohen and Welch, *Makin' Numbers*, 190.
22. Charles J. Bashe, Lyle R. Johnson, John H. Palmer, and Emerson W. Pugh, *IBM's Early Computers* (The MIT Press, 1986),30.
23. Charles Babbage, *Passages from the Life of a Philosopher* (Longman, Green, Longman, Roberts & Green, 1864), 450.
24. John Kobler, "You're Not Very Smart After All", The Saturday Evening Post, February 18, 1950.
25. Scott McCartney, *ENIAC: The Triumphs and Tragedies of the World's First Computer* (Walker and Company, 1999) 27.
26. Walter Isaacson, *The Innovators* (Simon and Shuster, 2014), 94.
27. "The Thinking Machine", Time Magazine, January 23, 1950.
28. Randy Alfred, "August 7, 1944: Harvard, IBM Dedicate Mark I Computer", Wired, August 7, 2008.
29. Nancy B. Stern, *From ENIAC to UNIVAC* (Digital Press, 1981), 111.
30. Bernard Cohen, "Howard H. Aiken's Mark I announced to the world the power of the computer", American Heritage Invention and Technology, Spring, 1999.
31. Herbert R. J. Grosch, *Computer: Bit Slices from a Life* (Underwood Books, 1991), 4.
32. Harry R. Lewis, "Computing's Cranky Pioneer", Harvard Magazine, May-June, 1999.

## COLOSSUS, THE SECRET DIGITAL CODE BREAKER

1. F. W. Winterbotham, *The Ultra Secret* (Harper and Row, 1974).
2. Colin B. Burke, "It Wasn't All Magic: The Early Struggle to Automate Cryptanalysis" (United States Cryptologic History, Special Series, Volume 6, National Security Agency, 2002), 130.
3. Alan M. Turing, "On Computable Numbers, with an Application to the Entscheidungsproblem", Proceedings of the London Mathematical Society, November 30, 1936.

4. Walter Isaacson, *The Innovators* (Simon and Shuster, 2014), 77.

5. Andrew Hodges, *Alan Turing: The Enigma*, (Princeton University Press, 1983), p. 218.

6. Isaacson, *The Innovators*, 79.

7. Linden Stead, "Tour of Bletchley Park", National Museum of Computing, Bletchley Park, retrieved February 6, 2023.

8. Stead, "Tour of Bletchley Park".

9. Paul Atkinson, *ENIAC versus Colossus and the early presentation of electronic computers* (University of Oxford, 2014).

10. Andy Clark, "Lorenz, Colossus, and the Dream of a Universal Machine for Cryptanalysis", The National Museum of Computing, 2015.

11. Atkinson," ENIAC versus Colossus".

12. Thomas H. Flowers, "The Design of Colossus", Annals of the History of Computing, 1983, p. 100.

13. Jane Smiley, *The Man Who Invented the Computer* (Doubleday, 2010), 105-106.

14. Tony Sale, "Other electronic code breaking machines", Tony Sale's Codes and Cyphers, retrieved January 18, 2022.

15. Smiley, *The Man Who Invent the Computer*, 107.

16. "The Modern History of Computing", Stanford Encyclopedia of Philosophy, retrieved from plato,Stanford.edu on October 5, 2023.

17. Alan Turing, *The Applications of Probability to Cryptography* (Hassel Street Press, 2021).

# BEYOND COLOSSUS

1. Paul E. Ceruzzi, *Reckoners: The Prehistory of the Digital Computer, From Relays to the Store Program Computer* (Greenwood Press, 1983), 25-44.

2. Raul Rojas, "The Z1: Architecture and Algorithms of Konrad Zuse's First Computer", Freie Universitat Berlin, June, 2014.

3. Jane Smiley, *The Man Who Invented the Computer* (Doubleday, 2010), 163.

4. Rojas, "The Z1: Architecture and Algorithms of Konrad Zuse's First Computer".

5. Colin B. Burke, "It Wasn't All Magic: The Early Struggle to Automate Cryptanalysis" (United States Cryptologic History, Special Series, Volume 6, National Security Agency, 2002), 198.

6. Eric Weiss, editor, "Biographies: Steven Warner Dunwell", *IEEE Annals of the History of Computing*, Vol. 16, No. 4, 1994.

7. Mina Rees, "The computing program of the Office of Naval Research, 1946-1953", *Communications of the ACM*, Vol. 30, Issue 10, October 1987.

8. Debbie Anderson, "Joseph R. Desch", retrieved from daytoncodebreakers.org on February 6, 2023.

9. Wenger, J. N. and H. T. Engstrom, R.I. Meader, "Memorandum for Director of Naval Communications: History of the Bombe Project", dated May 30, 1944, National Archives, Record Group 457 (Records of the National Security Agency).

10. Jim DeBrosse, "Dayton's Code Breakers", Dayton Daily News, February 25, 2001.

11. Jerry Russell, "Ultra and the Campaign Against the U-Boats in World War II", records of the NSA, May 20, 1980.

12. James Gleick, *Genius: The Life and Science of Richard Feynman* (Vintage Books, 1993), 137.

13. Gleick, *Genius*, 180.

14. Anne Fitzpatrick, "Igniting the Light Elements: The Los Alamos Thermonuclear Weapon Project, 1942-1952" (Los Alamos National Laboratory, 1999).

15. Richard Feynman, "Los Alamos From Below", lecture given at the University of California, Santa Barbara, 1975.

16. Jean Ford Brennan, *The IBM Watson Laboratory at Columbia University* (IBM, 1971), 14.

## WATSON'S RESPONSE TO HARVARD: THE SSEC

1. Larry Saphire, "Wallace Eckert interview", the Technical History Project, interviewed in July, 1967 at the New York City Watson Lab.

2. T. R. Kennedy, "Electronic Computer Flashes Answers, May Speed Engineering", New York Times, February 15, 1946.

3. Richard R. Mertz, "Oral History Herbert R. Grosch", interview on November 9, 1970, National Museum of American History.

4. Herbert R. J. Grosch, *Computer: Bit Slices from a Life* (Underwood Books, 1991), 10-11.

5. Bernard Cohen, *Howard Aiken, Portrait of a Computer Pioneer* (The MIT Press, 1999), 162.

6. Saphire, "Wallace Eckert interview".

7. John C. McPherson, Frank E. Hamilton, and Robert R. Seeber, "A Large-Scale, General-Purpose Electronic Digital Computer – The SSEC", from Jeffery R, Yost, *The IBM Century: Creating the IT Revolution* (IEEE Computer Society, 2011), 55-68.

8. Frank Da Cruz, "Columbia University Computing History", retrieved from Columbia.edu/computinghistory on February 6, 2023.

9. Richard R. Mertz, "Interview of Herbert R. Grosch", conducted November 1970 in Gaithersburg, Maryland.

10. Grosch, *Bit Slices from a Life*, 1.

11. Jennifer Baker, "The SSEC: IBM's Electronic Brain and Reaching the Moon", February 23, 2011, NC State University Libraries.

12. Seidel, Robert, "An Interview with Cuthbert C. Hurd", Charles Babbage Institute, conducted November 18, 1994 at Portola Valley, California.

13. Cohen, *Howard Aiken*, 50.

14. Steve Lohr, *Go To* (Basic Books, 2001), 16-17.

15. Jean Ford Brennan, *The IBM Watson Laboratory at Columbia University*, (IBM, 1971), 21.

16. Don Grice, "The Evolution of Technical Computing and Man-Machine Partnerships", NCAR Computational and Systems Lab, 2015.

17. Stern, Nancy, "An Interview with J. Presper Eckert", conducted on October 28, 1977, Charles Babbage Institute, University of Minnesota.

## THE VON NEUMANN MACHINES

1. Richard R. Mertz, "An Interview with Julian Bigelow", conducted on January 20, 1971, Smithsonian National Museum of American History.

2. Henry S. Tropp, "Interview with Byron E. Phelps and Werner Buchholz" conducted on July 20, 1973, National Museum of American History.

3. Jane Smiley, *The Man Who Invented the Computer* (Doubleday, 2010), 148-149.

4. Henry S. Tropp,, "I2728nterview with Nathaniel Rochester" July 24, 1973, Smithsonian Museum of American History.
5. Richard R. Mertz, "An Interview with Julian Bigelow", conducted on January 20, 1971, Smithsonian National Museum of American History.
6. Martin H. Weik, "A Survey of Domestic Electronic Digital Computing Systems", Aberdeen Proving Ground, December, 1955.
7. Pamela McCorduck, *Machines Who Think*, (A. K. Peters, 2004), p. 81.
8. Nicolas Lewis and Whitney Spivey, "Computing on the Mesa", National Security Science, Winter, 2020.
9. The BESM-1, completed in 1952, used 5,000 vacuum tubes and mercury delay storage. Performance was 8,000 to 10,000 operations per second, making it one of the fastest machines in Europe. Promising at the time, the Soviets also embraced differential analyzers at this late date, serving to stunt their progress in digital computers.
10. Maurice V. Wilkes, *Memoirs of a Computer Pioneer* (MIT Press, 1985), 108-109.
11. "J. Lyons and Co.", Wikipedia, retrieved February 6, 2023.
12. Wilkes, *Memoirs of a Computer Pioneer,* 127-142.
13. Wilkes, *Memoirs of a Computer Pioneer,* 164-165.
14. Wilkes, *Memoirs of a Computer Pioneer,* 145.
15. Wilkes, *Memoirs of a Computer Pioneer,* 142.
16. Fred Brooks, ICSE Plenary Sessions, May 27 – June 3, 2018, Gothenburg, Sweden.
17. Wilkes, *Memoirs of a Computer Pioneer,* 171.
18. Wilkes, *Memoirs of a Computer Pioneer,* 134-135.
19. "LEO: The Story of the World's First Business Computer", film produced by the Centre for Computer History, 2021.
20. Peter John Bird, *LEO: The First Business Computer* (Hasler, 1994), pp. 84,86, 228.
21. Simon Lavington, editor, *Alan Turing and his Contemporaries*, (BCS Learning and Development Limited, 2012), 39-41.
22. Byron Paul Mobley, "The ingenuity of common workmen: and the invention of the computer",( Iowa State University Capstones, Theses, and Dissertations, 2001), 265.
23. "The Modern History of Computing", Stanford Encyclopedia of Philosophy, retrieved from plato,Stanford.edu on October 5, 2023.
24. B. J. Archer, "The Los Alamos Computing Facility during the Manhattan Project", Los Alamos National Laboratory, 2021.
25. Smiley, *The Man Who Invented the Computer*, 84.

# THE RELUCTANT WATSON

1. Thomas J. Watson, Jr. and Peter Petre, *Father, Son, and Co.* (Bantam, 1990), 2.
2. Richard Tedlow, *The Watson Dynasty: The Fiery and Troubled Reign of IBM's Founding Father and Son* (Harper Business, 2003), 135.
3. Watson, *Father, Son, and Co.,* 85.
4. Scott S. Smith, "Thomas Watson Jr. Powered IBM to the Top of the Tech World", Investor's Business Daily, May 28, 2016.
5. Watson, *Father, Son, and Co.,* 93.
6. Ralph Watson MeElvenny and Marc Wortman, The Greatest Capitalist Who Ever Lived: Tom Watson, Jr. and the Epic Story of How IBM Created the Digital Age (Public Affairs, 2023), 123-152.
7. Watson, *Father, Son, and* Co., 106.

8. Ralph Watson McElvenny and Marc Wortman, *The Greatest Capitalist Who Ever Lived*, 156.
9. Watson, *Father, Son, and Co.*, 132.
10. Watson, *Father, Son, and Co.*, 136.
11. Watson, *Father, Son, and Co.*, 200.
12. Atsushi Akera, "IBM's Early Adaptation to Cold War Markets: Cuthbert Hurd and His Applied Science Field Men", The Business History Review, Vol. 76, No. 4 (Winter, 2002).
13. Watson, *Father, Son, and Co.*, 201.
14. Watson, *Father, Son, and Co.*, 59.
15. James T. Cortada, *IBM: The Rise and Fall and Reinvention of a Global Icon* (The MIT Press, 2019), 159.
16. Weiss, Eric A., "Arthur Lee Samuel", IEEE Annals of the History of Computing, Vol. 14, No. 3, 1992.
17. James W. Birkenstock, "Pioneering: On the Frontier of Electronic Data Processing, a Personal Memoir", from Jeffrey R. Yost, editor, *The IBM Century: Creating the IT Revolution* (IEEE Computer Society, 2011), 92-113.
18. Robert Stuewer and Erwin Tomash, "An Interview with James Birkenstock", conducted on August 12, 1980, Charles Babbage Institute.
19. "14K Days: A History of the Poughkeepsie Laboratory', IBM, November 30, 1984.
20. "The IBM 602 Calculating Punch", Columbia University Computing History, retrieved May 27, 2021.
21. Watson, *Father, Son, and Co.*, 136.
22. Ed Thelen, "IBM 604", retrieved from ed-thelen.org on February 6, 2023.
23. Emerson W. Pugh, *Building IBM: Shaping An Industry and Its Technology* (MIT Press, 1995), 183.
24. Pugh, Building IBM, 229.
25. Emerson W. Pugh, Lyle R. Johnson, and John H. Palmer, *IBM's 360 and Early 370 Systems*, (MIT Press, 1991), 15.
26. "Arthur Lee Samuel", IEEE Annals of the History of Computing, Vol. 14, No. 3, 1992.
27. "Arthur Lee Samuel".
28. "Arthur Lee Samuel".
29. Martin Campbell-Kelly, William Aspray, Nathan Ensmenger, and Jeffrey Yost, *Computer: A History of the Information Machine* (Westview Press, 2014), 104.
30. Charles J. Bashe, Lyle R. Johnson, John H. Palmer, and Emerson W. Pugh, *IBM's Early Computers* (The MIT Press, 1986),83.
31. Bashe, *IBM's Early Computers*, 70.
32. Herbert R. J. Grosch, *Computer: Bit Slices from a Life* (Underwood Books, 1991), 87.
33. Grosch, Computer: *Bit Slices from a Life*, 116.
34. George R. Trimble, "The IBM 650 Magnetic Drum Calculator", from Jeffery R, Yost, *The IBM Century: Creating the IT Revolution* (IEEE Computer Society, 2011), 138-147.
35. Fred J. Gruenberger, "A Short History of Digital Computing in Southern California", Annals of the History of Computing, Vol. 2, No. 3, July, 1980.
36. "Computation Seminar", seminar held at the Endicott Education Center in August, 1951, edited by IBM Applied Science Department, IBM Corporation, 1951.
37. Bobbi Mapstone and Henry Tropp, "Interview with Greg Toben, Jim Smith, Dave Montgomery, and Roy Harper", October 9, 1972, Smithsonian Institution.

38. Grosch, *Computer: Bit Slices from a Life*, 119.
39. Grosch, *Computer: Bit Slices from a Life*, 249.
40. Atsushi Akera, "IBM's Early Adaptation to Cold War Markets: Cuthbert Hurd and His Applied Science Field Men", The Business History Review, Vol. 76, No. 4 (Winter, 2002).

# THE RISE OF UNIVAC

1. Mina Rees, "The computing program of the Office of Naval Research, 1946-1953", *Communications of the ACM*, Vol. 30, Issue 10, October 1987.
2. Mina Rees, "The computing program of the Office of Naval Research, 1946-1953".
3. Byron Paul Mobley, "The ingenuity of common workmen: and the invention of the computer", Iowa State University Capstones, Theses, and Dissertations, 2001.
4. Nancy Stern, "An Interview with Isaac Auerbach", Charles Babbage Institute, April 10, 1978.
5. Stern, "An Interview with Isaac Auerbach." How the fortunes of UNIVAC might have turned out differently if John Mauchly had kept his security clearance and Eckert had listened to Auerbach.
6. "The BINAC: A Product of the Eckert-Mauchly Computer Corp", corporate brochure, 1949.
7. Mobley, "The ingenuity of common workmen", 11.
8. Scott McCartney, *ENIAC: The Triumphs and Tragedies of the World's First Computer* (Walker and Company, 1999), 163.
9. Stern, "An Interview with Isaac Auerbach."
10. Donald E. Eckdahl, Irving S. Reed, and Harold Sarkissian, "West Coast Contributions to the Development of the General Purpose Computer", IEEE Annals of the History of Computing, January-March, 2003.
11. Robina Mapstone, "Interview with W. W. Woodbury", conducted on January 15, 1973, Smithsonian National Museum of American History.
12. Maurice Wilkes, *Memoirs of a Computer Pioneer* (MIT Press, 1985), 138-139.
13. Mina Rees, "The computing program of the Office of Naval Research, 1946-1953".
14. Uta C. Merzbach, "Franz L. Alt interview", Smithsonian Institute Oral History Program, March 13, 1969.
15. Nancy Stern, "The Eckert-Mauchly Computers: Conceptual Triumphs, Commercial Tribulations." (Technology and Culture, vol. 23, no. 4, 1982), 569–582.
16. Mina Rees, "The computing program of the Office of Naval Research, 1946-1953".
17. Jane Smiley, *The Man Who Invented the Computer* (Doubleday, 2010), 149.
18. Matthew Lasar, "UNIVAC: the troubled life of America's first computer", Ars Technica, September 18, 2011.
19. "Manipulation Ban Put On J.H. Rand Jr., The New York Times, October 19, 1939.
20. John C. Dvorak, "Whatever Happened to the Seven Dwarfs", Dvorak News, 2008.
21. Roger Mills, moderator, "COT User Group Meeting", December 1, 1971, Los Angeles, California
22. "The Dawn of Commercial Computing in the 1950s", Carnegie Mellon University, School of Computer Science, 2017.
23. Watson: Thomas J. Watson, Jr. and Peter Petre, *Father, Son, and Co.* (Bantam, 1990), 227.
24. Ed LaHay, "Oral History of Burton Grad (IBM)", interviewed on November 28, 2007, Computer History Museum.

25. Ann Fitzpatrick, "Oral History: Robert Richtmyer", interviewed March 4, 1997, Boulder, Colorado.

## THE FIRST SUPERCOMPUTER

1. Calvin N. Mooers, "The Computer Project at the Naval Ordnance Laboratory", IEEE Annals of the History of Computing, April-June 2001.
2. Clark R. Mollenhoff, *Atanasoff: Forgotten Father of the Computer* (Iowa State University Press, 1988), 74.
3. Mollenhoff, *Atanasoff: Forgotten Father of the Computer*, 68.
4. Richard R. Mertz, "Interview of Herbert R. Grosch", conducted November 1970 in Gaithersburg, Maryland.
5. Mertz," Interview of Herbert R. Grosch".
6. Mertz, "Interview of Herbert R. Grosch".
7. William Aspray, "Oral History of John McPherson", conducted on April 29 and May 12, 1992 in Short Hills, New Jersey.
8. Kenneth Flamm, *Creating the Computer* (The Brookings Institute, 1988), 9.
9. Theresa Cramer, "The Naval Ordnance Research Computer (NORC)", DC Military, July 13, 2018.
10. "The IBM Naval Ordnance Research Computer", Columbia University Computing History, retrieved May 27, 2021.

## IBM'S 700 SERIES VAULTS TO #1

1. Martin Campbell, William Aspray, Nathan Ensmenger, and Jeffrey Yost, *Computer: A History of the Information Machine* (*Westview Press, 2014), 106.
2. Robert Stuewer and Erwin Tomash, "An Interview with James Birkenstock", conducted on August 12, 1980, Charles Babbage Institute.
3. Emerson W. Pugh, Building IBM: Shaping An Industry and Its Technology (MIT Press, 2009), 168..
4. Cuthbert C. Hurd, "IBM Early Computers", from the compilation by Jeffrey R. Yost, *The IBM Century: Creating the IT Revolution,* (*IEEE Computer Society, 2011), 71-90.
5. Ralph Watson McElvenny and Marc Wortman, *The Greatest Capitalist Who Ever Lived: Tom Watson Jr. and the Epic Story of How IBM Created the Digital Age* (Public Affairs, 2023), 384.
6. "Learson at IBM's Helm", Time Magazine, July 12, 1971.
7. Thomas J. Watson, Jr. and Peter Petre, *Father, Son, and Co.* (Bantam, 1990), 242.
8. Neil Hrab, "The Speechwriter's Life: Harrison B. Kinney", Pro Rhetoric, October 12, 2016.
9. Robert Garner, "Oral History of Charles (Chuck) Branscomb", Computer History Museum, November 10, 2009.
10. Arthur L. Norberg, "An Interview with Gene Amdahl", interviewed at Cupertino, on April 16, 1986, January 17 and April 5, 1989.
11. Angelo Donofrio, "No Stranger to the Helm", Think Magazine, July/August 1971.
12. Isabel S. Grossner, "Oral history interview with T. Vincent Learson", interviewed on August 6, 1982 in New York, Spencer Foundation, Columbia University Oral Histories, 1982.

13. Ralph Watson McElvenny and Marc Wortman, *The Greatest Capitalist Who Ever Lived*, 421-422.
14. Bob O. Evans, *The Genesis of the Mainfra*me, self-published memoir, 1999.
15. William D. Smith, "Non-Watson at Helm", The New York Times, June 30, 1971.
16. Robert Seidel, "An Interview with Cuthbert C. Hurd", Charles Babbage Institute, conducted November 18, 1994 at Portola Valley, California.
17. "The IBM 700 Series: Computing Comes to Business", IBM history archive on ibm.com, retrieved May 27, 2021.
18. Nathaniel Rochester, "The 701 Project as Seen by its Chief Architect", Annals of the History of Computing, Vol. 5, No. 2, April, 1983
19. Bobbi Mapstone and Henry Tropp, "Interview with Greg Toben, Jim Smith, Dave Montgomery, and Roy Harper", October 9, 1972, Smithsonian Institution.
20. Werner Buchholz, "The System Design of the IBM Type 701 Computer", Proceedings of the Institute of Radio Engineers, October, 1953.
21. Henry S. Tropp, "Interview with Nathaniel Rochester" July 24, 1973, Smithsonian Museum of American History.
22. William Lokke, "Early Computing and Its Impact on Lawrence Livermore National Laboratory", Department of Energy, March 19, 2007.
23. "Electronic Brain, '53 Model", The New York Times, April 8, 1953.
24. Eric A. Weiss, "Arthur Lee Samuel", IEEE Annals of the History of Computing (Vol. 14, No. 3, 1992)
25. Feng-hsiung Hsu, *Behind Deep Blue: Building the Computer that Defeated the World Chess Champion* (Princeton University Press, 2002), vii.
26. Claude Shannon, "Programming a Computer to Play Chess", Philosophical Magazine (Vol. 41, No. 314, 1950).
27. Arthur L. Samuel, "Some Studies in Machine Learning Using the Game of Checkers", IBM Journal, July 1959.
28. "Christopher Strachey", Computer History Museum, retrieved December 4, 2023.
29. Weiss, "Arthur Lee Samuel".
30. Bertram Raphael, The Thinking Computer: Mind Inside Matter (W. H. Freeman, 1976), 38.
31. "The IBM 700 Series: Computing Comes to Business", IBM history archive on ibm.com, retrieved May 27, 2021.
32. Weiss, "Arthur Lee Samuel".
33. Pamela McCorduck, *Machines Who Think* (A. K. Peters), p. 177.
34. Arthur L. Samuel, "Some Studies in Machine Learning Using the Game of Checkers", IBM Journal of Research and Development (Vol. 3, Is. 3, July, 1959).
35. Arthur L. Norberg, "An Interview with Gene Amdahl", interviewed at Cupertino, on April 16, 1986, January 17 and April 5, 1989.
36. William S. Anderson, "An Interview with Gene M. Amdahl", IEEE Solid-State Circuits Society News, Summer 2007.
37. Gene M. Amdahl, "Computer Architecture and Amdahl's Law" IEEE News, Summer 2007).
38. Anderson, "An Interview with Gene M. Amdahl".
39. C. J. Bashe, W. Buchholz, G. V. Hawkins, J. J. Ingram, and N. Rochester, "The Architecture of IBM's Early Computers" (*IBM Journal of Research and Development*, Vol. 25, No. 5, September, 1981)
40. Thomas Haigh, "Oral History of Robert L. Patrick", interviewed on February 16, 2006 in Mountain View, California.
41. Haigh, "Oral History of Robert L. Patrick".

42. Robert L. Patrick, "General Motors/North American Monitor For The IBM 704 Computer", The Rand Corporation, January, 1987.
43. Lokke, "Early Computing and Its Impact on Lawrence Livermore National Laboratory".

## PROGRAMMING: BEYOND MACHINE CODING

1. Grady Booch, "Oral History of John Backus", interviewed on September 5, 2006 at Ashland, Oregon, Computer History Museum.
2. Andrew Booth and Kathleen Britten, "Coding for A.R.C.", Institute for Advanced Study, Princeton, 1947.
3. Kurt W. Beyer, *Grace Hopper and the Invention of the Information Age* (MIT Press, 2009), 274.
4. John Backus, "The History of Fortran I, II, and III", from Jeffrey R. Yost, editor, *The IBM Century: Creating the IT Revolution* (IEEE Computer Society, 2011), 160. .
5. Steve Lohr, *Go To*, (Basic Books, 2001), 11-43. Lohr provides a comprehensive narrative of the development of FORTRAN.
6. Lois Haibt, "Oral History", interviewed by Janet Abbate for the IEEE History Center on August 2, 2001.
7. Mills, Roger, moderator, "COT User Group Meeting", December 1, 1971, Los Angeles, California.
8. Backus, "The History of FORTRAN I, II, and III", 166.
9. Backus, "The History of FORTRAN I, II, and III", 167
10. Steve Lohr, "Software: When Few Knew the Code, They Changed the Language", The New York Times, June 13, 2001.
11. Nathan Ensmenger, William Aspray and Thomas Misa, *The Computer Boys Take Over: Computers, Programmers, and the Politics of Technical Expertise*, (MIT Press, 2010), 93-101.

## THE IBM TAPE MACHINE – BETTER LATE THAN NEVER

1. "14K Days: A History of the Poughkeepsie Laboratory", IBM, November 30, 1984.
2. Charles J. Bashe, Lyle R. Johnson, John H. Palmer, and Emerson W. Pugh, *IBM's Early Computers* (The MIT Press, 1986), 108.
3. Bashe, *IBM's Early Computers*, 110-112.
4. Emerson W. Pugh, *Building IBM: Shaping An Industry and Its Technology* (MIT Press, 1995), 169.
5. Henry S. Tropp, "Oral History: Byron E. Phelps and Werner Buchholz", interviewed July 20, 1973.
6. Tropp, "Oral History: Byron E. Phelps and Werner Buchholz".
7. C. J. Bashe, W. Buchholz, G. V. Hawkins, J. J. Ingram, and N. Rochester, "The Architecture of IBM's Early Computers" (*IBM Journal of Research and Development*, Vol. 25, No. 5, September, 1981).
8. George Michael, "An Interview with Norman Hardy", interviewed on May23, 1994.
9. Mina Rees, "The computing program of the Office of Naval Research, 1946-1953", *Communications of the ACM*, Vol. 30, Issue 10, October 1987.

10. Brian Randell, "From Analytical Engine to Electronic Digital Computer: The Contributions of Ludgate, Torres, and Bush", Annals of the History of Computing, Vol. 4, No. 4, October, 1982.

11. Mina Rees, "The computing program of the Office of Naval Research, 1946-1953".

12. Wilkes, *Memoirs of a Computer Pioneer* (MIT Press, 1985), 176.

13. Emerson W. Pugh, Lyle R. Johnson, and John H. Palmer, *IBM's 360 and Early 370 Systems* (MIT Press, 1991), 706.

14. Bashe, *IBM's Early Computers*, 267.

15. James W. Birkenstock, "Pioneering: On the Frontier of Electronic Data Processing, a Personal Memoir", from Jeffrey R. Yost, editor, *The IBM Century: Creating the IT Revolution* (IEEE Computer Society, 2011), 117.

16. Bashe, *IBM's Early Computers*, 270.

17. Pugh, *IBM's 360 and Early 370 Systems*, 204.

18. Nathan Ensmenger, William Aspray and Thomas Misa, *The Computer Boys Take Over: Computers, Programmers, and the Politics of Technical Expertise* (MIT Press, 2010), 138.

19. Herbert F. Mitchell, "UNIVAC Marketing", personal remembrances in VIP Club, January 2007.

20. "UNIVAC", Computer History Museum, retrieved February 6, 2023.

21. Matthew Lasar, "UNIVAC: the troubles life of America's first computer", Ars Technica, September 18, 2011.

22. John Greenwald, "The Colossus That Works", Time, July 11, 1983.

23. Birkenstock, "Pioneering: On the Frontier of Electronic Data Processing, a Personal Memoir", from Jeffrey R. Yost, editor, *The IBM Century: Creating the IT Revolution*, 113..

# IBM 650: "THE MODEL T OF COMPUTERS"

1. Charles J. Bashe, Lyle R. Johnson, John H. Palmer, and Emerson W. Pugh, *IBM's Early Computers* (The MIT Press, 1986), 90.

2. Bashe, *IBM's Early Computers*, 101.

3. Robina Mapstone, "Interview with W. W. Woodbury", conducted on January 15, 1973, Smithsonian National Museum of American History.

4. Robina Mapstone, "Interview with W. W. Woodbury".

5. Robina Mapstone, "Interview with W. W. Woodbury".

6. Robina Mapstone, "Interview with W. W. Woodbury".

7. George R. Trimble, "The IBM 650 Magnetic Drum Calculator", from Jeffery R, Yost, *The IBM Century: Creating the IT Revolution* (IEEE Computer Society, 2011), 138-147.

8. Gustav Tauschek, "Electromagnetic memory for numbers, and other data" German Patent DE 643,803 (Filed: July 1, 1933 Issued: April 17, 1937)

9. J. A. N. Lee, "Clifford Edward Berry", IEEE Computer Society, 1995.

10. James W. Birkenstock, "Pioneering: On the Frontier of Electronic Data Processing, a Personal Memoir", from Jeffrey R. Yost, editor, *The IBM Century: Creating the IT Revolution* (IEEE Computer Society, 2011), 107.

11. Trimble, "The IBM 650 Magnetic Drum Calculator", 138-147.

12. Richard V. Andree, "Programming the IBM 650 Magnetic Drum Computer and Data Processing Machine", (Holt, Rinehart, and Winston, 1958), 5.

13. Cuthbert C. Hurd, "IBM Early Computers", from Yost, *The IBM Century: Creating the IT Revolution,*71.
14. James T. Cortada, *IBM: The Rise and Fall and Reinvention of a Global Icon* (The MIT Press, 2019), 165.
15. "IBM's Thomas J. Watson, Jr.", Time Magazine, vol. LXV no. 13, March 28, 1955.
16. Feigenbaum, Edward, "Donald Knuth Oral History", interviewed March 14, 2007 at Mountain View, California, Computer History Museum.

## THE FIRST DISK DRIVE, AKA "THE BOLOGNA SLICER"

1. Emerson W. Pugh, Lyle R. Johnson, and John H. Palmer, *IBM's 360 and Early 370 Systems* (MIT Press, 1991), 242.
2. Reynold Johnson, "Talk at the Data Storage '89 Conference", San Jose, September 19, 1989.
3. Charles J. Bashe, Lyle R. Johnson, John H. Palmer, and Emerson W. Pugh, *IBM's Early Computers* (The MIT Press, 1986), 274.
4. Albert S. Hoagland, *Magnetic Disk Storage: A Personal Memoir* (self-published, 2011), 13.
5. Jim Porter, "The Disk Drive Story", interviews conducted at the IBM Almaden Research Center, November, 2001 through March 22, 2002. The 1620 lacked an ALU (Arithmetic and Logic Unit). It's nickname was "CADET", meaning "Can't Add, Doesn't Even Try". This came directly from John Lentz's CADET personal computer program.
6. Hoagland, *Magnetic Disk Storage*, provides extensive of Hoagland's schooling and start with IBM's San Jose Lab.
7. Tom Gardner, "Oral History of William Crooks", interviewed on November 13, 2008 at Mountain View, California.
8. Gardner, "Oral History of William Crooks".
9. "World's First Magnetic Disk Drive", IBM fact sheet on the 350 in IBM Archives, retrieved February 6, 2023.
10. Reynold Johnson, "Dinner Talk at the Data Storage '89 Conference", San Jose, September 19, 1989.
11. Charles Branscomb, "Main 1401 Story", Computer History Museum, 2014.
12. Bashe, *IBM's Early Computers*, 290.

## WATSON JR. TAKES OVER

1. Thomas J. Watson, Jr. and Peter Petre, *Father, Son, and Co.* (Bantam, 1990), 172.
2. "The Brain Builders", Time Magazine, March 28, 1955.
3. "IBM's Thomas J. Watson, Jr.", Time Magazine, vol. LXV no. 13, March 28, 1955.
4. "Watson Yields IBM Helm at 82, The New York Times, May 9, 1956.
5. Watson, *Father, Son, and Co.*, 279.
6. Ralph Watson McElvenny and Marc Wortman, *The Greatest Capitalist Who Ever Lived: Tom Watson Jr. and the Epic Story of How IBM Created the Digital Age* (Public Affairs, 2023), 284.
7. Kevin Maney, *The Maverick and His Machine* (John Wiley and Sons, 2003), 442.
8. Bob O. Evans, *The Genesis of the Mainframe*, self-published memoir, 1999.
9. Watson, *Father, Son, and Co.*, 258.

10. Scott G. Knowles and Stuart W. Leslie, "Industrial Versailles: Eero Saarinen's Corporate Campuses for GM, IBM, and AT&T" , ISIS, vol. 92, no. 1, March, 2001.

11. Scott G. Knowles and Stuart W. Leslie, "Industrial Versailles: Eero Saarinen's Corporate Campuses for GM, IBM, and AT&T" (ISIS Journal, University of Chicago Press, Vol. 92, No. 1, March, 2001).

12. Randal Picker, The Arc of Monopoly: A Case Study in Computing", University of Chicago Law Review (2020).

13. James W. Birkenstock, "Pioneering: On the Frontier of Electronic Data Processing, a Personal Memoir", from Jeffrey R. Yost, editor, The IBM Century: Creating the IT Revolution (IEEE Computer Society, 2011), 105-106.

14. "Sperry Rand Sues IBM as Monopoly", The New York Times, December 28, 1955.

15. Bosworth, Edward L., "The Heritage of the IBM System/360", in Textbook for IBM Mainframe Assembler Language Course, 2009.

16. Cuthbert C. Hurd, "IBM Early Computers", from the compilation by Jeffrey R. Yost, The IBM Century: Creating the IT Revolution, (IEEE Computer Society, 2011), 71-90. The meetings with MIT's Lincoln Lab would include Watson, Jr., James Birkenstock, Cuthbert Hurd, Al Williams, Arthur Watson, Wally McDowell, Ralph Palmer, and sometimes Watson, Sr.

17. "SAGE: The First National Air Defense Network", ibm.com, retrieved February 6, 2023.

18. Robert Wendover, High Performance Hiring, (Thomas Crisp Learning, 2003), 179.

19. "14K Days: A History of the Poughkeepsie Laboratory", produced by the IBM Poughkeepsie Lab, 1984.

20. James T. Cortada, IBM: The Rise and Fall and Reinvention of a Global Icon (The MIT Press, 2019), 169.

21. Stan Augarten, Bit by Bit: An Illustrated History of Computers (Ticknor and Fields, 1984), 199.

22. Martin Campbell-Kelly, William Aspray, Nathan Ensmenger, and Jeffrey Yost, Computer: A History of the Information Machine (Westview Press, 2014), 150-152.

23. Emerson W. Pugh, Building IBM: Shaping An Industry and Its Technology (MIT Press, 1995), 212.

## AMERICAN AIRLINES AND THE SABRE SYSTEM

1. Robina Mapstone, "An Interview with R. Blair Smith", conducted on May 20, 1980, Charles Babbage Institute.

2. James McKenney, Waves of Change: Business Evolution Through Information Technology (Harvard University Press, 1995), 100.

3. Robert V. Head, "Getting SABRE Off the Ground", IEEE Annals of the History of Computing, vol. 24, no. 4 (Oct-Dec 2002), 32-39.

4. Martin Campbell-Kelly, William Aspray, Nathan Ensmenger, and Jeffrey Yost, Computer: A History of the Information Machine (Westview Press, 2014), 156.

5. Head, "Getting SABRE Off the Ground", 32-39.

6. Head, "Getting SABRE Off the Ground", 32-39.

7. Martin Campbell-Kelly and William Aspray, Computer: A History of the Information Machine (Harper Collins, 1996), 173.

8. Mapstone, "An Interview with R. Blair Smith".

9. Chris Bajorek and Tom Garner, "Oral History of Jack Harter with Denis Mee", interviewed at Mountain View, California, on May 30, 2007 and July 10, 2007.

10. George Rostky, "Disk Drives take eventful spin" , EE Times, July 13, 1998.
11. "R. Blair Smith", Oral history interview by Robina Mapstone, May 1980. Charles Babbage Institute, University of Minnesota, Minneapolis.
12. Michael Doran, "American Airlines Wins Major Ruling Amid Sabre Antitrust Case, Simply Flying, March, 2022.

## STRETCHING COMPUTER POWER

1. Mark Smotherman, "IBM Stretch (7030) – Aggressive Uniprocessor", January, 2016. Smotherman's analysis on Stretch and indeed all of IBM's early supercomputer efforts is exhaustive and compelling.
2. William Lokke, "Early Computing and Its Impact on Lawrence Livermore National Laboratory", December 15, 2006.
3. Lokke, "Early Computing and Its Impact on Lawrence Livermore National Laboratory".
4. "Engineering Department records, 1959-1974", Archive 1825-I, Sperry Rand Corporation, Univac Division records.
5. L. R. Johnson, "A Description of Stretch", IBM Research Report RC-160, December 10, 1959
6. Saul Rosen, "Electronic Computers: A Historical Survey", Computing Surveys, Vol. 1, No. 1, March, 1969.
7. Grady Booch, "Oral History of Fred Brooks", interviewed on September 16, 2007, Computer History Museum.
8. Gary Bishop, "Video Oral History with Frederick "Fred" Brooks", Association for Computer Machinery, March 12, 2020.
9. Brooch, "Oral History of Fred Brooks".
10. Walter Isaacson, The Innovators (Simon and Shuster, 2014), 151.
11. Robert Garner, "Oral History of Francis Underwood", interviewed for Computer History Museum, 2009.
12. Smotherman, "IBM Stretch (7030) – Aggressive Uniprocessor".
13. Robert Garner and Dag Spicer, "IBM 1401: The Legendary Data Processing System", Computer History Museum, 2009.
14. Ralph Watson McElvenny and Marc Wortman, The Greatest Capitalist Who Ever Lived: Tom Watson Jr. and the Epic Story of How IBM Created the Digital Age (Public Affairs, 2023), 327-328.
15. Garner and Spicer, "IBM 1401: The Legendary Data Processing System".
16. Garner and Spicer, "IBM 1401: The Legendary Data Processing System".
17. Werner Buchholz, editor, Planning a Computer System: Project Stretch (McGraw-Hill Book Company, 1962). This detailed, technical description of Stretch technologies grew out of the Datatron memos and the tracking of added system functions.
18. Buchholz, Planning a Computer System: Project Stretch, 40.
19. "Stretch Reunion Transcripts" (Computer History Museum, September 28, 2002).
20. Arthur L. Norberg, "An Interview with Gene Amdahl", interviewed at Cupertino, on April 16, 1986, January 17 and April 5, 1989.
21. Smotherman, "IBM Stretch (7030) – Aggressive Uniprocessor".
22. William Aspray, "An Interview with Steve Dunwell", conducted on February 13, 1989, Charles Babbage Institute.
23. Colin B. Burke, "It Wasn't All Magic: The Early Struggle to Automate Cryptanaly-

sis", United States Cryptologic History, Special Series, Volume 6, National Security Agency, 2002, 435-438. .

24. Burke, "It Wasn't All Magic", 453.
25. Burke, "It Wasn't All Magic", 456.
26. "Stretch Reunion Transcripts".
27. L. R. Johnson, "A Description of Stretch", IBM Research Report RC-160, December 10, 1959.
28. Smotherman, "IBM Stretch (7030) – Aggressive Uniprocessor".
29. Smotherman, "IBM Stretch (7030) – Aggressive Uniprocessor".
30. Edward K. Yasaki, "IBM Stretch", Datamation, January, 1982.
31. Michael Cowlishaw, "IBM Jargon and General Computing Dictionary Tenth Edition", 1990, retrieved December 19, 2022. IBM is duly noted for its insular jargon and acronyms, aided and abetted by the explosion of computing technology. Cowlishaw, a former IBMer noted for designing the REXX programming language and now a professor at the University of Warwick, set about capturing all of this lingo.
32. "Oral History of Michael J. Flynn", Interview by Bob Zeidman, Computer History Museum, December 17, 2012.
33. Charles J. Bashe, Lyle R. Johnson, John H. Palmer, and Emerson W. Pugh, IBM's Early Computers (The MIT Press, 1986), 457.
34. Smotherman, "IBM Stretch (7030) – Aggressive Uniprocessor".
35. Yasaki, "IBM Stretch".
36. "Stretch Reunion Transcripts", Computer History Museum.
37. James Strothman, "The Ancient History of System/360", Invention and Technology, Winter 1990, Volume 5, Issue 3.
38. Cuthbert C. Hurd, "IBM Early Computers", from Jeffrey R. Yost, editor, The IBM Century: Creating the IT Revolution (IEEE Computer Society, 2011), 87. .
39. Bob O. Evans, "System/360: A Retrospective View", from Yost, The IBM Century: Creating the IT Revolution, 173-197. .

## IBM IN THE 1950S

1. Thomas J. Watson, Jr. and Peter Petre, Father, Son, and Co. (Bantam, 1990), 311.
2. J. A. N. Lee, "Clifford Edward Berry", IEEE Computer Society, 1995.
3. John C. Dvorak, "Whatever Happened to the Seven Dwarfs", Dvorak News, 2008.
4. Scott McCartney, ENIAC: The Triumphs and Tragedies of the World's First Computer (Walker and Company, 1999), 171-172.
5. John C. Dvorak, "Whatever Happened to the Seven Dwarfs", Dvorak News, 2008.
6. Watson, Father, Son, and Co., 243.
7. Arthur L. Norberg, "An Interview with William Charles Norris", conducted July 28 and October 1, 1986 at Minneapolis, Charles Babbage Institute.

## THE SPERRY TRIAL: WHO INVENTED THE COMPUTER

1. Richard K. Richards, Electronic Digital Systems (Wiley and Son, New York, 1966).
2. Richard K. Richards, Arithmetic Operation in Digital Computers (Van Nostrand, 1955).
3. Geraldine M. Montag, "John Vincent Atanasoff, Inventor of the First Electronic Digital Computer", Iowa State University. .

4. Clark R. Mollenhoff, *Atanasoff: Forgotten Father of the Computer* (Iowa State University Press, 1988), 112. Mollenhoff was a reporter for the Des Moines Register.

5. Scott McCartney, *ENIAC: The Triumphs and Tragedies of the World's First Compute,* (Walker and Company, 1999), 177.

6. James W. Birkenstock, "Pioneering: On the Frontier of Electronic Data Processing, a Personal Memoir", from Jeffrey R. Yost, editor, *The IBM Century: Creating the IT Revolution* (IEEE Computer Society, 2011), 118.

7. McCartney, *ENIAC: The Triumphs and Tragedies of the World's First Computer,* 179.

8. The moderator was none other than Isaac Auerbach, the engineer who had prodded J. Presper Eckert to raise UNIVAC prices and dial down the clock speed of BINAC, and who left the company in frustration.

9. Mollenhoff, *Atanasoff,* 115-116.

10. Mollenhoff, *Atanasoff,* 133.

11. "Patent/Court Case: Honeywell v. Sperry Rand" , Iowa State University Department of Computer Science, 2011.

12. David W. Gardner, "The Computer: Born in a Tavern", The Washington Post, January 13, 1974.

13. Mollenhoff, *Atanasoff,* 6.

14. Alice R. Burks and Arthur W. Burks, *The First Electronic Computer: The Atanasoff Story* (University of Michigan Press, 1988), 252.

15. Mollenhoff, *Atanasoff,* 8.

16. Mollenhoff, *Atanasoff,* 213.

17. Mollenhoff, *Atanasoff,* 60.

18. Mollenhoff, *Atanasoff,* 217.

19. Mollenhoff, *Atanasoff,* 226.

20. Burks and Burks, *The First Electronic Computer: The Atanasoff Story,* 267.

21. Lee Loevinger, "The Invention and Future of the Computer" (*Journal of Computer and Information Law,* Vol. 15, Issue 1, Fall 1996).

22. Jane Smiley, *The Man Who Invented the Computer* (Doubleday, 2010), 197.

23. William C. Shaffer, *Shifting Gears: One Family's Journey Through the Automobile Age* (self-published, 2021), 74.

# THE WORLDWIDE ACCOUNTING MACHINE

1. Lars Heide, *Punched Card Systems and the Early Information Explosion* (Johns Hopkins University Press, 2009), 198-202.

2. Arthur Humphreys, "An Interview with Tom Watson, Jr. and James Birkenstock", conducted on April 25, 1985 in Armonk NY, Charles Babbage Institute.

3. Thomas J. Watson, Jr. and Peter Petre, *Father, Son, and Co.* (Bantam, 1990), 177.

4. Larry Saphire, "Oral History with Ralph Mork", interviewed at Harrison, NY on October 24, 1967, IBM World History of Computer Technology.

5. "Origins of Architecture and Design of the IBM 1401", retrieved from ibm-1401.info, April 26, 2021.

6. Eugene Estrems, "The IBM World Wide Accounting Machine", IBM Laboratory Publications Department No. PTP-88, June 20, 1956.

7. Robert Garner, "Oral History of Charles (Chuck) Branscomb", Computer History Museum, November 10, 2009.

8. Garner, "Oral History of Charles (Chuck) Branscomb".
9. Saphire, "Oral History with Ralph Mork."

## THE 1401: THE CHEVROLET OF COMPUTERS

1. Robert Garner, "Oral History of Francis Underwood", interviewed for Computer History Museum, 2009.
2. Charles J. Bashe, Lyle R. Johnson, John H. Palmer, and Emerson W. Pugh, *IBM's Early Computers* (The MIT Press, 1986), 470.
3. Larry Saphire, "Oral History with Karl Ganzhorn", interview TC-55 at Boblingen on February 5, 1968.
4. Garner, "Oral History of Francis Underwood".
5. Robert Garner, "Oral History of Charles (Chuck) Branscomb", Computer History Museum, November 10, 2009.
6. Larry Saphire, "Oral History with Ralph Mork", interviewed at Harrison, NY on October 24, 1967, IBM World History of Computer Technology.
7. Bashe, *IBM Early Computers*, 488-493.
8. Garner, Robert, and Frederick Dill, "The Legendary IBM 1401 Data Processing System", IEEE Solid-State Circuits Magazine, Winter 2010.
9. "FARGO for IBM 1401", IBM Systems Reference Library, IBM Publication C24-1464, 1964.
10. Alex Woodie, "Is it Time to Rename RPG", The Four Hundred, August 24, 2020.
11. Branscomb, "Main 1401 Story".
12. Kevin Maney, Steve Hamm, and Jeffrey O'Brien, *Making the World Work Better* (International Business Machines, 2011), 108.
13. Robert Garner and Dag Spicer, "IBM 1401: The Legendary Data Processing System", Computer History Museum, 2009.
14. Martin Campbell-Kelly, "ICL: Taming the R&D Beast", Business and Economic History, vol. 22, no. 1, (Cambridge University Press, 1993), 169-180.
15. Garner, "Oral History of Charles (Chuck) Branscomb".
16. Bashe, *IBM Early Computers*, 474-477.

## A VERITABLE MAZE OF COMPUTER MODELS

1. Fred J. Gruenberger, "A Short History of Digital Computing in Southern California", Annals of the History of Computing, vol. 2, no. 3, July, 1980.
2. Tom Burniece, "Oral History of Dal Allan", interview March 12, 2013, Mountain View, California.
3. Bob Bemer, "Birth of an Unwanted IBM Computer", Computer History Vignettes, retrieved December 9, 2021.
4. Chuck Boyer, "The 360 Revolution", IBM Corporation, 2004.
5. R. W. Avery and S. H. Blackford, L. McDonnell, "The IBM 7070 Data Processing System", International Workshop on Managing Requirements Knowledge, Philadelphia, 1958.

## THE "BET THE COMPANY" GAMBLE

1. Telex Corp. v. IBM, US District Court for the Northern District of Oklahoma, 1973.
2. T. A. Wise, "IBM's $5,000,000,000 Gamble", Fortune, September, 1966.
3. Bob O. Evans, *The Genesis of the Mainframe*, self-published memoir, 1999.
4. Grady Booch, "Oral History of Fred Brooks", interviewed on September 16, 2007, Computer History Museum.
5. Thomas J. Watson, Jr. and Peter Petre, *Father, Son, and Co.* (Bantam, 1990), 348.
6. Strothman, James, "The Ancient History of System/360", Invention and Technology, Winter 1990, volume 5, issue 3.
7. Strothman, "The Ancient History of System/360".
8. William Rodgers, *Think: A Biography of the Watsons and IBM* (Stein and Day, 1969), 288.
9. Brooch, "Oral History of Fred Brooks".
10. Grady Booch, "Oral History of Watts Humphrey", interviewed on June 17, 2009, Computer History Museum.
11. Mark Smotherman, "IBM Stretch (7030) – Aggressive Uniprocessor", January, 2016.
12. Dag Spicer, "Oral History of Jerome Svigals", interviewed June 19, 2007 at Mountain View, California, Computer History Museum.
13. Strothman, "The Ancient History of System/360".
14. Evans, *Genesis of the Mainframe*.
15. Paul E. Ceruzzi, *A History of Modern Computing* (MIT Press, 2003), 146.
16. Chuck Boyer, "The 360 Revolution" (IBM, 2004), 35.
17. Emerson W. Pugh, Lyle R. Johnson, and John H. Palmer, *IBM's 360 and Early 370 Systems* (MIT Press, 1991), 135.
18. Chuck Boyer, "The 360 Revolution", IBM Corporation, 2004.

## THE NEW PRODUCT LINE GETS UNDERWAY

1. "Historical Narrative: The 1960s", US vs. IBM Exhibit 14971, July 22, 1980.
2. Emerson W. Pugh, *Building IBM: Shaping An Industry and Its Technology* (MIT Press, 1995), 383-384.
3. "IBM System/360 Principles of Operation", publication A22-6821 (IBM Corporation, 1965).
4. David Gifford and Alfred Spector, "Case Study: IBM's System/360-370 Architecture" (*Communications of the ACM*, Volume 30, No. 4, April 1987).
5. Chuck Boyer, "The 360 Revolution", IBM Corporation, 2004.
6. Emerson W. Pugh, Lyle R. Johnson, and John H. Palmer, *IBM's 360 and Early 370 Systems* (MIT Press, 1991), 59.
7. Walter Isaacson, *The Innovators*, (Simon and Shuster, 2014), 168.
8. Eric Weiss, *Computer Usage Fundamentals* (McGraw-Hill, 1969), 55-56.

## THE SOFTWARE CHALLENGE

1. Chuck Boyer, "The 360 Revolution", IBM Corporation, 2004.
2. Emerson W. Pugh, Lyle R. Johnson, and John H. Palmer, *IBM's 360 and Early 370 Systems* (MIT Press, 1991), 302-303.

3. Burton Grad, "Oral History of Richard Case", interviewed on December 7, 2006, Westport, Connecticut.
4. Grad, "Oral History of Richard Case".
5. Pugh, *IBM's 360 and Early 370 Systems*, 310-312.
6. Grady Booch, "Oral History of Fred Brooks", interviewed on September 16, 2007, Computer History Museum.
7. T.A. Wise, "The Rocky Road to the Marketplace" (Fortune, October, 1966), 138-143.
8. Pugh, *IBM's 360 and Early 370 Systems*, 333.
9. William Rodgers, *Think: A Biography of the Watsons and IBM* (Stein and Day, 1969), 287-288.
10. Wilkes, *Memoirs of a Computer Pioneer* (MIT Press, 1985), 178.
11. Wilkes, *Memoirs of a Computer Pioneer* (MIT Press, 1985), 187.
12. Pugh, *IBM's 360 and Early 370 Systems*, 210-219.
13. "LEO: The Story of the World's First Business Computer", film produced by the Centre for Computer History.
14. Frederick P. Brooks, *The Design of Design: Essays from a Computer* Scientist (Addison-Wesley, 2010), 321.
15. T.A. Wise, "The Rocky Road to the Marketplace", 138-143.
16. "Historical Narrative: The 1960s", US vs. IBM Exhibit 14971, July 22, 1980.
17. Gordon Bell and Alan Newell, *Computer Structures: Principles and Examples* (McGraw-Hill, 1971), 856.

# DEEPENING TROUBLES

1. James W. Cortada, *The Rise and Fall and Reinvention of a Global Icon* (MIT, 2019), 222.
2. Emerson W. Pugh, Lyle R. Johnson, and John H. Palmer, *IBM's 360 and Early 370 Systems* (MIT Press, 1991), 313-314.
3. Pugh, *IBM's 360 and Early 370 Systems*, 305.
4. Pugh, *IBM's 360 and Early 370 Systems*, 328-329.
5. Jim Strickland, "Oral History of Fritz Trapnell", interviewed on May 27, 2014 at Mountain View, California.
6. Pugh, *IBM's 360 and Early 370 Systems*, 339-343.
7. Pugh, *IBM's 360 and Early 370 Systems*, 343.
8. Grady Booch, "Oral History of Watts Humphrey", interviewed on June 17, 2009, Computer History Museum.
9. T.A. Wise, "The Rocky Road to the Marketplace" (Fortune, October, 1966), 138-143.
10. Booch, "Oral History of Watts Humphrey".
11. Strickland, "Oral History of Fritz Trapnell".
12. Thomas J. Watson, Jr. and Peter Petre, *Father, Son, and Co.* (Bantam, 1990), 353.
13. Grady Booch, "Oral History of Fred Brooks", interviewed on September 16, 2007, Computer History Museum.
14. Richard S. Tedlow, "The IBM System/360: A Look Back at the Creation of a Computing History Giant", Core Magazine, 2014.
15. Frederick Trapnell, Interview by Jim Strickland on March 27, 2014 at Mountain View, California, Computer History Museum.

16. Frederick P. Brooks, Jr., *The Mythical Man-*Month (Addison Wesley Longman, 1995).
17. Brooks, Jr., *The Mythical Man-*Month, 17.
18. Daniel Roth, "Quoted Often, Followed Rarely", Fortune, December 12, 2005.
19. Pugh, *IBM's 360 and Early 370 Systems*, 344.

## THE SYSTEM/360 ARRIVES

1. James T. Cortada, *IBM: The Rise and Fall and Reinvention of a Global Icon* (The MIT Press, 2019), 214.
2. Bill Hansen, "Fifty Years Of Operating IBM Systems", The Four Hundred, March 11, 2019.
3. Henry B. Comstock, "Inside IBM's World's Fair Egg", Popular Science, July, 1964.
4. "The IBM Pavilion: New York World's Fair 1964-1965", IBM, 1964. This was an elaborate, four-color guide to the IBM space at the fair.
5. "IBM Creates an Information Machine", New York World's Fair Report, July, 1964.
6. Walter Carlson, "Eisenhower Gets Ovation at Fair", New York Times, July 9, 1964.
7. Ralph Watson MeElvenny and Marc Wortman, *The Greatest Capitalist Who Ever Lived: Tom Watson, Jr. and the Epic Story of How IBM Created the Digital Age* (Public Affairs, 2023), 177.
8. Emerson W. Pugh, Lyle R. Johnson, and John H. Palmer, *IBM's 360 and Early 370 Systems* (MIT Press, 1991), 171.
9. Ralph Watson MeElvenny and Marc Wortman, *The Greatest Capitalist Who Ever Lived*, 388.
10. T. A. Wise, "The Rocky Road to the Marketplace", (Fortune, October, 1966), 138.
11. Thomas J. Watson, Jr. and Peter Petre, *Father, Son, and Co.* (Bantam, 1990), 350.
12. T.A. Wise, "The Rocky Road to the Marketplace", 138-143.
13. Pugh, *IBM's 360 and Early 370 Systems* (MIT Press, 1991), 100-105.
14. "Historical Narrative: The 1960s", US vs. IBM Exhibit 14971, July 22, 1980.
15. Pugh, *IBM's 360 and Early 370 Systems*, 66. Jerry Haddad would observe regarding Gibson, "Components is the man's life."
16. Pugh, *IBM's 360 and Early 370 Systems*, 209.
17. Watson, *Father, Son, and Co.*, 359.
18. William D. Smith, "Arthur Watson of IBM, Ex-Envoy, Dies", The New York Times, July 27, 1974.
19. James T. Cortada, *IBM: The Rise and Fall and Reinvention of a Global Icon* (The MIT Press, 2019), 219.
20. Ken Shirriff, "IBM System/360 Model 30", Ken Shirriff's blog entry 2020, retrieved February 6, 2023..
21. Chuck Boyer, "The 360 Revolution", IBM Corporation, 2004.
22. Emerson W. Pugh, *Building IBM: Shaping An Industry and Its Technology* (MIT Press, 1995), 296.
23. Grady Booch, "Oral History of Fred Brooks", interviewed on September 16, 2007, Computer History Museum.
24. Cortada, *IBM: The Rise and Fall and Reinvention of a Global Icon*, 224.
25. Alex Taylor, Andrew Erdman, Justin Martin, and Tricia Welch, "US Cars Come Back", Fortune, November 16, 1992.
26. Richard S. Tedlow, "The IBM System/360: A Look Back at the Creation of a Computing History Giant, Core Magazine, 2014.

27. "System/360: From Computers to Computer Systems", from ibm.com/ibm/history, retrieved February 6, 2023.
28. Watson, *Father, Son, and Co.*, 349.
29. Pugh, *IBM's 360 and Early 370 Systems*, 109.
30. T. A. Wise, "The Rocky Road to the Marketplace" (Fortune, October 1966).

# A MISS AT THE HIGH END

1. William D. Nordhaus, "The Progress of Computing", (Yale University, 2001). The standard was the DEC VAX-11/780, rated at 1 MIP.
2. Thomas J. Watson, Jr. and Peter Petre, *Father, Son, and Co.* (Bantam, 1990), 383.
3. "US vs. IBM Exhibit 14971", dated July 2, 1980.
4. Charles J. Murray, The Supermen: The Story of Seymour Cray and the Technical Wizards behind the Supercomputer (John Wiley & Sons, 1997), 76.
5. Frederick P. Brooks, Jr., *The Design of Design: Essays from a Computer Scientist* (Addison-Wesley, 2010), 70.
6. Ernest von Simson, *The Limits of Strategy*, (iUniverse, 2009), 6.
7. Emerson W. Pugh, Lyle R. Johnson, and John H. Palmer, *IBM's 360 and Early 370 Systems* (MIT Press, 1991), 395,
8. Brooks, *The Design of Design*, 121.
9. T. Block, "Large Computer Systems and New Architectures" (*Computer Physics Communications*, vol. 26, 1982), 125-145.
10. William Hoffman, "Cray's Mark Remains Speed with Simplicity", University of Minnesota Update, Spring, 1993
11. Robert Johnson, "Cray's Condition Improves", Chippewa Herald, September 25, 1996.
12. "Historical Narrative: The 1960s", US vs. IBM Exhibit 14971, July 22, 1980.
13. Bob O. Evans, The Genesis of the Mainframe, self-published memoir, 1999.
14. Pugh, *IBM's 360 and Early 370 Systems*, 108.
15. Pugh, *IBM's 360 and Early 370 Systems*, 388.
16. Bob Zeidman, "Oral History of Michael J. Flynn", interviewed on December 17, 2012, Computer History Museum.
17. Pugh, *IBM's 360 and Early 370 Systems*, 395.
18. Frank Da Cruz, "The IBM 360/91", Columbia University Computing History, retrieved August 11, 2024.
19. R, T, Braden and W. F. Miller, "The SLAC Central Computer", SLAC Computation Group, October 25, 1967.
20. "The U.S. v. IBM Stipulation of Fact. Model 90 Program", February 11, 1975.
21. Pugh, *IBM's 360 and Early 370 Systems*, 417.
22. John P. Shen and Mikko H. Lipasti, *Modern Processor Design* (Wakeland Press, 2005), 373.
23. Robert Garner, "Oral History of Charles (Chuck) Branscomb", Computer History Museum, November 10, 2009.
24. Pugh, *IBM's 360 and Early 370 Systems*, 405.
25. Bob O. Evans, *The Genesis of the Mainframe*.

# A MISS WITH PLUG COMPATIBLES

1. "Historical Narrative: The 1960s", US vs. IBM Exhibit 14971, July 22, 1980.
2. T. Noyes, W.E. Dickinson, "Engineering Design of a Magnetic-Disk Random-Access Memory". Proceedings of the Western Joint Computer Conference, February 1952, 42-44.
3. Bill Carlson, "IBM 1311 (1962-75) and 2311 Disk Drives with 1316 Disk Pack", retrieved from cloudfront.net on February 6, 2023.
4. Emerson W. Pugh, Lyle R. Johnson, and John H. Palmer, *IBM's 360 and Early 370 Systems* (MIT Press, 1991), 490-491.
5. Pugh, Johnson, and Palmer, *IBM's 360 and Early 370 Systems*, 771. The twelve engineers were John Harmon, Jim Woo, Martin Halfhill, Russell Brunner, John McNulty, Robert Crouch, Frank Sordello, Rick Wilford, Steven MacArthur, Carlo, Westenskow, Stanley Brown, and Harold Yang.
6. Jim Porter, "Oral History of Jack Clemens", interview recorded July 9, 2007 at Mountain View California.
7. Porter, "Oral History of Jack Clemens".
8. Jim Porter, "Oral History of Ken Haughton", interviewed at Mountain View, California on February 14, 2007.
9. Pugh, Johnson, and Palmer, *IBM's 360 and Early 370 Systems*,494.
10. Chris Bajorek and Tom Garner, "Oral History of Jack Harker with Denis Mee".
11. Pugh, *IBM's 360 and Early 370 Systems*, 494.
12. Jim Porter, "Oral History of Jack Clemens", interview recorded July 9, 2007 at Mountain View, California.
13. Tom Burniece, "Oral History of Dal Allan", interview March 12, 2013, Mountain View, California.
14. Bajorek and Garner, "Oral History of Jack Harker with Denis Mee".
15. Porter, "Oral History of Ken Haughton". The core information on the development of the 3340 Winchester and 3370 drives comes from this interview.
16. Tom Gardner and M. Warner., "IBM 3340 Direct Access Storage Facility". CHM Storage SIG Research Notes, June, 2011.
17. Chris Bajorek and Tom Garner, "Oral History of Jack Harker with Denis Mee".
18. Tom Gardner, "Oral History of Jesse I. Aweida", interviews October 10, 2017 at Superior, Colorado, Computer History Museum.
19. Gardner, "Oral History of Jesse I. Aweida".

# A MISS WITH TIMESHARING

1. Bob O. Evans, The Genesis of the Mainframe, self-published memoir, 1999.
2. Ernest von Simson, *The Limits of Strategy* (iUniverse, 2009), 39.
3. Mike Murray and Dan Rockmore, "Birth of BASIC", video produced by the Trustees of Dartmouth College, 2014.
4. Paul E. Ceruzzi, *A History of Modern Computing* (MIT Press, 2003), 116.
5. Richard Feynman, "Los Alamos From Below", lecture given at the University of California, Santa Barbara, 1975.
6. Frank Da Cruz, "John G. Kemeny: BASIC and DTSS: Everyone a Programmer", retrieved from columbia.edu/cu/computinghistory on February 6, 2023.
7. Daniel Daily, "Interview of Thomas E. Kurtz". Conducted June 20, 2002 at Hanover, N.H.

8. Steve Lohr, *Go To* (Basic Books, 2001), 82.
9. Harry McCracken, "Fifty Years of BASIC, the Programming Language That Made Computers Personal", Time, April 28, 2014.
10. McCracken, "Fifty Years of BASIC".
11. Daniel Daily, "Interview of Thomas E. Kurtz".
12. Daniel Daily, "Interview of Thomas E. Kurtz".
13. Richard Wexelblat, editor, History of Programming Languages (Academic Press, 1981), 517.
14. Lohr, *Go To*, 85.
15. Daniel Daily, "Interview of Thomas E. Kurtz".
16. Daniel Daily, "Interview of Thomas E. Kurtz".

# EARLY EFFORTS IN EUROPE

1. Ernest von Simson, *The Limits of Strategy* (iUniverse, 2009), 233.
2. Ernest von Simson, The Limits of Strategy (iUniverse, 2009), 233.
3. Larry Saphire, "Oral History with Karl Ganzhorn", interview TC-55 at Boblingen on February 5, 1968. Ganzhorn's story is summarized here but his full interview covering war in North Africa, years in a POW camp, and rise to head IBM Europe's research and development laboratories is nothing short of amazing.
4. Bob Moore, "Unruly Allies: British Problems with the French Treatment of Axis Prisoners of War, 1943-1945" (War in History, Vol.7, No.2, April, 2000).
5. Saphire, "Oral History with Karl Ganzhorn".
6. Frederik Nebeker, "Oral History of Karl Ganzhorn", conducted in 1994, IEEE History Center, Piscataway, NJ, USA.
7. Larry Saphire, "Oral History with Ralph Mork", interviewed at Harrison, NY on October 24, 1967, IBM World History of Computer Technology.
8. Saphire, "Oral History with Ralph Mork".
9. Saphire, "Oral History with Karl Ganzhorn".
10. "Origins of Architecture and Design of the IBM 1401", retrieved from ibm-1401.info, April 26, 2021.
11. Emerson W. Pugh, Lyle R. Johnson, and John H. Palmer, *IBM's 360 and Early 370 Systems* (MIT Press, 1991), 446.
12. David Gifford and Alfred Spector, "Case Study: IBM's System/360-370 Architecture" (*Communications of the ACM*, Volume 30, No. 4, April 1987).
13. Ken Shirriff, "Iconic consoles of the IBM System/360 mainframes, 55 years old", Ken Shirriff's blog, retrieved February 6, 2023.
14. Saphire, "Oral History with Karl Ganzhorn".
15. "UNIVAC 9200 and 9300", Datapro Research Corporation, October 1971.
16. "Sperry Rand UNIVAC 9000 Series: Facts and Figures Guide", Sperry Rand Corporation, 1967.
17. Andrew Dalton, "GE Engineer, Father of Steven Spielberg, 103" Chicago Sun Times, August 27, 2020.
18. "Origin of the Level 64, the New Product Line of Honeywell in the 1970s", Bull-General Electric internal memo, February 20, 2003.

# THE ROAD TO SYSTEM/3

1. Jeffrey R. Yost, editor, *The IBM Century: Creating the IT Revolution* (IEEE Computer Society, 2011), 25.
2. David T. Bishop, "Why IBM came to Rochester", Albert Lea Tribune, November 5, 2011.
3. Arthur L. Norberg and Jeffrey R. Yost, "IBM Rochester @ 50 – 1956-2006", prepared by the Charles Babbage Institute, University of Minnesota.
4. Frank G. Soltis, *Fortress Rochester: The Inside Story of the IBM iSeries* (29th Street Press, 2001), 368.
5. Soltis, *Fortress Rochester: The Inside Story of the IBM iSeries*, 367.
6. Larry Saphire, "Oral History with Ralph Mork", interviewed at Harrison, NY on October 24, 1967, IBM World History of Computer Technology.
7. Walter D. Jones, "Personal Observations and Comments Concerning the History and Development of IBM: The Perspective of Walter D. Jones", from the compilation of Jeffrey R. Yost, *The IBM Century: Creating the IT Revolution* (IEEE Computer Society, 2011), 39-52.
8. Graham Pitcher, "Blasts from the Past", IBM Hursley Museum, April 28, 2015.
9. Andrew Lucas, "Comprehensible computer" (*Design Journal*, April, 1970), 60-63.
10. "New System for Small Business", The IBM Newsmagazine (August 11, 1969).
11. Ed LaHay, "Oral History of Burton Grad (IBM)", interviewed on November 28, 2007, Computer History Museum.
12. Ed Nanas, "New Punched Card, Monolithic Technology", press release from IBM, July 30, 1969.
13. "IBM System/3 History", posted on ibmsystem3.nl website, retrieved February 6, 2023.
14. "System/3, featuring new size punched card and low price disk capability, introduced for small business use", IBM Peach Letter, July 30, 1969.
15. Saphire, "Oral History with Ralph Mork".

# IBM IN THE SOUTH

1. Michael Cowlishaw, "IBM Jargon and General Computing Dictionary Tenth Edition", 1990, retrieved December 19, 2022.
2. Robert Garner, "Oral History of Charles (Chuck) Branscomb", Computer History Museum, November 10, 2009.
3. Gene Tharpe, "IBM Reaps Small Business Harvest, Atlanta Journal Constitution, October 9, 1972.
4. "IBM Frequently Asked Questions", retrieved from IBM.com on February 6, 2023.
5. "14K Days: A History of the Poughkeepsie Laboratory", IBM, November 30, 1984.

# THE FIRST PERSONAL COMPUTER

1. Angel West Pedrao, "Big Blue and the Concrete Wave: IBM Boca Raton and Marcel Breuer's Redefinition of Modernity", Modern and International Architecture Seminar, 2013.
2. Philip L. Frana, "An Interview with Glenn Henry", interviewed on August 7, 2001, Charles Babbage Institute.

3. Kevin Krewell, "Oral History of Glenn Henry", Computer History Museum, Interviewed Aug. 7, 2017 at Mountain View, CA..

4. Porter, James N., "The Disk Drive Story", interviews conducted at the IBM Almaden Research Center, November, 2001 through March 22, 2002.

5. Henry, Glenn, "The IBM System/36", Glenn's Computer Museum, posted 9/11/2014, retrieved February 6, 2023.

## GENERAL SYSTEMS GROWS UP

1. Alex Woodie, "In Orlando, Optimism Returns", The Four Hundred, May 10, 2010.

2. Thomas Haigh, "Oral History of Robert L. Patrick" , interviewed on February 16, 2006 in Mountain View, California.

3. "DRAM: The Invention of On-Demand Data", from IBM 100: Icons of Program, retrieved from https://www.ibm.com/ibm/history/ibm100/us/en/icons/dram/ on January 7, 2023.

4. Thomas J. Watson, Jr. and Peter Petre, *Father, Son, and Co.* (Bantam, 1990), 377.

5. Watson and Petre, *Father, Son, and Co.*, 380.

6. Sara Baase, "IBM: Producer or Predator" , Reason, April, 1974.

7. Watson and Petre, *Father, Son, and Co.*, 384.

8. Watson and Petre, *Father, Son, and Co.*, 381.

9. Ed LaHay, "Oral History of Burton Grad (IBM)", interviewed on November 28, 2007, Computer History Museum.

10. David Bird, "IBM Brief Disputed by U. S. as Distorted", The New York Times, August 28, 1979.

11. Ernest Holsendolph, "U.S. Settles Phone Suit, Drops IBM Case", The New York Times, January 9, 1982

12. Barnaby Feder, "IBM Says 13-Year Suite Didn't Affect Operations", The New York Times, January 10, 1982.

13. "U.S. vs. I.B.M.", The New York Times, February 15, 1981.

14. "Aero convicted of selling jet fighter parts to Iran", UPI, November 17, 1992.

15. Vince Ciotti, "H.I.S.-tory", presentations on Healthcare Industry history at histalk.com, retrieved February 6, 2023.

16. Emerson W. Pugh, Lyle R. Johnson, and John H. Palmer, *IBM's 360 and Early 370 Systems* (MIT Press, 1991), 514-520.

17. Chris Bajorek and Tom Garner, "Oral History of Jack Harker with Denis Mee", interviewed at Mountain View, California, on May 30, 2007 and July 10, 2007.

18. Chris Bajorek and Tom Garner, "Oral History of Jack Harker with Denis Mee".

## GLENN HENRY'S ENTRY SYSTEM

1. Philip L. Frana, "An Interview with Glenn Henry", interviewed on August 7, 2001, Charles Babbage Institute.

2. "International Competition in Electronics", Congress of the United States, Office of Technology Assessment, 1983.

3. Arthur L. Norberg, "Interview of David Schleicher" conducted on January 24, 2006 in Rochester, MN, Charles Babbage Institute, Center for the History of Information Processing, University of Minnesota.

4. Eric J. Wieffering, "The brave new world of IBM Rochester" (Post-Bulletin, May 23, 1992).

5. Frana, "An Interview with Glenn Henry".
6. Glenn Henry, "The IBM System/32: The Second IBM Personal Computer", Glenn's Computer Museum, 9/11/2014.
7. J. S. Heath, "Design of a Swinging Arm Actuator for a Disk File" IBM Journal of Research and Development, July 1976.
8. "International Competition in Electronics".
9. "International Competition in Electronics".
10. "1979: Hard Disk Diameter Shrinks to Eight Inches", Computer History Museum, retrieved from on January 7, 2023.
11. Andrew Pollack, "I.B.M. Introduces Computer to Replace System/34 Model", The New York Times, May 17, 1983.
12. "Big Chicken", Wikipedia.com, retrieved February 6, 2023.
13. Robert Snowdon Jones, "IBM has moved south, bringing business with it", Atlanta Journal Constitution, July 11, 1983.

## FUTURE SYSTEM (FS) CRASH AND BURN

1. "High Level System: Final Report", Machine Organization Concept Study Group, IBM, May 18, 1970.
2. John F. Sowa, "Memorandum No. 125", November 27, 1974. This is an extensive and cogent analysis of the progress of FS, done in real-time by one of the key participants.
3. Herbert Grosch, "IBM's John McPherson, 1908-1999", Columbia University Computing History, 2004.
4. John McPherson, "On Mechanical Tabulation of Polynomials" (*Annals of Mathematical Statistics*, Vol. 12, No. 3, September, 1941).
5. John F. Sowa, "Computer Systems", (self-published, 2000), retrieved at jfsowa.com on February 6, 2023..
6. Emerson W. Pugh, Lyle R. Johnson, and John H. Palmer, *IBM's 360 and Early 370 Systems* (MIT Press, 1991), 348.
7. Sowa, "Computer Systems".
8. "High Level System: Final Report".
9. Ralph Watson McElvenny and Marc Wortman, *The Greatest Capitalist Who Ever Lived: Tom Watson, Jr. and the Epic Story of How IBM Created the Digital Age* (Public Affairs, 2023), 418.
10. Sowa, "Memorandum No. 125".
11. Pugh, *IBM's 360 and Early 370 Systems*, 548.
12. Burton Grad, "Oral History of Richard Case", interviewed on December 7, 2006, Westport, Connecticut.
13. Sowa, "Memorandum No. 125".
14. Sowa, "Memorandum No. 125".
15. Frederick P. Brooks, *The Design of Design: Essays from a Computer Scientist* (Addison-Wesley, 2010), 248.
16. Sowa, "Memorandum No. 125".
17. Sowa, "Memorandum No. 125".
18. John F. Sowa, "Computer Systems", (Self-published, 2000).
19. Sowa, "Computer Systems".
20. Burton Grad, "Oral History of Richard Case".
21. Burton Grad, "Oral History of Richard Case".

22. Sowa, "Memorandum No. 125".
23. Burton Grad, "Oral History of Richard Case".
24. Sowa, "Memorandum No. 125".
25. Burton Grad, "Oral History of Richard Case".

## FROM THE ASHES COMES PACIFIC

1. Frank G. Soltis, *Inside the AS/400* (News/400 Books, 1997), 353.
2. Frank G. Soltis, "Automatic allocation of digital computer storage resources for time-sharing", doctorate dissertation, Iowa State University, 1968.
3. Soltis, *Inside the AS/400*, xxiv.
4. Stanley Mazor, "Fairchild Symbol Computer", IEEE Annals of the History of Computing, January-March, 2008.
5. Frank G. Soltis, *Fortress Rochester: The Inside Story of the IBM iSeries* (29th Street Press, 2001), 355.
6. Soltis, *Fortress Rochester: The Inside Story of the IBM iSeries*, 369.
7. Timothy Prickett Morgan, "A Frank Solstice" The Four Hundred, June 30, 2022.
8. Soltis, *Fortress Rochester*, 380.
9. Glenn Henry, conversation with the author on July 22, 2024. In the "single-level" store building, Henry and his software team had half with Ray Klotz and the hardware group the other half.
10. Glenn Henry, conversation with the author on July 22, 2024. Harry Tashjian, the lab director, had asked Glenn who he thought should be the System/38 system manager. Glenn had known Brian Utley from his San Jose days and did not care for him. Henry came to respect Utley, who took the fall for the system's delay. He believes Utley deserves much credit for the System/38.
11. Philip L. Frana, "An Interview with Glenn Henry", interviewed on August 7, 2001, Charles Babbage Institute.
12. Frank G. Soltis, *Inside the AS/400*, 361-363.
13. Soltis, *Fortress Rochester*, 378.
14. Frana, "An Interview with Glenn Henry".
15. Frana, "An Interview with Glenn Henry".
16. V. Zacharov, "Computer Systems and Networks: Status and Perspectives", lecture given at CERN School of Computing, September, 1980.
17. Glenn Henry, "Introduction to IBM System/38 Architecture", *IBM System/38 Technical Developments*, 1978.
18. Simon Lavington, *A History of Manchester Computers*, (Swindon, The British Computer Society, 1998), pp. 44-45.
19. Soltis, *Fortress Rochester*, 380-381.
20. Glenn Henry, from conversation on July 24, 2024. This subject is also covered by Wayne O. Evans in "Why MI? IBM AS/400 S/38 iSeries", June 13, 2016. Henry would note that Dick Bains leaned towards leaving and joining Ray Klotz on the hardware team. Bains would go on to champion the System/36 and Advanced/36.
21. Glenn Henry, July 24, 2024 conversation. Henry notes that George Radin came to Rochester to press the use of RISC architecture but the team already had enough on their plate.
22. E. F. Codd, "A Relational Model of Data for Large Shared Data Banks", IBM Research Laboratory, 1970.
23. Soltis, *Inside the AS/400*, 124.

24. Steve Lohr, *Go To* (Basic Books, 2001), 162-163.
25. Soltis, *Inside the AS/400*, 359.

## TROUBLE IN ROCHESTER

1. "IBM System/38", retrieved from the ibm.com archives on February 6, 2023.
2. Philip Frana, "An Interview with Dick Hedger", conducted in Minneapolis, Minnesota on May 17, 2001.
3. Philip Frana, "An Interview with Glenn Henry", interviewed on August 7, 2001, Charles Babbage Institute.
4. Glenn Henry, conversation with the author on July 22, 2024. Glenn called the System/38 "the hardest thing I've done" which says quite a bit considering his many other successes.
5. Burton Grad, "Oral History of Richard Case", interviewed on December 7, 2006, Westport, Connecticut.
6. Roger L. Taylor, "Low-End General-Purpose Systems", IBM Journal of Research and Development, Vol. 25, No. 5, September 1981.
7. Arthur L. Norberg, "Interview of David Schleicher" conducted on January 24, 2006 in Rochester, MN, Charles Babbage Institute, Center for the History of Information Processing, University of Minnesota.
8. Alex Woodie, "JD Edwards Co-Founder McVaney Leaves a Legacy" (The Four Hundred, September 2, 2020).

## FORT KNOX AND TOO MANY COMPUTERS

1. Frank G. Soltis, *Fortress Rochester: The Inside Story of the IBM iSeries* (29th Street Press, 2001), 380.
2. Frank G. Soltis, *Inside the AS/400* (News/400 Books, 1997), 369.
3. Arthur L. Norberg , "Interview of David Schleicher" conducted on January 24, 2006 in Rochester, MN, Charles Babbage Institute, Center for the History of Information Processing, University of Minnesota.
4. David E. Sanger, "The Moment of Truth for Big Blue", New York Times, January 3, 1988.
5. Sanger, "The Moment of Truth for Big Blue".
6. Rosemary Hamilton, "IBM 9370 user survives battle of misconceptions". Computerworld. February 5, 1990.
7. Frank G. Soltis, *Inside the AS/400* (News/400 Books, 1997), 370.
8. Norberg, "Interview of David Schleicher".
9. "The Origin of AS/400", AS/400 Magazine, June, 1998.
10. Arthur L. Norberg and Jeffrey R. Yost, "IBM Rochester @ 50 – 1956-2006", prepared by the Charles Babbage Institute, University of Minnesota.
11. Norberg, "Interview of David Schleicher.
12. Roy A. Bauer, Emilio Collar, and Victor Tang, *The Silverlake Project* (Oxford University Press, 1992), 49.
13. Norberg and Yost, "IBM Rochester @ 50 – 1956-2006".
14. Bauer, Collar, and Tang, *The Silverlake Project*, 53.
15. Bauer, Collar, and Tang, *The Silverlake Project*, 166.
16. "MAPICS, Inc. History", Funding Universe, retrieved June 2, 2022.

17. Stanley Gibson, "Splash! IBM midrange users in one pool", Computerworld, June 27, 1988.
18. "A Brief History of the IBM AS/400 and iSeries", retrieved from ibm.com on February 6, 2023.
19. Bauer, Collar, and Tang, The Silverlake Project, 177.
20. "A Brief History of the IBM AS/400 and iSeries", retrieved from ibm.com on February 6, 2023.
21. "Microsoft AS/400 corporate use", Stellar Solutions, September 22, 2014.
22. Norberg and Yost, "IBM Rochester @ 50 – 1956-2006".
23. Eric J. Wieffering, "The Brave New World of IBM Rochester", Rochester Post-Bulletin, May 23, 1992
24. Arthur L. Norberg, "Interview of David Schleicher" conducted on January 24, 2006 in Rochester, Minnesota, Charles Babbage Institute, Center for the History of Information Processing, University of Minnesota.
25. Peter S. DeLisi, "A Modern-Day Tragedy: The Digital Equipment Story", Journal of Management Inquiry (Vol. 7 No. 2, June, 1998).
26. Ernest von Simson, The Limits of Strategy (iUniverse, 2009), 109.

# THE RISC REVOLUTION

1. Jeff Kiger, "IBM Rochester, AS/400 Absorbed 10 Years Ago", Rochester Post Bulletin, April 2, 2018.
2. Frederick P. Brooks, Jr., The Design of Design: Essays from a Computer Scientist (Addison-Wesley, 2010), 249.
3. Cocke, John and V. Markstein, "The Evolution of RISC Technology at IBM", (IBM Journal of Research and Development, vol. 34, is. 1, January, 1990), 4-11.
4. Douglas Fairbairn and Robert Garner, "IBM 801 Microprocessor Oral History Panel, with Frank Carrubba, Peter Markstein, Richard Freitas, Rich Oehler, and Vicky Markstein", interviewed October 27, 2014 at Mountain View, California.
5. Steven Leibson, "Fifty (or Sixty) Years of Processor Development...for This?", Journal of Electronic Engineering, March 29, 2018.
6. Cocke and Markstein, "The Evolution of RISC Technology at IBM".
7. Charles H. Ferguson and Charles R. Morris, Computer Wars: The Fall of IBM and the Future of Global Technology (Times Books, 1993), 44-45.
8. Philip L. Frana, "An Interview with Glenn Henry", interviewed on August 7, 2001, Charles Babbage Institute.
9. Glenn Henry, conversation with the author on July 22, 2024. Henry received another large IBM cash award for this his development of AIX.
10. Kevin Krewell, "Oral History of Glenn Henry", Computer History Museum, Interviewed August 7, 2017, Mountain View, CA.
11. Cocke and Markstein, "The Evolution of RISC Technology at IBM".
12. Saul Austerlitz, "40 Years Ago, This Ad Changed the Super Bowl Forever", New York Times, February 9,2020. The author notes that Ridley Scott was paid nearly $1 million to direct the commercial.
13. Andrew Pollack, "I.B.M. Now Apple's Main Ally", The New York Times, October 3, 1991.
14. Tom Hormby, "A Brief Taligent History", Low End Mac, April 27, 2014.
15. Morley Safer, "The 60 Minutes Interview with Grace Hopper", CBS News, February 13, 2017.

16.  "Multi-Core Processor", retrieved from Wikipedia.com on October 8, 2023.

17.  Sowa, John F., "Computer Systems", 2000, self-published.

18.  Arthur L. Norberg, "Interview of David Schleicher" conducted on January 24, 2006 in Rochester, MN, Charles Babbage Institute, Center for the History of Information Processing, University of Minnesota.

19.  Frank G. Soltis, *Inside the AS/400* (News/400 Books, 1997), 27-28.

20.  William Berg, Marshall Cline, and Mike Girou, "Lessons Learned from the OS/400 OO Project" (*Communications of the ACM*, Vol. 38, No. 10, October, 1995).

21.  John L. Hennessy and David A. Patterson, "A New Golden Age for Computer Architecture" (*Communications of the ACM*, Vol. 62, No. 2, February, 2019).

22.  Soltis, *Inside the AS/400*, 58.

23.  "Assessing IBM's New RAMP-C Standard, TechMonitor, August 2, 1992.

24.  Timothy Prickett Morgan, "The More Things Change", The Four Hundred, September 20, 2010.

25.  Frank G. Soltis, *Fortress Rochester: The Inside Story of the IBM iSeries* (29th Street Press, 2001), xxv.

26.  "From Silverlake to IBM i, a Powerful Story of Evolution:, Software Engineering of America, July 26, 2018.

27.  IBM Rochester Timeline", IBM archives at ibm.com, retrieved February 6, 2023.

28.  Becky Bosch, "iSeries Sparklers", IBM internal presentation, July, 2004.

29.  Bob O. Evans, The Genesis of the Mainframe, self-published memoir, 1999.

30.  Norberg, "Interview of David Schleicher".

31.  Thomas J. Misa, *Digital State: The Story of Minnesota's Computing Industry* (University of Minnesota Press, 2013), 22.

# WATSON, LEARSON, AND SYSTEM/370

1.  Marjorie Williams, "Back to Siberia, With Gusto", The Washington Post, July 21, 1987.

2.  Peter E. Greulich, "Comparing IBM CEO Yearly Revenue Growth", discerning readers.com, August 5, 2021.

3.  "The Greatest Capitalist in History", Fortune Magazine, August 31, 1987.

4.  Thomas J. Watson, Jr., *A Business and its Beliefs* (McGraw-Hill, 2003).

5.  Williams, "Back to Siberia".

6.  "The Greatest Capitalist in History", Fortune Magazine, August 31, 1987.

7.  Thomas A. Stewart, Alex Taylor, Peter Petre, and Brent Schlender, "The Businessman of the Century", Fortune, vol. 140, no. 10, November 22, 1999.

8.  William Rodgers, *Think: A Biography of the Watsons and IBM* (New American Library, 1969), 273.

9.  "Learson at IBM's Helm", Time Magazine, July 12, 1971.

10.  A close third behind the Watson family and Sherman Fairchild was Lyle Spencer. Watson, Jr. asked James Birkenstock to negotiate the purchase of Science Research Associates (SRA) in 1963. The acquisition transferred 128,000 shares of IBM stock to Spencer, worth $62 million at IBM's lofty price of $569 per share. With this huge stake in the company, Spencer was added to the IBM board of directors. SRA underperformed Watson's expectations and was eventually sold in 1988. Learson recounted the story in 1982: Isabel Grossner: "Oral history interview with T. Vincent Learson", Interviewed on August 6, 1982 in New York, Spencer Foundation, Columbia University Oral Histories, 1982.

11. Marjorie Williams, "Back to Siberia, With Gusto".

12. Ralph Watson MeElvenny and Marc Wortman, *The Greatest Capitalist Who Ever Lived: Tom Watson, Jr. and the Epic Story of How IBM Created the Digital Age* (Public Affairs, 2023), 414.

13. Ralph Watson MeElvenny and Marc Wortman, *The Greatest Capitalist Who Ever Lived*, 440-441.

14. Chris Bajorek and Tom Garner, "Oral History of Jack Harker with Denis Mee", interviewed at Mountain View, California, on May 30, 2007 and July 10, 2007.

15. Bob O. Evans, The Genesis of the Mainframe, self-published memoir, 1999.

16. Ernest von Simson, *The Limits of Strategy* (iUniverse, 2009), 39.

17. "14K Days: A History of the Poughkeepsie Laboratory', IBM, November 30, 1984.

18. Mark Smotherman, "IBM Advanced Computing Systems – Timeline", updated December 30, 2016.

19. Steve Hawkins, "Gene Amdahl dies at 92; chief architect of IBM's mainframe computers", The Los Angeles Times, November 15, 2015.

20. William S. Anderson, "An Interview with Gene M. Amdahl", IEEE Solid-State Circuits Society News, Summer 2007.

21. Marvin Wolf, *The Japanese Conspiracy* (Hodder & Stoughton, 1985), 52.

22. Wolf, *The Japanese Conspiracy*, 42.

23. Simson, *The Limits of Strategy*, 67.

24. Robert Sobel, *IBM: Colossus in Transition* (Bantam Books, 1981), 318.

25. David T. Levy, "Short-term Leasing and Monopoly Power: The Case of IBM" (*Journal of Institutional and Theoretical Economics*, Vol. 144, No. 4, September, 1988).

## TALK TO FRANK

1. Thomas J. Watson, Jr. and Peter Petre, *Father, Son, and Co.* (Bantam, 1990), 395.

2. Thomas J. Watson, *Father, Son, and Co.*, 400.

3. Michael Cowlishaw, "IBM Jargon and General Computing Dictionary Tenth Edition", 1990, retrieved December 19, 2022.

4. Ernest von Simson, *The Limits of Strategy* (iUniverse, 2009), 31.

5. "Frequently Asked Questions: Legal", ibm.com/history, retrieved February 6, 2023.

6. "International Competition in Electronics", Congress of the United States, Office of Technology Assessment, 1983.

7. Connie Winkler, "Cary: IBM Vetoed Spin-Off for Small Machines", Computerworld, April 23, 1979.

8. "Are options the key to IBM buybacks?" The Globe and Mail, April 9, 2002.

9. Thomas J. Watson, *Father, Son, and Co.*, 408.

10. Thomas J. Watson, *Father, Son, and Co.*, 401.

## FRANK AND THE IBM PC

1. Frank Da Cruz, "The IBM 610 Auto-Point Computer", retrieved from Columbia University Computing History on February 6, 2024.

2. The original location of the Watson Lab was in the Delta Phi fraternity house. In 1953, IBM moved the lab to 612 W. 115th Street, a half block off campus. IBM ceded the 4-story building to Columbia in 1970. It is now Watson Hall.

3. Larry Saphire, "Oral History with Ralph Mork", interviewed at Harrison, New York on October 24, 1967, IBM World History of Computer Technology.
4. Paul Friedl, "SCAMP: The Missing Link in the PC Past?", PC Magazine, November, 1983, Vol. 2, No. 6.
5. Jeffrey Young, *Forbes Greatest Technology Stories: Inspiring Tales of the Entrepreneurs and Inventors Who Revolutionized Modern Business* (John Wiley and Sons, 1998), 218.
6. Robert Garner, "Oral History of Charles (Chuck) Branscomb", Computer History Museum, November 10, 2009.
7. Tim Danton, "The true story behind the IBM Personal Computer", IT Pro, December 3, 2021.
8. Patricia Sullivan, "Frank Cary", The Washington Post, January 6, 2006.
9. Robert Heller, *The Fate of IBM* (Warner Books, 1995), 51-52.
10. "Project Chess: The Story Behind the Original IBM PC", Techguide, retrieved February 24, 2022.
11. Bradley, David J., "The Creation of the IBM PC", Byte, September, 1990.
12. Carlos Frias, "In flash of keystrokes, former Boca Raton man Dave Bradley changed computer history", The Palm Beach Post, January 17, 2011..
13. James T. Cortada, *IBM: The Rise and Fall and Reinvention of a Global Icon* (The MIT Press, 2019), 386-387.
14. Bro Uttal, "Inside the Deal That Made Bill Gates $350,000,000", Fortune, July 21, 1986.
15. David Allison, "Video History Interview with Mr. William "Bill" Gates, National Museum of American History, interviewed in Bellevue, Washington on September 27, 1994.
16. Walter Isaacson, *The Innovators* (Simon and Shuster, 2014), 313-343.
17. Harry McCracken, "Fifty Years of BASIC, the Programming Language That Made Computers Personal", Time, April 28, 2014.
18. Steve Lohr, *Go To* (Basic Books, 2001), 83.
19. Brian Warner, "This Man Should Have Been Bill Gates", Celebrity Net Worth, December 8, 2014.
20. Isaacson, *The Innovators*, 358.
21. Michael Miller, "The Rise of DOS: How Microsoft Got the IBM PC OS Contract", PC Magazine, August, 12, 2021.
22. Tim Danton, "The true story behind the IBM Personal Computer" (IT Pro, December 3, 2021).
23. Cortada, *IBM: The Rise and Fall and Reinvention of a Global Icon*, 390.
24. Davie E. Sanger, "Philip Estridge Dies in Jet Crash", New York Times, August 5, 1985.

## JOHN OPEL AND IBM'S GREAT EXPANSION

1. William Lazonick, "Evolution of the New Economy Business Model", Business History Conference, 2006.
2. Thomas J. Watson, Jr. and Peter Petre, *Father, Son, and Co.* (Bantam, 1990), 408-409.
3. Bob O. Evans, The Genesis of the Mainframe, self-published memoir, 1999.
4. Robert X. Cringley, "Triumph of the Nerds", PBS series first aired April 14, 1996.

5. James Langdell, "Phoenix Says Its BIOS May Foil IBM's Lawsuits", PC Magazine, July 10, 1984.
6. Charles H. Ferguson and Charles R. Morris, *Computer Wars: The Fall of IBM and the Future of Global Technology* (Times Books, 1993), 54.
7. John Markoff, "Apple Reverses Strategy on Cloning", New York Times, September 3, 1997.
8. John Greenwald, "The Colossus That Works", Time, July 11, 1983.
9. Sandra Salmans, "Dominance Ended, IBM Fights Back", The New York Times, January 9, 1982l
10. Tom Peters, In Search of Excellence (Harper and Row, 1982).

## JOHN AKERS TACKLES AN OVERHEATED COMPANY

1. Rick Rojas and Steve Lohr, "John F. Akers, 79, Dies: Led IBM as PCs Ascended", The New York Times, August 23, 2014.
2. John A. Byrne, "Be nice to everybody", Forbes, November 5, 1984.
3. Carol J. Loomis, "Can John Akers Save IBM", Fortune, July 15, 1991.
4. Robert Snowdon Jones, "IBM to add building at Chattahoochee site", Atlanta Journal Constitution, March 8, 1983.
5. Sallye Salter, "IBM Tower to Grow Taller", The Atlanta Constitution, January 24, 1986.

## THE PC CHALLENGE

1. L. A. Lorek, "IBM's Boca Campus Was Cradle of PC Revolution", Sun Sentinel, December 11, 1999.
2. David Allison, "Video History Interview with Mr. William "Bill" Gates, National Museum of American History, interviewed in Bellevue, Washington on September 27, 1994.
3. Charles H. Ferguson and Charles R. Morris, *Computer Wars: The Fall of IBM and the Future of Global Technology* (Times Books, 1993), 72-75.
4. Ferguson and Morris, *Computer Wars: The Fall of IBM and the Future of Global Technology*, 76-77.
5. James Cortada, "IBM PC Turns 40", Senior Living, August 12, 2021.
6. James Cortada, IBM: The Rise and Fall and Reinvention of a Global Icon (MIT Press, 2019), 399.
7. Robert Sobel, *IBM: Colossus in Transition* (Bantam Books, 1981), 304.
8. Michael Killen, IBM: The Making of the Common View (Harcourt Brace Jovanovich, 1988), 107-134.
9. "Communiques". PC Magazine. June–July 1982, 27.
10. Anne R. Field, "Why the Hardware Giants are Hustling into Software", Business Week, July 27, 1987
11. Evelyn Richards, "Turning IBM Around", The Washington Post, November 24, 1991.
12. David E. Sanger, "The Moment of Truth for Big Blue", New York Times, January 3, 1988
13. Sanger, "The Moment of Truth for Big Blue".
14. "The Xerox PARC Visit", Making the Macintosh, retrieved from web.stanford.edu on February 6, 2023.

15. David C. Brock, "50 Years Later, We're Still Living in the Xerox Alto's World", IEEE Spectrum, March 1, 2023.

16. Jeremy Reimer, "Half an operating system: The triumph and tragedy of OS/2", Ars Technica, November 29, 2019.

17. John Burgess, "IBM Finishes One Race, Starts Another", The Washington Post, March 31, 1992.

18. Doug Garr, *IBM Redux: Lou Gerstner and the Business Turnaround of the Decade* (Harper Business, 1999), 189.

19. John Markoff, "Computer Feud Enters a New Phase", New York Times, May 26, 1992.

20. Bro Uttal, "Inside the Deal That Made Bill Gates $350,000,000", Fortune, July 21, 1986.

21. Casey Corr, "Microsoft's Next Move – The Brains Behind Windows NT Could Make Company A Pile of Money", Seattle Times, May 16, 1993.

22. David Allison, "Video History Interview with Mr. William "Bill" Gates, National Museum of American History, interviewed in Bellevue, Washington on September 27, 1994.

23. Frank Soltis, "AS/400: Designed for the Cloud", presentation at COMMON Europe, December 15, 2014.

24. Markoff, "Computer Feud Enters a New Phase".

25. Glenn Henry, conversation with the author on July 22, 2024. Glenn Henry was a good friend of Lee Reiswig. Lee tried to sell Henry on OS/2 but failed to sway him.

26. Markoff, "Computer Feud Enters a New Phase".

27. Brett D. Fleisch and Mark Allan, "Workplace microkernel and OS: a case study" (Software: Practice and Experience, Volume 28, Issue 6, January 1999).

28. Jason Pontin, "OS/2 for PowerPC release may be too little, too late", InfoWorld, January 15, 1996.

29. G. Glenn Henry, "From Mainframes to Microprocessors", IEEE Micro (Vol. 41, Is. 6, Nov-Dec 2021). Henry would have 50 engineers working on the L86 program.

30. Steve Lohr, "For Big Blue, the Ones Who Got Away", New York Times, January 9, 1994.

31. Glenn Henry, conversation with the author on July 22, 2024.

32. Philip L. Frana, "An Interview with Glenn Henry", interviewed on August 7, 2001, Charles Babbage Institute.

33. William Lazonick, "Evolution of the New Economy Business Model, Business History Conference, 2006.

34. Glenn Henry, conversation with the author on July 22, 2024. Henry and Michael Dell had a falling out.

35. Frana, "An Interview with Glenn Henry".

36. Dan Luu, "Glenn Henry Interview", interviewed in 1994, retrieved from linuxdevices.com on October 19, 2023.

37. Henry, "From Mainframes to Microprocessors", Henry knew what Dell was paying for Intel microprocessors and decided that he could build a "good enough" compatible microprocessor with a small team of talented engineers.

38. "CEO Interview: Glenn Henry, founder of VIA processor subsidiary Centaur", interview with linuxdevices.com on June 8, 2004.

39. Glenn Henry, "From Mainframes to Microprocessors", Henry did more than programming. With a patent suit lodged by Intel in 2001, he was involved in legal issues for 18 months.

# AKERS AND THE ROAD TO RUIN

1. Isabel S. Grossner, "Oral history interview with T. Vincent Learson", Interviewed on August 6, 1982 in New York, Spencer Foundation, Columbia University Oral Histories, 1982.
2. James T. Cortada, *IBM: The Rise and Fall and Reinvention of a Global Icon* (The MIT Press, 2019), 453.
3. Clinton Wilder, "IBM shuffles management", Computerworld, July 25, 1988.
4. Paul Carroll, *Big Blues: The Unmaking of IBM* (Crown Publishing, 1993), 265.
5. Robert Heller, *The Fate of IBM* (Warner Books, 1995), 47-48.
6. Glenn Rifkin, "Profile: It's Not IBM and It's Not Big, But He's Got Big Ideas", New York Times, February 27, 1994.
7. Rifkin, "Profile: It's Not IBM and It's Not Big, But He's Got Big Ideas".
8. Ernest von Simson, *The Limits of Strategy* (iUniverse, 2009), 312.
9. Carl J. Loomis and David Kirkpatrick, "The Hunt for Mr. X: Who Can Run IBM?", Fortune, February 22, 1993.
10. Carroll, *Big Blues*, 326-327.
11. Doug Garr, IBM Redux (Harper Business, 1999), 19.
12. Cortada, *IBM: The Rise and Fall and Reinvention of a Global Icon*, 453.
13. Simson, *The Limits of Strategy*, 313.
14. Robert McMillan, "Steve Jobs: HP Implosion was an iTragedy", Wired, October 5, 2011.

# ENTER THE "COOKIE MAN"

1. Patricia Sellers and David Kirkpatrick, "Can This Man Save IBM?, Fortune, April 19, 1993.
2. Doug Garr, *IBM Redux: Lou Gerstner and the Business Turnaround of the Decade* (Harper Business, 1999), 83.
3. Louis V. Gerstner, *Who Says Elephants Can't Dance* (Harper, 2002), 2-6.
4. Garr, *IBM Redux: Lou Gerstner and the Business Turnaround of the Decade*, 112..
5. Gerstner, *Who Says Elephants Can't Dance*, 4.
6. Gerstner, *Who Says Elephants Can't Dance*, 14.
7. Gerstner, *Who Says Elephants Can't Dance*, 16.
8. Duff McDonald, "The Godfather of CEO Megapay: McKinsey Consultant Arch Patton Didn't Invent Wealth Inequality", The Observer, August 13, 2013.
9. Garr, *IBM Redux: Lou Gerstner and the Business Turnaround of the Decade*, 27.
10. Steve Lohr, "On the Road with Chairman Lou", New York Times, June 26, 1994
11. Ernest von Simson, *The Limits of Strategy* (iUniverse, 2009), 322.
12. Gerstner, *Who Says Elephants Can't Dance*, 34-35.
13. Gerstner, *Who Says Elephants Can't Dance*, 204.
14. Gerstner, *Who Says Elephants Can't Dance*, 106.
15. Gerstner, *Who Says Elephants Can't Dance*, 43.
16. Sellers and Kirkpatrick, "Can This Man Save IBM?".
17. Paul Carroll, Big Blues: The Unmaking of IBM (Crown Publishing, 1993), 353-354.
18. Gerstner, *Who Says Elephants Can't Dance*, 152.
19. Gerstner, *Who Says Elephants Can't Dance*, 97.
20. D. Quinn Mills, "The Decline and Rise of IBM" (MIT Sloan Management Review, July 15, 1996).

21. Daniel Lyons, "IBM's Giant Gamble", Forbes, October 4, 1999.
22. Gary Hamel, "Waking Up IBM: How a Gang of Unlikely Rebels Transformed Big Blue", Harvard Business Review, July-August, 2000.
23. Gerstner, Who Says Elephants Can't Dance, 262.
24. Gerstner, *Who Says Elephants Can't Dance*, 182.
25. Laurence Zuckerman, "IBM's New Headquarters Reflects A Change of Corporate Style", Technology Cybertimes, September 17, 1997
26. Gerstner, *Who Says Elephants Can't Dance*, 48.
27. Simson, *The Limits of Strategy* (iUniverse, 2009), 331.
28. Feng-hsiung Hsu, Behind Deep Blue: Building the Computer That Defeated the World Chess Champion (Princeton University Press, 2002), vii.
29. Claude Shannon, "Programming a Computer to Play Chess", Philosophical Magazine, Ser.7, Vol. 41, No. 314 - March 1950.
30. Pamela McCorduck, Machines Who Think (A. K. Peters, 2004), pp. 184-187.
31. Garry Kasparov, Deep Thinking (Public Affairs, New York, 2017), pp. 54-55.
32. Hsu, Behind Deep Blue, 13.
33. Kasparov, Deep Thinking, 38.
34. Carol Hamilton, and S. R. Hedberg, "Modern Masters of an Ancient Game", AI Magazine, Vol. 18., No. 4, 1997.
35. Larry Kaufman, "Accuracy, Ratings, and Goats", chess.com, retrieved October 21, 2023.
36. Kasparov, Deep Thinking, 89.
37. Hsu, Behind Deep Blue, 25.
38. Dag Spicer, "Oral History of Feng-Hsiung Hsu", interviewed on February 14, 2005 at Mountain View, California.
39. Hsu, Behind Deep Blue,27.
40. Hsu, Behind Deep Blue, 24.
41. Dag Spicer, "Oral History of Feng-Hsiung Hsu".
42. Hsu, Behind Deep Blue, 93.
43. Hsu, Behind Deep Blue, 94.
44. Hsu, Behind Deep Blue, 111-116.
45. Irv Drasnin, "Kasparov versus Deep Thought Documentary", 1991 video.
46. Hsu, Behind Deep Blue, 132-135.
47. Hsu, Behind Deep Blue, 134 and 162-165.
48. Hsu, Behind Deep Blue, 167.
49. Kasparov, Deep Thinking, 132.
50. Kasparov, Deep Thinking, 154.
51. Hsu, Behind Deep Blue, 176.
52. Kasparov, Deep Thinking, 160.
53. Kasparov, Deep Thinking, 156.
54. "Deep Blue", IBM Research, retrieved from www.research.ibm/deepblue on October 5, 2023.
55. Kasparov, Deep Thinking, 165.
56. McCorduck, Machines Who Think, P. 481.
57. Hsu, Behind Deep Blue, 200.
58. Hsu, Behind Deep Blue, 217.
59. Kasparov, Deep Thinking, 160-163.
60. Kasparov, Deep Thinking, 133.
61. Kasparov, Deep Thinking, 182.
62. Kasparov, Deep Thinking, 190.

63. Hsu, Behind Deep Blue, 253.
64. Hsu, Behind Deep Blue, 257.
65. Bruce Weber, "Victory Eludes Champion After a Tense Fight with the Computer", The New York Times, May 8, 1997.
66. Kasparov, Deep Thinking, 112.
67. Hsu, Behind Deep Blue, 254.
68. Hsu, Behind Deep Blue, 263.
69. Kasparov, Deep Thinking, 218.
70. Hsu, Behind Deep Blue, 168.
71. Kasparov, Deep Thinking, 219.
72. Arthur L. Norberg and Jeffrey R. Yost, "IBM Rochester @ 50 – 1956-2006", prepared by the Charles Babbage Institute, University of Minnesota.
73. David Kirkpatrick, Suzanne Koudsi, Adam Lashinsky, Peter Lewis, Carol Lewis, Carol J. Loomis, Brent Schlender, Julie Schlosser, Fred Vogelstein, and Joshua Watson, "The Future Of IBM Lou Gerstner seems to have pulled off a miracle", Fortune, February 18, 2002.
74. "The Decline of IBM", The Market is Open, March 10, 2020.
75. Gerstner, Who Says Elephants Can't Dance, 166-167.
76. Gerstner, Who Says Elephants Can't Dance, 137.
77. Gerstner, Who Says Elephants Can't Dance, 140-145.

## GERSTNER TAKES ON THE PC

1. Doug Garr, IBM Redux: Lou Gerstner and the Business Turnaround of the Decade (Harper Business, 1999), 175.
2. Garr, IBM Redux: Lou Gerstner and the Business Turnaround of the Decade, 166.
3. James W. Cortada, I, (MIT Press, 2019), 405.
4. Louis V. Gerstner, Who Says Elephants Can't Dance (Harper, 2002), 185.
5. "IBM Tries its hand at Clone-Peddling with Launch of Individual Computer Products Line", Tech Monitor, June 4, 1992.
6. Tim Bajarin, "Why Lenovo's ThinkPad Still Matters 25 Years Later', Time Magazine, October 9, 2017.
7. Harry McCracken, "How IBM's ThinkPad Became A Design Icon", Fast Company, October 5, 2017.
8. Saul Hansell, "The Strategy for IBM: Loss: Leader PC Sales" The New York Times, October 25, 1999.
9. Garr, IBM Redux: Lou Gerstner and the Business Turnaround of the Decade, 193.
10. Gerstner, Who Says Elephants Can't Dance, 138.
11. Gerstner, Who Says Elephants Can't Dance, 143.
12. Hansell, "The Strategy for IBM: Loss Leader PC Sales",
13. Steve Lohr, "Microsoft, Google and Antitrust: Similar Legal Theories in a Different Era", The New York Times, September 11, 2023.
14. David McCabe and Cecilia Kang, "Microsoft CEO Testifies That Google Power in Search is Ubiquitous", The New York Times, October 2, 2023.
15. Lauren Feiner, "Google paid $26 billion in 2021 to become the default search engine on browsers and phones":, CNBC, October 27, 2023.
16. Jeremy Reimer, "Half an operating system: The triumph and tragedy of OS/2", Ars Technica, November 29, 2019.
17. Reimer, "Half an operating system: The triumph and tragedy of OS/2".

18. Steve Lohr, "He Loves to Win. At IBM, He Did", New York Times, March 10, 2002.
19. Paul B. Carrol and Chunka Mui, *Billion Dollar Lessons* (Penguin Group. 2008), p.222.
20. Mike Langbert, "Apple Walking the Wrong Path?", Mercury News, October 2, 1995.
21. Evan I. Schwartz, "30,000 Programmers Later, Software Gridlock", Information Processing, July 15, 1991.
22. Andrew Orlowski, "OS/2 a quarter century on: Why IBM lost out and how Microsoft won", The Register, November 27, 2012.
23. Ernie Smith, "A Kernel of Failure", Tedium, February 28, 2019. .
24. Paul E. Ceruzzi, *A History of Modern Computing* (MIT Press, 2003), 354-360.
25. Martin Campbell-Kelly and Daniel D. Garcia-Schwartz, "IBM's Love Affair with Open Source Software", University of Warwick, January 8, 2008.
26. Irving Wladawsky-Berger, "IBM's Linux Initiative", author's blog dated January 23, 2006
27. Lemos, Robert, "IBM gets $100,000 fine for 'Peace, Love and Linux' campaign", ZDNet, November 28, 2001.
28. Irving Wladawsky-Berger, "The Impressive Scope of the Linux Foundation in the 21st Century Digital Economy", author's blog dated June 30, 2022.

## "THE MAN WHO SAVED IBM"

1. Louis V. Gerstner, *Who Says Elephants Can't Dance* (Harper, 2002), 285.
2. Bethany McLean, "That old financial magic" Fortune, February 18, 2002, vol. 145, is. 4.
3. Lazonick, "Evolution of the New Economy Business Model".
4. William Lazonick, "Evolution of the New Economy Business Model", Business History Conference, 2006.
5. "Top Ten Largest Severance Packages of the Last Decade", Forbes, January 19, 2012.
6. Thomas Heath, "Louis V. Gerstner Jr. lays out his post-IBM life", The Washington Post, June 7, 2013.
7. Gerstner, *Who Says Elephants Can't Dance*, 254-256..

## PALMISANO AND THE ROADMAP ERA

1. David Kirkpatrick, Suzanne Koudsi, Adam Lashinsky, Peter Lewis, Carol Lewis, Carol J. Loomis, Brent Schlender, Julie Schlosser, Fred Vogelstein, and Joshua Watson, "The Future Of IBM Lou Gerstner seems to have pulled off a miracle" Fortune, February 18, 2002.
2. Matt Richtel, "A Gerstner Loyalist Cut From Quite Different Style" New York Times, January 30, 2002.
3. Michael Useem, "Sam Palmisano: Always Put the Enterprise Ahead of the Individual" interviewed January 13, 2012 at Wharton
4. David Kirkpatrick, "Inside Sam's $100 Billion Growth Machine", Fortune, June 14, 2004.
5. Steve Denning, "Why IBM is in decline", Forbes, May 30, 2014.
6. Louis V. Gerstner, *Who Says Elephants Can't Dance* (Harper, 2002), 97.

7. Useem, "Sam Palmisano: Always Put the Enterprise Ahead of the Individual".
8. Robert X. Cringley, "The Decline and Fall of IBM", NerdTV, 2014.
9. Kirkpatrick, "Inside Sam's $100 Billion Growth Machine".
10. Cringley, "The Decline and Fall of IBM".
11. Peter Greulich, "An IBM Case Study: Do Share Buybacks Work?", retrieved April 16, 2022 from discerningreaders.com. A comprehensive analysis of the scope and impact of Palmisano's program of using generated income to buy back shares.
12. Darryl Taft, "IBM's Palmisano Credits Culture for Big Blue's Success", eWeek, June 17, 2011.
13. "Songs of the IBM", IBM Corporation, 1937.
14. James T. Cortada, *IBM: The Rise and Fall and Reinvention of a Global Icon* (The MIT Press, 2019), 585.
15. William M. Bulkeley, "IBM Plots another Share Buyback", The Wall Street Journal, February 27, 2008.
16. Jessi Hempel, "IBM's Sam Palmisano: A super second act", Fortune, March 3, 2011
17. Denning, "Why IBM is in decline".
18. Lauren Ohnesorge, "IBM's former CEO getting million-dollar office", Triangle Business Journal, March 20, 2013.

# ROMETTY STAYS THE COURSE

1. James T. Cortada, *IBM: The Rise and Fall and Reinvention of a Global Icon* (The MIT Press, 2019), 554-556.
2. Jo Best, "IBM Watson: The inside story of how the Jeopardy-winning supercomputer was born, and what it wants to do next", Tech Republic, September 9, 2013.
3. Sissi Cao, "Can ChatGPT Really Think Like a Human?", The Observer, May 27, 2023.
4. Steve Lohr, "What Ever Happened to IBM's Watson", New York Times, July 16, 2021.
5. Lohr, "What Ever Happened to IBM's Watson".
6. Lex Fridman, "IBM Watson, Jeopardy!, and Deep Conversations with AI", an interview with David Ferrucci, October 11, 2019.
7. Fridman, "IBM Watson, Jeopardy!, and Deep Conversations with AI".
8. David Ferrucci, "Building Watson", IBM Corporation presentation, 2010.
9. John Markoff, "Computer Wins on Jeopardy!: Trivial, It's Not", The New York Times, Feb. 16, 2011.
10. Aleksey Tsalolikhin, "System Administration of the IBM Watson Supercomputer", Linux Journal, July 21, 2021.
11. Cortada, *IBM: The Rise and Fall and Reinvention of a Global Icon*, 572-574.
12. Best, "IBM Watson: The inside story of how the Jeopardy-winning supercomputer was born"..
13. Mac Schwerin, "America Forgot about IBM Watson. Is ChatGPT Next?", The Atlantic, May 5, 2023.
14. Jessi Hempel, "IBM's Massive Bet on Watson, Fortune, October 7, 2013, vol. 168 issue 6.
15. Adi Ignatius, "Don't Try to Protect the Past", Harvard Business Review, July-August 2017.
16. Michal Lev-ram, "Getting Past the Big Blues", Fortune, October 6, 2014, vol. 170 issue 5.

17. Lizzie O'Leary, "How IBM's Watson Went From the Future of Health Care to Sold Off for Parts", Slate, January 31, 2022 .
18. Matthew Herper, "MD Anderson Benches IBM Watson In Setback For Artificial Intelligence In Medicine", Forbes, February 19, 2017.
19. Ernest von Simson, *The Limits of Strategy*, (iUniverse, 2009), 8.
20. Robert Mullins, "IBM, Cray win DARPA supercomputer contracts", IDG News Service, November 22, 2006.
21. Tom Papatheodore, "Summit System Overview", Oak Ridge National Laboratory, June 1, 2018.
22. Doug Black, "Where Have You Gone, IBM", InsideHPC, October 14, 2020.

## THE CLOUD COMPUTING WAKE-UP CALL

1. Kevin McLaughlin, "Amazon Wins $600 million Cloud Deal as IBM Withdraws Protest", CRN, October 30, 2013.
2. Frank Konkel, "CIA Awards Secret Multibillion-Dollar Cloud Contract", Nextgov, November 20, 2020.
3. Irving Wladawsky-Berger, "The Transformative Power of Cloud Computing", author's blog dated December 12, 2016.
4. Leena Rao, "A Leader with His Head in the Cloud", Fortune, July 1, 2015.
5. Frank Konkel, "The Details about CIA's Deal with Amazon", Atlantic Magazine, July 17, 2014.
6. Aran Ali, "AWS: Powering the Internet and Amazon's Profits", Visual Capitalist, July 10, 2022.
7. Beth Kindig, "Azure Growth Proves Resilient", Forbes, August 8, 2022 .
8. Chris Kanaracus, "The roots of Oracle's cloud evolution: A 20-year review", Tech Target, September 11, 2019.
9. Michal Lev-ram, "Getting Past the Big Blues", Fortune, October 6, 2014, Vol. 170 Issue 5.
10. Tom Krazit, "How IBM lost the cloud", September 29, 2021, Protocol Media.
11. Krazit, "How IBM lost the cloud".
12. Richard Leiby, "A Chain Of Riches: Making Money Was a Snap. Making It Count Is Another Story", The Washington Post, March 15, 2000.
13. Leiby, "A Chain Of Riches: Making Money Was a Snap. Making It Count Is Another Story".
14. Timothy Pickett Morgan, "Millions Pay AWS to Give Amazon an Insurmountable IT Advantage", The Next Platform, February 4, 2022.

## BYE BYE ROAD MAP

1. James T. Cortada, *IBM: The Rise and Fall and Reinvention of a Global Icon* (The MIT Press, 2019), 561-562.
2. Tiernan Ray, "IBM: No Targets? That's Somewhat by Design, Says CFO", Barron's March 9, 2018. Kavanaugh went on to say "I learned a tough lesson" in the interview but neglected to survey the damage that IBM road maps wrought.
3. Steve Lohr, "Weak Results at IBM as Its Strategy Shifts", New York Times, October 20, 2014.
4. Lohr, "Weak Results at IBM as Its Strategy Shifts."

5. Timothy Prickett Morgan, "IBM Chips in to Drive 2 Nanometer Semiconductor Manufacturing", The Next Platform, May 6, 2021.

6. Joab Jackson, "IBM's Rometty Defends Bumpy Financial Ride as Company Shifts Strategy", Computerworld, October 20, 2014.

7. Jonathan Vanian, "IBM CEO Ginni Rometty to step down" (Fortune, January 30, 2020).

8. Steven Zolman, "Shrinking Value, Shrinking Sales, Sinking Ship: The Story of IBM's Ginni Rometty", retrieved from netnetweb.com on October 8, 2023.

9. Jennifer Saba and Antony Currie, "Will IBM be the next Microsoft or Nokia", Reuters, January 30, 2020.

10. "Revenue per employee of Selected Tech and Internet Companies in 2022", Statista.com, 2023.

11. Peter Greulich, A View From Beneath the Dancing Elephant, (MBI Concepts, 2014), 107.

12. Steven Zolman, "Shrinking Value, Shrinking Sales, Sinking Ship".

13. "Fortune 500 List", Fortune, May, 2022.

14. Mike Wheatley, "IBM's 2015 roadmap following a very uncertain path", Silicon Angle, May 26, 2014.

15. Noam Scheiber, "Making Dinobabies Extinct: IBM's Push for a Younger Work Force", New York Times, February 12, 2022.

16. "World's Most Admired Companies", Fortune, October 18, 2022.

17. "Ginni Rometty, The Trailblazing Tech Icon and CEO of IBM is Stepping Down" Forbes, January 31, 2020.

18. Michael Hiltzik, "Ginni Rometty's lousy record, but excellent pay, at IBM", Los Angeles Times, January 31, 2020. .

19. Steven Zolman, "Shrinking Value, Shrinking Sales, Sinking Ship".

20. "Dow Jones Industrial Average", retrieved from Wikipedia.com on February 6, 2023.

21. Sean Williams, "If Amazon and Alphabet Join the Dow, These Stocks May be Kicked Out", The Motley Fool, March 15, 2022.

22. Peter E. Greulich, "An IBM Case Study: Do Share Buybacks Work?", retrieved April 16, 2022 from discerningreaders.com, published April 2, 2022.

23. "List of companies by research and development spending", Wikipedia with source data from Strategy+Business Magazine, retrieved February 6, 2023.

## CAN THIS MAN SAVE IBM, PART TWO

1. Matthew Burbridge, "What's next for Jim Whitehurst?", The Margin, October 20, 2021.

2. "Who will get custody of Rochester in the IBM split", Rochester Post-Bulletin, September 7, 2021.

3. Paul Kunert, "20,000-plus staff could face the chop in spin-off of IBM's IT outsourcing biz", OnPrem, October 14, 2020.

4. Tobias Mann, "IBM sell off cloud business, yes we mean weather.com", The Register, August 22, 2023.

5. "IBM Releases Fourth Quarter Results, IBM Corporation, retrieved from ibm.com, January 24, 2022.

6. Ryan DeRousseau, "Pondering the Puzzle of IBM", Fortune, July 1, 2016.

7. Eric J. Savitz, "IBM Earnings Edge Estimates. The AI Bet is Starting to Pay Off, CEO Says", Barron's, October 25, 2023.

8. "The Decline of IBM", The Market is Open, March 10, 2020.

9. Matthew Gooding, "Hyperscalers clean up in Pentagon's $9bn multi-cloud Jedi replacement" Tech Monitor, December 8, 2022.

10. Kevin Roose, "The Brilliance and Weirdness of ChatGPT", The New York Times, December 5, 2022.

11. Naveen Balani, "From IBM Watson to ChatGPT: AI Chatbots an Limitations", Medium, January 5, 2023.

12. Chris Paine, "Do You Trust This Computer?", Busy Child LLC, 2018 film.

13. Paul Elias, "IBM's Watson rebooted as a secure AI alternative", SC Media, July 20, 2023.

14. "IBM is Pitching the AI Story, Again", Seeking Alpha, April 21, 2023, retrieved August 30, 2023.

15. Pamela McCorduck, Machines Who Think (A. K Peters) pp. 111-118.

16. Steve Lohr, "One Man's Dream of Fusing AI with Common Sense", The New York Times, September 8, 2022.

17. Garry Kasparov, Deep Thinking: Where Machine Intelligence Ends and Human Creativity Begins (Public Affairs, New York, 2017), pp. 70-72.

18. Hayden Field, "The scientist behind IBM Watson has raised $60 million for his AI startup in New York", CNBC Report, August 17, 2023.

19. Paul Smith-Goodson, "IBM WatsonX Empowers Business to Build, Tune, and Deploy Reliable Generative AI Models", Forbes, July 7, 2023.

20. Marla Diaz, "AI threatens 7,800 jobs as IBM pauses hiring", Zdnet, May 3, 2023.

21. Kyle Wiggers, "IBM taps AI to translate COBOL code to Java", TechCrunch, August 22, 2023.

22. John Simons, IBM Was Once the King of AI", Time Magazine, April 17, 2022.

23. Jeremy Kahn, "The Quest for Human-Level AI", Fortune, February, 2020.

24. Lex Fridman, "IBM Watson, Jeopardy, and Deep Conversations with AI", interview with David Ferrucci, October 11, 2019.

25. Karl Freund, "IBM Launches Quantum System Two and a Roadmap to Quantum Advantage", Forbes, December 4, 2023.

26. Saba Ali, "IBM debuts z16 mainframe built in Poughkeepsie", Poughkeepsie Journal, April 5, 2022.

27. Brian Anderson and Hemanth Jagannathan, "IBM and Samsung Unveil Semiconductor Breakthrough That Defies Conventional Design", IBM.com, December 14, 2021.

28. Jeff Sommer, "Stock Buybacks and Dividends Become a $1.5 Trillion Political Target, New York Times, August 13, 2022.

29. John Loeffler, "No more transistors: The end of Moore's law", Interesting Engineering, February 14, 2022.

30. John L. Hennessy and David A. Patterson, "A New Golden Age for Computer Architecture" (Communications of the ACM, Vol. 62, No. 2, February, 2019).

## EPILOGUE, A SMALLER IBM

1. Grace Chen, "Georgia Schools: Posh New High School Serving Atlanta", Public School Review, October 29, 2022.

2. Jeff Kiger, "Rochester's Big Blue workers to split between IBM, Kyndryl", Post Bulletin, September 20, 2021.

3. David Kirkpatrick, "Inside Sam's $100 Billion Growth Machine", Fortune, June 14, 2004.

4. Duska Anastasijevic, "Mayo Clinic selects Google as strategic partner for health care innovation, cloud computing", Mayo Clinic news release, September 10, 2019.

5. "From Silver Lake to IBM I, A Powerful Story of Evolution", July 26, 2018, Software Engineering of America.

6. "TIOBE Index for December 2022", www.tiobe.com, retrieved December 26, 2022.

7. John Rockwell, "IBM's AS400 – A Simple Analogy Explains Everything", January24, 2020.

8. Ed Aswad and Suzanne M. Meredith, "IBM in Endicott", self-published, 2003.

9. Chris Potter, "By the numbers: Breaking down the sale of the old IBM campus in Endicott", Press Connect, October 7, 2021.

10. Jim Ehmke, "Final Days for IBM in Endicott", Nexstar Media, August 30, 2023.

11. Alexandra Clough, "Former IBM Campus in Boca: Once Tech Hub, Now Set for Housing?" Palm Beach Post, November 9, 2017.

12. Kim Vo, "Historic IBM Building 25 in San Jose destroyed by fire", The Mercury News, March 8, 2008.

13. "IBM opens $30 million manufacturing facility in Poughkeepsie", RP News Wires, May, 2022.

## FINAL COMMENTS

1. William Rodgers, *Think: A Biography of the Watsons and IBM* (Stein and Day, 1969),304.